Geophysical Monograph Series

Including

IUGG Volumes
Maurice Ewing Volumes
Mineral Physics Volumes

Geophysical Monograph Series

Geophysical Monograph 134

The North Atlantic Oscillation
Climatic Significance and Environmental Impact

James W. Hurrell
Yochanan Kushnir
Geir Ottersen
Martin Visbeck
Editors

American Geophysical Union
Washington, DC

Library of Congress Cataloging-in-Publication Data

The North Atlantic oscillation : climatic significance and environmental impact / James
W. Hurrell … [et al.], editors.
p. cm. -- (Geophysical monograph ; 134)
Includes bibliographical references.
ISBN 0-87590-994-9
1. North Atlantic oscillation--Environmental aspects. 2. Atmospheric circulation--North Atlantic Ocean. 3. North Atlantic Ocean--Climate. I. Hurrell, James W., 1962- II. Series.

QC880.4.A8 N67 2003
551.5'24633--dc21 2002038309

ISSN 0065-8448
ISBN 0-87590-994-9

Printed in the United States of America.

CONTENTS

Acknowledgments

The editors wish to thank the contributing authors for their dedicated effort and patience in seeing this project to completion. Their involvement has extended over nearly two years. We hope they are as proud of the volume as we are.

The idea for the monograph evolved from the well-attended and successful American Geophysical Union (AGU) Chapman Conference in Ourense, Spain over 28 November—1 December, 2000. We wish to acknowledge the University of Vigo, and we extend special thanks to the team of Spanish scientists who worked hard to host the meeting and contributed greatly to its success: R. García, L. Gimeno, E. Hernández, and P. Ribera.

We sincerely thank the referees of individual chapters. They provided careful and thought-provoking reviews that considerably improved the volume as a whole. They are: M. Alexander, J. Alheit, C. Appenzeller, D. Battisti, K. Brander, G. Branstator, K. Briffa, A. Broccoli, T. Callaghan, C. Cassou, A. Czaja, C. Deser, C. Folland, T. Haine, R. Harris, M. Hoerling, P. Jones, T. Joyce, P. Lamb, S. Lee, A. Loison, J. Luterbacher, J. Magnuson, J. Marshall, B. Planque, W. Randel, W. Robinson, J. Rogers, R. Saravanan, R. Seager, D. Shindell, N. Stenseth, R. Sutton, K. Trenberth, and J. Wallace. In addition, comments on the original book proposal from seven anonymous referees substantially refined the organization of the volume.

We are appreciative of the technical assistance of Gaylynn Potemkin and Liz Rothney from NCAR, both of whom helped prepare the camera-ready manuscripts, as well as Allan Graubard, Marie Poole, and Bethany Matsko from the Books Department of AGU. Bethany, in particular, provided valuable guidance and advice as the volume evolved.

Finally, financial backing from agencies on both sides of the Atlantic supported much of the work presented in this volume. The volume itself would not have materialized if not for the funding of the Chapman Conference provided by the National Science Foundation, the National Oceanic and Atmospheric Administration's Office of Global Programs, and the University of Vigo.

This volume represents a current and authoritative survey of the ever-growing body of literature on the North Atlantic Oscillation. It is unique: no other such volume exists. As such, we, the editors, hope it is a valuable resource for students and researchers alike.

PREFACE

Over the middle and high latitudes of the Northern Hemisphere the most prominent and recurrent pattern of atmospheric variability is the North Atlantic Oscillation (NAO). The NAO refers to swings in the atmospheric sea level pressure difference between the Arctic and the subtropical Atlantic that are most noticeable during the boreal cold season (November-April) and are associated with changes in the mean wind speed and direction. Such changes alter the seasonal mean heat and moisture transport between the Atlantic and the neighboring continents, as well as the intensity and number of storms, their paths, and their weather. Significant changes in ocean surface temperature and heat content, ocean currents and their related heat transport, and sea ice cover in the Arctic and sub-Arctic regions are also induced by changes in the NAO. Such climatic fluctuations affect agricultural harvests, water management, energy supply and demand, and fisheries yields. All these effects have led to many studies of the phenomenon; yet, despite this interest, unanswered questions remain regarding the climatic processes that govern NAO variability, how the phenomenon has varied in the past or will vary in the future, and whether it is at all predictable.

Focusing exclusively on the NAO and its impacts, this monograph brings together for the first time atmospheric scientists, oceanographers, paleoclimatologists, and biologists to present a state-of-the-art assessment of the current understanding of this important climate phenomenon and its environmental and societal consequences. Indeed, the outstanding feature of the monograph is its multidisciplinary content presented in 12 papers thematically organized. Each paper provides a thorough overview of different facets of the NAO phenomenon and contains new research as well.

The NAO is one of the oldest known world weather patterns, as some of the earliest descriptions of it were from seafaring Scandinavians several centuries ago. The monograph presents a stimulating account of the major scientific landmarks of NAO research through time while also stressing the present renewed interest in the phenomenon. The NAO and its time dependence, for instance, appear central to the current global climate change debate. The monograph assesses whether recent changes in the atmospheric circulation of the North Atlantic are beyond natural variability by synthesizing a diverse body of literature dealing with how the NAO might change in response to increasing concentrations of greenhouse gases. The relationship between the NAO and anthropogenic climate change has also made it critical to better understand how the NAO and its influence on surface climate have varied in the past. These issues are addressed in separate papers using long instrumental records and paleoclimate proxies.

Another reason for recent invigorated interest in the NAO is that the richly complex and differential responses of the surface-, intermediate- and deep-layers of the ocean to NAO-induced forcing are becoming better documented and understood. The state of knowledge about the oceanic response to changes in NAO forcing is reviewed from theoretical, numerical experimentation and observational perspectives.

That the ocean may play an active role in determining the evolution of the NAO is one of the most debated aspects of this climatic phenomenon, and it is thoroughly investigated in the monograph. If such a role can be delineated and understood, it may also provide one pathway to predictability. New statistical analyses have revealed patterns in North Atlantic sea surface temperatures that precede specific phases of the NAO by 6-9 months, a link that most likely involves the remarkable tendency of the extratropical ocean to preserve its thermal state throughout the year. These results are reviewed and evaluated in a separate paper using observations and global ocean-atmosphere model data.

A second pathway that offers hope for NAO predictions involves links to the stratosphere. A wintertime statistical connection between the month-to-month variability of the strength of the stratospheric polar vortex and the NAO was established several years ago; more recently, it has been argued that large amplitude anomalies in the former precede anomalous behavior of the NAO by 1-2 weeks. The monograph describes these results together with the general state of knowledge on the complex dynamical processes within the atmosphere necessary to understand the NAO behavior.

Renewed interest in the NAO has also come from the biological community. Variations in climate have a profound influence on a variety of ecological processes and, consequently, patterns of species abundance and dynamics. Fluctuations in temperature and salinity, vertical mixing, circulation patterns and ice formation of the North Atlantic Ocean influenced by variations in the NAO have a demonstrated effect on marine biology and fish stocks through both direct and indirect pathways. Responses of terrestrial ecosystems to NAO fluctuations have also been documented. Likewise, the NAO affects lake temperature profiles, lake ice phenology, river runoff, lake water levels, and ultimately the population dynamics of freshwater organisms.

The monograph derives from the American Geophysical Union Chapman Conference on the NAO convened in Ourense, Spain, in the fall of 2000. Each paper was subjected to critical peer review and was revised accordingly. A total of 42 specialists participated in writing the material for the book, and 36 expert referees made substantial contributions to the overall quality and content of the monograph. It is the first time that the NAO phenomenon is addressed in such a comprehensive manner, providing a current and authoritative survey of the ever-growing body of literature on the NAO. The monograph offers extensive information on different levels that can be useful to the students and scientists of climate and environmental studies as well as to non-scientists who are interested in this topic.

James W. Hurrell
National Center for Atmospheric Research

Yochanan Kushnir
Lamont-Doherty Earth Observatory

Geir Ottersen
Institute of Marine Research

Martin Visbeck
Lamont-Doherty Earth Observatory

An Overview of the North Atlantic Oscillation

James W. Hurrell[1], Yochanan Kushnir[2], Geir Ottersen[3], and Martin Visbeck[2]

The North Atlantic Oscillation (NAO) is one of the most prominent and recurrent patterns of atmospheric circulation variability. It dictates climate variability from the eastern seaboard of the United States to Siberia and from the Arctic to the subtropical Atlantic, especially during boreal winter, so variations in the NAO are important to society and for the environment. Understanding the processes that govern this variability is, therefore, of high priority, especially in the context of global climate change. This review, aimed at a scientifically diverse audience, provides general background material for the other chapters in the monograph, and it synthesizes some of their central points. It begins with a description of the spatial structure of climate and climate variability, including how the NAO relates to other prominent patterns of atmospheric circulation variability. There is no unique way to define the spatial structure of the NAO, or thus its temporal evolution, but several common approaches are illustrated. The relationship between the NAO and variations in surface temperature, storms and precipitation, and thus the economy, as well as the ocean and ecosystem responses to NAO variability, are described. Although the NAO is a mode of variability internal to the atmosphere, indices of it exhibit decadal variability and trends. That not all of its variability can be attributed to intraseasonal stochastic atmospheric processes points to a role for external forcings and, perhaps, a small but useful amount of predictability. The surface, stratospheric and anthropogenic processes that may influence the phase and amplitude of the NAO are reviewed.

1. INTRODUCTION

Over the middle and high latitudes of the Northern Hemisphere (NH), especially during the cold season months (November-April), the most prominent and recurrent pattern of atmospheric variability is the North Atlantic Oscillation (NAO). The NAO refers to a redistribution of atmospheric mass between the Arctic and the subtropical Atlantic, and swings from one phase to another produce large changes in the mean wind speed and direction over the Atlantic, the heat and moisture transport between the Atlantic and the neighboring continents, and the intensity and number of storms, their paths, and their weather. Agricultural harvests, water management, energy supply and demand, and yields from fisheries, among many other things, are directly affected by the NAO. Yet, despite this pronounced influence, many open issues remain about which climate processes govern NAO variability, how the phenomenon has varied in the past or will vary in the future, and whether it is at all predictable. These and other topics are dealt with in detail in the following chapters. Our intent is to provide general background material for these chapters, as well as synthesize some of the central points made by other authors.

The NAO is one of the oldest known world weather patterns, as some of the earliest descriptions of it were from seafaring Scandinavians several centuries ago. The history of scientific research on the NAO is rich, and *Stephenson et al.* [this volume] present a stimulating account of the major scientific landmarks of NAO research through time. They also note that, today, there is considerable renewed interest

[1]National Center for Atmospheric Research, Boulder, Colorado
[2]Lamont-Doherty Earth Observatory, Columbia University, Palisades, New York
[3]Institute of Marine Research, Bergen, Norway. Current address: Department of Biology, Division of Zoology, University of Oslo, Oslo, Norway

The North Atlantic Oscillation:
Climatic Significance and Environmental Impact
Geophysical Monograph 134

10.1029/134GM01

in the phenomenon. The NAO and its time dependence, for instance, appear central to the global change debate. Surface temperatures over the NH are likely warmer now than at any other time over the past millennium [*Mann et al.*, 1999; *Jones et al.*, 2001], and the rate of warming has been especially high (~ 0.15°C decade^{-1}) over the past 40 years or so [*Folland et al.*, 2001; *Hansen et al.*, 2002]. A substantial fraction of this most recent warming is linked to the behavior of the NAO [*Hurrell,* 1996; *Thompson et al.*, 2000; also section 5.1], in particular a trend in its index from large amplitude anomalies of one phase in the 1960s to large amplitude anomalies of the opposite phase since the early 1980s. This change in the atmospheric circulation of the North Atlantic accounts for several other remarkable alterations in weather and climate over the extratropical NH as well, and it has added considerably to the debate over our ability to detect and distinguish between natural and anthropogenic climate change. Improved understanding of the relationship between the NAO and anthropogenic climate change has emerged as a key goal of modern climate research [*Gillett et al.*, this volume]. It has also made it critical to better understand how the NAO and its influence on surface climate has varied naturally in the past, either as measured from long instrumental records [*Jones et al.*, this volume] or estimated through multi-century multi-proxy reconstructions [*Cook*, this volume].

While it has long been recognized that the North Atlantic Ocean varies appreciably with the overlying atmosphere [*Bjerknes*, 1964], another reason for invigorated interest in the NAO is that the richly complex and differential responses of the surface-, intermediate- and deep-layers of the ocean to NAO forcing are becoming better documented and understood [*Visbeck et al.*, this volume]. The intensity of wintertime convective renewal of intermediate and deep waters in the Labrador Sea and the Greenland-Iceland-Norwegian (GIN) Seas, for instance, is not only characterized by large interannual variability, but also by interdecadal variations that appear to be synchronized with fluctuations in the NAO [e.g., *Dickson et al.*, 1996]. These changes in turn affect the strength and character of the Atlantic thermohaline circulation (THC) and the horizontal flow of the upper ocean, thereby altering the oceanic poleward heat transport and the distribution of sea surface temperature (SST).

On seasonal time scales, the upper North Atlantic Ocean varies primarily in response to changes in the surface winds, air-sea heat exchanges and freshwater fluxes associated with NAO variations [*Cayan,* 1992a,b]. This does not mean, however, that the extratropical interaction is only one-way. The dominant influence of the ocean on the overlying atmosphere is to reduce the thermal damping of atmospheric variations, and this influence becomes greater on longer time scales. The extent to which the influence of the ocean extends beyond this local thermodynamic coupling to affect the evolution and dynamical properties of the atmospheric flow is probably small, but the effect is non-zero [*Robinson,* 2000; *Kushnir et al.*, 2002]. The role of ocean-atmosphere coupling in determining the overall variability of the NAO is, therefore, a topic of much interest and ongoing research [*Czaja et al.*, this volume].

That the ocean may play an active role in determining the evolution of the NAO is also one pathway by which some limited predictability might exist [*Rodwell*, this volume]. New statistical analyses have revealed patterns in North Atlantic SSTs that precede specific phases of the NAO by up to 9 months, a link that likely involves the remarkable tendency of the extratropical ocean to preserve its thermal state throughout the year [*Kushnir et al.*, 2002]. On longer time scales, recent modeling evidence suggests that the NAO responds to slow changes in global ocean temperatures, with changes in the equatorial regions playing a central role [*Hoerling et al.*, 2001].

A second pathway that offers hope for improved predictability of the NAO involves links through which changes in stratospheric wind patterns might exert some downward control on surface climate [*Thompson et al.*, this volume]. A statistical connection between the month-to-month variability of the NH stratospheric polar vortex and the tropospheric NAO was established several years ago [e.g., *Perlwitz and Graf*, 1995], and more recently it has been documented that large amplitude anomalies in the wintertime stratospheric winds precede anomalous behavior of the NAO by 1-2 weeks [*Baldwin and Dunkerton*, 2001], perhaps providing some useful extended-range predictability. The mechanisms are not entirely clear, but likely involve the effect of the stratospheric flow on the refraction of planetary waves dispersing upwards from the troposphere [e.g., *Hartmann et al.*, 2000]. Similarly, processes that affect the stratospheric circulation on longer time scales, such as reductions in stratospheric ozone and increases in greenhouse gases, could factor into the trend in Atlantic surface climate observed over the past several decades [*Gillett et al.*, this volume]. Regardless of whether predictability arises from the influence of the ocean or from processes internal to the atmosphere, the salient point is that relatively little attention was paid to the NAO, until recently, because changes in its phase and amplitude from one winter to the next were considered unpredictable. The possibility that a small, but useful, percentage of NAO variance is predictable has motivated considerable recent research.

Finally, renewed interest in the NAO has come from the biological community. Variations in climate have a profound influence on a variety of ecological processes and, consequently, patterns of species abundance and dynamics. Fluctuations in temperature and salinity, vertical mixing, circulation patterns and ice formation of the North Atlantic Ocean induced by variations in the NAO [*Visbeck et al.*, this volume] have a demonstrated influence on marine biology and fish stocks through both direct and indirect pathways [*Drinkwater et al.*, this volume]. This includes not only longer-term changes associated with interdecadal NAO variability, but interannual signals as well. Responses of terrestrial ecosystems to NAO fluctuations have also been documented [*Mysterud et al.*, this volume]. In parts of Europe, for example, many plant species have been blooming earlier and longer because of increasingly warm and wet winters, and variations in the NAO are also significantly correlated with the growth, development, fertility and demographic trends of many land animals. The NAO has a demonstrated influence on the physics, hydrology, chemistry and biology of freshwater ecosystems across the NH as well [*Straile et al.*, this volume]. Increasing awareness among and interactions between biologists and climate scientists will undoubtedly further our insights into the critical issue of the response of ecosystems to climate variability and climate change, and mutual interest in the NAO as a dominant source of climate variability is serving as an impetus for this interdisciplinary research.

For many reasons, then, there is broad and growing interest in the NAO. Improved understanding of the physical mechanisms that govern the NAO and its intraseasonal-to-interdecadal variability, and how modes of natural variability such as the NAO may be influenced by anthropogenic climate change, are research questions of critical importance. Setting the stage for the following more detailed review chapters, we begin with a description of the spatial structure of climate and climate variability, including a brief discussion of how the NAO is defined and how it relates to other, prominent patterns of atmospheric circulation variability. The impacts of the NAO on surface temperature, precipitation, storms, the underlying ocean and sea ice, and the local ecology are also briefly described, as are the mechanisms that most likely govern NAO variability. We conclude by expressing our thoughts on outstanding issues and future challenges.

2. THE SPATIAL STRUCTURE OF CLIMATE AND CLIMATE VARIABILITY

Climate variability is usually characterized in terms of "anomalies", where an anomaly is the difference between the instantaneous state of the climate system and the climatology (the mean state computed over many years representative of the era under consideration). Since the spatial structure of climate variability in the extratropics is strongly seasonally dependent [*Wallace et al.*, 1993], it is useful to briefly examine the seasonal evolution of the mean state upon which the climate variations are superimposed.

2.1. The Mean State and Planetary Waves

Large changes in the mean distribution of sea level pressure (SLP) over the NH are evident from boreal winter (December-February) to boreal summer (June-August, Figure 1). Perhaps most noticeable are those changes over the Asian continent related to the development of the Siberian anticyclone during winter and the monsoon cyclone over Southeast Asia during summer. Over the northern oceans, subtropical anticyclones dominate during summer, with the Azores high-pressure system covering nearly all of the North Atlantic. These anticyclones weaken and move equatorward by winter, when the high-latitude Aleutian and Icelandic low-pressure centers predominate.

Because air flows counterclockwise around low pressure and clockwise around high pressure in the NH, westerly flow across the middle latitudes of the Atlantic sector occurs throughout the year. The vigor of the flow is related to the meridional pressure gradient, so the surface winds are strongest during winter when they average near 5 m s^{-1} from the eastern United States across the Atlantic onto northern Europe (Figure 2). These middle latitude westerlies extend throughout the troposphere and reach their maximum (up to 40 m s^{-1}) at a height of about 12 km. This "jet stream" roughly coincides with the path of storms (atmospheric disturbances operating on time scales of days) traveling between North America and Europe. Over the subtropical Atlantic the prevailing surface northeasterly trade winds are relatively steady but strongest during boreal summer.

In the middle troposphere (~ 5-6 km), the boreal winter map of the geopotential height field reveals a westward tilt with elevation of the high latitude surface cyclones and anticyclones (Figure 3). There is a clear "wavenumber two" configuration with low-pressure troughs over northeastern Canada and just east of Asia, and high-pressure ridges just to the west of Europe and North America. These strong zonal asymmetries reflect the so-called "stationary waves" that are forced primarily by the continent-ocean heating contrasts and the presence of the Rocky and Himalayan mountain ranges. In summer the flow is much weaker and more symmetric, consistent with a much more uniform equator-to-pole distribution of solar radiation.

Although the planetary-scale wave patterns (Figure 3) are geographically anchored, they do change in time either because the heating patterns in the atmosphere vary or

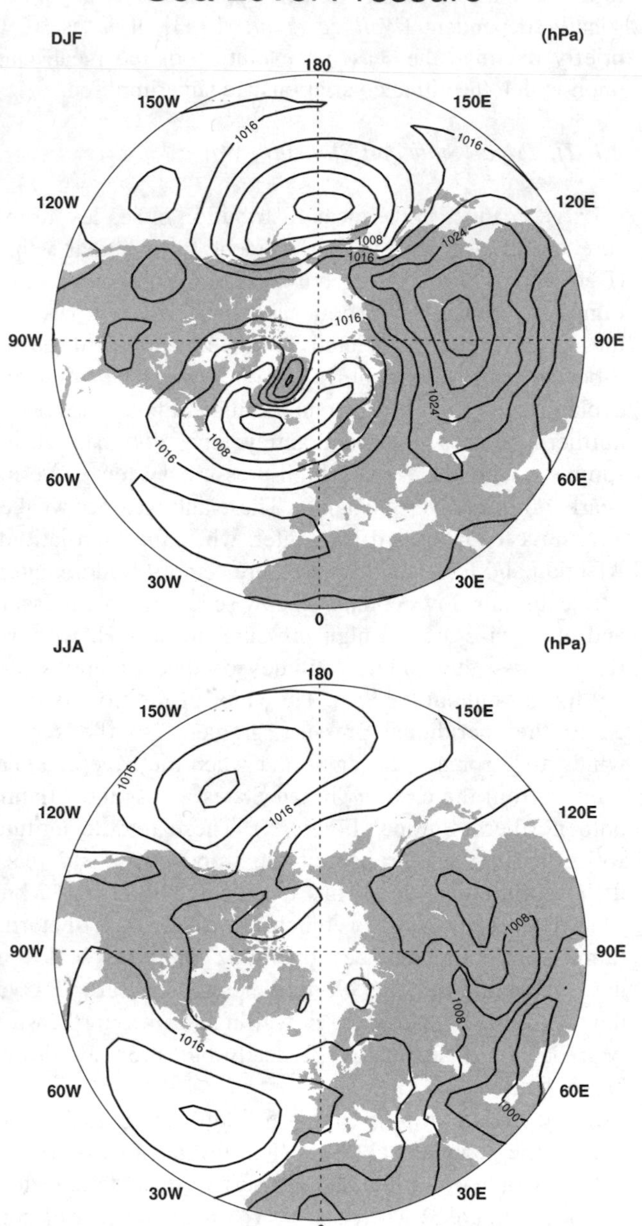

Figure 1. Mean sea level pressure for (top) boreal winter (December-February) and (bottom) boreal summer (June-August). The data come from the NCEP/NCAR reanalysis project over 1958-2001 [*Kalnay et al.,* 1996], and the contour increment is 4 hPa.

because of internal (chaotic) processes. The amplitude and structure of the variability of the seasonal mean 500 hPa geopotential height field (Figure 4) is characterized by a strong longitudinal dependence with maximum temporal variance over the northern oceans, especially during boreal winter. The frequency dependence of the winter pattern is subtle: maps of the variability of monthly mean data, or data filtered to retain fluctuations within specific frequency bands (e.g., 60-180 days), also exhibit distinct variance maxima at 500 hPa over the Atlantic and Pacific Oceans, although the longitudinal contrasts become increasingly apparent as longer time scales are examined [*Kushnir and Wallace*, 1989]. In comparison, throughout most of the NH, the standard deviations of boreal summer 500 hPa heights are only about half as large as those of the wintertime means (Figure 4) [see also *Wallace et al.*, 1993].

2.2. *Teleconnections: The PNA and the NAO*

A consequence of the transient behavior of the atmospheric planetary-scale waves is that anomalies in climate on seasonal time scales typically occur over large geographic regions. Some regions may be cooler or perhaps drier than average, while at the same time thousands of kilometers away, warmer and wetter conditions prevail. These simultaneous variations in climate, often of opposite sign, over distant parts of the globe are commonly referred to as "teleconnections" in the meteorological literature [*Wallace and Gutzler*, 1981; *Esbensen*, 1984; *Barnston and Livezey*, 1987; *Kushnir and Wallace*, 1989; *Trenberth et al.*, 1998]. Though their precise nature and shape vary to some extent according to the statistical methodology and the data set employed in the analysis, consistent regional characteristics that identify the most conspicuous patterns emerge.

Arguably the most prominent teleconnections over the NH are the NAO and the Pacific-North American (PNA) patterns. Both patterns are of largest amplitude during the boreal winter months, and their mid-tropospheric spatial structure is illustrated most simply through one-point correlation maps (Figure 5). These maps are constructed by correlating the 500 hPa height time series at a "reference gridpoint" with the corresponding time series at all gridpoints [e.g., *Wallace and Gutzler*, 1981]. That these two patterns "stand out" above a background continuum comprised of a complete (hemispheric) set of one-point correlation maps is, of course, subjective [*Wallace,* 1996], but the strong consensus is that they do [e.g., *Kushnir and Wallace*, 1989].

The PNA teleconnection pattern has four centers of action. Over the North Pacific Ocean, geopotential height fluctuations near the Aleutian Islands vary out-of-phase with those to the south, forming a seesaw pivoted along the mean position of the Pacific subtropical jet stream (Figure 2). Over North America, variations in geopotential height over western Canada and the northwestern U.S. are negatively correlated with those over the southeastern U.S., but are positively correlated with the subtropical Pacific center. The significance of the locations and the respective phases

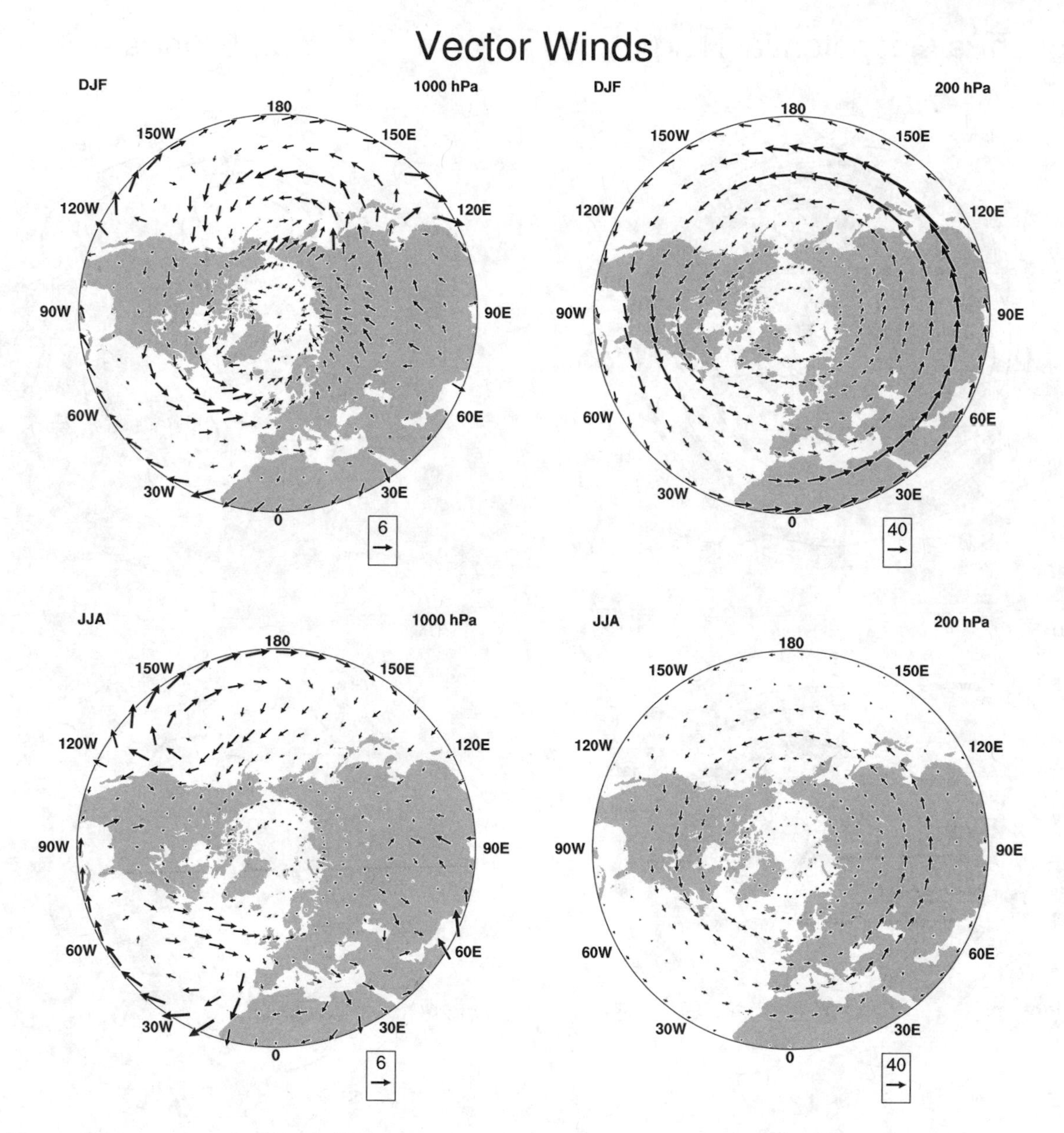

Figure 2. Mean vector winds for (top) boreal winter (December-February) and (bottom) boreal summer (June-August) for (left) 1000 hPa and (right) 200 hPa over 1958-2001. The scaling vectors are indicated in the boxes and are given in units of m s^{-1}.

of the four centers of the PNA is their relation to the mean atmospheric circulation (Figure 3). As stated by *Kushnir* [2002], variations in the PNA pattern "represent variations in the waviness of the atmospheric flow in the western half-hemisphere and thus the changes in the north-south migration of the large-scale Pacific and North American air masses and their associated weather".

On interannual time scales, atmospheric circulation anomalies over the North Pacific, including the PNA, are linked to changes in tropical Pacific sea surface temperatures associated with the El Niño/Southern Oscillation (ENSO) phenomenon. This association reflects mainly the dynamical teleconnection to higher latitudes forced by deep convection in the tropics [see *Trenberth et al.*, 1998 for a review]. The PNA pattern is sometimes viewed, then, as the extratropical arm of ENSO, as is the similar Pacific South American (PSA) teleconnection pattern in the Southern Hemisphere [SH; *Kiladis and Mo,* 1998]. Significant variability of the PNA occurs

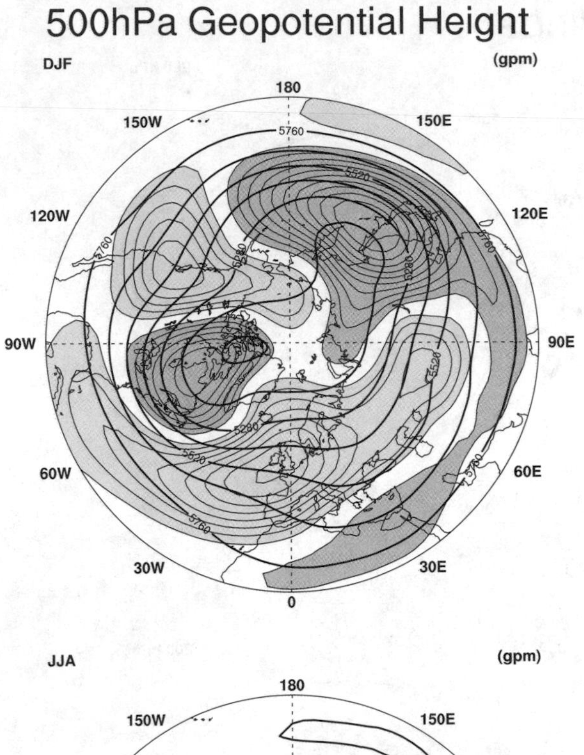

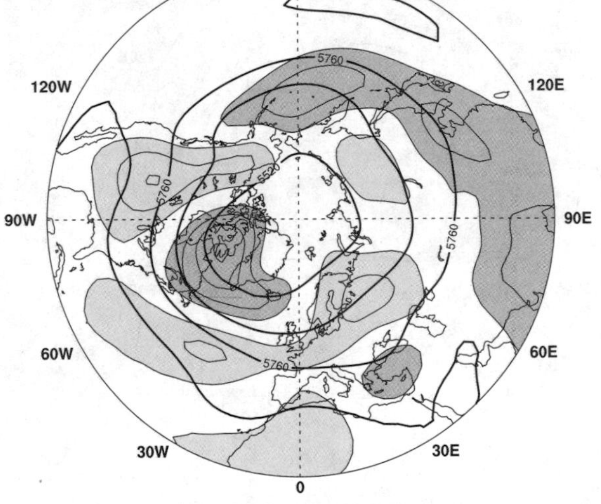

Figure 3. Mean 500 hPa geopotential height for (top) boreal winter (December-February) and (bottom) boreal summer (June-August), indicated by the thick contours every 120 gpm, over 1958-2001. The thin contours (every 20 gpm, zero contour excluded) indicate departures from the zonal average: negative (positive) departures are indicated by dark (light) shading.

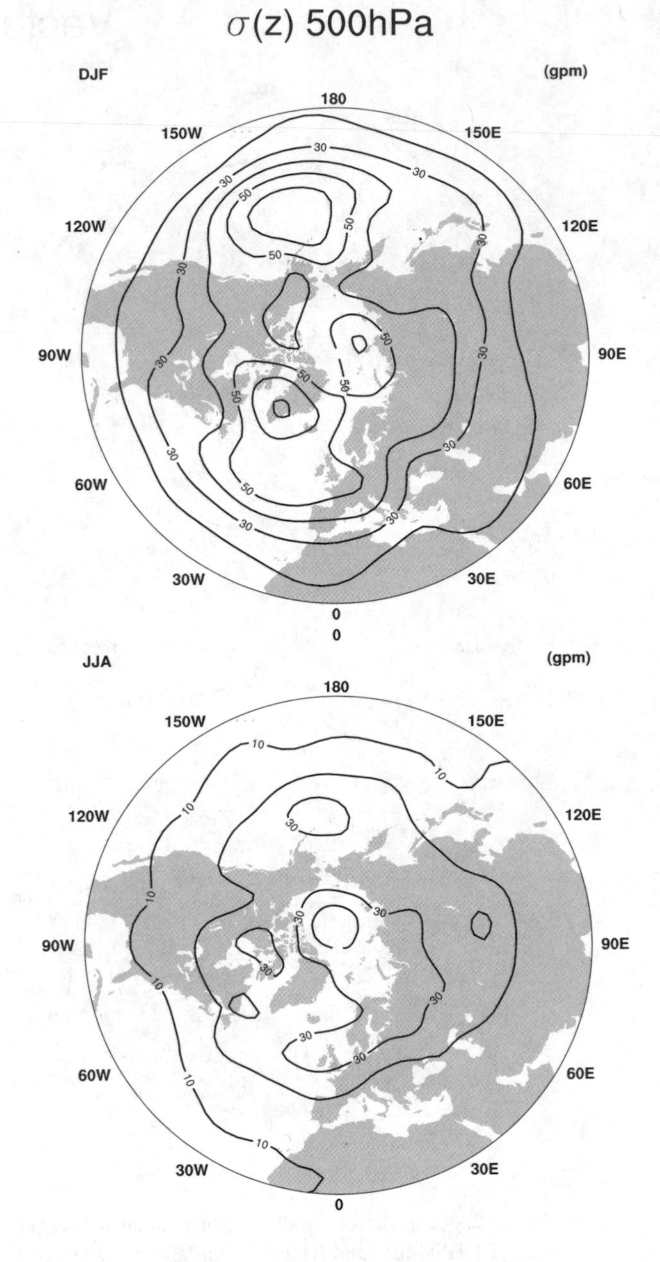

Figure 4. Interannual variability of 500 hPa geopotential height for (top) boreal winter (December-February) and (bottom) boreal summer (June-August) over 1958-2001. The contour increment is 10 gpm.

even in the absence of ENSO, however, indicating that the PNA is an "internal" mode of atmospheric variability.

Similarly, the NAO does not owe its existence to coupled ocean-atmosphere-land interactions [*Thompson et al.*, this volume; *Czaja et al.*, this volume], as is evident from observations and climate model experiments that do not include SST, sea ice or land surface variability (see section 6.1 and Figure 19). In contrast to the wave-like appearance of the PNA, the NAO is primarily a north-south dipole characterized by simultaneous out-of-phase height anomalies between temperate and high latitudes over the Atlantic sector (Figure 5; section 3). Both the NAO and PNA are also reflected in the spatial patterns of the two leading empirically-determined orthogonal functions (EOFs) of NH bore-

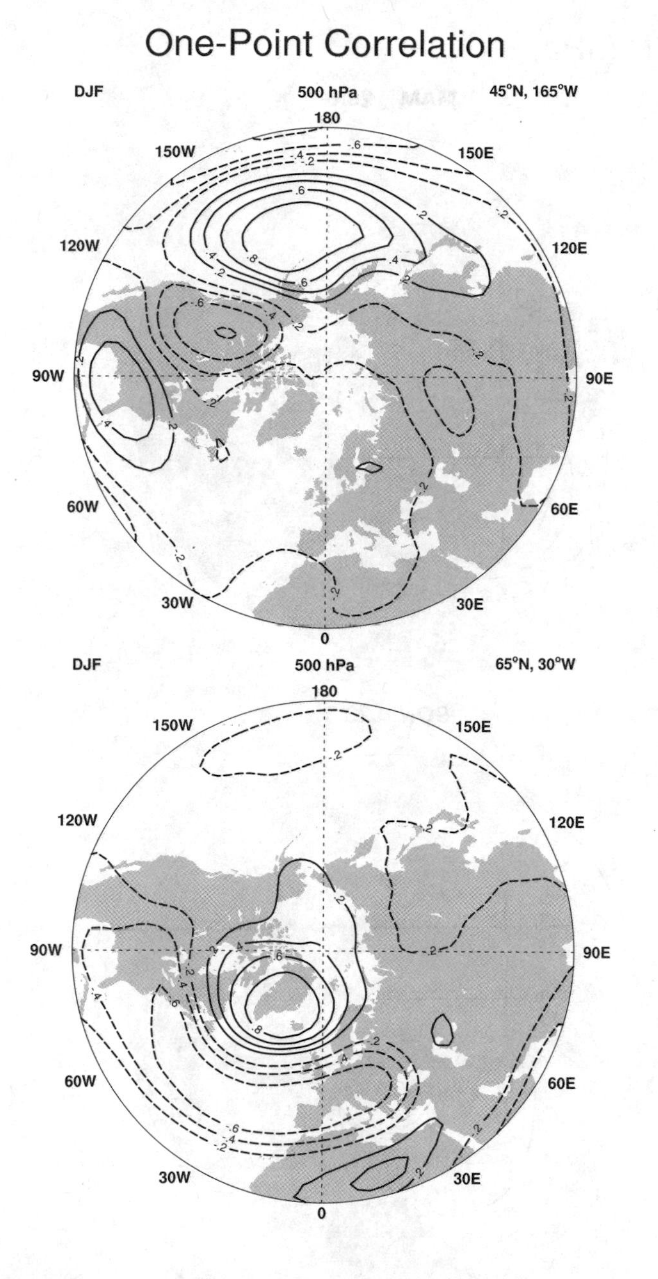

Figure 5. One-point correlation maps of 500 hPa geopotential heights for boreal winter (December-February) over 1958-2001. In the top panel, the reference point is 45°N, 165°W, corresponding to the primary center of action of the PNA pattern. In the lower panel, the NAO pattern is illustrated based on a reference point of 65°N, 30°W. Negative correlation coefficients are dashed, the contour increment is 0.2, and the zero contour has been excluded.

al winter 500 hPa height (not shown), but in order to see them clearly it is necessary to rotate (i.e., to form linear combinations of) the EOFs in a manner that tends to simplify their spatial structure [e.g., *Barnston and Livezey*, 1987; *Kushnir and Wallace*, 1989]. This is less of an issue at the surface, however, where the NAO dominates the leading EOF of the NH SLP field [section 3.2; see also *Kutzbach*, 1970; *Rogers*, 1981; *Trenberth and Paolino*, 1981; *Thompson et al.*, this volume]. Analyzing SLP also allows for the longer-term behavior of the NAO to be evaluated, as a long series of SLP charts over the NH begin in 1899 [*Trenberth and Paolino*, 1980], in contrast to 500 hPa height charts that are confined to after 1947. Moreover, even longer instrumental records of SLP variations are available, especially from European stations [*Jones et al.*, this volume]. Thus, in the following, we examine the spatial structure and time evolution of the NAO in more detail from SLP records.

3. THE SPATIAL SIGNATURE OF THE NAO

There is no single way to "define" the NAO. One approach is through conceptually simple one-point correlation maps (e.g., Figure 5), identifying the NAO by regions of maximum negative correlation over the North Atlantic [*Wallace and Gutzler*, 1981; *Kushnir and Wallace,* 1989; *Portis et al.*, 2001]. Another technique is EOF (or principal component) analysis. In this approach, the NAO is identified from the eigenvectors of the cross-covariance (or cross-correlation) matrix, computed from the time variations of the gridpoint values of SLP or some other climate variable. The eigenvectors, each constrained to be spatially and temporally orthogonal to the others, are then scaled according to the amount of total data variance they explain. This linear approach assumes preferred atmospheric circulation states come in pairs, in which anomalies of opposite polarity have the same spatial structure. In contrast, climate anomalies can also be identified by cluster analysis techniques, which search for recurrent patterns of a specific amplitude and sign. Clustering algorithms identify weather or climate "regimes", which correspond to peaks in the probability density function of the climate phase space [*Lorenz*, 1963]. Interest in this nonlinear interpretation of atmospheric variability has been growing, and recently has found applications within the climate framework [e.g., *Palmer,* 1999; *Corti et al.*, 1999; *Cassou and Terray*, 2001a,b; see also *Monahan et al.*, 2000; 2001]. In the following, we compare the spatial patterns of the NAO as estimated from both traditional EOF and clustering techniques.

3.1. EOF Analysis of North Atlantic SLP

The leading eigenvectors of the cross-covariance matrix calculated from seasonal (3-month average) SLP anomalies in the North Atlantic sector (20°-70°N; 90°W-40°E) are illustrated in Figure 6. The patterns are very similar if based on the cross-correlation matrix (not shown). The patterns are

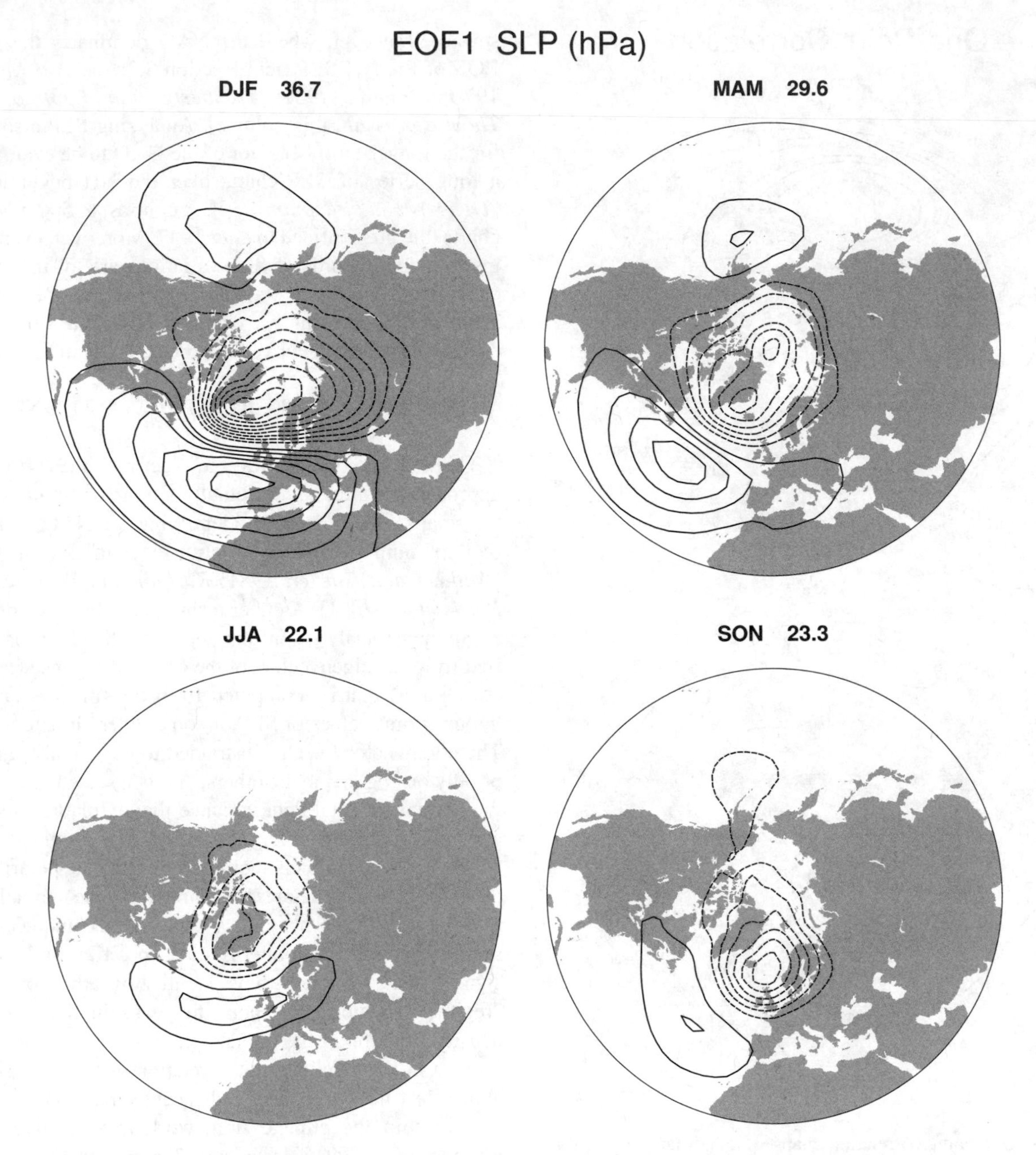

Figure 6. Leading empirical orthogonal functions (EOF 1) of the seasonal mean sea level pressure anomalies in the North Atlantic sector (20º-70ºN, 90ºW-40ºE), and the percentage of the total variance they explain. The patterns are displayed in terms of amplitude (hPa), obtained by regressing the *hemispheric* sea level pressure anomalies upon the leading principal component time series. The contour increment is 0.5 hPa, and the zero contour has been excluded. The data cover 1899-2001 [see *Trenberth and Paolino,* 1980].

displayed in terms of amplitude, obtained by regressing the hemispheric SLP anomalies upon the leading principal component (PC) time series from the Atlantic domain.

The largest amplitude anomalies in SLP occur during the boreal winter months; however, throughout the year the leading pattern of variability is characterized by a surface pressure dipole, and thus may be viewed as the NAO, although the spatial pattern is not stationary [*Barnston and Livezey*, 1987; *Hurrell and van Loon*, 1997; *Portis et al.*, 2001]. Since the eigenvectors are, by definition, structured to explain maximum variance, it is expected that the "centers of action" of the leading EOFs will coincide with the regions of strongest variability, and the movement of those regions through the annual cycle is reflected in Figure 6.

The NAO is the *only* teleconnection pattern evident throughout the year in the NH [*Barnston and Livezey*, 1987]. During the winter season (December-February), it accounts for more than one-third of the total variance in SLP over the North Atlantic, and appears with a slight northwest-to-southeast orientation. In the so-called positive phase (depicted), higher-than-normal surface pressures south of 55°N combine with a broad region of anomalously low pressure throughout the Arctic to enhance the climatological meridional pressure gradient (Figure 1). The largest amplitude anomalies occur in the vicinity of Iceland and across the Iberian Peninsula. The positive phase of the NAO is associated with stronger-than-average surface westerlies across the middle latitudes of the Atlantic onto Europe, with anomalous southerly flow over the eastern U.S. and anomalous northerly flow across the Canadian Arctic and the Mediterranean (Figure 7).

The NAO is well separated (and thus less likely to be affected by statistical sampling errors) in all seasons from

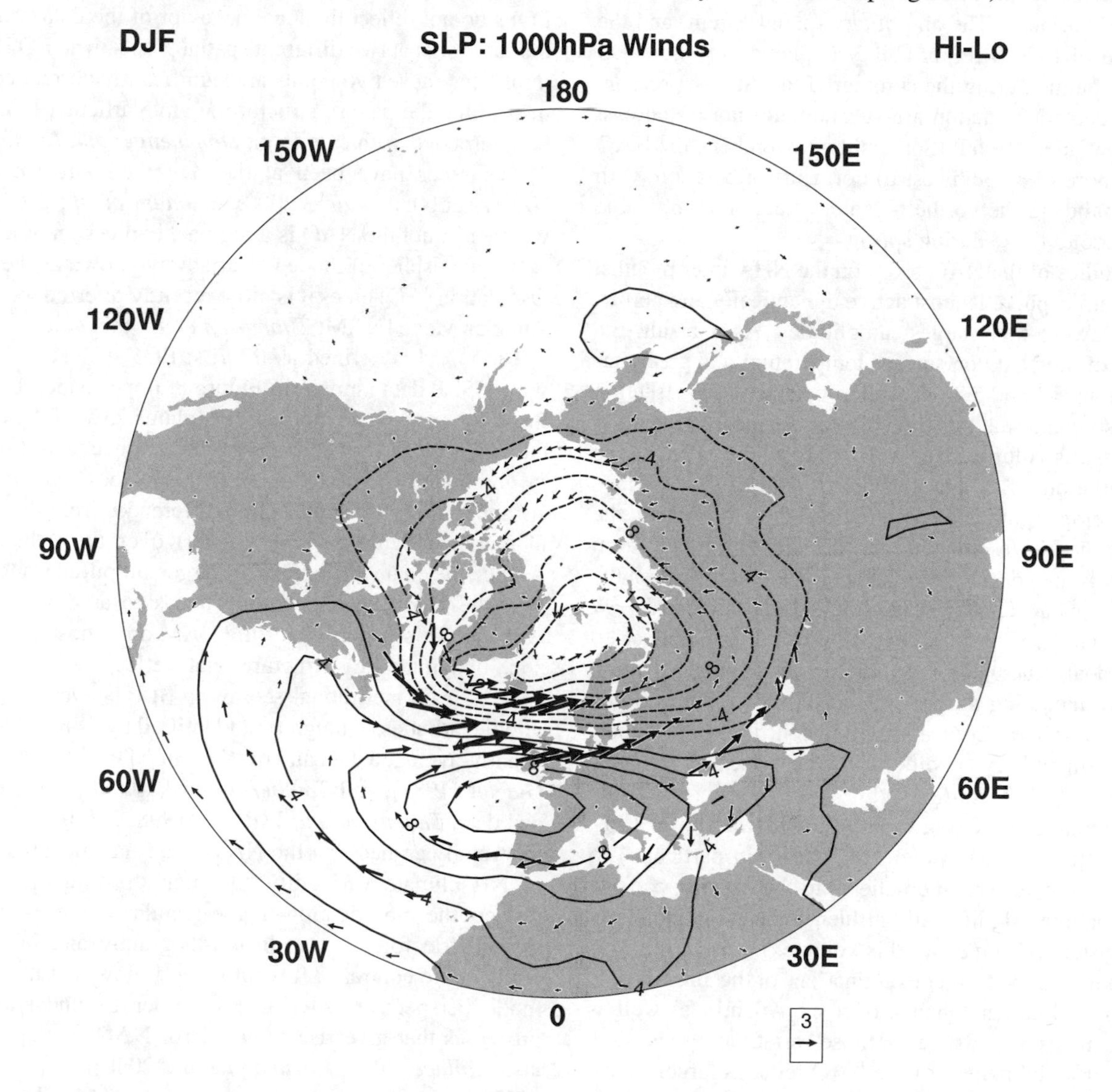

Figure 7. The difference in boreal winter (December-February) mean sea level pressure and 1000 hPa vector winds between positive (hi) and negative (lo) index phases of the NAO. The composites are constructed from winter data (the NCEP/NCAR reanalyses over 1958-2001) when the magnitude of the NAO index (defined as the principal component time series of the leading empirical orthogonal function of Atlantic-sector sea level pressure, as in Figures 6 and 10) exceeds one standard deviation. Nine winters are included in each composite. The contour increment for sea level pressure is 2 hPa, negative values are indicated by the dashed contours, and the zero contour has been excluded. The scaling vector is 3 m s^{-1}.

the second eigenvector, according to the criterion of *North et al.* [1982]. The second EOF, which resembles the East Atlantic (EA) pattern during the winter and spring months [*Wallace and Gutzler,* 1981; *Barnston and Livezey*, 1987], generally accounts for about 15% of the total SLP variance (not shown). By boreal spring (March-May), the NAO appears as a north-south dipole with a southern center of action near the Azores. Both the spatial extent and the amplitude of the SLP anomalies are smaller than during winter, but not by much, and the leading EOF explains 30% of the SLP variance. The amplitude, spatial extent, and the percentage of total SLP variability explained by the NAO reach minimums during the summer (June-August) season, when the centers of action are substantially north and east relative to winter. By fall (September-November), the NAO takes on more of a southwest-to-northeast orientation, with SLP anomalies in the northern center of action comparable in amplitude to those during spring.

Most studies of the NAO focus on the NH winter months, when the atmosphere is most active dynamically and perturbations grow to their largest amplitudes. As a result, the influence of the NAO on surface temperature and precipitation (sections 5.1 and 5.2), as well as on ecosystems (section 5.4), is also greatest at this time of year. As most of the other chapters in this volume do as well, we focus hereafter on the winter variations. But that coherent fluctuations of surface pressure, temperature and precipitation occur throughout the year over the North Atlantic, and decadal and longer-term variability is not confined to winter, should not be lost on the reader. For instance, *Hurrell et al.* [2001; 2002] and *Hurrell and Folland* [2002] document significant interannual to multi-decadal fluctuations in the summer NAO pattern (Figure 6), including a trend toward persistent anticyclonic flow over northern Europe that has contributed to anomalously warm and dry conditions in recent decades [see also *Sexton et al.*, 2002; *Rodwell*, this volume]. Moreover, the vigorous wintertime NAO can interact with the slower components of the climate system (the ocean, in particular) to leave persistent surface anomalies into the ensuing parts of the year that may significantly influence the evolution of the climate system [*Czaja et al.*, this volume; *Rodwell*, this volume]. Undoubtedly, further examinations of the annual cycle of climate and climate change over the Atlantic, as well as the mechanisms responsible for those variations, are needed.

That the spatial pattern of the NAO remains largely similar throughout the year does not imply that it also tends to persist in the same phase for long. To the contrary, it is highly variable, tending to change its phase from one month to another (section 4), and its longer-term behavior reflects the combined effect of residence time in any given phase and its amplitude therein.

3.2. EOF Analysis of Northern Hemisphere SLP

A well-known shortcoming of EOF analysis is that eigenvectors are mathematical constructs, constrained by their mutual orthogonality and the maximization of variance over the entire analysis domain. There is no guarantee, therefore, that they represent physical/dynamical modes of the climate system. An EOF analysis, for instance, will not clearly reveal two patterns that are linearly superposed if those patterns are not orthogonal. Moreover, the loading values of EOFs do not reflect the local behavior of the data: values of the same sign at two different spatial points in an EOF do not imply that those two points are significantly correlated. This means that the pattern structure of any particular EOF must be interpreted with care [e.g., *Dommenget and Latif*, 2002]. These issues have been at the center of a recent debate [*Deser*, 2000; *Wallace*, 2000; *Ambaum et al.*, 2001] over whether or not the NAO is a regional expression of a larger-scale (hemispheric) mode of variability known as the Arctic Oscillation (AO) or, as it is more recently referred to, the NH Annular Mode [NAM; *Thompson et al.*, this volume].

The NAM is defined as the first EOF of NH (20°-90°N) winter SLP data (shown in Figure 8, upper panel, based on the cross-covariance matrix). It explains 23% of the extended winter mean (December-March) variance, and it is clearly dominated by the NAO structure in the Atlantic sector. Although there are some subtle differences from the regional pattern (Figure 8, lower panel) over the Atlantic and Arctic, the main difference is larger amplitude anomalies over the North Pacific of the same sign as those over the Atlantic. This feature gives the NAM an almost annular (or zonally-symmetric) structure that reflects a more hemispheric-scale meridional seesaw in SLP between polar and middle latitudes. Though first identified by *Lorenz* [1951] in zonally-averaged data and by *Kutzbach* [1970], *Wallace and Gutzler* [1981], and *Trenberth and Paolino* [1981] in gridded data, *Thompson and Wallace* [1998; 2000] have recently strongly argued that the NAM is a fundamental structure of NH climate variability, and that the "regional" NAO reflects the modification of the annular mode by zonally-asymmetric forcings, such as topography and land-ocean temperature contrasts. It would then follow that the annular mode perspective is critical in order to understand the processes that give rise to NAM (or NAO) variations [see also *Wallace*, 2000; *Hartmann et al.*, 2000].

The arguments for the existence of the NAM, described in much more detail by *Thompson et al.* [this volume], include the following: (1) the zonally-symmetric component of the NAM is evident in the leading EOFs of heights and winds from the surface through the stratosphere, with variability in the latter region being dominated by a truly annular mode;

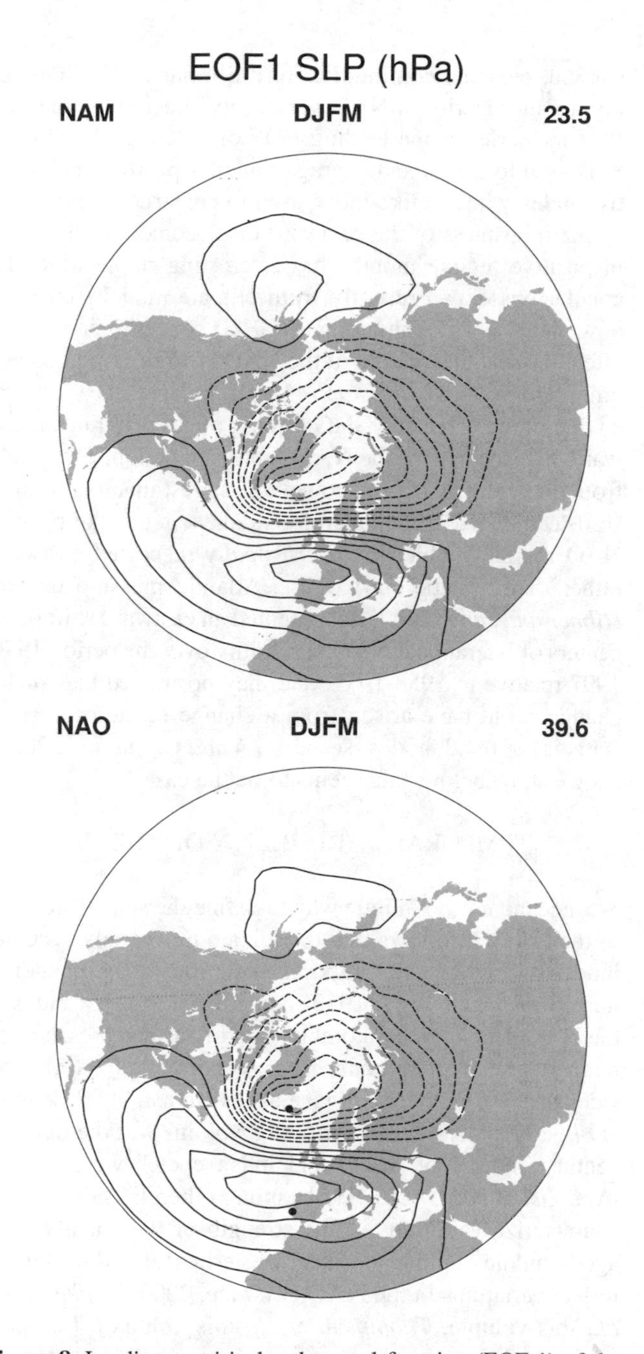

Figure 8. Leading empirical orthogonal function (EOF 1) of the winter (December-March) mean sea level pressure anomalies over (top) the Northern Hemisphere (20º-90ºN) and (bottom) the North Atlantic sector (20º-70ºN, 90ºW-40ºE), and the percentage of the total variance they explain. The patterns are displayed in terms of amplitude (hPa), obtained by regressing the *hemispheric* sea level pressure anomalies upon the leading principal component time series. The contour increment is 0.5 hPa, and the zero contour has been excluded. The data cover 1899-2001 [see *Trenberth and Paolino,* 1980]. The dots in the bottom panel represent the locations of Lisbon, Portugal and Stykkisholmur, Iceland used in the station based NAO index of *Hurrell* [1995a] (see Figure 10).

(2) the strong similarity of the NAM to the spatial pattern of circulation variability in the SH, known as the Southern Annular Mode (SAM); (3) the "signature" of the NAM in the meridional profiles of the month-to-month variance of the zonally-averaged circulation; and (4) that the NAM seems to orchestrate weather and climate over the hemisphere, not just the Atlantic sector, on time scales from weeks to decades. This point of view clearly suggests that the NAM reflects dynamical processes that transcend the Atlantic sector. It is not a view that is universally accepted, however [*Kerr*, 1999].

Deser [2000] has argued that the NAM is not a teleconnection pattern in the sense that there are only weak correlations between the Atlantic and Pacific middle latitude centers on both intraseasonal (month-to-month) and interannual time scales. In addition, while interannual fluctuations in SLP over the Arctic and Atlantic centers of action are significantly (negatively) correlated (e.g., Figure 5), the Arctic and Pacific centers are not. This leads her to conclude "the annular character of the AO is more a reflection of the dominance of its Arctic center of action than any coordinated behavior of the Atlantic and Pacific centers". *Ambaum et al.* [2001] reach a similar conclusion, but also based on an assessment of the physical consistency between the NAM and NAO structures in SLP and the leading patterns of variability in other, independent climate variables. In particular, they show that leading EOFs of SLP, lower tropospheric winds and temperature over the Atlantic sector are dynamically related and are clear representations of the NAO, while the same analysis applied to the hemispheric domain yields very different results and patterns that are not obviously related. Rather, over the Pacific sector, they show that dynamical consistency among fields emerges for the PNA. *Ambaum et al.* [2001] also note that NAM variability is superposed upon a strongly zonally asymmetric climatology (Figures 1-3; note that the Icelandic and Aleutian low pressure centers occupy different latitudes), so that it does not correspond to a uniform modulation of the climatological features. In the positive NAM phase (depicted in Figure 8), the North Atlantic tropospheric subtropical and polar jets (Figure 2) are strengthened, but the subtropical jet in the Pacific is weakened.

While the above arguments suggest that the NAO paradigm may be more robust and physically relevant for NH variability, the debate is not over. Recently, for instance, *Wallace and Thompson* [2002] suggest that the lack of teleconnectivity between the Atlantic and Pacific sectors is consistent with the NAM if a second mode is present that favors out-of-phase behavior between these sectors. They suggest this mode could be the PNA. Regardless, the important point is that the physical mechanisms associated with annular mode behavior may be very relevant to understanding the existence of the NAO, regardless of the robustness of the NAM paradigm. For instance, as previously noted, the leading wintertime pattern of variability in the lower stratosphere is clearly annular, but the SLP anomaly pattern that is associated with it is confined almost entirely to the Arctic and Atlantic sectors and coincides with the spatial structure of the NAO [e.g., *Perlwitz and Graf*, 1995; *Kodera et al.*, 1996; *Thompson and Wallace*, 1998; *Deser*, 2000]. *Thompson et al.* [this volume] present a thorough overview of the dynamics governing annular mode behavior, including a discussion of the mechanisms by which annular variability in the stratosphere might drive NAO-like variations in surface climate.

3.3. Cluster Analysis of North Atlantic SLP

The dynamical signature of interannual variability in the North Atlantic domain can also be examined through non-linear approaches, such as cluster analysis or non-linear principal component analysis [*Monahan et al.*, 2000; 2001]. Here we apply the former to 100 years of December-March monthly SLP data using the procedures of *Cassou and Terray* [2001a,b], which are based on the clustering algorithm of *Michelangeli et al.* [1995]. The solutions are robust among different algorithms and SLP data sets (not shown).

The clustering algorithm applied over the Atlantic domain (20°-70°N; 90°W-40°E) identifies four winter climate regimes in SLP (Figure 9). Two of them correspond to the negative and positive phases of the NAO, while the third and fourth regimes display a strong anticyclonic ridge and trough, respectively, off western Europe and bear some resemblance to the EA teleconnection pattern [*Wallace and Gutzler*, 1981; *Barnston and Livezey*, 1987]. Both the ridge and negative NAO regimes occur in about 30% of all winter months since 1900, while both the positive NAO and trough regimes occur in about 20% of all winter months. These numbers are sensitive to the period of analysis, reflecting that the dominance of certain regimes over others varies over time (section 4).

In contrast to the typical NAO pattern identified through linear approaches (e.g., Figures 5 and 6), some interesting spatial asymmetries are evident in Figure 9. Most striking is the difference in the position of the middle latitude pressure anomalies between the two NAO regimes: in particular, the eastward shift (by ~30° longitude) in the positive relative to the negative regime. The main difference in the northern center is the northeastward extension of SLP anomalies during positive NAO regime months. These spatial asymmetries are not dependent on the analysis period: they are evident in subperiods of the ~100-year long SLP data set [C. Cassou, personal communication]. Similar results, indicating a non-linearity in NAO variability, are found when the PC time series of the leading EOF of Atlantic SLP (Figure 8) is used to define and average together positive and negative index winters (like those used to construct Figure 7).

The robustness of the eastward displacement of the NAO in positive regime months has interesting implications for conclusions drawn recently from climate model studies on how increasing greenhouse gas (GHG) concentrations might affect the spatial structure of the NAO [*Gillett et al.*, this volume]. *Ulbrich and Christoph* [1999], for instance, concluded that future enhanced GHG forcing might result in an eastward displacement of the NAO centers of action. The results from the regime analysis, however, suggest that longitudinal shifts could arise from the preferential excitement of positive NAO regimes, which are *intrinsically* displaced eastward, rather than a static shift of the Atlantic pressure centers. *Hilmer and Jung* [2000] documented an eastward shift of the centers of interannual NAO variability over the period 1978-1997 relative to 1958-1977, and they postulated that such a change could have arisen from a change in the occupation statistics of fixed modes [see also *Lu and Greatbatch,* 2002]. As we show below, this seems to be the case.

4. TEMPORAL VARIABILITY OF THE NAO

Since there is no unique way to define the spatial structure of the NAO, it follows that there is no universally accepted index to describe the temporal evolution of the phenomenon. *Walker and Bliss* [1932] constructed the first index of the NAO using a linear combination of surface pressure and temperature measurements from weather stations on both sides of the Atlantic basin [see also *Wallace*, 2000; *Wanner et al.*, 2001; *Stephenson et al.*, this volume]. In the mid-20th century, indices of the "zonal index cycle" were popular [*Namias*, 1950; *Lorenz*, 1951 among others]. These indices characterize variations in the strength of the zonally averaged middle latitude surface westerlies and thus largely reflect variations in the NAO [*Wallace*, 2000; *Stephenson et al.*, this volume; *Thompson et al.*, this volume]. European scientists have introduced many others, all also strongly related to the NAO but generally not well known. *Stephenson et al.* [this volume] describe several of them. An example is the "westerly index" of *Lamb* [1972], one of several indices associated with a set of circulation types relevant to the climate of the United Kingdom that are still used in research today [C. Folland, personal communication].

Most modern NAO indices are derived either from the simple difference in surface pressure anomalies between various northern and southern locations, or from the PC time series of the leading (usually regional) EOF of SLP.

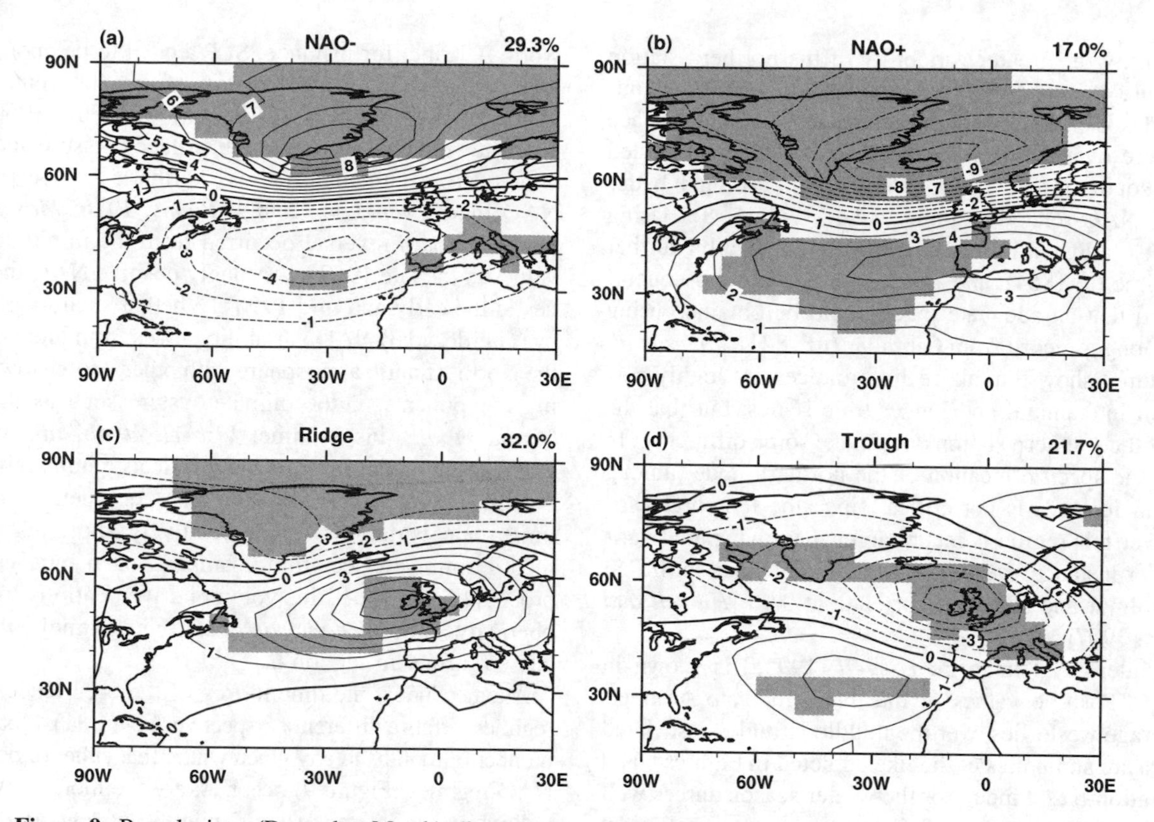

Figure 9. Boreal winter (December-March) climate regimes in sea level pressure (hPa) over the North Atlantic domain (20°-70°N, 90°W-40°E) using monthly data over 1900-2001. Shaded areas exceed the 95% confidence level using T and F statistics [see *Cassou*, 2001]. The percentage at the top right of each panel expresses the frequency of occurrence of a cluster out of all winter months since 1900. The contour interval is 1 hPa.

Many examples of the former exist, usually based on instrumental records from individual stations near the NAO centers of action [e.g., *Rogers*, 1984; *Hurrell*, 1995a; *Jones et al.*, 1997; *Slonosky and Yiou*, 2001], but sometimes from gridded SLP analyses [e.g., *Portis et al.*, 2001; *Luterbacher et al.*, 2002]. *Jones et al.* [this volume] discuss and compare various station-based indices in detail. They note that a major advantage of most of these indices is their extension back to the mid-19th century or earlier, and they even present a new instrumental NAO index from London and Paris records dating back to the late 17th century [see also *Slonosky et al.*, 2001].

A disadvantage of station-based indices is that they are fixed in space. Given the movement of the NAO centers of action through the annual cycle (Figure 6), such indices can only adequately capture NAO variability for parts of the year [*Hurrell and van Loon*, 1997; *Portis et al.*, 2001; *Jones et al.*, this volume]. Moreover, individual station pressures are significantly affected by small-scale and transient meteorological phenomena not related to the NAO and, thus, contain noise [see *Trenberth*, 1984]. *Hurrell and van Loon* [1997] showed, for instance, that the signal-to-noise ratio of commonly-used winter NAO station-based indices is near 2.5, but by summer it falls to near unity.

An advantage of the PC time series approach is that such indices are more optimal representations of the full NAO spatial pattern; yet, as they are based on gridded SLP data, they can only be computed for parts of the 20th century, depending on the data source. Below we compare a station-based index to the PC time series of the leading EOF (PC1) of both Atlantic-sector and NH SLP. We also present the time history of occurrence of the NAO regimes identified in Figure 9. All comparisons are for the winter (December-March) season. *Osborn et al.* [1999], *Wallace* [2000], *Wanner et al.* [2001], *Portis et al.* [2001], and *Jones et al.* [this volume] present quantitative comparisons of these and other NAO-related indices, the latter two papers for other seasons as well.

4.1. Time Series

Rogers [1984] simplified the NAO index of *Walker and Bliss* [1932] by examining the difference in normalized SLP anomalies from Ponta Delgada, Azores and Akureyri, Iceland. Normalization is used to avoid the series being dominated by the greater variability of the northern station (e.g., Figure 4). *Hurrell* [1995a] analyzed the important coupled modes of wintertime variability in SLP and surface temperature over the North Atlantic sector, and concluded that the southern-node station of Lisbon, Portugal better captured NAO-related variance (e.g., Figure 8). Using Lisbon also allowed him to extend the record a bit further back in time (to 1864), and *Jones et al.* [1997] subsequently showed that an adequate index could be obtained using the even longer record from Gibraltar (to 1821). *Jones et al.* [this volume] show that all of these indices are highly correlated on interannual and longer time scales, but that the choice of the southern station does make some difference. In contrast, the specific location of the northern node (among stations in Iceland) is not critical since the temporal variability over this region is much larger than the spatial variability. For instance, December-March anomalies in SLP at Stykkisholmur and Akureyri correlate at 0.98 [*Hurrell and van Loon*, 1997].

The winter-mean index of *Hurrell* [1995a] is shown in Figure 10. Positive values of the index indicate stronger-than-average westerlies over the middle latitudes associated with pressure anomalies of the like depicted in Figures 6 and 8. The station-based index for the winter season agrees well with PC1 of Atlantic-sector SLP. The correlation coefficient between the two is 0.92 over the common period 1899-2002, indicating that the station-based index adequately represents the time variability of the winter-mean NAO spatial pattern. Moreover, it correlates with PC1 of NH SLP [the NAM index of *Thompson and Wallace*, 1998] at 0.85, while the correlation of the two PC1 time series is 0.95. These results again emphasize that the NAO and NAM reflect essentially the same mode of tropospheric variability. When intraseasonal anomalies are considered by stringing together the individual winter months, the correlation coefficient between the two PC1 time series is reduced slightly to 0.89, but the correlations involving the station-based index remain unchanged.

One conclusion from Figure 10 is that there is little evidence for the NAO to vary on any preferred time scale. Large changes can occur from one winter to the next, and there is also a considerable amount of variability within a given winter season [*Nakamura*, 1996; *Feldstein*, 2000; see also Figure 11]. This is consistent with the notion that much of the atmospheric circulation variability in the form of the NAO arises from processes internal to the atmosphere [section 6.1; *Thompson et al.*, this volume], in which various scales of motion interact with one another to produce random (and thus unpredictable) variations. There are, however, periods when anomalous NAO-like circulation patterns persist over quite a few consecutive winters. In the subpolar North Atlantic, for instance, SLP tended to be anomalously low during winter from the turn of the 20th century until about 1930 (positive NAO index), while the 1960s were characterized by unusually high surface pressure and severe winters from Greenland across northern Europe [negative NAO index; *van Loon and Williams*, 1976; *Moses et al.*, 1987]. A sharp reversal occurred from the minimum index values in the late 1960s to strongly positive NAO index values in the early and mid 1990s. Whether such low frequency (interdecadal) NAO variability arises from interactions of the North Atlantic atmosphere with other, more slowly varying components of the climate system such as the ocean [*Czaja et al.*, this volume; *Visbeck et al.*, this volume], whether the recent upward trend reflects a human influence on climate [*Gillett et al.*, this volume], or whether the longer time scale variations in the relatively short instrumental record simply reflect finite sampling of a purely random process [*Czaja et al.*, this volume] driven entirely by atmospheric dynamics [*Thompson et al.*, this volume] will be discussed further in section 6.

Another index, the time history of the occurrence of NAO regimes, offers a different perspective (Figure 11). Plotted is the number of months in any given winter that either or both of the NAO regimes (Figure 9) occur. As for the more conventional indices, strong interannual variability is evident, and there are periods when one NAO regime occurs almost to the exclusion of the other. For instance, very few positive NAO regime months are found during the 1960s, while very few negative regime months have been observed recently, consistent with the upward trend in the indices of Figure 10 and the "eastward shift" of the NAO centers of action in recent decades [Figure 9; *Hilmer and Jung*, 2000; *Lu and Greatbatch*, 2002].

The regime analysis also illustrates two other important points. First, there is a large amount of within-season variance in the atmospheric circulation of the North Atlantic. Most winters are not dominated by any particular regime; rather, the atmospheric circulation anomalies in one month might resemble the positive index phase of the NAO, while in another month they resemble the negative index phase or some other pattern altogether. Over the ~100-year record, for example, all four-winter months are classified as the negative NAO regime for only four winters (1936, 1940, 1969, and 1977, the year given by January), only two winters (1989 and 1990) have more than two months classified under the positive NAO regime, and there are nine winters during which neither NAO regime can be identified in any months. This is also a reminder of the second point: although the NAO is the dominant pattern of atmospheric circulation variability over the North Atlantic, it explains only a fraction of the total variance, and most winters cannot be characterized by the canonical NAO pattern in Figures 6 and 8. One notable difference

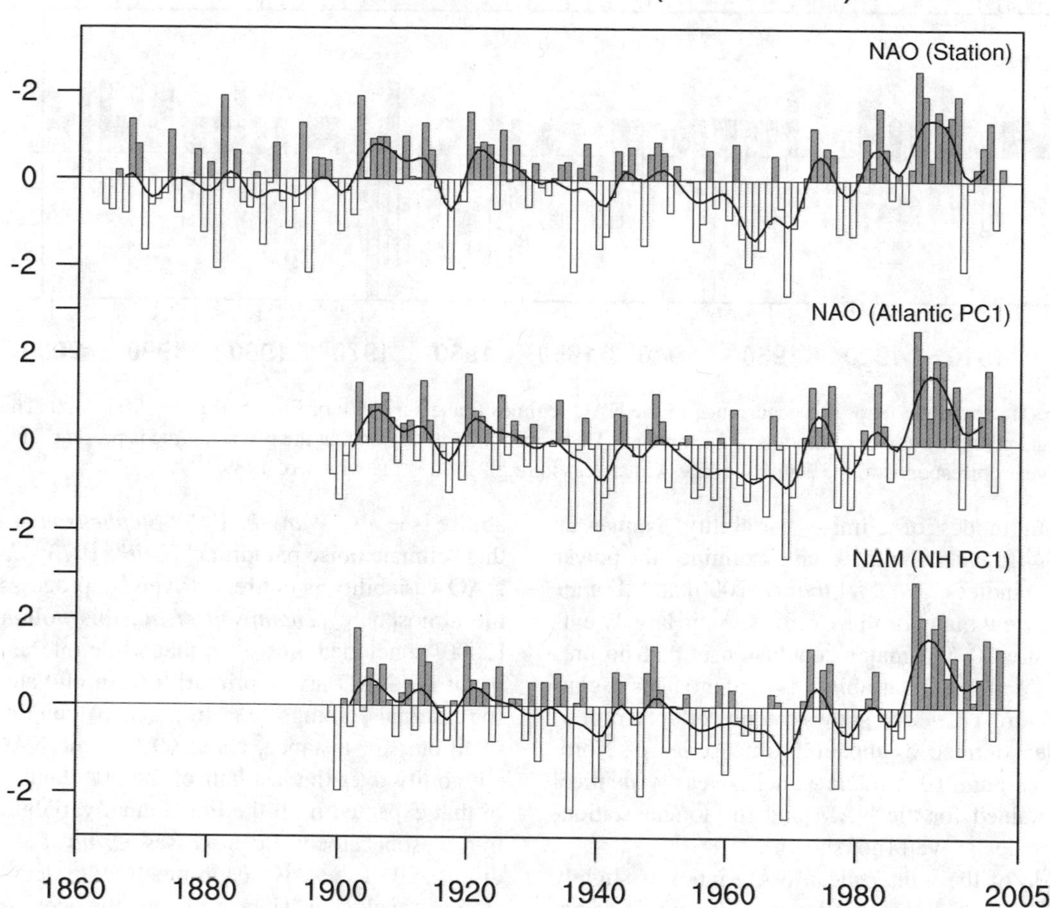

Figure 10. Normalized indices of the mean winter (December-March) NAO constructed from sea level pressure data. In the top panel, the index is based on the difference of normalized sea level pressure between Lisbon, Portugal and Stykkisholmur/Reykjavik, Iceland from 1864 through 2002. The average winter sea level pressure data at each station were normalized by division of each seasonal pressure by the long-term mean (1864-1983) standard deviation. In the middle panel, the index is the principal component time series of the leading empirical orthogonal function (EOF) of Atlantic-sector sea level pressure (bottom panel of Figure 8). In the lower panel, the index is the principal component time series of the leading EOF of Northern Hemisphere sea level pressure (top panel of Figure 8). The heavy solid lines represent the indices smoothed to remove fluctuations with periods less than 4 years. The indicated year corresponds to the January of the winter season (e.g., 1990 is the winter of 1989/1990). See http://www.cgd.ucar.edu/~jhurrell/nao.html for updated time series.

between the time history of NAO regime occurrences and more conventional NAO indices occurs early in the 20th century. All three conventional indices have generally positive values from 1900 until about 1930 (Figure 10). However, the ridge and trough regimes (Figure 9) were more dominant than the NAO regimes over this period (as can be deduced by the small occupancy rates in Figure 11; see *Cassou* [2001] for more discussion). This is also consistent with *van Loon and Madden* [1983], who show that the early 20th century was characterized by a southward displacement of the maximum North Atlantic SLP variance from the Irminger Sea to near Ireland (their Figures 1 and 3). That the ridge regime projects upon the positive index phase of the NAO (Figure 9) helps explain the strong positive values of the conventional indices, but it also warns that reducing the complexities of the North Atlantic atmospheric circulation to one simple index can be misleading.

4.2. Power Spectrum

Spectral analysis is used to quantify periodicities in a time series and to gain insights into the dynamical processes

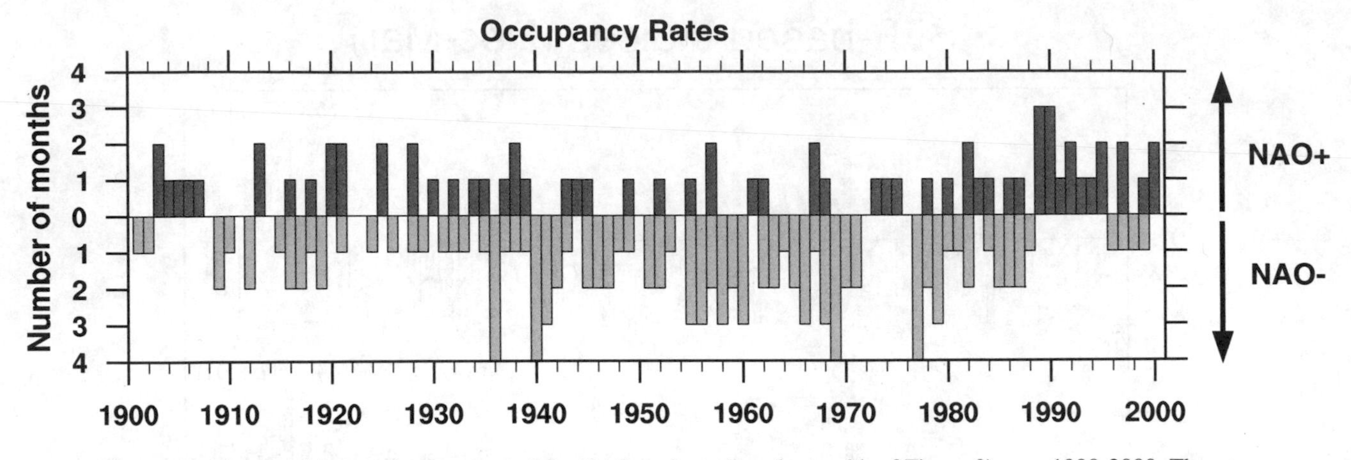

Figure 11. The time history of occurrence of the NAO regimes (panels a and b of Figure 9) over 1900-2000. The vertical bars give the number of months in each winter (December-March) season that the given regime is present. The indicated year corresponds to the January of the winter season (e.g., 1990 is the winter of 1989/1990).

associated with modes of climate variability. Numerous authors using different techniques have examined the power spectra of NAO indices, and *Greatbatch* [2000] and *Wanner et al.* [2001] review many of these efforts. As is largely evident from Figure 10, the major conclusion is that no preferred time scale of NAO variability is evident. This is consistent with Figure 12, which shows the power spectrum of the NAO index defined as the PC1 time series of North Atlantic SLP (Figure 10, middle panel). Nearly identical spectra are obtained for the NAM and the longer station-based NAO indices as well (not shown).

The spectrum of the winter-mean NAO index is slightly "red", with power increasing with period. It reveals somewhat enhanced variance at quasi-biennial periods, a deficit in power at 3 to 6 year periods, and slightly enhanced power in the 8-10 year band, but no significant peaks. The power evident at the lowest frequencies reflects the trends evident in Figure 10. *Hurrell and van Loon* [1997] examined the time evolution of these signals using a station-based index, and found that the quasi-biennial variance was enhanced in the late 19th and early 20th centuries, while the 8-10 year variance was enhanced over the latter half of the 20th century. They also noted similar behavior in the spectra of European surface temperature records over the same period of analysis. *Appenzeller et al.* [1998], *Higuchi et al.* [1999], and *Wanner et al.* [2001] among others have also noted that apparent "bumps" in the NAO spectra come and go over time.

Feldstein [2000] examined the spectral characteristics of the NAO using daily data, and concluded that its temporal evolution is largely consistent with a stochastic (Markov, or first-order autoregressive) process with a fundamental time scale of about 10 days. This then means that observed interannual and longer time scale NAO fluctuations (Figure 12) could entirely be a remnant of the energetic weekly variability [see also *Wunsch*, 1999; *Stephenson et al.*, 2000]. In this "climate noise paradigm" [*Leith*, 1973; *Madden*, 1976], NAO variability is entirely driven by processes intrinsic to the atmosphere [*Thompson et al.*, this volume]. *Feldstein* [2000] concluded, however, that while interannual variability of the NAO arises primarily from climate noise, a role for external forcings (e.g., the ocean) couldn't be entirely ruled out; for instance, about 60% of the NAO interannual variability over the last half of the 20th century is in excess of that expected if all the interannual variability was due to intraseasonal stochastic processes. *Czaja et al.* [this volume] discuss all of these issues in much more detail, and reach a similar conclusion. They argue that the ocean can be expected to modulate NAO variability on interannual and longer time scales [see also *Visbeck et al.*, this volume], and that spectral analyses of dynamical NAO indices as in Figure 12 are not optimal for detecting the impact of the ocean.

5. IMPACTS OF THE NAO

The NAO exerts a dominant influence on wintertime temperatures across much of the NH. Surface air temperature and SST across wide regions of the North Atlantic Ocean, North America, the Arctic, Eurasia and the Mediterranean are significantly correlated with NAO variability. These changes, along with related changes in storminess and precipitation, ocean heat content, ocean currents and their related heat transport, and sea ice cover have significant impacts on a wide range of human activities as well as on marine, freshwater and terrestrial ecosystems. In the following, we present a brief overview of these impacts. More detailed discussions can be found particularly in *Jones et al.* [this volume], *Visbeck et al.* [this volume], *Mysterud et al.* [this volume], *Drinkwater et al.* [this volume], and *Straile et al.* [this volume].

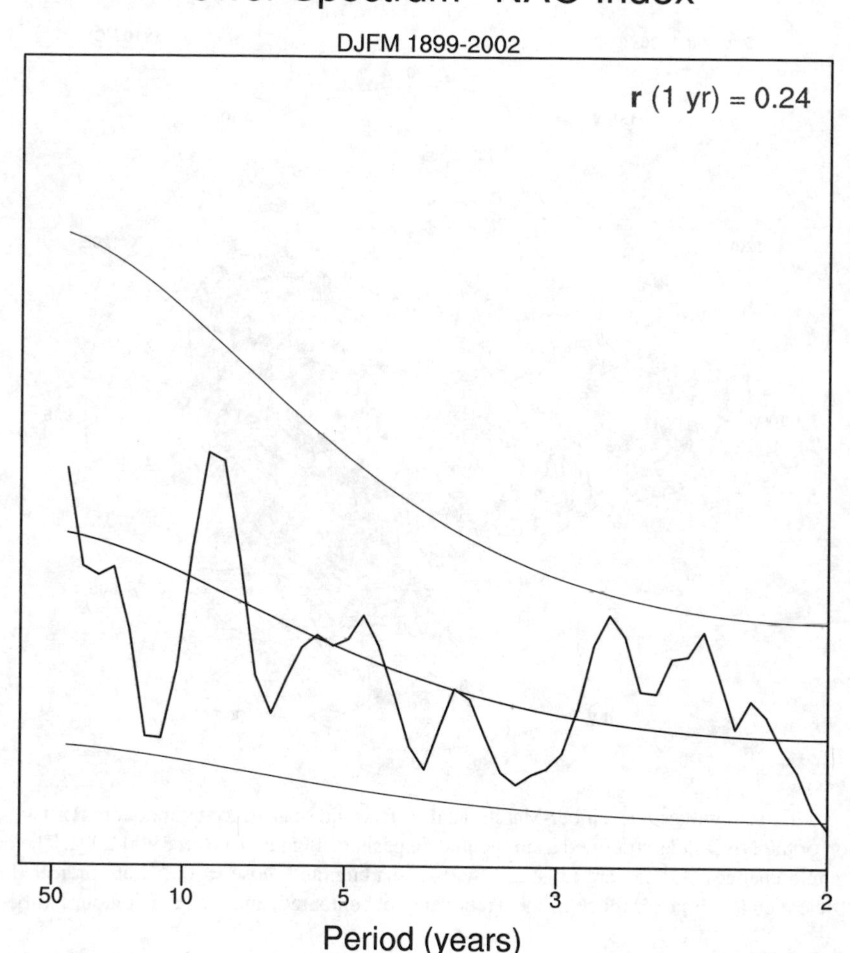

Figure 12. Power spectrum of the mean winter (December-March) NAO index over 1899-2002, defined as in the middle panel of Figure 10. Also shown is the corresponding red noise spectrum with the same lag one autocorrelation coefficient (0.24) and the 5 and 95% confidence limits.

5.1. Surface Temperature

When the NAO index is positive, enhanced westerly flow across the North Atlantic during winter (Figure 7) moves relatively warm (and moist) maritime air over much of Europe and far downstream, while stronger northerly winds over Greenland and northeastern Canada carry cold air southward and decrease land temperatures and SST over the northwest Atlantic (Figure 13). Temperature variations over North Africa and the Middle East (cooling), as well as North America (warming), associated with the stronger clockwise flow around the subtropical Atlantic high-pressure center are also notable.

This pattern of temperature change is important. Because the heat storage capacity of the ocean is much greater than that of land, changes in continental surface temperatures are much larger than those over the oceans, so they tend to dominate average NH (and global) temperature variability [e.g., *Wallace et al.*, 1995; *Hurrell and Trenberth*, 1996]. Given the especially large and coherent NAO signal across the Eurasian continent from the Atlantic to the Pacific, it is not surprising that NAO variability contributes significantly to interannual and longer-term variations in NH surface temperature during winter. *Jones et al.* [this volume] show that the strength of this relationship can change over time, both locally and regionally. This aspect has implications for proxy-based reconstructions of past NAO variability [*Cook*, this volume].

Much of the warming that has contributed to the often-cited global temperature increases of recent decades has occurred during winter and spring over the northern continents [*Folland et al.*, 2001]. Since the early 1980s, winter

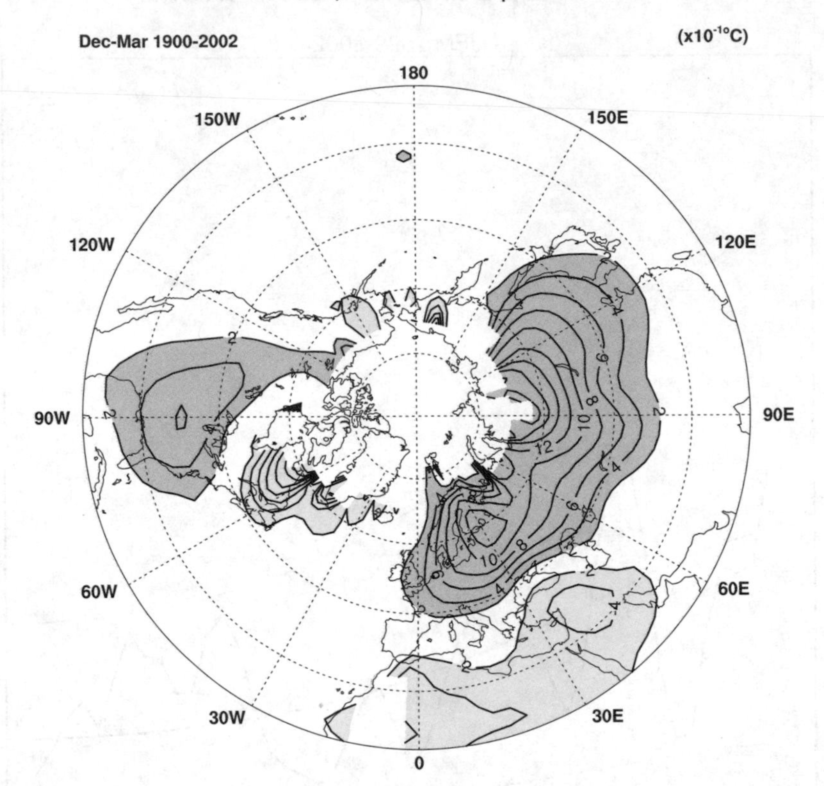

Figure 13. Changes in mean winter (December-March) land surface and sea surface temperatures (x 10^{-1}°C) corresponding to a unit deviation of the NAO index (defined as in the middle panel of Figure 10) over 1900-2002. The contour increment is 0.2°C. Temperature changes > 0.2°C are indicated by dark shading, and those < -0.2°C are indicated by light shading. Regions of insufficient data (e.g., over much of the Arctic) are not contoured, and the zero contour has been excluded.

temperatures have been 1-2°C warmer-than-average over much of North America and from Europe to Asia, while temperatures over the northern oceans have been slightly colder-than-average (Figure 14, upper panel). This pattern is strongly related to changes in the atmospheric circulation, which are reflected by lower-than-average SLP over the middle and high latitudes of the North Pacific and North Atlantic, as well as over much of the Arctic, and higher-than-average SLP over the subtropical Atlantic (Figure 14, lower panel). The Atlantic sector SLP changes clearly reflect the predominance of the positive index phase of the NAO over this period (Figure 10), while the North Pacific changes correspond to an intensification of the Aleutian low-pressure system (and an enhancement of the middle tropospheric PNA pattern) driven, at least in part, by decadal variations in ENSO [e.g., *Trenberth and Hurrell*, 1994].

Hurrell [1996] used multivariate linear regression to quantify the influence of atmospheric circulation variability associated with the NAO and ENSO on NH winter-mean surface temperatures. He showed that much of the local cooling in the northwest Atlantic and the warming across Europe and downstream over Eurasia (Figure 14) resulted directly from decadal changes in the North Atlantic atmospheric circulation in the form of the NAO [see also *Thompson et al.*, 2000], and that the NAO (ENSO) accounted for 31% (16%) of the wintertime interannual variance of NH extratropical temperatures over the latter half of the 20th century. Moreover, changes in the atmospheric circulation associated with the NAO and ENSO accounted (linearly) for much, but not all, of the hemispheric warming through the mid-1990s [*Hurrell*, 1996]. The warming of the most recent winters, however, is beyond that that can be linearly explained by changes in the NAO or ENSO (not shown). Over 1999-2002, for instance, record warmth was recorded while generally cold conditions prevailed in the tropical Pacific and NAO-related circulation anomalies were weak.

5.2. Storms and Precipitation

Changes in the mean circulation patterns over the North Atlantic associated with the NAO are accompanied by changes in the intensity and number of storms, their paths,

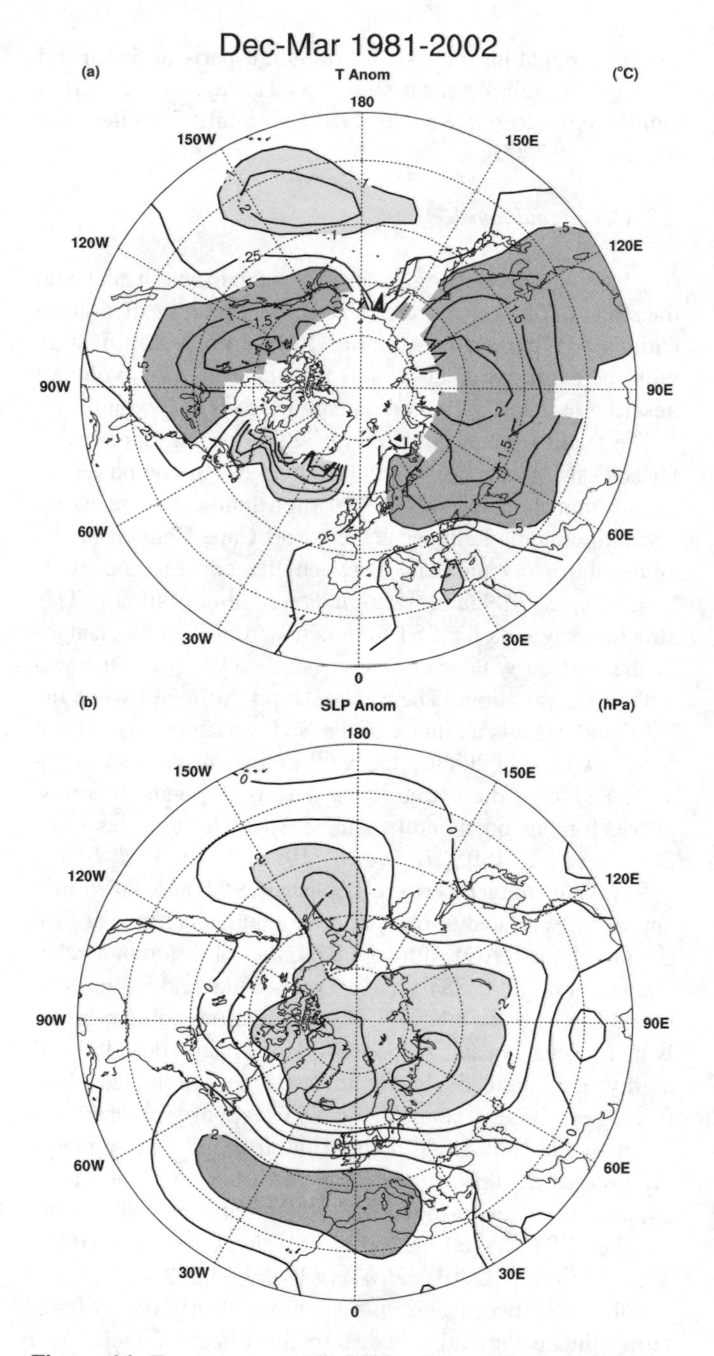

Figure 14. Twenty-two (1981-2002) winter (December-March) average (a) land surface and sea surface temperature anomalies and (b) sea level pressure anomalies expressed as departures from the 1951-1980 means. Temperature anomalies > 0.25°C are indicated by dark shading, and those < -0.10°C are indicated by light shading. The contour increment is 0.1°C for negative anomalies, and the 0.25, 0.5, 1.0, 1.5, and 2.0°C contours are plotted for positive anomalies. Regions with insufficient temperature data are not contoured. The same shading convention is used for sea level pressure, but for anomalies greater than 2 hPa in magnitude. The contour increment in (b) is 1 hPa.

and their weather. During winter, a well-defined storm track connects the North Pacific and North Atlantic basins, with maximum storm activity over the oceans (Figure 15). The details of changes in storminess differ depending on the analysis method and whether one focuses on surface or upper-air features. Generally, however, positive NAO index winters are associated with a northeastward shift in the Atlantic storm activity (Figure 15) with enhanced activity from Newfoundland into northern Europe and a modest decrease in activity to the south [*Rogers*, 1990, 1997; *Hurrell and van Loon,* 1997; *Serreze et al.*, 1997; *Alexandersson et al.*, 1998]. Positive NAO index winters are also typified by more intense and frequent storms in the vicinity of Iceland and the Norwegian Sea [*Serreze et al.*, 1997; *Deser et al.*, 2000].

The ocean integrates the effects of storms in the form of surface waves, so that it exhibits a marked response to long lasting shifts in the storm climate. The recent upward trend toward more positive NAO index winters has been associated with increased wave heights over the northeast Atlantic and decreased wave heights south of 40°N [*Bacon and Carter*, 1993; *Kushnir et al.*, 1997; *Carter*, 1999]. Such changes have consequences for the regional ecology, as well as for the operation and safety of shipping, offshore industries such as oil and gas exploration, and coastal development.

Changes in the mean flow and storminess associated with swings in the NAO index are also reflected in pronounced changes in the transport and convergence of atmospheric moisture and, thus, the distribution of evaporation (E) and precipitation (P) [*Hurrell,* 1995a; *Dickson et al.*, 2000]. Evaporation exceeds precipitation over much of Greenland and the Canadian Arctic during high NAO index winters (Figure 16), where changes between high and low NAO index states are on the order of 1 mm day^{-1}. Drier conditions of the same magnitude also occur over much of central and southern Europe, the Mediterranean and parts of the Middle East, whereas more precipitation than normal falls from Iceland through Scandinavia [*Hurrell*, 1995a; *Dai et al.*, 1997; *Dickson et al.*, 2000; *Visbeck et al.*, this volume].

This spatial pattern, together with the upward trend in the NAO index since the late 1960s (Figure 10), is consistent with recent observed changes in precipitation over much of the Atlantic basin. One of the few regions of the world where glaciers have not exhibited a pronounced retreat over the past several decades is in Scandinavia [e.g., *Hagen*, 1995; *Sigurdsson and Jonsson*, 1995) where more than average amounts of precipitation have been typical of many winters since the early 1980s. In contrast, over the Alps, snow depth and duration in many recent winters have been among the lowest recorded this century, and the retreat of Alpine glaciers has been widespread [e.g., *Frank*, 1997].

Mean Storm Track (DJFM) 1958-1998

$(Z'^2)^{1/2}$ 300 hPa (gpm)

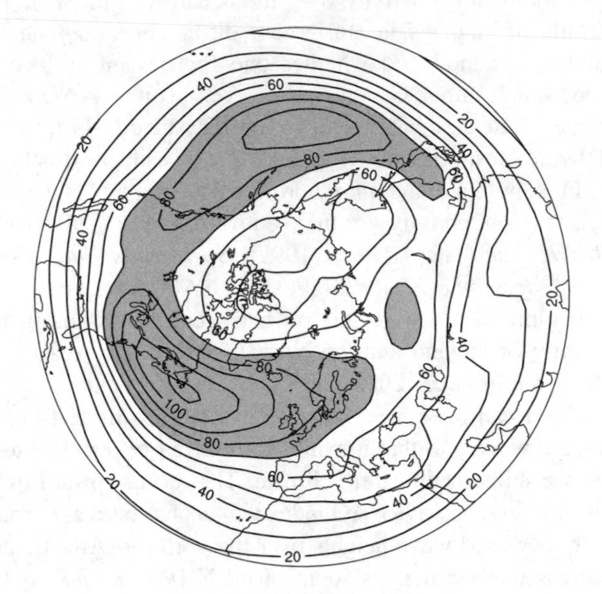

Regression onto NAO Index (gpm)

Figure 15. In the top panel, mean storm tracks for 1958-1998 winters (December-March) as revealed by the 300 hPa root mean square transient geopotential height (gpm) bandpassed to include 2-8 day period fluctuations. Values greater than 70 gpm are shaded and the contour increment is 10 gpm. In the lower panel, anomalies are expressed in terms of amplitude (gpm) by regression onto the NAO index (defined as in the middle panel of Figure 10). The contour increment is 2 gpm, and anomalies greater than 4 gpm in magnitude are shaded. The data come from the NCEP/NCAR reanalyses.

Severe drought has persisted throughout parts of Spain and Portugal as well. As far eastward as Turkey, river runoff is significantly correlated with NAO variability [*Cullen and deMenocal*, 2000].

5.3. Ocean and Sea Ice

It has long been recognized that fluctuations in SST and the strength of the NAO are related, and there are clear indications that the North Atlantic Ocean varies significantly with the overlying atmosphere. *Visbeck et al.* [this volume] describe in detail the oceanic response to NAO variability.

The leading pattern of SST variability during boreal winter consists of a tri-polar structure marked, in one phase, by a cold anomaly in the subpolar North Atlantic, a warm anomaly in the middle latitudes centered off Cape Hatteras, and a cold subtropical anomaly between the equator and 30°N [e.g., *Cayan*, 1992a,b; *Visbeck et al.,* this volume]. This structure suggests the SST anomalies are driven by changes in the surface wind and air-sea heat exchanges associated with NAO variations. The relationship is strongest when the NAO index leads an index of the SST variability by several weeks, which highlights the well-known result that large-scale SST over the extratropical oceans responds to atmospheric forcing on monthly and seasonal time scales [e.g., *Battisti et al.*, 1995; *Delworth*, 1996; *Deser and Timlin*, 1997]. Over longer periods, persistent SST anomalies also appear to be related to persistent anomalous patterns of SLP (including the NAO), although a number of different mechanisms can produce SST changes on decadal and longer time scales [e.g., *Kushnir,* 1994]. Such fluctuations could primarily be the local oceanic response to atmospheric decadal variability. It is quite likely, for instance, that sustained NAO forcing results in a hemispheric SST response, in which the northern and subtropical parts of the tri-polar pattern merge [*Visbeck et al.,* this volume]. On the other hand, non-local dynamical processes in the ocean could also be contributing to the SST variations [e.g., *Visbeck et al.,* 1998; *Krahmann et al.,* 2001; *Eden and Willebrand,* 2001].

Subsurface ocean observations more clearly depict long-term climate variability, because the effect of the annual cycle and month-to-month variability in the atmospheric circulation decays rapidly with depth. These measurements are much more limited than surface observations, but over the North Atlantic they too indicate fluctuations that are coherent with the low frequency winter NAO index to depths of 400 m [*Curry and McCartney*, 2001].

The oceanic response to NAO variability is also evident in changes in the distribution and intensity of winter convective activity in the North Atlantic. The convective renewal of intermediate and deep waters in the Labrador Sea and the

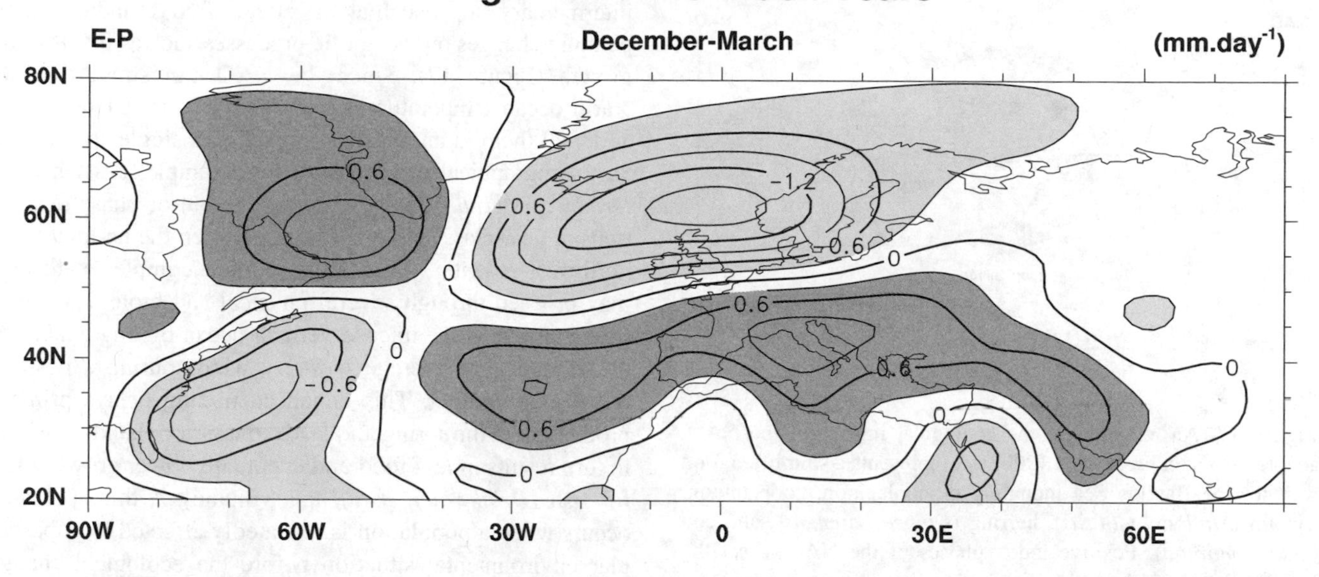

Figure 16. Difference in mean winter (December-March) evaporation (E) minus precipitation (P) between years when the NAO index exceeds one standard deviation. The NAO index is defined as in the middle panel of Figure 10, and nine winters enter into both the high index and the low index composites. The E-P field is obtained as a residual of the atmospheric moisture budget [see *Hurrell*, 1995a]. The calculation was based on the NCEP/NCAR reanalyses over 1958-2001, and truncated to 21 wavenumbers. The contour increment is 0.3 mm day^{-1}, differences greater than 0.3 mm day^{-1} (E exceeds P) are indicated by dark shading, and differences less than -0.3 mm day^{-1} (P exceeds E) are indicated by light shading.

GIN Seas contribute significantly to the production and export of North Atlantic Deep Water and, thus, help to drive the global thermohaline circulation. The intensity of winter convection at these sites is not only characterized by large interannual variability, but also interdecadal variations that appear to be synchronized with variations in the NAO [*Dickson et al.*, 1996]. Deep convection over the Labrador Sea, for instance, was at its weakest and shallowest in the postwar instrumental record during the late 1960s. Since then, Labrador Sea Water has become progressively colder and fresher, with intense convective activity to unprecedented ocean depths (> 2300 m) in the early 1990s [*Visbeck et al.,* this volume; their Figure 10]. In contrast, warmer and saltier deep waters in recent years are the result of suppressed convection in the GIN Seas, whereas tracer evidence suggests that intense convection likely occurred during the late 1960s [*Schlosser et al.,* 1991].

Some global warming scenarios have suggested that the next decades might show a preferred positive index phase of the NAO [*Gillett et al.*, this volume]. This would lead to increased deep water formation in the Labrador Sea region which might offset, or at least delay, the buildup of fresh water, which in many models leads to a sudden reduction of the thermohaline circulation [*Delworth and Dixon*, 2000; *Cubasch et al.,* 2001].

For this reason there has also been considerable interest in the past occurrences of low salinity anomalies that propagate around the subpolar gyre of the North Atlantic. The most famous example is the Great Salinity Anomaly (GSA) [*Dickson et al.*, 1988]. The GSA formed during the extreme negative index phase of the NAO in the late 1960s (Figure 10), when clockwise flow around anomalously high pressure over Greenland fed record amounts of freshwater from the Arctic Ocean through the Fram Strait into the Nordic Seas. From there some of the fresh water passed through the Denmark Strait into the subpolar North Atlantic Ocean gyre. There have been other similar events as well, and statistical analyses have revealed that the generation [*Belkin et al.*, 1998] and termination [*Houghton and Visbeck*, 2002] of these propagating salinity modes are closely connected to a pattern of atmospheric variability strongly resembling the NAO.

The strongest interannual variability of Arctic sea ice occurs in the North Atlantic sector. The sea ice fluctuations display a seesaw in ice extent between the Labrador and Greenland Seas. Strong interannual variability is evident in the sea ice changes, as are longer-term fluctuations including a trend over the past 30 years of diminishing (increasing) ice concentration during boreal winter east (west) of Greenland. Associated with the sea ice fluctuations are

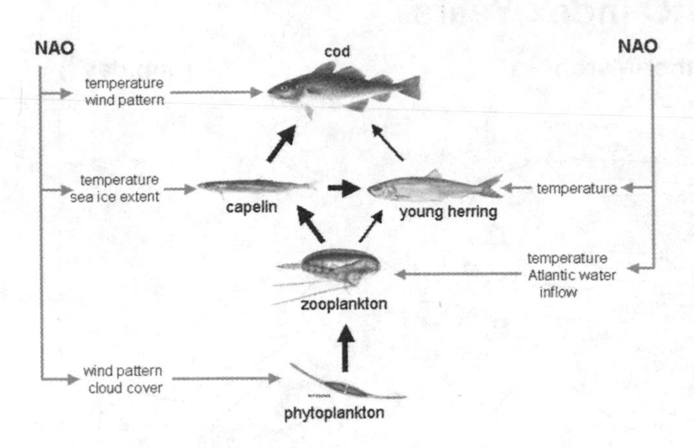

Figure 17. An example of the ecological impact of the NAO, adapted from *Stenseth et al.* [2002]. It represents a simplified food web for the Barents Sea including phytoplankton, zooplankton, capelin (*Mallotus villosus*), herring (*Clupea harengus*), and cod (*Gadus morhua*). Positive index phases of the NAO affect the Barents Sea through increasing volume flux of warm water from the southwest, cloud cover and air temperature, all leading to increased water temperature, which influences fish growth and survival both directly and indirectly [*Ottersen and Stenseth*, 2001].

large-scale changes in SLP that closely resemble the NAO [*Deser et al.*, 2000].

When the NAO is in its positive index phase, the Labrador Sea ice boundary extends farther south while the Greenland Sea ice boundary is north of its climatological extent. This is qualitatively consistent with the notion that the atmosphere directly forces the sea ice anomalies, either dynamically via wind-driven ice drift anomalies, or thermodynamically through surface air temperature anomalies. The relationship between the NAO index and an index of the North Atlantic ice variations is strong, although that it does not hold for all individual winters [*Deser et al.*, 2000; *Hilmer and Jung,* 2000; *Lu and Greatbatch*, 2002] illustrates the importance of the regional atmospheric circulation in forcing the extent of sea ice.

5.4. Ecology

Over the last couple of years interest in the ecological impacts of NAO variability has increased markedly [e.g., *Ottersen et al.*, 2001; *Walther et al.*, 2002; *Stenseth et al.*, 2002]. *Drinkwater et al.* [this volume], *Mysterud et al.* [this volume] and *Straile et al.* [this volume] show the NAO affects a broad range of marine, terrestrial and freshwater ecosystems across large areas of the NH, diverse habitats and different trophic levels. Although such effects are far-reaching, the nature of the impacts varies considerably.

Ottersen et al. [2001] attempted to systematize the ecological effects of NAO variability, and they identified three possible pathways. The first is relatively simple with few intermediary steps, such as the effect of NAO-induced temperature changes on metabolic processes such as feeding and growth (Figure 17). Since the NAO can simultaneously warm ocean temperatures in one part of the Atlantic basin and cool them in another, its impact on a single species can vary geographically. An interesting example, described by *Drinkwater et al.* [this volume], is the out-of-phase fluctuations in year-class strength of cod between the northeast and northwest Atlantic. Alternatively, more complex pathways may proceed through several physical and biological steps. One example is the intense vertical ocean mixing generated by stronger-than-average westerly winds during a positive NAO index winter. This enhanced mixing delays primary production in the spring and leads to less zooplankton, which in turn results in less food and eventually lower growth rates for fish [*Drinkwater et al.,* this volume]. A third pathway occurs when a population is repeatedly affected by a particular environmental situation before the ecological change can be perceived (biological inertia), or when the environmental parameter affecting the population is itself modulated over a number of years [physical inertia; *Heath et al.,* 1999].

Mysterud et al. [this volume] demonstrate how the NAO influences a wide range of terrestrial animals and plants, including the intriguing example of red deer on the west coast of Norway. These animals stay in low land regions during winter, while in summer they forage at higher elevations. Altitude is a key factor determining whether precipitation comes as rain or snow, and it thereby explains why positive NAO index conditions are favorable for these red deer populations. Two separate mechanisms operate. First, warm and rainy conditions in the low-elevation wintering areas decrease energetic costs of thermoregulation and movement while they increase access to forage in the field layer during winter. Second, more winter snowfall at high elevations leads to a prolonged period of access of high quality forage during summer.

Although research on the influence of the NAO on freshwater ecosystems is still in its early stages, *Straile et al.* [this volume] show that a pronounced effect on the physics, chemistry and biology of many NH lakes and rivers is apparent. To a large extent, the strong, coherent impact of the NAO on European lakes through the year is set up in winter and early spring. This time of year is critical to lakes, as both spring turnover and the onset of stratification occur. Thus, strong variations in climate driven by the NAO exert a major impact on the distribution and seasonal development of temperature and nutrients, as well as influence the time of onset, and the rate, of plankton succession. But although the NAO strongly influences a diversity of freshwater ecosystems, the actual effects differ with altitude, latitude, size, and depth of a lake.

The studies of the NAO impact on terrestrial [*Mysterud et al.*, this volume] and freshwater [*Straile et al.*, this volume] ecosystems are mainly from the eastern side of the Atlantic, and few results are available on several large groups of animals and plants. Since the study of the ecological impacts of the NAO is still relatively new and conducted by only a few scientists, it is not yet possible to determine if the NAO influences on ecology are more pronounced over Europe, or if the studies reflect more suitable European data sets and greater interest initially by European scientists. Nevertheless, interest will no doubt continue to grow leading to many new insights [*Stenseth et al.*, 2002].

5.5. Economy

Significant changes in winter rainfall and temperature driven by the NAO have the potential to affect the economy. Anecdotal evidence of this is plentiful, ranging from the economic costs of increased protection of coastal regions and the redesign of off shore oil platforms to cope with the recent NAO-driven increases in significant wave heights [e.g., *Kushnir et al.*, 1997] to altered tourism associated with changed snow conditions in the Alps and winter weather in northern Mediterranean countries (section 5.2). Some studies have also suggested the likelihood of major hurricanes land falling on the east coast of the U.S. depends on the phase of the NAO [e.g., *Elsner and Kocher*, 2000; *Elsner et al.*, 2000].

Cullen and deMenocal [2000] studied the connection between the NAO and stream flow of the Euphrates and Tigris rivers. A major issue in this region involves water supply shortages and surpluses for irrigation farming in the Middle East. Decreases in rainfall associated with the long term trend in the NAO index have had catastrophic effects on crop yields and have contributed to high level political disputes on water withdrawals from the rivers between Turkey, where most of the rain falls, and Syria, a downstream riparian neighbor.

A more recent study [J. Cherry, H. Cullen and others, personal communication] investigated the economic impact of NAO-induced changes in temperature and precipitation over Scandinavia. Norway, and its energy trade with Sweden, was examined because the Norwegian energy sector has qualities that are particularly useful for analyzing the relationship between climate and energy commodities. Norway is the leading Organization of Economic Cooperation and Development (OECD) exporter of oil and gas, and it is the second largest oil exporter in the world. Norway's climate and topography also make it ideal for hydroelectric power generation. More than 99% of electricity generation in Norway comes from hydropower, and Norway has the highest electricity consumption per capita in the world. Sweden, however, generates only 47% of its electricity from hydropower, and another 47% comes from nuclear facilities. Sweden normally adjusts its nuclear electricity production downward during the period from March to October because electricity consumption is low and water reservoirs produce plenty of hydropower. Both Norway and Sweden underwent market deregulation and increased privatization during the 1990s. This made international trade in electricity feasible in Scandinavia. In 1993, Nord Pool was established as the world's first multinational exchange for electric power trading.

In Norway the reservoir influx and level show a pronounced seasonal cycle with much month-to-month variability (Figure 18). The peak inflow in summer is associated with snowmelt in the mountain ranges. At the inflow maximum in June 1995, enough water to generate 10,000 GWh per week of hydroelectric power flowed into Norwegian reservoirs, while only enough water to generate about 5,000 GWh per week flowed into Swedish reservoirs. At the summer inflow peak in 1996, the numbers dropped to 7,000 and 4,500 GWh per week for Norway and Sweden, respectively.

This dramatic reduction in Norwegian precipitation is reflected in the strongly negative NAO index value of the December 1995-March 1996 winter, which followed a long run of wet winters (and hydropower surpluses) beginning in the late 1980s associated with the positive index phase of the NAO (Figures 10 and 16). There is a strong correlation between the winter NAO index and the amount of rainfall over western Norway (Figure 18), for instance $r \sim 0.8$ at Bergen [*Hurrell*, 1995a]. Rainfall anomalies are clearly evident in reservoir levels, which for instance were 40% below average in 1996 (Figure 18), and ultimately in hydroelectric power generation. The approximate 6-month time lag between the winter precipitation and reservoir level anomalies illustrates that much of the freshwater is stored in the winter snow pack.

The large swing in the phase of the NAO between 1995 and 1996 brought international attention to the physical connection between the NAO and the availability of water in Scandinavia for hydropower generation. Norwegian power producers sold contracts during 1995 to end-users both locally and abroad. But, during the relatively dry conditions of 1996, these contracts could not be met with Norwegian hydropower alone: power had to be purchased on short-term markets at high cost to the industry and consumers. Still embroiled in its energy crisis early in 1997, Norway imported a large amount of electricity from coal-fired power plants in Denmark, straining the commitment both countries made under the Kyoto Protocol.

Because the climates in Norway and Sweden are impacted somewhat differently by the NAO, and the electricity generation sources are distributed differently between hydropow-

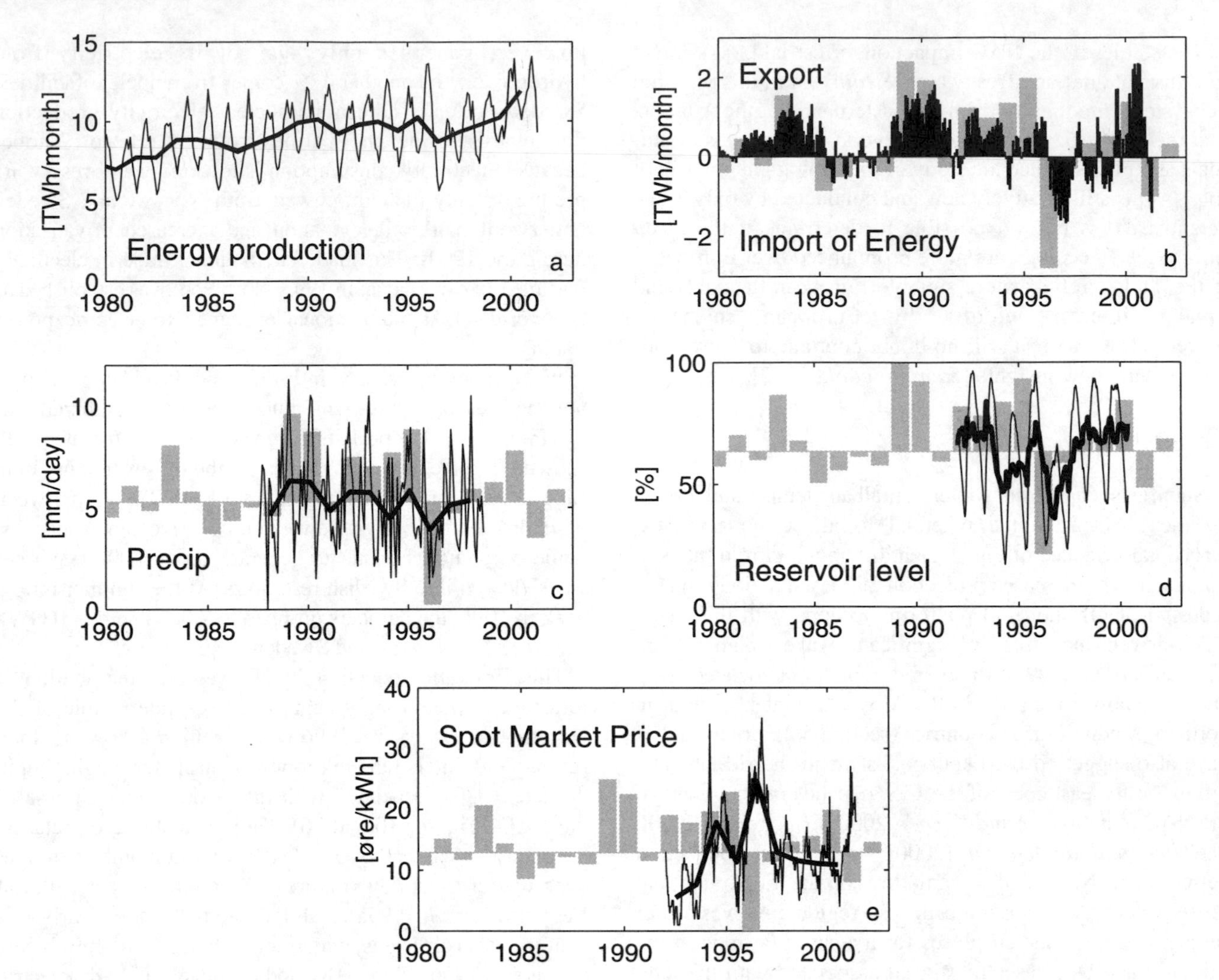

Figure 18. a) Monthly hydroelectric energy production in TWh per month for Norway (thin line). The heavy line denotes the annual average. b) Difference between hydroelectric energy production and electric energy consumption in Norway. Positive values represent times when Norway is able to export energy to Sweden and negative values indicate a shortage in hydropower production. The heavy gray bars in the background show the winter mean NAO index for this and all of the following graphs. c) Monthly time series of the first empirical orthogonal function of rainfall obtained from a gridded station based data set for all of Scandinavia (thin line). Norway has the largest spatial loading in the area where most of the hydroelectric dams are located. The heavy line denotes the annual average. d) Monthly average reservoir level for all hydroelectric dams in Norway in % (thin line). The heavy line denotes the departure from the mean annual cycle with the total time mean added back. e) Monthly price of electric energy on the Scandinavian energy spot market (thin line). The solid represents the annual average.

er, nuclear power, and fossil fuels, each country may have a natural competitive advantage under particular climate events. If the NAO is in a negative index phase and winter precipitation is relatively low, power producers in Norway can meet supply contracts by buying nuclear power from Sweden. In contrast, when the NAO is in its positive index phase and hydroelectric power is plentiful in Norway, Sweden may find it cheaper to buy hydropower from Norway than to produce that power itself. The existence of electricity trade means that Norway can maintain its heavy reliance on hydropower instead of burning fossil fuels. Figure 18 shows how, in recent years, the NAO has influenced both the trade amount and price of energy on the spot market.

6. MECHANISMS

Although the overwhelming indication is that the NAO is a mode of variability internal to the atmosphere, there is some evidence that external factors such as volcanic aerosols, anthropogenic influences on the atmospheric composition, and variations in solar activity can influence its phase and amplitude. Moreover, it has been argued that interactions between the atmosphere and the underlying surface, or between the troposphere and stratosphere, can lend a "low-frequency" component to the NAO variability, such that limited prediction is plausible. At present there is no consensus on the relative roles such processes play in NAO variability, especially where long (interdecadal) times scales are concerned. Considering the significant impact the NAO exerts on the climate of the NH, understanding the mechanisms that control and affect the NAO is therefore crucial to the current debate on climate variability and change. *Thompson et al.* [this volume], *Czaja et al.* [this volume], and *Gillett et al.* [this volume] provide more detailed discussions of external physical processes thought to affect the NAO in the context of thorough overviews of the dynamical factors that give rise to the horizontal and vertical structure of NAO variability, as well as its amplitude and time scales. Here we provide a quick look at these mechanisms and the debate surrounding their importance.

6.1. Atmospheric Processes

Atmospheric general circulation models (AGCMs) provide strong evidence that the basic structure of the NAO arises from the internal, nonlinear dynamics of the atmosphere. The observed spatial pattern and amplitude of NAO anomalies are well simulated in AGCMs forced with climatological annual cycles (no interannual variations) of all forcings "external" to the atmosphere, like insolation, SST, sea ice, snow cover, and land surface moisture, as well as fixed atmospheric trace-gas composition [e.g., *Barnett*, 1985; *James and James*, 1989; *Kitoh et al.*, 1996; *Osborn et al.*, 1999]. The results from one such integration are illustrated in Figure 19. *Thompson et al.* [this volume] discuss in depth the governing atmospheric dynamical mechanisms, of which interactions between the time-mean flow and the departures from that flow (the so-called transient eddies) are central and give rise to a fundamental time scale for NAO fluctuations of about 10 days [*Feldstein*, 2000]. Such intrinsic atmospheric variability exhibits little temporal coherence (Figure 19), mostly consistent with the time scales of observed NAO variability (Figures 10 and 12) and the climate noise paradigm discussed earlier (section 4.2).

A possible exception to this reference is the enhanced NAO variability over the latter half of the 20th century [*Feldstein*, 2000], including the apparent upward trend in the boreal winter NAO index (Figure 19). *Thompson et al.* [2000] concluded that the linear component of the observed upward trend of the index is statistically significant relative to the degree of internal variability that it exhibits, and *Gillett et al.* [2001] and *Feldstein* [2002] showed that the upward trend is statistically significant compared to appropriate red noise models [*Trenberth and Hurrell*, 1999; although see *Wunsch,* 1999]. *Osborn et al.* [1999] showed, furthermore, that the observed trend in the winter NAO index is outside the 95% range of internal variability generated in a 1,400 year control run with a coupled ocean-atmosphere climate model, and *Gillett et al.* [this volume] reach the same conclusion based on their examination of multi-century control runs from seven different coupled climate models. This indicates that either the recent climate change is due in part to external forcing, or all of the models are deficient in their ability to simulate North Atlantic interdecadal variability (although the simulated variability was similar to that observed in the instrumental record prior to 1950). Comparisons to NAO indices reconstructed from proxy data have also concluded the recent behavior is unusual, although perhaps not unprecedented [*Jones et al.* 2001; *Cook*, this volume]; however, such extended proxy records are liable to considerable uncertainties [e.g., *Schmutz et al.,* 2000; *Cook*, this volume].

A possible source of the trend in the winter NAO index could be external processes that affect the strength of the atmospheric circulation in the lower stratosphere on long time scales, such as increases in GHG concentrations [*Gillett et al.*, this volume]. In contrast to the mean flow (Figures 1-3), the NAO has a pronounced "equivalent barotropic" structure (i.e., it does not display a westward tilt with elevation), and its anomalies increase in amplitude with height in rough proportion to the strength of the mean zonal wind [*Thompson et al.*, this volume]. In the lower stratosphere, the leading pattern of geopotential height variability is characterized by a much more annular (zonally symmetric) structure than in the troposphere. When heights over the polar region are lower than normal, heights at nearly all longitudes in middle latitudes are higher than normal, and *vice versa*. In the former phase, the stratospheric westerly winds that encircle the pole are enhanced and the polar vortex is "strong" and anomalously cold. Simultaneously at the surface, the NAO tends to be in its positive index phase [e.g., *Baldwin*, 1994; *Perlwitz and Graf*, 1995; *Kitoh et al.*, 1996; *Kodera et al.*, 1996; *Baldwin and Dunkerton*, 1999].

The trend in the NAO phase and strength during the last several decades has been associated with a stratospheric trend toward much stronger westerly winds encircling the pole and anomalously cold polar temperatures [e.g., *Randel and Wu*, 1999; *Thompson et al.*, 2000]. There is a considerable body of research and observational evidence to support the notion that variability in the troposphere can drive variability in the stratosphere, but it now appears that some stratospheric control of the troposphere may also occur [e.g., *Baldwin and Dunkerton*, 2001]. *Thompson et al.* [this volume] review the evidence for such "downward control" and evaluate the possible mechanisms, which likely involve the effect of the stratospheric flow on the refraction of planetary waves dispersing upwards from the troposphere [e.g., *Hartmann et al.*, 2000; *Shindell et al.*, 2001; *Ambaum and Hoskins*, 2002], although more direct momentum forcing could also be important [e.g., *Haynes et al.*, 1991; *Black*, 2002].

The atmospheric response to strong tropical volcanic eruptions provides some evidence for a stratospheric influence on the surface climate [*Gillett et al.*, this volume]. Volcanic aerosols act to enhance north-south temperature gradients in the lower stratosphere by absorbing solar radiation in lower latitudes, which produces warming. In the troposphere, the aerosols exert only a very small direct influence [*Hartmann et al.*, 2000]. Yet, the observed response following eruptions does not include only lower geopotential heights over the pole with stronger stratospheric westerlies, but also a strong, positive NAO-like signal in the tropospheric circulation [e.g., *Robock and Mao*, 1992; *Kodera*, 1994; *Graf et al.*, 1994; *Kelley et al.*, 1996].

Reductions in stratospheric ozone and increases in GHG concentrations also appear to enhance the meridional temperature gradient in the lower stratosphere, via radiative cooling of the wintertime polar regions. This change implies a stronger polar vortex. It is possible, therefore, that the upward trend in the boreal winter NAO index in recent decades (Figure 10) is associated with trends in either or both of these trace-gases quantities. In particular, a decline in the amount of ozone poleward of 40°N has been observed during the last two decades [*Randel and Wu*, 1999].

Shindell et al. [1999] subjected a coupled ocean-atmosphere model to realistic increases in GHG concentrations and found a trend toward the positive index phase of the NAO in the surface circulation. In similar experiments with different coupled models, *Ulbrich and Christoph* [1999] and *Fyfe et al.* [1999] also found an increasing trend in the NAO index, although *Osborn et al.* [1999] and *Gillett et al.* [2000] did not. This led *Cubasch et al.* [2001] to conclude that there is not yet a consistent picture of the NAO response to increasing concentrations of greenhouse gases. *Gillett et al.* [this volume], however, examine 12 coupled ocean-atmosphere models and find 9 show an increase in the boreal winter NAO index in response to increasing GHG levels, although the results are somewhat sensitive to the definition of the NAO index used (section 3). This led them to conclude that increasing GHG concentrations have contributed to a strengthening of the North Atlantic surface pressure gradient [see also *Osborn*, 2002]. Forcing from stratospheric ozone depletion has generally been found to have a smaller effect than GHG changes on the NAO [e.g., *Graf et al.*, 1998; *Shindell et al.*, 2001], although *Volodin and Galin* [1999] claim a more significant influence. *Gillett et al.* [this volume] synthesize the diverse and growing body of literature dealing with how the NAO might change in response to several anthropogenic forcings.

6.2. The Ocean's Influence on the NAO

In the extratropics, the atmospheric circulation is the dominant driver of upper ocean thermal anomalies [section 5.3; *Visbeck et al.*, this volume]. A long-standing issue, however, has been the extent to which the forced, anomalous extratropical SST field feeds back to affect the atmosphere [*Kushnir et al.*, 2002]. Most evidence suggests this effect is quite small compared to internal atmospheric variability [e.g., *Seager et al.*, 2000]. Nevertheless, the interaction between the ocean and atmosphere could be important for understanding the details of the observed amplitude of the NAO and its longer-term temporal evolution [*Czaja et al.*, this volume], as well as the prospects for meaningful predictability [*Rodwell et al.*, this volume].

The argument for an oceanic influence on the NAO goes as follows: while intrinsic atmospheric variability exhibits temporal incoherence, the ocean tends to respond to it with marked persistence or even oscillatory behavior. The time scales imposed by the heat capacity of the upper ocean, for example, leads to low frequency variability of both SST and lower tropospheric air temperature [*Frankignoul and Hasselmann*, 1977; *Manabe and Stouffer*, 1996; *Barsugli and Battisti*, 1998]. Other studies suggest that basin-wide, spatially-coherent atmospheric modes such as the NAO interact with the mean oceanic advection in the North Atlantic to preferentially select quasi-oscillatory SST anomalies with long time scales [*Saravanan and McWilliams*, 1998] and even excite selected dynamical modes of oceanic variability that act to redden the SST spectrum [*Griffies and Tziperman*, 1995; *Frankignoul et al.*, 1997; *Capotondi and Holland*, 1997; *Saravanan and McWilliams*, 1997; 1998; *Saravanan et al.*, 2000]. These theoretical studies are supported by observations that winter SST anomalies, born in the western subtropical gyre, spread eastward along the path of the Gulf Stream and North Atlantic Current with a transit time of roughly a decade [*Sutton and Allen*, 1997; *Krahmann et al.*, 2001]. *Czaja et al.* [this volume] offer a detailed examination of all these possibilities of ocean-atmosphere interaction.

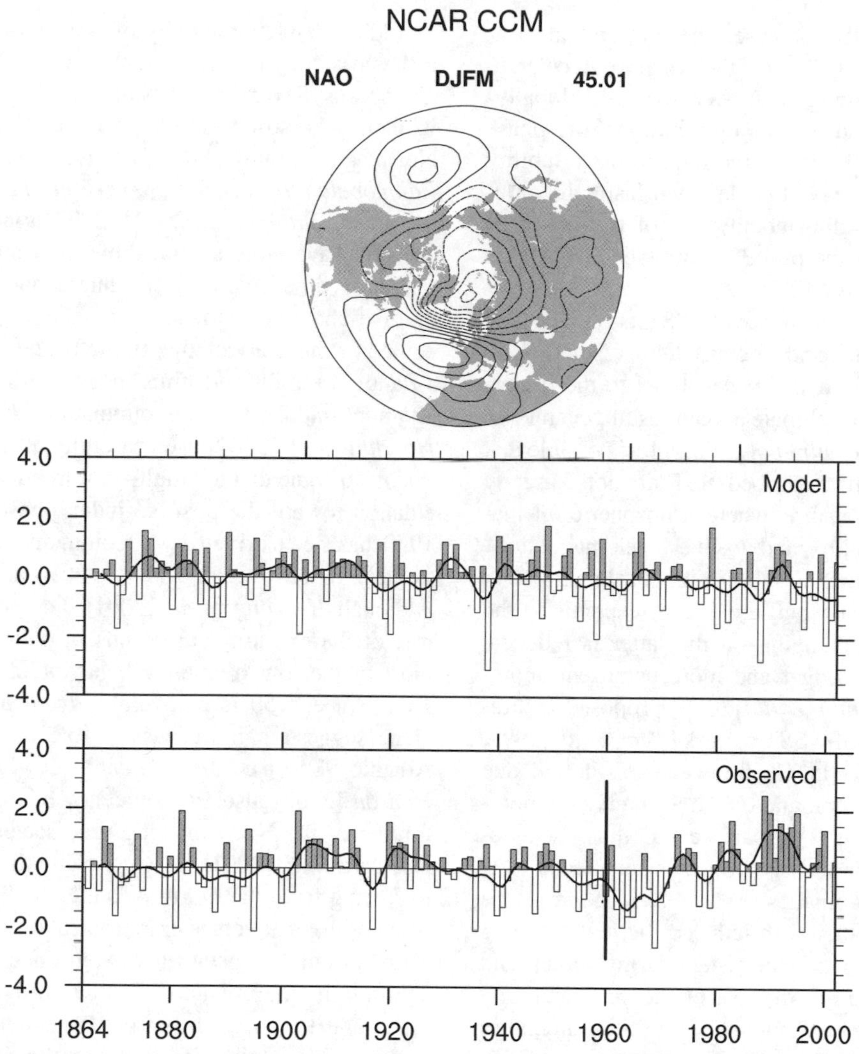

Figure 19. Leading empirical orthogonal function (EOF 1) of the winter (December-March) mean sea level pressure anomalies over (top) the North Atlantic sector (20°-70°N, 90°W-40°E), and the percentage of the total variance it explains in a 200 year integration with the NCAR Community Climate Model (CCM) with climatological annual cycles of all forcings external to the atmosphere. The pattern is displayed in terms of amplitude (hPa), obtained by regressing the hemispheric sea level pressure anomalies upon the leading principal component time series. A subset of 139 years of the latter is plotted in the middle panel for comparison to the station-based NAO index (lower panel) of equivalent length (upper panel of Figure 10). Both the model and observed NAO indices have been normalized. In the top panel, the contour increment is 0.5 hPa, and the zero contour has been excluded.

A key question in this debate is the sensitivity of the middle latitude atmosphere, away from the surface, to changes in SST (and other surface boundary conditions including sea-ice and land surface snow cover). This issue has been addressed in numerous studies, many of them based on AGCM experiments with prescribed SST anomalies. *Palmer and Sun* [1985] and *Peng et al.* [1995], for instance, showed that a stationary, warm SST anomaly south of Newfoundland, in the Gulf Stream Extension region, forces a high-pressure response over the North Atlantic, but in spatial quadrature with the NAO pattern. When a more realistic reproduction of the entire basin SST response to NAO variability [*Visbeck et al.*, this volume] is prescribed, a realistic atmospheric NAO pattern emerges as a response, but whether the forcing comes from the tropical or extratropical part of the ocean has not been unequivocally resolved [*Venzke et al.*, 1999; *Sutton et al.*, 2001; *Peng et al.*, 2002]. *Robertson et al.* [2000] reported that changing the SST distribution in the North Atlantic affects the frequency of occurrence of different regional low-frequency atmospheric modes and substantially increases the interannual variability of the NAO simulated by their AGCM. *Rodwell et al.* (1999) show that by forcing their AGCM with globally-varying observed SST and sea ice distributions, the phase (though not the full amplitude) of the long-term variability in the observed, wintertime NAO index over last half century can be captured, including about 50% of the observed strong upward trend over the past 30 years [see also *Mehta et al.*, 2000; *Hoerling et al.*, 2001].

The weakness of AGCM responses to SST anomalies and the occasional confusing and inconsistent results [e.g., *Kushnir and Held,* 1996] created a debate as to the importance of oceanic forcing for climate anomalies in general and the NAO in particular [*Kushnir et al.*, 2002]. It is possible that AGCM experiments with prescribed SST do not correctly represent the processes in nature, where atmospheric fluctuations cause the SST variability, and it is the "back interaction" or feedback of the SST anomalies that is sought. *Barsugli and Battisti* [1998] argued that when the ocean responds to the atmosphere, the thermal damping on the latter is reduced, thereby contributing to stronger and more persistent anomalies in both media. *Kushnir et al.* [2002] proposed a paradigm, based on North Pacific SST studies [*Peng et al.,* 1997; *Peng and Whitaker*, 1999], in which the changes in SST due to large scale circulation anomalies (such as the NAO) modify the temperature gradient at the surface and, thus, the associated baroclinic transient activity, which in turn feeds back on the large scale flow anomalies [e.g., *Hurrell* 1995b]. This mechanism may act together with reduced thermal damping to explain why the NAO is more persistent during winter (and well into March) than during the rest of the year, and why there is a high correlation between late spring SST anomalies and the NAO state in the ensuing fall and early winter [*Czaja and Frankignoul*, 2002; *Rodwell*, this volume].

In their review, *Czaja et al.* [this volume] explore the implication of the reduced thermal damping paradigm, as well as evidence from coupled general circulation model experiments that the climate system exhibits quasi-oscillatory behavior due to the long-term (multi-year) response of the ocean circulation to atmospheric forcing [e.g., *Latif and Barnett*, 1996; *Grötzner et al*., 1998; *Timmermann et al.*, 1998]. These latter studies argue that changes in the oceanic heat transport (perhaps driven by NAO variability) can lead to a negative feedback on the atmosphere and thus reverse the phase of the atmospheric forcing phenomenon. Such behavior is obviously conditioned on the ability of the atmosphere to respond to upper ocean heat content anomalies (see discussion above). The possibility that this delayed circulation response will lead to unstable, oscillatory modes of the climate system has also been explored in simplified models [e.g., *Jin* 1997; *Goodman and Marshall*, 1999; *Weng and Neelin*, 1998].

Adding to the complexity of ocean-atmosphere interaction is the possibility of remote forcing of the NAO from the tropical oceans. Several recent studies have concluded that NAO variability is closely tied to SST variations over the tropical North and South Atlantic [*Xie and Tanimoto*, 1998; *Rajagopalan et al.*, 1998; *Venzke et al.*, 1999; *Robertson et al., 2000*; *Sutton et al.*, 2001]. SST variations in the tropical Atlantic have large spatial dimension and occur on a wide range of time scales, from interannual to decadal. They involve changes in the meridional SST gradient across the equator, which affect the strength and location of tropical Atlantic rainfall and thus possibly influence the North Atlantic middle latitude circulation. On the other hand, *Hoerling et al.* [2001] used carefully designed AGCM experiments to argue that the multi-year trend in North Atlantic circulation toward the positive index phase of the NAO since 1950 has been driven by a commensurate warming of the tropical Indian Ocean and western Pacific surface waters. Although *Hoerling et al.* [2001] did not directly assess the role of the extratropical oceans in their paper, the fact that most of the low frequency behavior of the observed NAO index since 1950 is recoverable from tropical SST forcing alone suggests a more passive role for extratropical North Atlantic SSTs [see also *Peterson et al.*, 2002]. *Sutton and Hodson* [2002] also find evidence of tropical Indian Ocean forcing of the NAO on long time scales, but in contrast to *Hoerling et al.* [2001], they conclude this effect is secondary to forcing from the North Atlantic itself. Clearly, the importance of tropical versus extratropical ocean-atmosphere interaction has not yet been fully determined.

Similarly, the impact of ENSO on the North Atlantic climate, and the NAO in particular, remains open to debate, although most evidence suggests the effects are small but non-trivial. *Rogers* [1984] concluded from an analysis of historical SLP data that simultaneous occurrences of particular modes of ENSO and the NAO "seem to occur by chance", and this is consistent with the fact that there is no significant correlation between indices of the NAO and ENSO on interannual and longer time scales. *Pozo-Vázquez et al.* [2001], however, composited extreme ENSO events and found for cold conditions in the tropical Pacific a statistically significant boreal winter SLP anomaly pattern resembling the positive index phase of the NAO. *Cassou and Terray* [2001b] also argued for an influence of La Niña on the North Atlantic atmosphere, although they found an atmospheric response that more closely resembles the ridge regime of Figure 9. *Sutton and Hodson* [2002] point out, importantly, that the influence of ENSO most likely depends on the state of the North Atlantic itself [see also *Venzke et al.,* 1999; *Mathieu et al.*, 2002]. Moreover, the possibility of an *indirect* link between ENSO and the NAO is suggested by numerous studies that illustrate a direct impact of ENSO on tropical North Atlantic SSTs [e.g., *Enfield and Mayer*, 1997; *Klein et al.*, 1999; *Saravanan and Chang*, 2000; *Chiang et al.*, 2000; 2002]. If this were a strong effect, however, it would be detected by statistical analyses.

6.3. Sea Ice and Land Snow Cover

The role of sea ice and land snow cover in affecting atmospheric variability has received very little attention, especially relative to the role of ocean anomalies. Here, too, the issue is whether changes in surface properties due to the NAO are able to modify its phase and amplitude in turn. As discussed in section 5.3, changes in sea-ice cover in both the Labrador and Greenland Seas as well as over the Arctic are well correlated with NAO variations [*Deser et al.*, 2000]. Since changes in ice cover produce large changes in sensible and latent heat fluxes, it is reasonable to ask if there is a subsequent feedback onto the atmospheric circulation anomalies. *Deser et al.* [2000] suggest from observations that a local response of the atmospheric circulation to reduced sea ice cover east of Greenland in recent years is apparent. However, more recent AGCM experiments, with imposed ice cover anomalies consistent with the observed trend of diminishing (increasing) ice concentration during winter east (west) of Greenland, suggest a weak, negative feedback onto the NAO [C. Deser, personal communication].

Watanabe and Nitta [1999] suggested that land processes are responsible for decadal changes in the NAO. They find that the change toward a more positive wintertime NAO index in 1989 was accompanied by large changes in snow cover over Eurasia and North America. Moreover, the relationship between snow cover and the NAO was even more coherent when the preceding fall snow cover was analyzed, suggesting that the atmosphere may have been forced by surface conditions over the upstream land mass. *Watanabe and Nitta* [1998] reproduce a considerable part of the atmospheric circulation changes by prescribing the observed snow cover anomalies in an AGCM. The debate on the importance of NAO interactions with the earth land and ocean surface is thus far from over.

7. CONCLUSIONS AND CHALLENGES

The NAO is a large-scale atmospheric phenomenon primarily arising from stochastic interactions between atmospheric storms and both the climatological stationary eddies and the time mean jet stream. As such, the month-to-month and even year-to-year changes in the phase and amplitude of the NAO are largely unpredictable. But that external forces might nudge the atmosphere to assume a high or low NAO index value over a particular month or season is important: even a small amount of predictability could be useful considering the significant impact the NAO exerts on the climate of the NH, and a better understanding of how the NAO responds to external forcing is crucial to the current debate on climate variability and change.

A number of different mechanisms that could influence the detailed state of the NAO have been proposed. Within the atmosphere itself, changes in the rate and location of tropical heating have been shown to influence the atmospheric circulation over the North Atlantic and, in particular, the NAO. Tropical convection is sensitive to the underlying SST distribution, which exhibits much more persistence than SST variability in middle latitudes. This might lead, therefore, to some predictability of the NAO phenomenon.

Interactions with the lower stratosphere are a second possibility. This mechanism is of interest because it might also explain how changes in atmospheric composition influence the NAO. For example, changes in ozone, GHG concentrations and/or levels of solar output affect the radiative balance of the stratosphere that, in turn, modulates the strength of the winter polar vortex. Given the relatively long time scales of stratospheric circulation variability (anomalies persist for weeks), dynamic coupling between the stratosphere and the troposphere via wave mean flow interactions could yield a useful level of predictive skill for the wintertime NAO. Such interactions have also been used to rationalize the recent trend in the winter NAO index in terms of global warming.

A third possibility is that the state of the NAO is influenced by variations in heat exchange between the atmosphere and the ocean, sea-ice and/or land systems. A significant amount of numerical experimentation has been done to test the influence of tropical and extratropical SST anomalies on the NAO, and these experiments are now beginning to lead to more conclusive and coherent results.

One of the most urgent challenges is to advance our understanding of the interaction between GHG forcing and the NAO. It now appears as though there may well be a deterministic relationship, which might allow for moderate low frequency predictability and thus needs to be studied carefully. Also, while the predictability of seasonal to interannual NAO variability will most likely remain low, some applications may benefit from the fact that this phenomenon leaves long-lasting imprints on surface conditions, in particular over the oceans. At the same time, the response of marine and terrestrial ecosystems to a shift in the NAO index might enhance or reduce the atmospheric carbon dioxide levels and thus provide a positive or negative feedback.

Acknowledgments. The authors wish to thank C. Cassou for his helpful comments, suggestions and use of his clustering algorithms, A. Phillips for his assistance with many of the figures, and

L. Rothney for formatting the manuscript. The National Center for Atmospheric Research is sponsored by the National Science Foundation. Some of the work was supported by NOAA OGP CLIVAR Atlantic grant NOAA-NA06GP0394.

REFERENCES

Alexandersson, H., T. Schmith, K. Iden, and H. Tuomenvirta, Long-term variations of the storm climate over NW Europe, *The Global Ocean Atmosphere System, 6*, 97–120, 1998.

Ambaum, M. H., and B. J. Hoskins, The NAO troposphere-stratosphere connection, *J. Climate, 15*, 1969–1978, 2002.

Ambaum, M. H. P., B. J. Hoskins, and D. B. Stephenson, Arctic Oscillation or North Atlantic Oscillation?, *J. Climate*, *14*, 3495–3507, 2001.

Appenzeller, C., T. F. Stocker, and M. Anklin, North Atlantic Oscillation dynamics recorded in Greenland ice cores, *Science*, *282*, 446–449, 1998.

Bacon, S., and D. J. T. Carter, A connection between mean wave height and atmospheric pressure gradient in the North Atlantic, *Int. J. Climatol., 13*, 423–436, 1993.

Baldwin, M. P., X. Cheng, and T. J. Dunkerton, Observed correlations between winter-mean tropospheric and stratospheric circulation anomalies, *Geophys. Res. Lett., 21*, 1141–1144, 1994.

Baldwin, M. P., and T. J. Dunkerton, Propagation of the Arctic Oscillation from the stratosphere to the troposphere, *J. Geophys. Res.*, *104*, 30937–30946, 1999.

Baldwin, M. P., and T. J. Dunkerton, Stratospheric harbingers of anomalous weather regimes, *Science*, *294*, 581–584, 2001.

Barnett, T. P., Variations in near-global sea level pressure, *J. Atmos. Sci.*, *42*, 478–501, 1985.

Barnston, A. G., and R. E. Livezey, Classification, seasonality and persistence of low frequency atmospheric circulation patterns, *Mon. Wea. Rev.*, *115*, 1083–1126, 1987.

Barsugli, J. J., and D. S. Battisti, The basic effects of atmosphere-ocean thermal coupling on midlatitude variability, *J. Atmos. Sci.*, *55*, 477–493, 1998.

Battisti, D. S., U. S. Bhatt, and M. A. Alexander, A modeling study of the interannual variability in the wintertime North Atlantic Ocean, *J. Climate, 8*, 3067–3083, 1995.

Belkin, I. M., S. Levitus, J. I. Antonov, and S.-A. Malmberg, "Great Salinity Anomalies" in the North Atlantic, *Prog. Oceanogr.*, *41*, 1–68, 1998.

Bjerknes, J., Atlantic air-sea interactions, *Advances in Geophysics*, *10*, 1–82, 1964.

Black, R. X., Stratospheric forcing of surface climate in the Arctic Oscillation, *J. Climate*, *15*, 268–277, 2002.

Capotondi, A., and W. R. Holland, Decadal variability in an idealized ocean model and its sensitivity to surface boundary conditions, *J. Phys. Oceanogr.*, *27*, 1072-1093, 1997.

Carter, D. J. T., Variability and trends in the wave climate of the North Atlantic: a review, in *Proc. 9th ISOPE Conf.*, Brest Vol. III, pp 12–18, 1999.

Cassou, C., Role de l'ocean dans la variabilite basse frequence de l'atmosphere sur la region Nord Atlantique-Europe. Ph.D dissertation, University of Paul Sabatier, Toulouse, France, 280 pp., 2001.

Cassou, C., and L. Terray, Oceanic forcing of the wintertime low frequency atmospheric variability in the North Atlantic European sector: a study with the ARPEGE model, *J. Climate*, *14,* 4266–4291, 2001a.

Cassou, C., and L. Terray, Dual influence of Atlantic and Pacific SST anomalies on the North Atlantic/Europe winter climate, *Geophys. Res. Lett.*, *28*, 3195–3198, 2001b.

Cayan, D. R., Latent and sensible heat flux anomalies over the northern oceans: Driving the sea surface temperature, *J. Phys. Oceanogr.*, *22*, 859–881, 1992a.

Cayan, D. R., Latent and sensible heat flux anomalies over the northern oceans: The connection to monthly atmospheric circulation, *J. Climate*, *5*, 354–369, 1992b.

Chiang, J. C. H., Y. Kushnir, and A. Giannini, Deconstructing Atlantic intertropical convergence zone variability: Influence of the local cross-equatorial sea surface temperature gradient and remote forcing from the eastern equatorial Pacific, *J. Geophys. Res.*, *107*, 10.1029/2000JD000307, 2002.

Chiang, J. C. H., Y. Kushnir, and S. E. Zebiak, Interdecadal changes in eastern Pacific ITCZ variability and its influence on the Atlantic ITCZ, *Geophys. Res. Lett.*, *27*, 3687–3690, 2000.

Cook, E. R., Multi-proxy reconstructions of the North Atlantic Oscillation (NAO) index: A critical review and a new well-verified winter index reconstruction back to AD 1400, this volume.

Corti, S., F. Molteni, and T. N. Palmer, Signature of recent climate change in frequencies of natural atmospheric circulation regimes, *Nature*, *398*, 799–802, 1999.

Cubasch, U., and co-authors, Projections of future climate change, in *Climate Change 2001, The Scientific Basis*, J. T. Houghton, Y. Ding, D. J. Griggs, M. Noguer, P. J. van der Linden and D. Xiaosu, Eds., pp. 525–582, Cambridge Univ. Press, 2001.

Cullen, H., and P. B. deMenocal, North Atlantic influence on Tigris-Euphrates streamflow, *Int. J. Climatol.*, *20*, 853–863, 2000.

Curry, R. G., and M. S. McCartney, Ocean gyre circulation changes associated with the North Atlantic Oscillation, *J. Phys. Oceanog.*, *31*, 3374–3400, 2001.

Czaja A., and C. Frankignoul, Observed impact of North Atlantic SST anomalies on the North Atlantic Oscillation, *J. Climate*, *15*, 606–623, 2002.

Czaja, A., A. W. Robertson, and T. Huck, The role of Atlantic ocean-atmosphere coupling in affecting North Atlantic Oscillation variability, this volume.

Dai, A., I. Y. Fung, and A. D. Del Genio, Surface observed global land precipitation variations during 1900-88, *J. Climate*, *10*, 2943–2962, 1997.

Delworth, T. L., North Atlantic interannual variability in a coupled ocean-atmosphere model, *J. Climate*, *9*, 2356–2375, 1996.

Delworth, T. L., and K.W. Dixon, Implications of the recent trend in the Arctic/North Atlantic Oscillation for the North Atlantic thermohaline circulation, *J. Climate*, *13*, 3721–3727, 2000.

Deser, C., On the teleconnectivity of the Arctic Oscillation, *Geophys. Res. Lett., 27*, 779–782, 2000.

Deser, C., and M. S. Timlin, Atmosphere-ocean interaction on weekly time scales in the North Atlantic and Pacific. *J. Climate*, *10*, 393-408,1997.

Deser, C., J. E. Walsh, and M. S. Timlin, Arctic sea ice variability in the context of recent atmospheric circulation trends, *J. Climate*, *13*, 617–633, 2000.

Dickson, R. R., J. Lazier, J. Meincke, P. Rhines, and J. Swift, Long-term co-ordinated changes in the convective activity of the North Atlantic, *Prog. Oceanogr.*, *38*, 241–295, 1996.

Dickson, R. R., J. Meincke, S. A. Malmberg, and A. J. Lee, The "Great Salinity Anomaly" in the northern North Atlantic 1968–1982, *Prog. Oceanogr.*, *20*, 103–151, 1988.

Dickson, R. R., T. J. Osborn, J. W. Hurrell, J. Meincke, J. Blindheim, B. Adlandsvik, T. Vigne, G. Alekseev, and W. Maslowski, The Arctic Ocean response to the North Atlantic Oscillation, *J. Climate*, *13*, 2671–2696, 2000.

Dommenget D., and M. Latif, A cautionary note on the interpretation of EOFs, *J. Climate*, 15, 216-225, 2002.

Drinkwater, K. F., and co-authors, The response of marine ecosystems to climate variability associated with the North Atlantic Oscillation, this volume.

Eden, C., and J. Willebrand, Mechanism of interannual to decadal variability of the North Atlantic circulation, *J. Climate*, *14*, 2266–2280, 2001.

Elsner, J. B., and B. Kocher, Global tropical cyclone activity: A link to the North Atlantic Oscillation, *Geophs. Res. Lett.*, *27*, 129–132, 2000.

Elsner, J. B., T. Jagger, and X. F. Niu, Changes in the rates of North Atlantic major hurricane activity during the 20th century, *Geophys. Res. Lett.*, *27*, 1743–1746, 2000.

Enfield, D. B., and D. A. Mayer, Tropical Atlantic sea surface temperature variability and its relation to El Niño-Southern Oscillation, *J. Geophys. Res.*, *102*, 929–945, 1997.

Esbensen, S. K., A comparison of intermonthly and interannual teleconnections in the 700 mb geopotential height field during the Northern Hemisphere winter. *Mon. Wea. Rev.*, *112*, 2016–2032, 1984.

Feldstein, S. B., Teleconnections and ENSO: The timescale, power spectra, and climate noise properties, *J. Climate*, *13*, 4430–4440, 2000.

Feldstein, S. B., The recent trend and variance increase of the Annular Mode, *J. Climate*, *15*, 88–94, 2002.

Folland, C. K., and co-authors, Observed climate variability and change, in *Climate Change 2001, The Scientific Basis*, J. T. Houghton, Y. Ding, D. J. Griggs, M. Noguer, P. J. van der Linden and D. Xiaosu, Eds., pp. 99–181, Cambridge Univ. Press, 2001.

Frank, P., Changes in the glacier area in the Austrian Alps between 1973 and 1992 derived from LANDSAT data. Max Planck Institute Report 242, 21 pp., 1997.

Frankignoul, C., and K. Hasselmann, Stochastic climate models, part II: application to sea-surface temperature variability and thermocline variability, *Tellus*, *29*, 289–305, 1977.

Frankignoul, C., P. Muller, and E. Zorita, A simple model of the decadal response of the ocean to stochastic wind forcing, *J. Phys. Oceanogr.*, *27*, 1533–1546, 1997.

Fyfe, J. C., G. J. Boer, and G. M. Flato, The Arctic and Antarctic Oscillations and their projected changes under global warming, *Geophys. Res. Lett.*, *26*, 1601–1604, 1999.

Gillett, N. P., M. P. Baldwin, and M. R. Allen, Evidence for non-linearity in observed stratospheric circulation changes, *J. Geophys. Res.*, *106*, 7891–7901, 2001.

Gillett, N. P., H. F. Graf, and T. J. Osborn, Climate change and the North Atlantic Oscillation, this volume.

Gillett, N. P., G. C. Hegerl, M. R. Allen, and P. A. Stott, Implications of changes in the Northern Hemisphere circulation for the detection of anthropogenic climate change, *Geophys. Res. Lett.*, *27*, 993–996, 2000.

Goodman J., and J. Marshall, A model of decadal middle-latitude atmosphere-ocean coupled modes, *J. Climate*, *12*, 621–641, 1999.

Graf, H. F., I. Kirchner, and J. Perlwitz, Changing lower stratospheric circulation: The role of ozone and greenhouse gases, *J. Geophys. Res.*, *103*, 11,251–11,261, 1998.

Graf, H. F., J. Perlwitz, and I. Kirchner, Northern Hemisphere tropospheric midlatitude circulation after violent volcanic eruptions, *Contrib. Atmos. Phys.*, *67*, 3–13, 1994.

Greatbatch, R. J., The North Atlantic Oscillation, *Stochastic and Environmental Risk Assessment*, *14*, 213–242, 2000.

Griffies, S. M., and E. Tziperman, A linear thermohaline oscillator driven by stochastic atmospheric forcing, *J. Climate*, *8*, 2440–2453, 1995.

Grötzner, A., M. Latif, and T. P. Barnett, A decadal climate cycle in the North Atlantic Ocean as simulated by the ECHO coupled GCM, *J. Climate*, *11*, 831–847, 1998.

Hagen, J. O., Recent trends in the mass balance of glaciers in Scandinavia and Svalbard, in *Proceedings of the international symposium on environmental research in the Arctic*, Tokyo, Japan, National Institute of Polar Research, 343–354, 1995.

Hansen, J., R. Ruedy, M. Sato, and K. Lo, Global warming continues, *Science*, *295,* 275, 2002.

Hartmann, D. L., J. M. Wallace, V. Limpasuvan, D. W. J. Thompson, and J. R. Holton, Can ozone depletion and greenhouse warming interact to produce rapid climate change?, *Proc. Nat. Acad. Sci.*, *97*, 1412–1417, 2000.

Haynes, P. H., C. J. Marks, M. E. McIntyre, T. G. Shepherd, and K. P. Shine, On the "downward control" of extratropical diabatic circulations by eddy-induced mean zonal forces, *J. Atmos. Sci.*, *48*, 651–678, 1991.

Heath M.R., and co-authors, Climate fluctuations and the spring invasion of the North Sea by *Calanus finmarchicus, Fish. Ocean.*, *8,* 163–176, 1999.

Higuchi, K., J. Huang, and A. Shabbar, A wavelet characterization of the North Atlantic oscillation variation and its relationship to the North Atlantic sea surface temperature, *Int. J. Climatol.*, *19,* 1119–1129.

Hilmer, M., and T. Jung, Evidence for a recent change in the link between the North Atlantic Oscillation and Arctic sea ice export, *Geophy. Res. Lett.*, *27*, 989–992, 2000.

Hoerling, M. P., J. W. Hurrell, and T. Xu, Tropical origins for recent North Atlantic climate change, *Science*, *292*, 90–92, 2001.

Houghton, B., and M. Visbeck, Quasi-decadal salinity fluctuations in the Labrador Sea, *J. Phys. Oceanogr.*, *32*, 687–701, 2002.

Hurrell, J. W., Decadal trends in the North Atlantic oscillation: Regional temperatures and precipitation, *Science*, *269*, 676–679, 1995a.

Hurrell, J. W., An evaluation of the transient eddy forced vorticity balance during northern winter, *J. Atmos. Sci.*, *52*, 2286–2301, 1995b.

Hurrell, J. W., Influence of variations in extratropical wintertime teleconnections on Northern Hemisphere temperature, *Geophys. Res. Lett.*, *23*, 665–668, 1996.

Hurrell, J. W., and C. K. Folland, A change in the summer atmospheric circulation over the North Atlantic, *CLIVAR Exchanges*, *25*, 52–54, 2002.

Hurrell, J. W., and K. E. Trenberth, Satellite versus surface estimates of air temperature since 1979. *J. Climate*, *9*, 2222–2232, 1996.

Hurrell, J. W., and H. van Loon, Decadal variations in climate associated with the North Atlantic Oscillation, *Climatic Change*, *36*, 301–326, 1997.

Hurrell, J. W., M. P. Hoerling, and C. K. Folland, Climatic variability over the North Atlantic. *Meteorology at the Millennium*. R. Pearce, Ed., Academic Press, London, 143–151, 2001.

Hurrell, J. W., C. Deser, C. K. Folland, and D. P. Rowell, The relationship between tropical Atlantic rainfall and the summer circulation over the North Atlantic. Proc. U. S. CLIVAR Atlantic meeting, D. Legler, Ed., Boulder, CO., 108–110, 2002.

James, I. N., and P. M. James, Spatial structure of ultra-low frequency variability of the flow in a simple atmospheric circulation model, *Quart. J. Roy. Meteor. Soc.*, *118*, 1211–1233, 1992.

Jin F., A theory of interdecadal climate variability of the North Pacific Ocean-atmosphere system, *J. Climate*, *10*, 1821–1835, 1997.

Jones, P. D., T. Jónsson, and D. Wheeler, Extension to the North Atlantic Oscillation using early instrumental pressure observations from Gibraltar and south-west Iceland, *Int. J. Climatol.*, *17*, 1433–1450, 1997.

Jones, P. D., T. J. Osborn, and K. R. Briffa, The evolution of climate over the last millennium, *Science*, *292*, 662–667, 2001.

Jones, P. D., T. J. Osborn, and K. R. Briffa, Pressure-based measures of the North Atlantic Oscillation (NAO): A comparison and an assessment of changes in the strength of the NAO and in its influence on surface climate parameters, this volume.

Kalnay, E., et al. The NCEP/NCAR 40-year reanalysis project, *Bull. Amer. Meteor. Soc.*, *77*, 437-471, 1996.

Kelly, P. M., P. D. Jones, and J. I. A. Pengquin, The spatial response of the climate system to explosive volcanic eruptions, *Intl. J. Climatol.*, *16*, 537–550, 1996.

Kerr, R. A., A new force in high-latitude climate *Science*, *284*, 241–242, 1999.

Kiladis, G. N., and K. C. Mo, Interannual and intraseasonal variability in the Southern Hemisphere, in *Meteorology of the Southern Hemisphere*, D. J. Karoly and D. G. Vincent, Eds., Meteorological Monographs, 49, American Meteorological Society, Boston, MA, 1–46, 1998.

Kitoh, A., H. Doide, K. Kodera, S. Yukimoto, and A. Noda, Interannual variability in the stratospheric- tropospheric circulation in a coupled ocean-atmosphere GCM. *Geophys. Res. Lett.*, *23*, 543-546, 1996.

Klein, S. A., B. J. Soden, and N.-C. Lau, Remote sea surface temperature variations during ENSO: Evidence for a tropical atmospheric bridge, *J. Climate*, *12*, 917–932, 1999.

Kodera, K., Influence of volcanic eruptions on the troposphere through stratospheric dynamical processes in the northern hemisphere winter, *J. Geophys. Res.*, *99*, 1273–1282, 1994.

Kodera, K., M. Chiba, H. Koide, A. Kitoh, and Y. Nikaidou Y, Interannual variability of the winter stratosphere and troposphere in the Northern Hemisphere, *J. Meteorol. Soc. Jpn.*, *74*, 365–382, 1996.

Krahmann, G., M. Visbeck, and G. Reverdin, Formation and propagation of temperature anomalies along the North Atlantic Current, *J. Phys. Oceanogr.*, *31*, 1287–1303, 2001.

Kushnir, Y., Interdecadal variations in North Atlantic sea surface temperature and associated atmospheric conditions, *J. Climate*, *7*, 142–157, 1994.

Kushnir, Y., Pacific North American pattern. In *Encyclopedia of Environmental Change*, in press, 2002.

Kushnir, Y., and I. Held, Equilibrium atmospheric responses to North Atlantic SST anomalies, *J. Climate*, *9*, 1208–1220, 1996.

Kushnir, Y., and J. M. Wallace, Low frequency variability in the Northern Hemisphere winter: Geographical distribution, structure and time dependence, *J. Atmos. Sci.*, *46*, 3122–3142, 1989.

Kushnir, Y., V. J. Cardone, J. G. Greenwood, and M. Cane, On the recent increase in North Atlantic wave heights, *J. Climate*, *10*, 2107–2113, 1997.

Kushnir, Y., W. A. Robinson, I. Bladé, N. M. J. Hall, S. Peng, and R. T. Sutton, Atmospheric GCM response to extratropical SST anomalies: Synthesis and evaluation, *J. Climate*, *15*, 2233–2256, 2002.

Kutzbach, J. E., Large-scale features of monthly mean Northern Hemisphere anomaly maps of sea-level pressure *Mon. Wea. Rev.*, *98*, 708–716, 1970.

Lamb, H. H., British Isles weather types and a register of the daily sequence of circulation patterns, *Geophysical Memoirs*, Vol. 16, 116, 85 pp, 1972.

Latif, M., and T. P. Barnett, Decadal climate variability over the North Pacific and North America: dynamics and predictability, *J. Climate*, *9*, 2407–2423, 1996.

Leith, C. E., The standard error of time-averaged estimates of climatic means, *J. Appl. Meteor.*, *12*, 1066–1069, 1973.

Lorenz, E. N., Seasonal and irregular variations of the Northern Hemisphere sea-level pressure profile, *J. Meteorol.*, *8*, 52–59, 1951.

Lorenz, E. N., Deterministic nonperiodic flow, *J. Atmos. Sci.*, *20*, 130–141, 1963.

Lu, J., and R. J. Greatbatch, The changing relationship between the NAO and Northern Hemisphere climate variability, *Geophys. Res. Lett.*, *29*, 10.1029/2001GL014052, 2002.

Luterbacher, J., et al. Extending North Atlantic Oscillation reconstructions back to 1500, *Atmos. Sci. Lett.*, *2*, 114-124, 10.1006/asle.2001.0044, 2002.

Madden, R. A., Estimates of the natural variability of time-averaged sea-level pressure, *Mon. Wea. Rev.*, *104*, 942–952, 1976.

Manabe, S., and R. Stouffer, Low frequency variability of surface air temperature in a 1000-year integration of a coupled ocean-atmosphere-land model, *J. Climate*, *9*, 376–393, 1996.

Mann, M. E., R. S. Bradley, and M. K. Hughes, Northern Hemisphere temperatures during the past millennium: Inferences, uncertainties, and limitations, *Geophys. Res. Lett.*, *26*, 759–762, 1999.

Mathieu, P.-P., R. T. Sutton, and B. W. Dong, Impact of individual El Niño events on the North Atlantic European region, *CLIVAR Exchanges*, *25*, 49–51, 2002.

Mehta, V., M. Suarez, J. V. Manganello, and T. D. Delworth, Oceanic influence on the North Atlantic Oscillation and associated Northern Hemisphere climate variations: 1959–1993, *Geophys. Res. Lett.*, *27*, 121–124, 2000.

Michelangeli, P. A., R. Vautard, and B. Legras, Weather regime occurrence and quasi stationarity, *J. Atmos. Sci.*, *52*, 1237–1256, 1995.

Monahan, A. H., J. C. Fyfe, and G. M. Flato, A regime view of northern hemisphere atmospheric variability and change under global warming, *Geophys. Res. Lett.*, *27*, 1139–1142, 2000.

Monahan, A. H., L. Pandolfo, and J. C. Fyfe, The preferred structure of variability of the Northern Hemisphere atmospheric circulation, *Geophys. Res. Lett.*, *28*, 1019–1022, 2001.

Moses, T., G. N. Kiladis, H. F. Diaz, and R. G. Barry, Characteristics and frequency reversals in mean sea level pressure in the North Atlantic sector and their relationships to long-term temperature trends, *J. Climatol.*, *7*, 13–30, 1987.

Mysterud, A., N. C. Stenseth, N. G. Yoccoz, G. Ottersen, and R. Langvatn, The response of terrestrial ecosystems to climate variability associated with the North Atlantic Oscillation, this volume.

Nakamura, H., 1996: Year-to-year and interdecadal variability in the activity of intraseasonal fluctuations in the Northern Hemisphere wintertime circulation. *Theor. Appl. Climatol.*, *55*, 19-32.

Namias, J., The index cycle and its role in the general circulation, *J. Meteorol.*, *7*, 130–139, 1950.

North, G. R., T. L. Bell, R. F. Cahalan, and F. J. Moeng, Sampling errors in the estimation of empirical orthogonal functions, *Mon. Wea. Rev.*, *110*, 699–706, 1982.

Osborn, T. J., The winter North Atlantic oscillation: roles of internal variability and greenhouse gas forcing, *CLIVAR Exchanges*, *25*, 54–57, 2002.

Osborn, T. J., K. R. Briffa, S. F. B. Tett, P. D. Jones, and R. M. Trigo, Evaluation of the North Atlantic oscillation as simulated by a climate model, *Clim. Dyn.*, *15*, 685–702, 1999.

Ottersen, G., B. Planque, A. Belgrano, E. Post, P. C. Reid, and N. C. Stenseth, Ecological effects of the North Atlantic Oscillation, *Oecologia*, *128*, 1–14, 2001.

Ottersen, G., and N. C. Stenseth, Atlantic climate governs oceanographic and ecological variability in the Barents Sea, *Limn. Ocean.*, *46*, 1774-1780, 2001.

Palmer, T. N., A nonlinear dynamical perspective on climate prediction, *J. Climate*, *12*, 575–591, 1999.

Palmer, T. N., and Z. Sun, A modeling and observational study of the relationship between sea surface temperature in the north west Atlantic and the atmospheric general circulation, *Quart. J. Roy. Meteor. Soc.*, *111*, 947–975, 1985.

Peng, S., and J. S. Whitaker, Mechanisms determining the atmospheric response to midlatitude SST anomalies, *J. Climate*, *12*, 1393–1408, 1999.

Peng, S., A. Mysak, H. Ritchie, J. Derome, and B. Dugas, The difference between early and middle winter atmospheric response to sea surface temperature anomalies in the northwest Atlantic, *J. Climate*, *8*, 137–157, 1995.

Peng, S., W. A. Robinson, and M. P. Hoerling, The modeled atmospheric response to midlatitude SST anomalies and its dependence on background circulation states, *J. Climate*, *10*, 971–987, 1997.

Peng, S., W. A. Robinson, and S. Li, North Atlantic SST forcing of the NAO and relationships with intrinsic hemispheric variability, *Geophys. Res. Lett.*, *29*(8), 10.1029/2001GL014043, 2002.

Perlwitz, J., and H. -F. Graf, The statistical connection between tropospheric and stratospheric circulation of the Northern Hemisphere in winter, *J. Climate*, *8*, 2281–2295, 1995.

Peterson, K. A., R. J. Greatbatch, J. Lu, H. Lin, and J. Derome, Hindcasting the NAO using diabatic forcing of a simple AGCM, *Geophys. Res. Lett.*, *29*, 10.1029/2001GL014502, 2002.

Portis, D. H., J. E. Walsh, M. El Hamly, and P. J. Lamb, Seasonality of the North Atlantic Oscillation, *J. Climate, 14*, 2069-2078, 2001.

Pozo-Vázquez, D., M. J. Esteban-Parra, F. S. Rodrigo, and Y. Castro-Díez, The association between ENSO and winter atmospheric circulation and temperature in the North Atlantic region, *J. Climate*, *14*, 3408–3420, 2001.

Rajagolapan, B., Y. Kushnir, and Y. Tourre, Observed decadal midlatitude and tropical Atlantic climate variability, *Geophys. Res. Lett.*, *25*, 3967–3970, 1998.

Randel, W. J., and F. Wu, Cooling of the Arctic and Antarctic polar stratospheres due to ozone depletion, *J. Climate*, *12*, 1467-1479, 1999.

Robertson, A. W., C. R. Mechoso, and Y. J. Kim, The influence of Atlantic sea surface temperature anomalies on the North Atlantic Oscillation, *J. Climate*, *13*, 122–138, 2000.

Robinson, W. A, Review of WETS-The workshop on extra-tropical SST anomalies. *Bull. Am. Meteorol. Soc.*, *81*, 567–577, 2000.

Robock, A., and J. P. Mao, Winter warming from large volcanic eruptions, *Geophys. Res. Lett.*, *19*, 2405–2408, 1992.

Rodwell, M. J., On the predictability of North Atlantic climate, this volume.

Rodwell, M. J., D. P. Rowell, and C. K. Folland, Oceanic forcing of the wintertime North Atlantic Oscillation and European climate, *Nature*, *398*, 320–323, 1999.

Rogers, J. C., Spatial variability of seasonal sea-level pressure and 500-mb height anomalies, *Mon. Wea. Rev.*, *109*, 2093–2106, 1981.

Rogers, J. C., The association between the North Atlantic Oscillation and the Southern Oscillation in the Northern Hemisphere, *Mon. Wea. Rev.*, *112*, 1999–2015, 1984.

Rogers, J. C., Patterns of low-frequency monthly sea level pressure variability (1899-1986) and associated wave cyclone frequencies, *J. Climate*, *3*, 1364–1379, 1990.

Rogers, J. C., North Atlantic storm track variability and its association to the North Atlantic Oscillation and climate variability of Northern Europe, *J. Climate*, *10*, 1635–1645, 1997.

Saravanan, R., and P. Chang, Interaction between tropical Atlantic variability and El Niño-Southern Oscillation, *J. Climate*, *13*, 2177–2194, 2000.

Saravanan, R., and J. C. McWilliams, Stochasticity and spatial resonance in interdecadal climate fluctuations, *J. Climate*, *10*, 2299-2320, 1997.

Saravanan, R., and J. C. McWilliams, Advective ocean-atmosphere interaction: an analytical stochastic model with implications for decadal variability, *J. Climate*, *11*, 165–188, 1998.

Saravanan, R., G. Danabasoglu, S. C. Doney, and J. C. Mc Williams, Decadal variability and predictability in the midlatitude ocean-atmosphere system, *J. Climate*, *13*, 1073–1097, 2000.

Schlosser, P., G. Bonisch, M. Rhein, and R. Bayer, Reduction of deepwater formation in the Greenland Sea during the 1980's: Evidence from tracer data, *Science*, *251*, 1054–1056, 1991.

Schmutz, C., J. Luterbacher, D. Gyalistras, E. Xoplaki, and H. Wanner, Can we trust proxy-based NAO index reconstructions?, *Geophy. Res. Lett.*, *27*, 1135–1138, 2000.

Seager, R., Y. Kushnir, M. Visbeck, N. Naik, J. Miller, G. Krahmann, and H. Cullen, Causes of Atlantic Ocean climate variability between 1958 and 1998, *J. Climate*, *13*, 2845–2862, 2000.

Serreze, M. C., F. Carse, R. G. Barry, and J. C. Rogers, Icelandic low activity: climatological features, linkages with the NAO, and relationships with recent changes in the Northern Hemisphere circulation. *J. Climate*, *10*, 453–464, 1997.

Sexton, D. M. H., D. E. Parker, and C. K. Folland, Natural and human influences on Central England temperature, *J. Climate*, in press, 2002.

Shindell, D. T., R. L. Miller, G. Schmidt, and L. Pandolfo, Simulation of recent northern winter climate trends by greenhouse-gas forcing, *Nature*, *399*, 452–455, 1999.

Shindell, D. T., G. A. Schmidt, R. L. Miller, and D. Rind, Northern hemisphere winter climate response to greenhouse gas, ozone, solar, and volcanic forcing, *J. Geophys. Res.*, *106*, 7193–7210, 2001.

Siggurdson, O., and T. Jonsson, Relation of glacier variations to climate changes in Iceland, *Annals of Glaciology*, *21*, 263-270, 1995.

Slonosky, V. C., and P. Yiou, Secular changes in the North Atlantic Oscillation and its influence on 20th century warming, *Geophys. Res. Lett.*, *28*, 807–810, 2001.

Slonosky, V. C., P. D. Jones, and T. D. Davies, Instrumental pressure observations from the seventeenth and eighteenth centuries: London and Paris, *Int. J. Climatol.*, *21*, 285–298, 2001.

Stephenson, D. B., V. Pavan, and R. Bojariu, Is the North Atlantic Oscillation a random walk?, *Int. J. Climatol.*, *20*, 1–18, 2000.

Stenseth, N. C., A. Mysterud, G. Ottersen, J. W. Hurrell, K.-S. Chan, and M. Lima, Ecological effects of climate fluctuations, *Science*, *297*, 1292–1296, 2002.

Stephenson, D. B., H. Wanner, S. Brönnimann, and J. Luterbacher, The history of scientific research on the North Atlantic Oscillation, this volume.

Straile, D., D. M. Livingstone, G. A. Weyhenmeyer, and D. G. George, The response of freshwater ecosystems to climate variability associated with the North Atlantic Oscillation, this volume.

Sutton R. T., and M. R. Allen, Decadal predictability of North Atlantic sea surface temperature and climate, *Nature*, *388*, 563–567, 1997.

Sutton, R. T., and D. L. R. Hodson, Influence of the ocean on North Atlantic climate variability 1871-1999, *J. Climate*, in press, 2002.

Sutton, R. T., W. A. Norton, and S. P. Jewson, The North Atlantic Oscillation - What role for the Ocean? *Atm. Sci. Lett.*, *1, 89-100, 10.1006/asle.2000.0018*, 2001.

Thompson, D. W. J, and J. M. Wallace, The Arctic Oscillation signature in the wintertime geopotential height and temperature fields, *Geophys. Res. Lett.*, *25*, 1297–1300, 1998.

Thompson, D. W. J., and J. M. Wallace, Annular modes in the extratropical circulation, Part I: Month-to-month variability, *J. Climate*, *13*, 1000–1016, 2000.

Thompson, D. W. J., J. M. Wallace, and G. C. Hegerl, Annular modes in the extratropical circulation, Part II: Trends, *J. Climate*, *13*, 1018–1036, 2000.

Thompson, D. W. J., S. Lee, and M. P. Baldwin, Atmospheric processes governing the Northern Hemisphere Annular Mode/North Atlantic Oscillation, this volume.

Timmerman, A., M. Latif, R. Voss, and A. Grötzner, Northern hemispheric interdecadal variability: a coupled air-sea mode, *J. Climate*, *11*, 1906–1931, 1998.

Ting, M., and S. Peng, Dynamics of early and middle winter atmospheric responses to northwest Atlantic SST anomalies, *J. Climate*, *8*, 2239–2254.

Trenberth, K. E., Signal versus noise in the Southern Oscillation, *Mon. Wea. Rev.*, *112*, 326–332, 1984.

Trenberth, K. E., and J. W. Hurrell, Decadal atmosphere-ocean variations in the Pacific, *Clim. Dyn.*, *9*, 303–319, 1994.

Trenberth, K. E., and J. W. Hurrell, Comments on "The interpretation of short climate records with comments on the North Atlantic and Southern Oscillations," *Bull. Am. Meteorol. Soc.*, *80*, 2721–2722, 1999.

Trenberth, K. E., and D. A. Paolino, The Northern Hemisphere sea level pressure data set: Trends, errors and discontinuities, *Mon. Wea. Rev.*, *108*, 855–872, 1980.

Trenberth, K. E., and D. A. Paolino, Characteristic patterns of variability of sea level pressure in the Northern Hemisphere, *Mon. Wea. Rev.*, *109*, 1169–1189, 1981.

Trenberth, K. E., G. W. Branstator, D. Karoly, A. Kumar, N.-C. Lau, and C. Ropelewski, Progress during TOGA in understanding and modeling global teleconnections associated with tropical sea surface temperatures. *J. Geophys. Res.*, *103*, 14291-14324, 1998.

Ulbrich, U., and M. Christoph, A shift of the NAO and increasing storm track activity over Europe due to anthropogenic greenhouse gas forcing, *Clim. Dyn.*, *15*, 551–559, 1999.

van Loon, H., and R. A. Madden, Interannual variations of mean monthly sea-level pressure in January, *J. Clim. Appl. Meteor.*, *22*, 687–692, 1983.

van Loon, H., and J. Williams, Connection between trends of mean temperature and circulation at the surface. 1. Winter, *Mon. Wea. Rev.*, *104*, 365–380, 1976.

Venzke, S., M. R. Allen, R. T. Sutton, and D. P. Rowell, The atmospheric response over the North Atlantic to decadal changes in sea surface temperatures, *J. Climate*, *12*, 2562–2584, 1999.

Visbeck, M., H. Cullen, G. Krahmann, and N. Naik, An ocean model's response to North Atlantic Oscillation like wind forcing. *Geophys. Res. Lett.*, *25*, 4521–4524, 1998.

Visbeck, M., E. .P. Chassignet, R. G. Curry, T. L. Delworth, R. R. Dickson, and G. Krahmann, The ocean's response to North Atlantic Oscillation variability, this volume.

Volodin, E. M., and V. Y. Galin, Interpretation of winter warming on Northern Hemisphere continents in 1977-94, *J. Climate*, *12*, 2947–2955, 1999.

Walker, G. T., and E. W. Bliss, World Weather V, *Mem. Roy. Met. Soc.*, *4*, 53–84, 1932.

Wallace, J. M., Observed climate variability: Spatial structure, in *Decadal Variability and Climate*, D.L.T. Anderson and J. Willebrand, Eds., Springer, Berlin, 31–81, 1996.

Wallace, J. M., North Atlantic Oscillation/annular mode: Two paradigms - one phenomenon, *Quar. J. Roy. Meteorol. Soc.*, *126*, 791–805, 2000.

Wallace, J. M., and D. S. Gutzler, Teleconnections in the geopotential height field during the Northern Hemisphere winter, *Mon. Wea. Rev.*, *109*, 784–812, 1981.

Wallace, J. M., and D. W. J. Thompson, The Pacific center of action of the Northern Hemisphere Annular Mode: Real or artifact?, *J. Climate*, *15*, 1987-1991, 2002.

Wallace, J. M., Y. Zhang, and K.-H. Lau, Structure and seasonality of interannual and interdecadal variability of geopotential height and temperature fields in the Northern Hemisphere troposphere, *J. Climate*, *6*, 2063–2082, 1993.

Wallace, J. M., Y. Zhang, and J. A. Renwick, Dynamic contribution to hemispheric mean temperature trends, *Science*, *270*, 780–783, 1995.

Walther, G.-R., E. Post, P. Convey, A. Menzel, C. Parmesan, T. J. C. Beebee, J.-M. Fromentin, O. Hoegh-Guldberg, and F. Bairlein, Ecological responses to recent climate change, *Nature*, *416*, 389–395, 2002.

Wanner, H., S. Brönnimann, C. Casty, D. Gyalistras, J. Luterbacher, C. Schmutz, D. B. Stephenson, and E. Xoplaki, North Atlantic Oscillation- concepts and studies, *Survey Geophys.*, *22*, 321–381, 2001.

Watanabe, M. and T. Nitta, Relative impact of snow and sea surface temperature anomalies on an extreme phase in the winter atmospheric circulation, *J. Climate*, *11*, 2837-2857, 1998

Watanabe, M. and T. Nitta, Decadal changes in the atmospheric circulation and associated surface climate variations in the Northern Hemisphere winter, *J. Climate*, *12*, 494-510, 1999.

Weng, W., and J. D. Neelin, On the role of ocean-atmosphere interaction in midlatitude interdecadal variability, *Geophys. Res. Lett.*, *25*, 167–170, 1998.

Wunsch C., The interpretation of short climate records, with comments on the North Atlantic Oscillation and Southern Oscillations, *Bull. Am. Meteorol. Soc.*, *80*, 245–255, 1999.

Xie, S.-P., and Y. Tanimoto, A pan-Atlantic decadal climate oscillation, *Geophys. Res. Lett.*, *25*, 2185–2188, 1998.

James W. Hurrell, National Center for Atmospheric Research, Climate Analysis Section, P. O. Box 3000, Boulder, CO 80307-3000, USA
jhurrell@ucar.edu

Yochanan Kushnir, Lamont-Doherty Earth Observatory of Columbia University, 61 Route 9W, Palisades, New York, 10964, USA
kushnir@ldeo.columbia.edu

Geir Ottersen, Institute of Marine Research, P.O. Box 1870 Nordnes, 5817 Bergen, Norway
Current address:
Department of Biology, Division of Zoology, University of Oslo, P.O. Box 1050 Blindern, N-0316 Oslo, Norway
geir.ottersen@bio.uio.no

Martin Visbeck, Lamont-Doherty Earth Observatory of Columbia University, Department of Earth and Environmental Sciences, 204D Oceanography, 61 Route 9W, Palisades, New York, 10964, USA
visbeck@ldeo.columbia.edu

The History of Scientific Research on the North Atlantic Oscillation

David B. Stephenson[1], Heinz Wanner[2,3], Stefan Brönnimann[2,4], Jürg Luterbacher[2,3]

The North Atlantic Oscillation is one of the most dominant and oldest known world weather patterns. This article will briefly review the intriguing discovery and history of scientific research of this fascinating phenomenon. By reviewing previous scientific investigations, it can be seen that several contemporary themes have recurred ever since the early scientific studies. The review will conclude with a brief speculation on how these themes are likely to develop with future research.

1. INTRODUCTION

It is not possible in a short article like this to present a fully comprehensive review of all research on the North Atlantic Oscillation (NAO) – an extremely complex phenomenon that has been the subject of scientific attention for more than two centuries. Instead, we hope to present a stimulating account of the major scientific landmarks, and refer readers to other studies for more detailed historical accounts [e.g., *Wallace*, 2000; *Stephenson et al.*, 2000; *Wanner et al.*, 2001]. By reviewing the earlier scientific literature, it can be noted that many of the contemporary issues concerning the North Atlantic Oscillation have been around for a long time. It is our hope that future historical reviews will be able to say that many of these issues were satisfactorily resolved in the 21st century!

2. DISCOVERY OF NORTH ATLANTIC TELECONNECTIONS AD 1000-1780

Perhaps because of the volatility of their local weather, the people of Northern Europe have always had a deep fascination for weather and climate. In the "ragnarok" (twilight of the Gods) legend, Norse mythology predicts that the end of the world will begin with the occurrence of a severe "fimbul-winter", when "snow drives from all quarters, the frosts are so severe, the winds so keen and piercing, that there is no joy in the sun. There will be three such winters in succession, without any intervening summer" [*Sturluson*, 1984]. This long-term prognosis could be taken to imply a run of three years in which the phase of the North Atlantic Oscillation was strongly negative.

It is not surprising, therefore, that the earliest descriptions of North Atlantic Oscillation were first noted by seafaring Scandinavians. Because of their excursions to Greenland, they were well aware of the relationship between climate in different regions of the North Atlantic and surrounding landmasses. The history of Greenland and its climate, fauna, and flora were carefully documented by the Danish missionary Hans Egede Saabye [*Egede*, 1745]. Figure 1 shows a map of Greenland published in an English translation of this book showing the main settlements and coastal features. As described in *van Loon and Rogers* [1978], Hans Egede Saabye also made the following revealing remarks in a diary that he kept in Greenland:

"In Greenland, all winters are severe, yet they are not alike. The Danes have noticed that when the winter in Denmark was severe, as we perceive it, the winter in Greenland in its manner was mild, and conversely."

The fact that this relationship was common knowledge at the time was commented upon by the editor of Saabye's diary [*Ostermann*, 1942]. It was also known and discussed by physical geographers in 19th century Germany [*Gilbert*, 1819].

It is natural to speculate whether earlier visitors such as the Norse colonizers of Greenland might also have noticed this teleconnection between climates in different regions of the North Atlantic basin. Since their first settlement of

[1]Department of Meteorology, University of Reading, UK

[2]Institute of Geography, University of Bern, Bern, Switzerland

[3]National Center of Competence in Research (NCCR) in Climate, University of Bern, Bern, Switzerland

[4]Lunar and Planetary Laboratory, University of Arizona, Tucson, Arizona

The North Atlantic Oscillation:
Climatic Significance and Environmental Impact
Geophysical Monograph 134

10.1029/134GM02

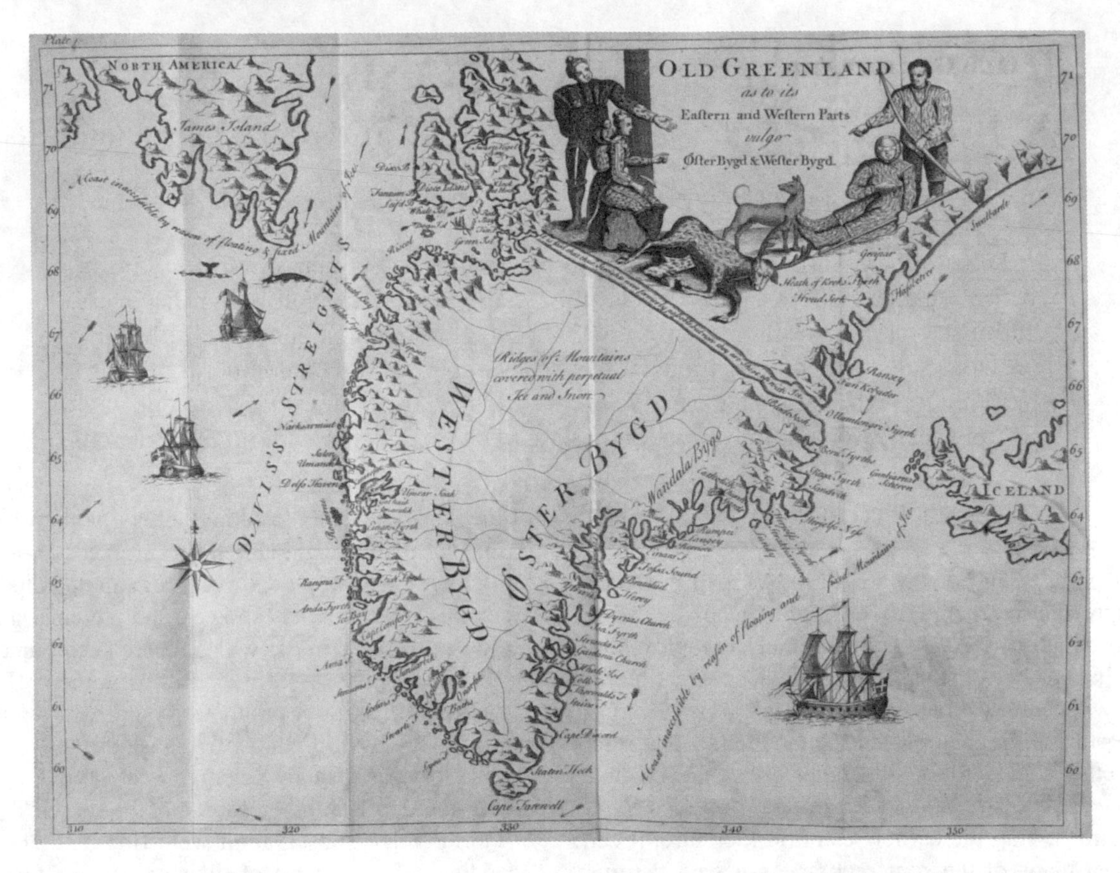

Figure 1. Vintage map of Old Greenland circa 1745 from an English translation of Hans Egede Saabye's History of Greenland. The original color map of Greenland drawn by Egede in 1737 is on display in the Royal Library in Denmark and a copy of it can be seen at http://www.kb.dk/kultur/expo/klenod/.

Greenland in AD 985, the Norse made regular summer transatlantic sailings between Norway and Greenland until around AD 1370, and as such were keen observers of North Atlantic weather. Around AD 1230, a remarkable book appeared in Norway known as the "King's Mirror" [*Anonymous*, 1917] – alternatively sometimes referred to as "Speculum Regale" (Latin) or "Konungs skuggsjá" (Old Norse). It provides fascinating descriptions and insight into several scientific topics including the Northern Lights, the spherical geometry of the Earth, and the climate of Greenland. The book is presented in the form of a dialogue between a son and his father, and one of the most revealing meteorological replies by the father is as follows:

"In reply to your remark about the climate of Greenland, that you think it strange that it is called a good climate, I shall tell you something about the nature of the land. When storms do come, they are more severe than in most other places, both with respect to keen winds and vast masses of ice and snow. But usually these spells of rough weather last only a short while and come at long intervals only. In the meantime the weather is fair, though the cold is intense. For it is in the nature of the glacier to emit a cold and continuous breath which drives the storm clouds away from its face so that the sky above is usually clear. But the neighbouring lands often have to suffer because of this; for all the regions that lie near get severe weather from this ice, inasmuch as all the storms that the glacier drives away from itself come upon others with keen blasts."

This passage demonstrates that the Norse knew from their observations that colder than normal conditions in Greenland were associated with more storminess elsewhere – an important aspect of the North Atlantic Oscillation [*Rogers*, 1997]. Their great expertise in transatlantic navigation and keen sense of observation made it possible for the Norse to be able to discover nonlocal relationships between weather in different parts of the North Atlantic. The Norse colony had been in Greenland almost 300 years before the King's Mirror book was written, which is much longer than it took the later Danish colonizers to note the seesaw relationship based on qualitative observations.

3. EARLY SCIENTIFIC EXPLORATION AD 1811-1905

The temperature seesaw between winters in Greenland and Germany was first documented in tables of above and below normal winters (1709-1800) published by *Gronau* [1811] – see *van Loon and Rogers* [1978] for a copy of one of these tables. The anomalous temperatures in Greenland were not actually measured but were inferred from general impressions of the state of the sea ice, which was important for the then flourishing whaling and sealing industries in Greenland waters. These qualitative observations were used in several seesaw studies [e.g., *Gronau*, 1811; *Dove*, 1839; *Dannmeyer*, 1948].

Some of the earliest regular observations were first taken at Ny Herrnhut near Godthaab on the west coast of Greenland in 1767-8 [*Loewe*, 1966]. Ny Herrnhut was established by the "herrnhuts", a catholic Moravian mission from Germany, who had been given a permit by the Danish king in 1733 to assist Hans Egede in his conversion of the Greenlanders [*Crantz*, 1765]. Temperature observations made at Greenland stations were used in the first meteorological observing network with uniform instrumentation – the Societas Meteorologica Palatina established in 1780 [*Loewe*, 1966]. During 1783, the Moravian brothers made meteorological observations at Godthaab (Nuuk). Later, from 1806 to 1813, Dr. Giesecke at Godthaab (personal communication, Dr Trausti Jonsson, Iceland Met. Office) made regular observations three times a day. The availability of long records of observed temperature measurements made it possible for 19th century climatologists to start scientifically exploring the spatial and temporal variations in climates of different regions.

Dove [1839; 1841] investigated 60 temperature time series of up to 40-yr length from all over the Northern Hemisphere, and noted that east-west variations in temperatures were often more pronounced than north-south variations. He noted an opposition of the monthly and seasonal temperature anomalies of northern Europe with respect to both North America and Siberia, and thereby scientifically confirmed the statement made by Hans Egede Saabye. The famous Austrian climatologist, *Julius Hann* [1890], later demonstrated the east-west temperature seesaw by using 42 years of monthly mean temperatures from Jakobshavn on the west coast of Greenland (69°N, 51°W) and Vienna, Austria (48°N, 16°E). Later studies used Oslo, Norway (60°N, 11°E) instead of Vienna [*Hann*, 1906].

A major stimulus for the research came from severe climate events in Europe such as the anomalously cold winter of 1879/80. In a pioneering study, *Teisserenc de Bort* [1883] compared European climate during different anomalous winters. He investigated the positions of large pressure centers (which he called "centres d'action"), and distinguished five types of anomalous winters according to the position of the Azores High and the Russian High and to some extent also the Icelandic Low. He suggested that surface influences (such as Eurasian snow cover) were possibly responsible for these displacements. Figure 2 shows some of the pressure maps for various winters published in *Teisserenc de Bort* [1883].

Inspired by the concept of "centres of action" introduced by Teisserenc de Bort, *Hildebrandsson* [1897] then investigated sea-level pressure time series from different sites and found a distinct inverse relation between the pressure at Iceland and the Azores. He also noted that series from the Azores and Siberia ran "parallel", whereas Alaska and Siberia showed an opposite behavior. This study was the forerunner for all future studies that have used sea-level pressures at Iceland and Azores to characterise the North Atlantic Oscillation.

Several authors of the late 19th century also addressed the relation between ocean and atmosphere in the North Atlantic region. In a study of sea surface temperatures near Norwegian lighthouses, *Pettersson* [1896] noted that milder winters were associated with warmer sea surface temperatures, and speculated that the climate of Western Europe was influenced by the Gulf Stream. *Meinardus* [1898] suggested that interannual fluctuations in the Gulf Stream might be responsible for anomalous winters, and that these fluctuations could affect the weather in western Iceland and Greenland in the opposite way than in Europe. With the benefit of hindsight, it is now known that coastal sea surface temperature anomalies are a local wind-induced response rather than the cause of variations in the North Atlantic Oscillation. A comprehensive review of these early theories of North Atlantic climate variability is given by *Helland-Hansen and Nansen* [1920].

4. DESCRIPTIVE CORRELATION STUDIES AD 1908-1937

The early explorations based on visual inspection of time series led to many mechanisms being proposed for explaining climate variations, for example, solar cycles, number of icebergs advected by the Gulf stream, etc. [*Helland-Hansen and Nansen*, 1920]. Many of these relationships were spurious and had arisen purely due to sampling uncertainty caused by the small length of available time series.

Felix Exner's comprehensive and accurate 1913 study of Northern Hemisphere sea-level pressure anomalies went one step further by producing the first correlation map showing the spatial structure of the North Atlantic

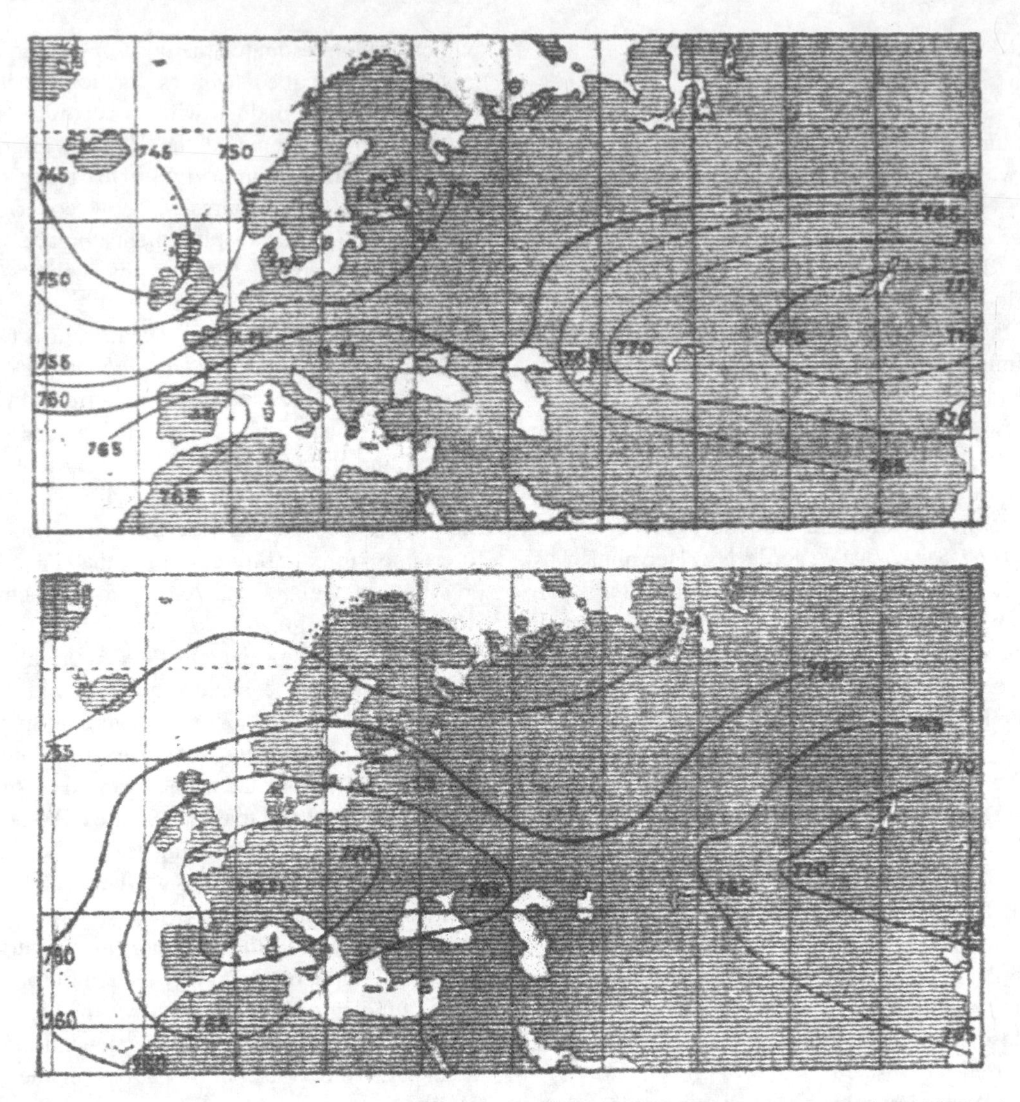

Figure 2. Vintage map showing early monthly mean sea level pressure (SLP) maps for December 1868 (top) and for December 1879 (bottom) (units in Torr; 1 Torr = 1.3332 hPa). From *Teisserenc de Bort* [1883].

Oscillation in the Northern Hemisphere. This map is reproduced in Figure 3a and shows the correlation between the monthly mean pressure anomalies at the North Pole (approximated by the mean of three time series from Greenland, northern Norway, and Northern Siberia) and some 50 sites in the Northern Hemisphere. Exner emphasized the annular appearance of the pattern and the strong signature in the North Atlantic and Mediterranean areas. In fact, his correlation pattern closely resembles the Northern Annular Mode (or Arctic Oscillation) found using principal component analysis of sea-level pressure [*Wallace*, 2000; *Deser*, 2000; *Ambaum et al.*, 2001; *Thompson et al.*, this volume]. Figure 3b shows Exner's later map of correlations of pressure anomalies between Stykkisholmur and some 70 sites [*Exner*, 1924]. This pattern is more regionally confined to the North Atlantic sector and resembles more closely the North Atlantic Oscillation pattern.

Felix Exner's studies used the statistical technique of correlation analysis introduced into climate research by Gilbert Walker [*Walker*, 1909; 1910]. As director of the Indian Meteorological Department in Pune, Walker was confronted with the problem of how to seasonally forecast ("foreshadow" in his words) the Indian summer monsoon and the flooding of the Nile. To do this he made expert use of the "regression" and "correlation" concepts that had been recently conceived by Francis Galton in 1877 [*Galton*, 1888]. It is possible that Walker learnt about these statistical techniques either directly from Galton or one of Galton's close acquaintances such as Richard Strachey (1817-1908)

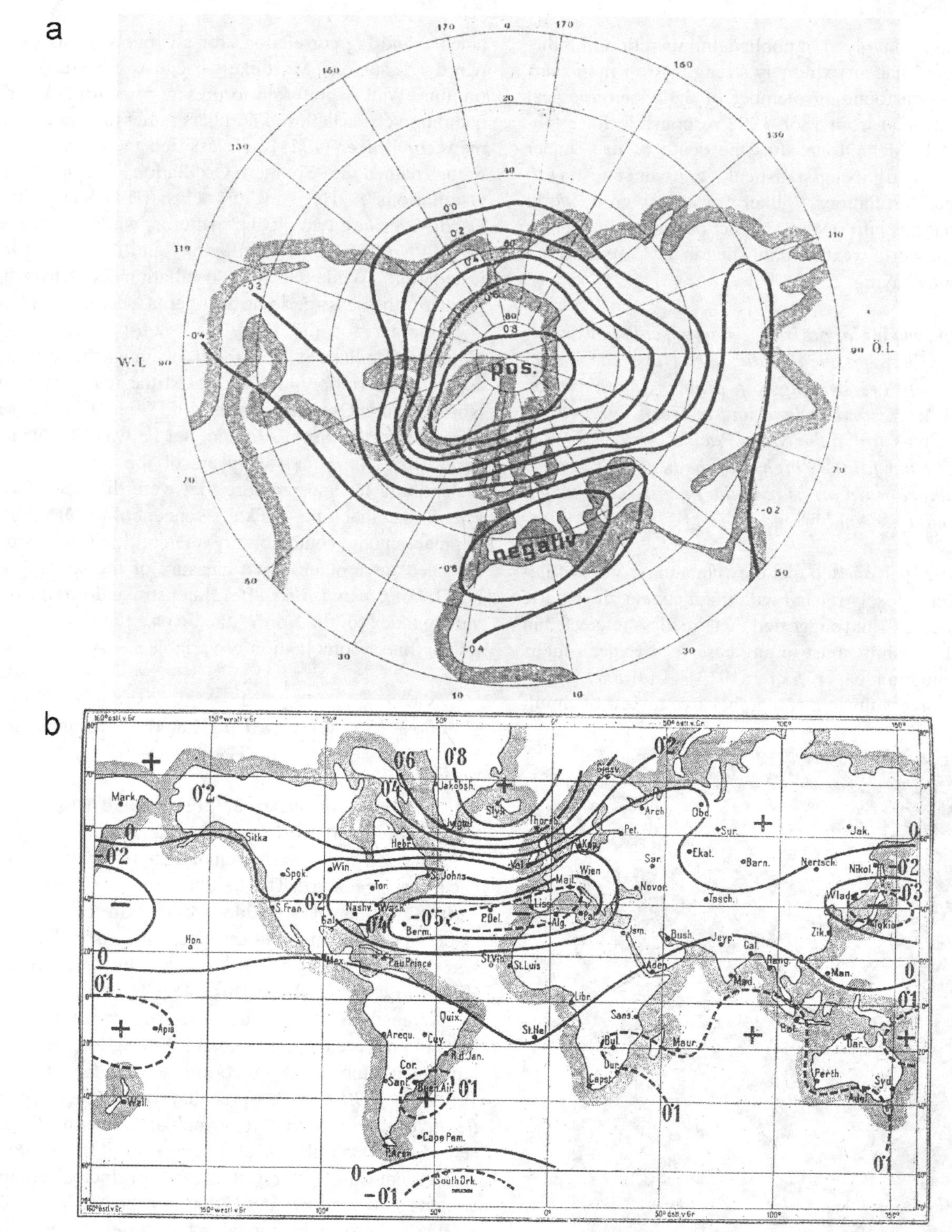

Figure 3. (a) Map of the correlation between monthly anomalies of "polar pressure" (average of three stations in northern Greenland, northern Norway, and northern Siberia, respectively) and pressure at around 50 sites of the Northern Hemisphere from 1887 to 1906 [from *Exner*, 1913]. (b) Map of the correlation between monthly anomalies of pressure at Stykkisholmur and pressure at about 70 sites for winter months (September to March) from 1887 to 1916 [from *Exner*, 1924].

who was heavily involved in public administration in India. Galton, himself, had an extremely strong interest in meteorology and was a founding member of the Meteorological Committee/Council from 1868-1902 responsible for establishing the UK Met Office after the death of its founder Admiral Fitzroy. By using statistical significance testing to reject spurious correlations, Walker was able to group world weather variations into several distinct patterns. His preliminary study of world weather published in 1923 summed up the situation by saying:

"there is a swaying of pressure on a big scale backwards and forwards between the Pacific ocean and the Indian ocean; and there are swayings, on a much smaller scale, between the Azores and Iceland and between the areas of high and low pressure in the North Pacific: further, there is a marked tendency for the "highs" of the last two swayings to be accentuated when pressure in the Pacific is raised and that in the Indian ocean lowered."

Walker [1923] also noted that the "Iceland Azores oscillation is not very closely related with that between the Pacific and Indian oceans" and suggested that "readers interested in northern relationships must in any case read Exner's interesting and important paper [*Exner*, 1913]". In *Walker* [1924] he extended his results to a comprehensive review of simultaneous and lag correlations for all four seasons for many worldwide stations. Most likely because of his mathematical training, Walker preferred to present his correlation results quantitatively as tables of numbers rather than as correlation maps. In *Walker* [1924] he classified the correlations into groups named the "Southern Oscillation" and the "Northern Oscillations". He further classified the Northern Oscillations into two distinct patterns, which he referred to as the "North Atlantic Oscillation" and the "North Pacific Oscillation". While the word "oscillation" is perhaps appropriate for the quasi-periodic Southern Oscillation, it remains a misnomer for the noisy and non-periodic (in time) Northern Oscillations – Northern Swayings might be a more accurate description! He discussed the relationship of the North Atlantic Oscillation to the Gulf Stream and the sea-ice dynamics in the North Atlantic, but he was skeptical about periodicities of 2 and 4.5 years of the sea-ice extent off Iceland and Iceland pressure that were discussed by other scientists at that time. Walker's concept of the NAO became popular among contemporary meteorologists and created the need for a quantitative measure of the strength of the NAO. *Walker and Bliss* [1932] constructed a robust multivariate index for the North Atlantic Oscillation using the following linear combination of variables

Figure 4. Photograph of Gilbert Thomas Walker (courtesy of R. W. Katz, NCAR).

$$P_{Vienna} + 0.7P_{Bermuda} - P_{Stykkisholmur} - P_{Ivigtut} + T_{Bodö} + T_{Stornoway} + 0.7(T_{Hatteras} + T_{Washington})/2 - 0.7T_{Godthaab}$$

where P stands for surface air pressure and T for surface air temperature averaged over the winter period December to February (all series were standardized to have zero mean and a variance of 20). The weights were discovered using an early selection of variables regression procedure. The Azores pressure time series gave only a small coefficient less than 0.5 and so was not retained in the expression by *Walker and Bliss* [1932]. This rather surprising rejection of the Azores predictor is due to its strong collinearity with Iceland pressure etc. rather than it having too weak a correlation with the NAO. According to *Wallace* [2000], the *Walker and Bliss* [1932] procedure may be considered an iterative approximation to Principal Component Analysis (PCA). However, this is not the case since the Azores pressure is not similarly rejected in the principal component weights shown by *Wallace* [2000] for the Arctic Oscillation. Walker's approach is more akin to a "selection of variables" approach as often used to select explanatory variables for multiple regression. A photograph of Gilbert Walker is shown in Figure 4.

Defant [1924] also published an original study of the monthly pressure anomaly fields over the North Atlantic from 1881 to 1905. He distinguished two pairs (four types) of anomalies, where the first pair (83% of all months) corresponds to the NAO-pattern and the second pair to a strong anomaly at 55°N and a weak opposite anomaly between 10° and 30°N. By subjectively attributing to each month an anomaly type and strength and applying a weighting procedure he was able to draw annual time series. He considered these variations to be internal oscillations of the climate system disturbed by sea ice extent off Iceland and volcanic eruptions. *Defant* [1924] also pointed to possible relations between the North Atlantic climate and the "heat engine" of the tropical Atlantic, taking up older, speculative ideas of *Shaw* [1905] and *Hann* [1906].

Although Defant went further than Exner and Walker in his search for dynamical causes for the NAO-like anomalies, his study still remained primarily exploratory. Other descriptive studies based on longer time series were published in the 1930s and 40s on the temperature seesaw between Northern Europe and Greenland [*Angström*, 1935; *Loewe*, 1937; *Dannmeyer* 1948]. *Angström* [1935] invented the word "teleconnection" to describe the association of climatic variations between different regions.

5. MODERN STUDIES

Theoretical developments in understanding the dynamics of large-scale planetary waves opened up a new way of looking at climate patterns. A number of theoretically motivated studies about the interaction of the zonal circulation and pressure centers were published by a group of leading meteorologists that included Rossby, Willett, Namias, Lorenz and others [*Lorenz*, 1967]. *Rossby et al.* [1939] studied the structure and dynamics of the planetary waves in the presence of disturbances and deduced an influence of the strength of the zonal circulation on the temporal behavior of the quasi-stationary centers of action. Apparently unaware of Exner's and Walker's climatological studies, they introduced a "zonal index" defined as the zonally averaged zonal wind at 45°N as a measure for the strength of the polar vortex in the free atmosphere to the north. *Rossby and Willett* [1948] focussed their studies on the polar vortex and addressed the issue of stratosphere-troposphere coupling. Rossby's zonal index became very popular, and many climatological studies of the zonal circulation were performed. *Namias* [1950], with a clear focus towards the improvement of forecasts, recognized the importance of latitudinal shifts in the zonal mean zonal wind. *Lorenz* [1951] studied the variability of the zonal mean circulation and the oscillations in the distribution of atmospheric mass. He introduced a new zonal index based on the zonal mean meridional pressure gradient at 55°N.

Another driving force in modern NAO studies has been the insight provided by the application of more advanced multivariate statistical techniques to grid point data sets. After visually exploring the correlation matrix between zonal mean sea-level pressure and zonal wind at various latitudes in *Lorenz* [1951], *Lorenz* [1956] went on to construct a less subjective zonal index by using "Empirical Orthogonal Functions" (EOFs) of the zonal means. EOFs are eigenvectors of the sample covariance matrix and were previously used in a weather forecasting study by *Fukuoka* [1951]. EOF analysis is equivalent to the widely used statistical technique known as Principal Component Analysis (PCA) invented by Karl Pearson in 1903. It is a descriptive multivariate technique for obtaining linear combinations of the variables (Principal Component indices) that explain maximum variance. The principal component time series are each associated with a set of constant weights, which in the case of grid point variables, define a spatial EOF pattern. With access to increased amounts of computer power, *Kutzbach* [1970] was able to demonstrate the power of PCA for studying large-scale circulation anomalies in two dimensional gridded pressure data. Many later studies have since adopted the same approach for pressure as well as for other fields e.g., *Trenberth and Paolino* [1980], *Barnett* [1985], and many subsequent studies. The intention of many of these studies was to identify the leading "modes" of the low-frequency atmospheric circulation. In a comprehensive review, *Wallace and Gutzler* [1981] noted the dominance of a zonally symmetric, global-scale seesaw between polar and temperate latitudes in Northern Hemisphere sea-level pressure, as well as to the more regional-scale pattern resembling the so-called Pacific North American (PNA) and Western Atlantic (WA) pressure patterns at mid-tropospheric levels. By applying similar techniques to a 700-hPa geopotential height, *Barnston and Livezey* [1987] demonstrated that the North Atlantic Oscillation is the only low-frequency circulation pattern, which is found in every month of the year. These studies confirmed that NAO is one of the most dominant and robust large-scale patterns of natural climate variability.

Many European studies relevant to NAO research were published in the latter half of the 20th century but, unfortunately, they remain relatively unknown and rarely cited in the NAO literature. Two main reasons for this obscurity were the preference for using (perceived to be less old-fashioned) expressions such as "westerly flow type" instead of "North Atlantic Oscillation", and their publication in not widely accessible international journals. A good example is provided by the studies of English climatologist, Hubert Lamb, who was fascinated by the variations in westerly flow and blocking that had greatly affected the UK and its society over the previous century and more – many interesting discussions and references can be found in his fascinating books on climate [*Lamb*, 1972; *Lamb*, 1977; *Lamb*, 1995]. Lamb [1972] devised a set of circulation types and associated indices for the UK and the nearby Atlantic that were later updated [*Hulme and Barrow*, 1977] and are still used today. One of Lamb's indices, the "westerly index", is very closely related to the NAO index and its variability since 1860 was discussed in detail in *Lamb* [1972] and *Hulme and Barrow* [1977]. Other westerly flow indices were also constructed such as the "PSCM" atmospheric circulation indices [*Murray and Benwell*, 1970] and the 500 hPa zonal index in the Atlantic/ European sector by H. Trenkle shown in Fig 7.9, p271 of *Lamb* [1972] – many of these and other zonal flow indices are reviewed in *Steinrücke* [1999]. An early study on the negative (blocked) phase of the NAO was published by *Sumner* [1959]. Several pioneering long-range forecasting studies at the UK Met Office attempted to use the NAO (westerly flow) to predict future circulation, temperatures, and precipitation in following seasons for the UK and the Atlantic region [*Hay*, 1967; *Murray*, 1972; *Murray and Lankester*, 1974]. *Miles* [1977] contains an interesting discussion of the annual cycle in the NAO.

In an attempt to find mechanisms explaining year-to-year and longer period variations in North Atlantic climate, *Bjerknes* [1964] used the pressure difference between Iceland and the Azores as a simple measure of westerly flow strength. This is in fact a simple North Atlantic Oscillation index, although Bjerknes used the term "zonal index". His studies were inspired by earlier investigations such as those of *Pettersson* [1896] *and Helland-Hansen and Nansen* [1920], and involved the dynamical analysis of past climate variability following *Lamb and Johnson* [1961]. Bjerknes noted that short-term variations in the zonal index were associated with latent cooling of North Atlantic sea surface temperatures (the tripole pattern) whereas long-term trends had a more complicated temperature pattern that he speculated were related to oceanic heat advection.

Several scientists in the 1970s attempted to exploit sea surface temperatures in order to make empirical long-range forecasts of European climate. *Ratcliffe and Murray* [1970] and *Ratcliffe* [1971, 1973] suggested that SST anomalies south of Newfoundland play a crucial role in this context and developed a classification, which could be used for long-range forecasts during winter. Cooler than normal SSTs south of Newfoundland were found to be followed one month later by positive SLP anomalies over the North Sea and the northern North Atlantic and negative SLP anomalies over the Atlantic south of 40°N (i.e. negative NAO index conditions). In a prescient seasonal forecast study of the causes of the extremely cold European winter of 1962/63, *Rowntree* [1976] investigated the influence of SST anomalies in the tropical Atlantic on the circulation of the northern extratropics by means of a set of climate model simulations including random perturbation experiments ("ensembles"). He prescribed a positive SST anomaly off the west coast of Africa at 16°N – 20°N, similar to that observed in the winter of 1962/1963, and then analyzed days 41 to 80 of a 80 day model runs (seasonal hindcasts). Although the different experiments gave different results north of 50°N and in the western North Atlantic, all experiments consistently showed a negative SLP anomaly west of the Iberian Peninsula. This finding was in agreement with the SLP anomalies observed in the 1962/1963 winter as well as with composites of other SLP anomalies for anomalously warm winters in the tropical North Atlantic. *Folland* [1983] later confirmed that SSTs in the eastern tropical Atlantic (as estimated by Cap Verde surface air temperatures) were significantly negatively correlated with the westerly flow over the North Atlantic [see *Wanner et al.*, 2001, for a comprehensive description of Atlantic SST anomalies related to the NAO]. To summarise, these early studies discovered the three main nodes of what is now known to be the North Atlantic SST tripole pattern associated with NAO [*Visbeck et al.*, this volume].

The papers by Fritz Loewe are worthy of further mention in that they provide an important historical bridge between the early German work and modern NAO studies. A polar explorer and meteorologist, Fritz Loewe experienced the severity of Greenland's temperatures first hand on the fateful expedition to Greenland in 1930, which led to the death of Alfred Wegener. Based on earlier German studies, he wrote about the temperature seesaw between Western Greenland and Europe in 1937, and then again in his retirement in 1966. Both *Loewe* [1937] and *Loewe* [1966] provided an important historical basis for the later studies of van Loon and coworkers in the late 1970s. With access to much longer historical data records, these studies were able to re-examine the winter temperature seesaw between Greenland and Northern Europe. *Loewe* [1966] confirmed beyond doubt that the seesaw in temperatures between western Greenland and Europe had been well established (stable) over a period of more than 200 years. The dynamical causes for the seesaw were then investigated in a series of pivotal studies by Harry van Loon and his PhD and masters students in Colorado [*van Loon and Rogers*, 1978; *Rogers and van Loon*, 1979; *Meehl and van Loon*, 1979]. They found significant correlations between atmospheric circulation and North Atlantic surface temperatures, and investigated the teleconnections with the Pacific region and with the tropical climate system. These studies significantly "shaped" the current concept of the NAO as a large-scale climate mode in the North Atlantic region with important impacts on European climate.

Rogers [1984] used a simple two-station NAO index based on sea-level pressures at Ponta Delgada, Azores and Akureyri, Iceland to assess the association between NAO and the Southern Oscillation. He confirmed *Walker's* [1924] earlier findings that the two phenomena are not strongly associated. *Rogers* [1984] also speculated on the long-term decreasing secular trend in the NAO index, which has since vanished due to the subsequent runs of positive NAO index winters. *Lamb and Peppler* [1987] successfully invoked the North Atlantc Oscillation for explaining the 1979-1984 droughts in Moroccan precipitation. This led to the (negative phase of the) North Atlantic Oscillation phenomenon being renamed "Al Moubarak" (the bountiful) by His Majesty King Hassan II of Morocco (personal communication, Dr El Hamly).

6. THE LATE 20TH CENTURY RENAISSANCE

Since 1990, there has been a remarkable growth in scientific research on the North Atlantic Oscillation. This can be clearly seen in Figure 5, which shows the number of scientific articles in the Institute for Scientific Information (ISI, see http://www.wos.mimas.ac.uk) citation database with the expression "North Atlantic Oscillation" in either the title or the abstracts published between 1985 and 2001. Various factors contribute to this renaissance of interest; in particular, the abundance of positive NAO index winters since 1970 that are associated with recent European climate variations and associated impacts [*Hurrell*, 1995]. In the year 2000, no fewer than 129 scientific papers were published on aspects of the North Atlantic Oscillation. In addition to these articles, a similar search for articles with titles and abstracts containing the words "Northern Annular Mode" and "Arctic Oscillation" revealed 1, 5, 22, and 41 articles published in the years 1998 to 2001. This great interest in the subject will undoubtedly lead to significant advances in the future. The rest of this section will briefly sketch some of the main developments that have occurred in this period, and we refer the reader to other articles in this volume for more in-depth reviews of the recent literature.

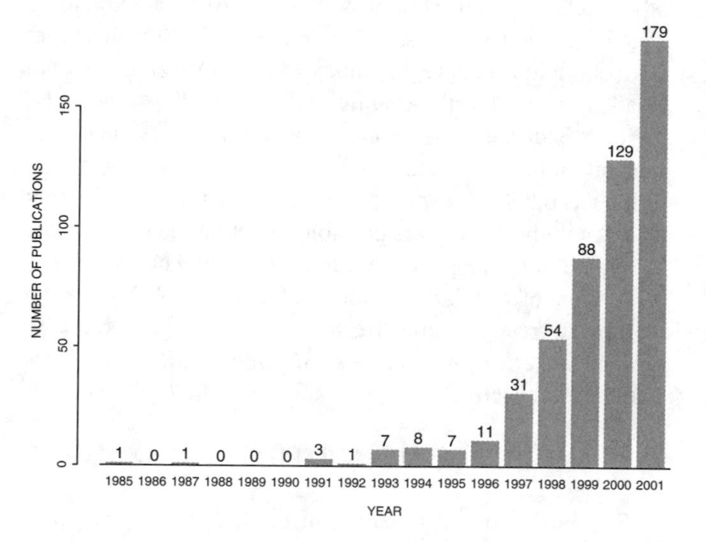

Figure 5. The increasing number of published articles containing the expression "North Atlantic Oscillation" either in the title or in the abstract during the period January 1985 and November 2001. Source: web of science bibliographic database.

Hurrell [1995] investigated the influence of the NAO on interannual and decadal temperature and pressure variations over the European continent. He demonstrated that recent trends in the NAO could help explain climate trends over Europe. *Hurrell* [1996] used multiple regression to show that the NAO explained 31% of the total variance in wintertime mean surface temperatures averaged from 20°N to 90°N over the period 1935-94. In other words, a substantial fraction of recent extratropical warming in the Northern Hemisphere can be attributed to the recent upward trend in the NAO index from 1965 to 1995. However, using longer NAO indices shows that the 31% explained variance reduces to zero over 1895-1920 and is about 25% over the period 1851-1894 [*Osborn et al.*, 1999]. A more comprehensive review of these findings was presented in *van Loon and Hurrell* [1997]. *Gillett et al.* [this volume] discuss likely changes to NAO variability due to anthropogenic forcing in more detail.

Hurrell [1995] defined a new NAO index as the difference between the standardized station pressure series of Lisbon, Portugal minus Stykkisholmur, Iceland. This index has become one of the most commonly used NAO indices in climate research. *Jones et al.* [1997] further extended an instrumental NAO index back to 1821 by using station pressure observations from Gibraltar and the Reykjavik area [see also *Jones et al.*, this volume]. Despite their obvious advantages for historical studies, two-point station indices are not particularly robust to changes in the NAO spatial pattern [*Kapala et al.*, 1998; *Wanner et al.*, 1997]. They are also less representative and noisy for sampling the NAO over periods less than one season [*Stephenson et al.*, in preparation]. For daily NAO studies, it is necessary to define the NAO index using data from the whole North Atlantic region [*Stephenson et al.,* in preparation]. Refer to *Jones et al.* [this volume] for more discussion of the recent work on historical NAO indices.

For studying long-term trends in the NAO, it is necessary to extend the index further back beyond the pre-instrumental period. Much current research has focussed on constructing "proxy" NAO indices using early instrumental pressure, temperature and precipitation station series and environmental proxy and documentary proxy data [e.g., *Cook*, this volume; *Appenzeller et al.*, 1998; *Luterbacher et al.*, 1999, 2002; *Glueck and Stockton,* 2001]. An important caveat of the proxy approach is that proxies can only be used to infer NAO by its assumed response rather than measure NAO directly via dynamical quantities such as pressure. Refer to *Cook* [this volume] for more discussion of the recent work on proxy indices.

In the 1990s, climate modellers also began to investigate the NAO in their model simulated climates. *Glowienka-Hense* [1990] showed that NAO variability was realistically simulated by an atmosphere-only general circulation model, confirming that atmospheric processes fundamentally cause the NAO. Several studies have also shown that NAO can be captured quite realistically by coupled ocean-atmosphere models [*Pittawala and Hameed*, 1991; *Davies et al.*, 1997; *Osborn et al.*, 1999]. *Davies et al.* [1997] showed that the HadAM1 atmospheric model gave low but significant predictability for the NAO from North Atlantic SST in the model in winter and spring with likely El Niño/Southern Oscillation (ENSO) influences on at least the western half the NAO pattern. By comparing coupled simulations made using different ocean models, *Christoph et al.* [2000] demonstrated that the NAO in sea-level pressure variations simulated by the ECHAM4 coupled model was not sensitive to the choice of ocean model. *Stephenson and Pavan* [2002] reviewed the ability of 17 coupled models to simulate the NAO temperature pattern, and found a wide range of different model behavior. Although the spatial NAO pattern was found to be generally well simulated, the simulated NAO time series was too strongly correlated with ENSO for several of the coupled models. More recently, *Hoerling et al.* [2001] have used a GCM to trace the origins of decadal NAO predictability to the tropical Indian Ocean, thereby extending the earlier work of *Shaw* [1905], *Hann* [1906], *Defant* [1924], and *Rowntree* [1976].

Stimulated by the progress made in understanding the coupled processes controlling ENSO, several studies in the 1990s started to investigate North Atlantic ocean-atmosphere processes in more detail. The NAO was related to interdecadal variability of latent and sensible heat flux anomalies of the North Atlantic and the oceanic circulation [e.g. *Cayan*, 1992a, b; *Deser and Blackmon*, 1993; *Kushnir*, 1994]. More recent research with atmosphere-only models forced by observed prescribed sea surface temperatures demonstrated a small amount of ex-ante skill at hindcasting post 1950 decadal trends in the NAO [*Rodwell et al.*, 1999; *Mehta et al.*, 2000]. Whether this apparent skill is relevant to the fully coupled system is still a serious matter for debate [*Bretherton and Battisti*, 2000]. See *Rodwell* [this volume] and *Cjaza et al.* [this volume] for more discussion of recent research on these matters.

The 1990s have seen the emergence of a debate on the spatial structure of the North Atlantic Oscillation. Based on principal component analysis of hemispheric sea-level pressure north of 20°N, *Thompson and Wallace* [1998; 2000; 2001] isolated a large-scale pattern they referred to as the "Arctic Oscillation (AO)" or "Northern Annular Mode (NAM)". The AO/NAM pattern has striking resemblances to the NAO in the Atlantic sector, but is more zonally symmetric with a low pressure center of action over the Arctic region and an annular high pressure band in the subtropics. The AO/NAM index has strong correlations with the stratosphere [*Baldwin and Dunkerton*, 1999 and references therein], and shows more marked decadal trends than does the NAO index. However, according to *Deser* [2000] and *Ambaum et al.* [2001], the annular appearance of the AO/NAM is caused by the Arctic center of action, while there is no co-ordinated behavior of the Atlantic and Pacific centers of action. *Ambaum et al.* [2001] showed that the NAO reflects the correlations between the surface pressure variability at all of its centers of action whereas this is not the case for the AO/NAM. The only significant correlation between centers of action in the AO/NAM pattern is between the Iceland and the Azores [*Ambaum et al.*, 2001]. For more discussion of this ongoing debate refer to these cited articles and to *Thompson et al.* [this volume], and *Hurrell et al.* [this volume].

A major contribution to the recent growth in publications on the NAO has been the rising awareness by researchers in other fields of the usefulness of the NAO for explaining weather related impacts. NAO provides a convenient (yet far from unique) aggregate index for summarizing seasonal weather in the North Atlantic and surrounding land mass regions. Because of its strong association with temperature, precipitation, and surface winds, the NAO has a strong influence on the biosphere. By relating impacts to recent changes in the NAO, it is possible to get an idea of how the biosphere may respond to future climate change. For a discussion of the large amount of recent NAO work on marine, terrestrial, and freshwater ecosystems refer to *Drinkwater et al.*, *Mysterud et al.*, and *Straile et al.* [this volume], respectively.

7. COMMON THREADS AND FUTURE DIRECTIONS

Several important threads can be discerned in previous research on the NAO.

Firstly, NAO is without doubt one of the most ancient, robust, and omnipresent climate teleconnections known to mankind. Scientific research has confirmed (and will undoubtedly continue to demonstrate) the role of NAO as a very useful aggregate factor for describing the variations of many diverse phenomena. NAO provides a macro-scale index that encapsulates a lot of information about climate and weather variations in the North Atlantic region.

Secondly, the overall spatial extent of the NAO pattern is still a debatable topic. Is the NAO primarily located in the North Atlantic region (and surrounding rim) or is it really part of a more global hemispheric pattern? In the absence of any dynamical arguments, the pattern has always been isolated using descriptive statistical techniques such as correlation mapping or principal component analysis [*Hurrell et al.*, this volume]. These methods do not provide any definitive answer to what is the most "physically" correct pattern for the NAO. In some ways, the recent debate on NAO versus AO [see *Ambaum et al.*, 2001] has similarities with the debate in the first half of the 20th century on Spearman's g factor (leading principal component) for intelligence. After much heated debate, Spearman finally conceded that neither his principal component nor the alternative rotated principal components provided a definitive factor for intelligence – the problem of intelligence had to be considered multidimensional [see *Gould*, 1981 for a very interesting account]. Despite the obvious temptation, we should avoid falling into a similar trap by remembering that other teleconnection patterns also need to be carefully considered in order to fully understand the many facets of climate variability.

A third recurrent theme of NAO research has been concerned with understanding underlying long-term climate trends. Despite being quite noisy from year to year, the NAO index has also exhibited substantial long-term trends (both up and down) in the past. These trends are the result of runs of similar sign in the NAO index. Various mechanistic explanations have been sought for these trends such as oceanic forcing, and more recently stratospheric forcing. However, the trends might also be stochastic trends resulting from natural variability containing long-range dependence [*Wunsch*, 1999; *Stephenson et al.*, 2000]. In order to discriminate likely future trends in the NAO induced by anthropogenic effects, it is necessary that we understand the behavior and causes of the natural long-term variations in the NAO. Longer time series from both climate model simulations and proxy reconstruction are likely to help enormously in this endeavor.

A fourth theme in NAO research has been the search for statistically significant periodicities. This is a particularly difficult exercise that is prone to error due to the presence of noise and trends in the NAO time series [see *Wunsch*, 1999 for some warning remarks]. Perhaps the most significant periodicity (if any) in the NAO is the quasi-biennial one remarked upon by *Walker* [1924] and reviewed in more detail in *Stephenson et al.* [2000]. *Coughlin and Tung* [2001] discuss the earlier QBO literature reviewed by *Stephenson et al.* [2000] and present a reanalysis of this robust feature of the NAO. It should be remembered, however, that NAO is a broad band phenomena with an almost white noise power spectrum and so, unlike ENSO, individual periodicities do not account for large fractions of the total variance.

Perhaps the most challenging thread in NAO research relates to our ability to understand its underlying dynamical mechanisms. Although we can describe its features, we still have very little conceptual understanding of why this particular pattern emerges out of weather variations or how it evolves prognostically. Without a dynamical understanding of how it ticks, it is difficult to define the "mode" unambiguously and thereby assess its predictability. To make progress, it will be necessary to develop conceptual understanding of the key mechanisms controlling the evolution of the North Atlantic Oscillation as has been achieved spectacularly in recent decades for the Southern Oscillation.

Rather than being merely a recent bandwagon phenomenon, the history of the last few centuries shows that NAO research waxes and wanes but always seems to come back again. There is no doubt that in the future the complex NAO phenomenon will continue to attract the curiosity of scientists as it has repeatedly done so in the past.

Acknowledgements. The authors wish to thank Chris Folland, Mike Wallace, and Phil Jones for reviewing this article and offering useful suggestions for improvements. DBS is indebted to all the people who kindly supplied historical information for this review, in particular, Dr Nils Gunnar Kvamstø who managed to find an old copy of Helland-Hansen and Nansen's book in the dusty archives of the Geophysical Institute, Bergen. HW, SB, and JL are indebted to the Swiss National Science Foundation, as represented by the National Centre of Competence in Research (NCCR) in Climate and FLOODRISK (SNF no. 15-079).

REFERENCES

Anonymous, The King's Mirror (Speculum Regale – Konungs skuggsjá)', in *Scandinavian Monographs*, translated by L. M. Larson, 3, 388 pp. The American Scandinavian Foundation, Harvard University Press, 1917.

Ambaum, M.H.P., B. Hoskins, and D. B. Stephenson, Arctic Oscillation or North Atlantic Oscillation?, *J. Climate, 14*, 3495-3507, 2001.

Ångström, A., Teleconnections of climatic changes in present time, *Geogr. Ann., 17*, 242-258, 1935.

Appenzeller, C., T. F. Stocker, and M. Anklin, North Atlantic Oscillation dynamics recorded in Greenland ice cores, *Science 282*, 446-449, 1998.

Baldwin, M. P., and T. J. Dunkerton, Propagation of the Arctic Oscillation from the stratosphere to the troposphere, *J. Geophys. Res. 104*, 30937-30946, 1999.

Barnett, T. P., Variations in near-global sea level pressure, *J. Atmos. Sci. 42*, 478-501, 1985.

Barnston, A. G., and R. E. Livezey, Classification, seasonality and persistence of low frequency atmospheric circulation patterns, *Mon.Wea. Rev., 115,* 1083-1126, 1987.

Bjerknes, J., Atlantic air-sea interaction, *Adv. Geophys., 10*, 1-82, 1964.

Bretherton, C. S., and D. S. Battisti, An interpretation of the results from atmospheric general circulation models forced by the time history of the observed sea surface temperature distribution, *Geophys. Res. Lett., 27*, 767-770, 2000.

Cayan, D. R., Latent and sensible heat flux anomalies over the Northern Oceans: Driving the sea surface temperature, *J. Phys. Ocean., 22*, 859-881, 1992a.

Cayan, D. R., Latent and sensible heat flux anomalies over the Northern Oceans: The connection to monthly atmospheric circulation, *J. Climate, 5*, 354-369, 1992b.

Christoph, M., U. Ulbrich, J. M. Oberhuber, and E. Roeckner, The role of ocean dynamics for low-frequency fluctuations of the NAO in a coupled ocean- atmosphere GCM, *J. Climate, 13*, 2536-2549, 2000.

Cook, E. R., R. D'Arrigo, and K. Briffa, A reconstruction of the North Atlantic Oscillation using tree-ring chronologies from North America and Europe, *The Holocene, 8*, 9-17, 1998.

Cook, E. R., Multi-proxy reconstructions of the North Atlantic Oscillation (NAO) index: A critical review and a new well-verified winter NAO index reconstruction back to AD 1400, this volume.

Coughlin, K., and K.-K. Tung, QBO signal found at the extratropical surface through Northern Annular Modes, *Geophys. Res. Lett., 28*, 4563, 2001.

Crantz, D., *The History of Greenland; including an Account of the Mission Carried on by the United Brethren in that Country*, 2 volumes, xi, 359 pp., vi, 323 pp., Longman, Hurst, Rees, Orme and Brown, 1820, London, 1765.

Czaja, A., A. W. Robertson, and T. Huck, The role of North Atlantic ocean-atmosphere coupling in affecting North Atlantic Oscillation variability, this volume.

Dannmeyer, F., Zur Frage der Gegensätzlichkeit der kalten Winter in Grönland zu den warmen Wintern in Deutschland, *Polarforschung*, *2*, 29, 1948.

Davies, J. R., D. P. Rowell, and C. K. Folland, North Atlantic and European seasonal predictability using an ensemble of multidecadal AGCM simulations, *Int. J. Climatol., 12*, 1263-1284, 1997.

Defant, A., Die Schwankungen der atmosphärischen Zirkulation über dem Nordatlantischen Ozean im 25-jährigen Zeitraum 1881-1905, *Geogr. Ann., 6*, 13-41, 1924.

Deser, C., On the teleconnectivity of the Arctic Oscillation, *Geophys. Res. Lett., 27*, 779-782, 2000.

Deser, C., and M. L. Blackmon, Surface climate variations over the North Atlantic Ocean during winter: 1900-1989, *J. Climate, 6*, 1743-1753, 1993.

Dove, H. W., Über die geographische Verbreitung gleichartiger Witterungserscheinungen. Erste Abhandlung: Über die nicht periodischen Änderungen der Temperaturvertheilung auf der Oberfläche der Erde, *Abhandlungen der Königlichen Akademie der Wissenschaften in Berlin 1838,* 287-415, 1839.

Dove, H. W., Über die nicht periodischen Änderungen der Temperaturvertheilung auf der Oberfläche der Erde, *Abhandlungen der Königlichen Akademie der Wissenschaften in Berlin 1839,* 305-440, 1841.

Drinkwater, K., A. Belgrano, A. Borja, A. Conversi, M. Edwards, C. H. Greene, G. Ottersen, A. J. Pershing, and H. Walker, The response of marine ecosystems to climate variabiliity associated with the North Atlantic Oscillation, this volume.

Egede, H., *History of Greenland – A description of Greenland: shewing the natural history, situation, boundaries, and face of the country; the nature of the soil; the rise and progress of the old Norwegian colonies; the ancient and modern inhabitants; their genius and way of life, and produce of the soil; their plants, beasts, fishes, etc.*, 220 pp., (translated from the Danish), Pickering Bookseller, Picadilly, London, 1745.

Exner, F. M., Übermonatliche Witterungsanomalien auf der nördlichen Erdhälfte im Winter, *Sitzungsberichte d. Kaiserl, Akad. der Wissenschaften, 122*, 1165-1241, 1913.

Exner, F. M., Monatliche Luftdruck- und Temperaturanomalien auf der Erde, *Sitzungsberichte d. Kaiserl, Akad. der Wissenschaften 133*, 307-408, 1924.

Folland, C. K., Regional-scale interannual variability of climate - a north west European perspective, *Met. Mag., 112*, 163-183, 1983.

Fukuoka, A., A study of 10-day forecast (a synthetic report), *Geophys. Mag. 22*, 177-208, 1951.

Galton, F., Co-relations and their Measurement, chiefly from Anthropometric Data, *Proceedings of the Royal Society of London 45*, 135-145, 1888.

Gilbert, L. W., Physikalisch Geographische Nachrichten aus dem nördlichen Polarmeer, Als Anhang zu den Aufsätzen im vorigen Hefte, *Gilbert's Annalen (Annalen der Physik), 62*, 137-166, 1819.

Gillett, N. P., H. F. Graf, and T. J. Osborn, Climate Change and the North Atlantic Oscillation, this volume.

Glowienka-Hense, R., The North Atlantic Oscillation in the Atlantic-European SLP, *Tellus, 42A*, 497–507, 1990.

Glueck, M. F., and C. W. Stockton, Reconstruction of the North Atlantic Oscillation, 1429-1983, *Int. J. Climatol., 21*, 1453-1465, 2001.

Gould, S. J., The Mismeasure of Man, 352 pp., W.W. Norton, New York, 1981.

Gronau, K. L., Das Klima der Polarländer, in H.F. Flörke, *Repertorium des Neuesten und Wissenwürdigsten aus der gesamten Naturkunde*, Berlin, 340-354, 1811.

Hann, J., Zur Witterungsgeschichte von Nord-Grönland, Westküste, *Meteor. Zeitschrift, 7*, 109-115, 1890.

Hann, J., Der Pulsschlag der Atmosphäre, *Meteor. Zeitschrift 23*, 82-86, 1906.

Hay, R. F. M., The association between autumn and winter circulations near Britain, *Met. Mag., 96*, 167-178, 1967.

Hoerling, M. P., J. W. Hurrell, and T. Xu, Tropical Origins for Recent North Atlantic Climate Change, *Science, 292*, 90-92, 2001.

Helland-Hansen, B., and F. Nansen, Temperature variations in the North Atlantic ocean and in the atmosphere, *Smithsonian Misc. Collections, 70(4)*, 406, 1920.

Hildebrandsson, H. H., Quelques recherches sur les centres d'action de l'atmosphère, *Kongl. Svenska Vetenskaps-akad. Handl., 29, (3)*, 1897.

Hulme, M., and E. Barrow (Eds), *Climates of the British Isles, past, present and future*, 454 pp., Routledge, London, 1997.

Hurrell, J. W., Decadal trends in the North Atlantic oscillation: Regional temperatures and precipitation, *Science, 269*, 676-679, 1995.

Hurrell, J. W., Influence of variations in extratropical wintertime teleconnections on Northern Hemisphere temperature, *Geophys. Res. Lett., 23*, 665-668, 1996.

Hurrell, J. W. and H. van Loon, Decadal variations in climate associated with the North Atlantic Oscillation, *Clim. Change 36*, 301-326, 1997.

Hurrell, J. W., Y. Kushnir, G. Ottersen, and M. Visbeck, An overview of the North Atlantic Oscillation, this volume.

Jones, P. D., T. Jónsson, and D. Wheeler, Extension to the North Atlantic Oscillation using early instrumental pressure observations from Gibraltar and south-west Iceland, *Int. J. Climatol., 17*, 1433-1450, 1997.

Jones, P. D., T. J. Osborn, and K. R. Briffa, Pressure-based measures of the North Atlantic Oscillation (NAO): A comparison and an assessment of changes in the strength of the NAO and in its influence on surface climate parameters, this volume.

Kapala, A. H., H. Mächel, and H. Flohn, Behaviour of the centres of action above the Atlantic since 1881, Part II: Associations with regional climate anomalies, *Int. J. Climatol., 18*, 23-36, 1998.

Kushnir, Y., Interdecadal variations in North Atlantic sea surface temperature and associated atmospheric conditions, *J. Climate, 7*, 142-157, 1994.

Kutzbach, J. E., Large-scale features of monthly mean Northern Hemisphere anomaly maps of sea-level pressure, *Mon. Wea. Rev., 98*, 708-716, 1970.

Lamb, H. H., and A. I. Johnson, Climatic variations and observed changes in the general circulation, III. Investigation of long series of observations and circulation changes in July, *Geogr. Ann., 18*, 363-400, 1961.

Lamb, H. H., British Isles weather types and a register of the daily sequence of circulation patterns, *Geophysical Memoirs*, Vol. 16, 116, 85 pp, 1972.

Lamb, H. H., *Climate: Present, Past, and Future – Volume 1: Fundamentals and Climate Now*, 613 pp, Methuen, London, 1972.

Lamb, H. H., *Climate: Present Past and Future – Volume 2: Climatic History and the Future*, 835 pp, Methuen, London, 1977.

Lamb, H. H., *Climate, History, and the Modern World*, 433 pp. Rouledge, London, 1995.

Lamb, P. J., and R. A. Peppler, North Atlantic Oscillation: Concept and application, *Bull. Amer. Meteor. Soc. 68*, 1217-1225, 1987.

Loewe, F., A period of warm winters in Western Greenland and the temperature see-saw between Western Greenland and Central Europe, *Q. J. R. Meteorol. Soc., 63*, 365-371, 1937.

Loewe, F., The temperature see-saw between western Greenland and Europe, *Weather, Vol. XXI, No. 7*, 241-246, 1966.

Lorenz, E. N., Seasonal and irregular variations of the Northern Hemisphere sea-level pressure profile, *J. Meteorol., 8*, 52-59, 1951.

Lorenz, E. N., Empirical orthogonal functions and statistical weather prediction, Technical Report 1, Statistical Forecasting Project, 48 pp., Department of Meteorology, MIT, Cambridge, 1956.

Lorenz, E. N., The Nature and Theory of the General Circulation of the Atmosphere, 161 pp., World Meteorological Organization (WMO), Geneva, 1967.

Luterbacher, J., C. Schmutz, D. Gyalistras, E. Xoplaki, and H. Wanner, Reconstruction of monthly NAO and EU indices back to AD 1675, *Geophys. Res. Lett., 26*, 2745-2748, 1999.

Luterbacher, J., et al., Extending North Atlantic Oscillation reconstructions back to 1500, *Atmos. Sci. Lett., 2,* 114-124, doi:10.1006/asle.2001.0044, 2002.

Meehl, G. A., and H. van Loon, The seesaw in winter temperatures between Greenland and Northern Europe, Part III: Teleconnections with lower latitudes, *Mon. Wea. Rev., 107*, 1095-1106, 1979.

Mehta, V. M., M. J. Suarez, J. Y. Manganello, and T. L. Delworth, Oceanic influence on the North Atlantic Oscillation and associated Northern Hemisphere climate variations: 1959-1993, *Geophys. Res. Lett., 27*, 121-124, 2000.

Meinardus, W., Der Zusammenhang des Winterklimas in Mittel- und Nordwest-Europa mit dem Golfstrom, *Z. d. Ges. f. Erdkunde in Berlin*, 23, 183-200, 1898.

Miles, M. K., The annual course of some indices of the zonal and meridional circulation in middle latitudes of the Northern Hemisphere, *Met. Mag., 106*, 52-66, 1977.

Murray, R., and P. R. Benwell, PSCM indices in synoptic meteorology and long range forecasting, *Met. Mag., 99*, 232-245, 1970.

Murray, R., An objective method of foreshadowing winter rainfall and temperature for England and Wales, *Met. Mag., 101*, 97-110, 1972.

Murray, R., and J. D. Lankester, Central England temperature quintiles and associated pressure analyses on a monthly time scale, *Met. Mag., 103*, 3-14, 1974.

Mysterud, A., N. C. Stenseth, N. G. Yoccoz, G. Ottersen, and R. Langvatn, The response of terrestrial ecosystems to climate variability associated with the North Atlantic Oscillation, this volume.

Namias, J., The index cycle and its role in the general circulation, *J. Meteorol. 7*, 130-139, 1950.

Osborn, T. J., K. R. Briffa, S. F. B. Tett, P. D. Jones, and R. M. Trigo, Evaluation of the North Atlantic oscillation as simulated by a climate model, *Clim. Dyn., 15*, 685-702, 1999.

Ostermann, H., (editor), Brudstykker af en Dagbog holden I Groenland I Aarene 1770-1778, *Medd. Groenland, 129, (2)*, 1-103, 1942.

Pettersson, O., Über die Beziehungen zwischen hydrographischen und meteorologischen Phänomenen, *Meteorol. Z., 13*, 285-321, 1896.

Pittawala, I. I., and S. Hameed, Simulation of the North Atlantic Oscillation in a General Circulation Model, *Geophys. Res. Lett., 18*, 841-844, 1991

Ratcliffe, R. A. S., and R. Murray, New lag association between North Atlantic sea temperature and European pressure applied to long-range weather forecasting, *Q. J. R. Meteorol. Soc., 96*, 226-246, 1970.

Ratcliffe, R. A. S., North Atlantic sea temperature classifications 1877-1970, *Met. Mag., 100*, 225-232, 1971.

Ratcliffe, R. A. S., Recent work on sea surface temperature anomalies related to long-range forecasting, *Weather, 28*, 106-117, 1973.

Rodwell, M. J., D. P. Rowell, and C. K. Folland, Oceanic forcing of the wintertime North Atlantic Oscillation and European climate, *Nature, 398*, 320-323, 1999.

Rodwell, M. J., On the predictability of North Atlantic climate, this volume.

Rogers, J. C., The association between the North Atlantic Oscillation and the Southern Oscillation in the Northern Hemisphere, *Mon. Wea. Rev., 112*, 1999-2015, 1984.

Rogers, J. C., North Atlantic storm track variability and its association to the North Atlantic Oscillation and climate variability of Northern Europe, *J. Climate 10*, 1635-1645, 1997.

Rogers, J. C., and H. van Loon, The seasaw in winter temperature between Greenland and Northern Europe, Part II: Some oceanic and atmospheric effects in middle and high latitudes, *Mon. Wea. Rev., 107*, 509-519, 1979.

Rossby, C.-G., and collaborators, Relations between variations in the intensity of the zonal circulation of the atmosphere and the displacements of the semipermanent centers of action, *J. Mar. Res., 3*, 38-55, 1939.

Rossby, C. G., and H. C. Willett, The circulation of the upper troposphere and lower stratosphere, *Science*, *108*, 643-652, 1948.

Rowntree, P. R., Response of the atmosphere to a tropical Atlantic ocean temperature anomaly, *Q. J. R. Meteorol. Soc., 102*, 607-626, 1976.

Shaw, W. N., The pulse of the atmospheric circulation, *Nature, 73*, 175-176, 1905.

Steinrücke, J., Changes in the Northern-Hemispheric zonal circulation and their relationship to precipitation frequencies in Europe, *Bochumer Geogr. Arbeiten, 65*, 1999.

Stephenson, D. B., V. Pavan, and R. Bojariu, Is the North Atlantic Oscillation a random walk?, *Int. J. Climatol., 20*, 1-18, 2000.

Stephenson, D. B., and V. Pavan, The North Atlantic Oscillation in coupled climate models: a CMIP1 evaluation, *Climate Dynamics,* in press, 2002.

Straile, D., D. M. Livingstone, G. A. Weyhenmeyer, and D. G. George, The response of freshwater ecosystems to climate variability associated with the North Atlantic Oscillation, this volume.

Sturluson, S., *Prose Edda of Snorri Sturluson: Tales from Norse Mythology*, edited by J. Young, University of California Press, 1984.

Sumner, E. J., Blocking anticyclones in the Atlantic-European sector of the Northern Hemisphere, *Met. Mag., 88*, 300-311, 1959.

Teisserenc de Bort, L. P., Etude sur l'hiver de 1879-80 et recherches sur l'influence de la position des grands centres d'action de l'atmosphère dans les hivers anormaux, *Ann. Soc. Météor. France, 31*, 70-79, 1883.

Thompson, D. W. J., and J. M. Wallace, The Arctic Oscillation signature in the wintertime geopotential height and temperature fields, *Geophs. Res. Lett., 25*, 1297-1300, 1998.

Thompson, D. W. J., and J. M. Wallace, Annular modes in the extratropical circulation. Part I: Month-to-month variability, *J. Climate, 13*, 1000-1016, 2000.

Thompson, D. W. J., and J. M. Wallace, Regional Climate Impacts of the Northern Hemisphere Annular Mode, *Science, 293*, 85-89, 2001.

Thompson, D. W. J., J. M. Wallace, and G. C. Hegerl, Annular modes in the extratropical circulation. Part II: Trends, *J. Climate, 13*, 1018-1036, 2000.

Thompson, D. W. J., S, Lee, and M. P. Baldwin, Atmospheric Processes governing the Northern Hemisphere Annular Mode/North Atlantic Oscillation, this volume.

Trenberth, K. E., and D. A. Paolino, The Northern Hemisphere sea level pressure data set: Trends, errors and discontinuities, *Mon. Wea. Rev., 108*, 855-872, 1980.

van Loon, H., and J. C. Rogers, The seesaw in winter temperatures between Greenland and northern Europe. Part I: General descriptions, *Mon. Wea. Rev., 106*, 296-310, 1978.

Visbeck, M., E. Chassignet, R. Curry, T. Delworth, B. Dickson, and G. Krahmann, The ocean's response to North Atlantic Oscillation variability, this volume.

Walker, G. T., Correlation in seasonal variation of climate, *Mem. Ind. Met. Dept., 20*, 122, 1909.

Walker, G. T., Correlation in seasonal variation of weather, VIII, a preliminary study of world weather, *Mem. Ind. Met. Dept., 24*, 75-131, 1923.

Walker, G. T., Correlation in seasonal variation of weather, IX *Mem. Ind. Met. Dept., 25*, 275-332, 1924.

Walker, G. T., and E. W. Bliss, World Weather V, *Mem. Roy. Met. Soc., 4*, 53-84, 1932.

Wallace, J. M., North Atlantic Oscillation/annular mode: Two paradigms - one phenomenon, *Q. J. R. Meteorol. Soc., 126*, 791-805, 2000.

Wallace, J. M., and D. S. Gutzler, Teleconnections in the geopotential height field during the Northern Hemisphere winter, *Mon. Wea. Rev., 109*, 784-812, 1981.

Wanner, H., R. Rickli, E. Salvisberg, C. Schmutz, and M. Schüepp, Global climate change and variability and its influence on Alpine climate - concepts and observations, *Theor. Appl. Climatol., 58*, 221-243, 1997.

Wanner, H., S. Brönnimann, C. Casty, D. Gyalistras, J. Luterbacher, C. Schmutz, D. B. Stephenson, and E. Xoplaki, North Atlantic Oscillation- concepts and studies, *Survey Geophys., 22*, 321-381, 2001.

Willett, H. C., Long-period fluctuations of the general circulation of the atmosphere, *J. Meteorol., 6*, 34-50, 1949.

Wunsch, C., The interpretation of short climate records, with comments on the North Atlantic and Southern Oscillations, *Bull. Am. Met. Soc. 80*, 245-255, 1999.

Stefan Brönnimann, Institute of Geography, University of Bern, Hallerstrasse 12, CH-3012 Bern, Switzerland.
broenn@giub.unibe.ch

Jürg Luterbacher, Institute of Geography, University of Bern, Hallerstrasse 12, CH-3012 Bern, Switzerland.
juerg@giub.unibe.ch

David B. Stephenson, Department of Meteorology, University of Reading, Earley Gate, PO Box 243, Reading RG6 6BB, U.K.
d.b.stephenson@reading.ac.uk

Heinz Wanner, Institute of Geography, University of Bern, Hallerstrasse 12, CH-3012 Bern, Switzerland.
wanner@giub.unibe.ch

Pressure-Based Measures of the North Atlantic Oscillation (NAO): A Comparison and an Assessment of Changes in the Strength of the NAO and in its Influence on Surface Climate Parameters

Philip D. Jones, Timothy J. Osborn, Keith R. Briffa

Climatic Research Unit, University of East Anglia, Norwich, U.K.

The North Atlantic Oscillation (NAO) is the principal mode of wintertime variability in surface pressure over the North Atlantic/European sector of the Northern Hemisphere (NH). Europe has long instrumental records and an index of the NAO can be derived back to the early 1820s using monthly pressure data from Reykjavik in Iceland and Gibraltar in southern Spain. The specific location of the northern node in Iceland is not critical, but the location of the southern node (Ponta Delgada, Lisbon and Cadiz are among other possibilities) determines the nature of the seasonal importance (southwestern European locations provide the longest records, but are only useful during the winter part of the year). Various instrumental indices are intercompared and the strengths and variability of their relationships with wintertime temperature and precipitation patterns over Europe are assessed. The influence of the NAO is shown to vary, sometimes significantly, over the last 150-200 years. The influence is strongest recently and during the late 19th and early 20th centuries and weaker at other times. Proxy reconstructions of the NAO based on regression, where the predictors are drawn from natural and written documentary archives, assume stationarity in the teleconnection patterns of precipitation and temperature with the NAO. Changes in the influence of the NAO will also be captured in these reconstructions. The performance of most proxy-based reconstructions, produced to date, reduces before the 1870s. We conclude by demonstrating the potential of creating a new index based on the two probable longest pressure series in Europe, Paris and London (back to the 1670s). Considerable effort will be required, however, to locate and homogenize the raw pressure measurements before the 1760s.

1. INTRODUCTION

The strong anti-phase relationship between monthly pressure series from Iceland and the Azores was first referred to by *Walker* [1924] as the North Atlantic Oscillation (NAO). A commonly used index has been calculated from the difference between the normalized monthly pressures at Ponta Delgada (Azores) and Stykkisholmur (Iceland). These locations were chosen originally as they were the sites with the longest readily available series of pressure observations, from near the two centers of action that represent the phenomenon.

The winter NAO value is a measure of the westerly wind strength over much of western, central and northern Europe. Positive values of the NAO index, implying stronger westerlies, bring milder weather to Europe, north of about 40°N and vice versa [see *Hurrell*, 1995; *Osborn et al.*, 1999; Slonosky et al., 2000]. Figure 1 shows correlations between a NAO index for the winter season (December to February) and surface temperature and precipitation for two periods 1901-50 and 1951-2000. The relationship with surface tem-

The North Atlantic Oscillation:
Climatic Significance and Environmental Impact
Geophysical Monograph 134

10.1029/134GM03

perature is strongest over southern Fennoscandia. South of 40°N, the relationship is inverse, particularly over the southern Balkans, Turkey and parts of the Middle East and over southern Spain and northwestern Africa. Positive NAO values are also associated with higher amounts of winter precipitation north of 45°N and reduced precipitation to the south [see also, *Hurrell*, 1995].

Figure 1 also illustrates that the winter NAO relationships with surface climate vary significantly depending on time period. All the main centers of influence over the North Atlantic, Europe and eastern North America are more coherent and stronger during 1951-2000 than 1901-50. Whilst some of the lesser features in other parts of the Northern Hemisphere (NH) may be affected by the quality of the surface climate data, the quality for the main centers is high. Thus, the marked reduction in strength and extent over much of the United States is a real feature. Lack of surface data for parts of northern Russia does not allow the extent of the positive influence during the 1901-50 period to be adequately assessed. The precipitation influence extends over a much smaller region, being limited to northern Europe, the Mediterranean and North Africa. Correlations with precipitation are also weaker during the 1901-50 period.

The NAO is, therefore, an important index determining winter severity in northern Europe and the precipitation amount during the winter rainy season in the Mediterranean region. The NAO also influences the climate over eastern North America, with positive values associated with warmer winters over the southeastern United States and cooler winters over eastern Canada and Greenland. The four main centers of influence have been extensively studied by *Hurrell* [1995] and referred to as a quadropole pattern by *Slonosky and Yiou* [2001]. The NAO has also been shown to influence a whole range of climatic and non-climatic variables [see the extensive discussions in recent reviews by *Stephenson et al.*, this volume; *Wanner et al.*, 2001].

Principal Component Analysis (PCA) of winter monthly-mean sea-level pressure (MSLP) for the Atlantic half of the NH identifies the NAO as the most important mode of pressure variability [see, e.g., *Osborn et al.*, 1999; *Slonosky et al.*, 2000]. The NAO is less strongly characterized in the other seasons, explaining a smaller percentage of the total pressure variance, but it is still the dominant pattern [*Hurrell and van Loon*, 1997]. The first principal component (PC1) of the winter (DJF) analysis locates the northern node over Iceland, but the southern node encompasses a large area of the tropical Atlantic, from near Bermuda to the Iberian Peninsula. In analyses of summer (JJA) MSLP data, the northern node moves to the southwest of Iceland, almost to the Labrador Sea, while the southern node is less extensive than in winter and is centered over the Azores [*Hurrell and van Loon*, 1997].

The availability of pressure data in the original definition of the index enables a series to be developed back to 1865. This is optimal for much of the year, and is only slightly unsatisfactory in summer. This is because the two sites are located near the two centers of action in each season. In the winter half of the year, the southern node extends north-eastwards to encompass parts of the Iberian Peninsula. This allows longer pressure records from the Iberian Peninsula [e.g. Gibraltar, Lisbon or Cadiz, see also *Hurrell*, 1995; *Jones et al.*, 1997] to extend the index back to the early 1820s.

In this paper we extend the analyses of Figure 1 by comparing various 'station-pair' NAO indices and assess the strengths of their relationships with surface climate variability over Europe, using as long a period of instrumental data as possible. The influence of the NAO on European climate is shown to vary on decadal and longer timescales over the last two centuries. This non-stationarity is particularly important when longer reconstructions of the NAO are presented based upon tree-ring and ice-core indicators, which respond to local temperatures and precipitation amounts. Rather than measuring the MSLP index itself, reconstructions made using natural archives are clearly also dependent upon how temperature and precipitation respond to the NAO. The possibility that regional teleconnection patterns between the NAO and temperature and precipitation variability, experienced across western Europe, vary over time should be recognized. Similar arguments can be made about the 'instrumental' NAO reconstructions of *Luterbacher et al.* [1999, 2002]. Before the late 18th century, these reconstructions are based on a mix of instrumental (principally temperature and precipitation series, but including a few early pressure series) and documentary series. They do not, therefore, strictly represent long-term changes in pressure-based indices of the NAO, but rather a combination of NAO variability and changes in the influence of the NAO.

2. DATA

2.1. Instrumental MSLP

Table 1 lists the lengths and sources of MSLP data for sites commonly used to derive 'station-pair' NAO indices. Reykjavik is the only one of these series offering considerable potential for further extension. About two thirds of the months between 1780 and 1820 have daily instrumental pressure observations somewhere in Iceland [*Jones et al.*, 1997].

Earlier instrumental MSLP data are available for several sites in western Europe [see e.g. *Jones*, 2001]. Zonal pressure indices have been derived back to the mid-18th century using combinations of long records, from Trondheim, Lund,

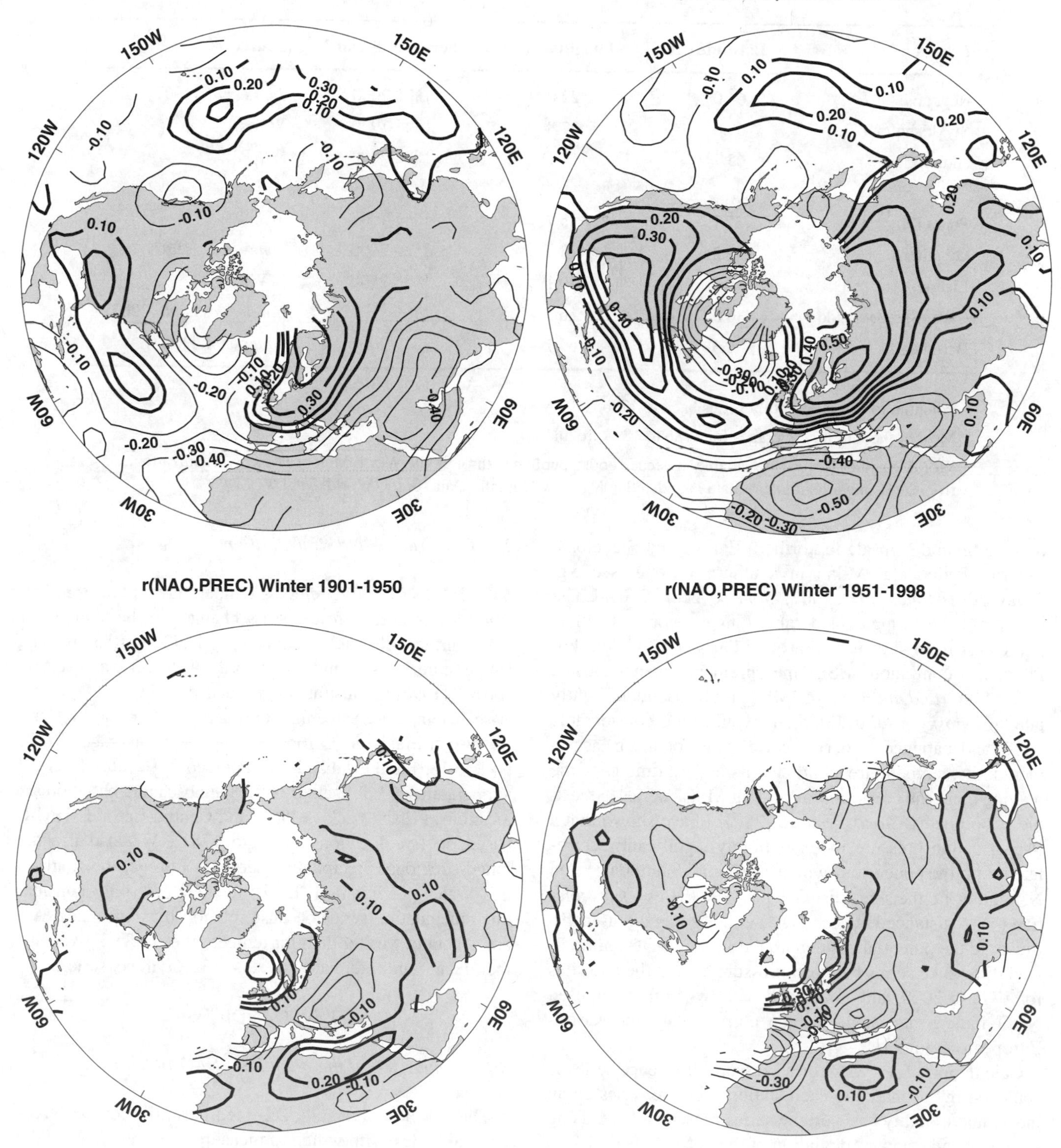

Figure 1. Correlations between the winter (December to February) NAO (based on Ponta Delgada and Reykjavik) and surface temperature and precipitation for two periods 1901-50 and 1951-2000. The temperature data come from *Jones et al.* [1999a] and the precipitation data from *Hulme* [1994].

Table 1. MSLP Data Sources for the Northern and Southern Nodes of the NAO

Location	Latitude (°N)	Longitude (°W)	Period of Record	Source
Reykjavik	64.0	22.0	1821-2000[1]	*Jones et al.* [1997]
Stykkisholmur	65.0	22.8	1846-1990	WWR[3]
Akureyri	65.7	18.1	1882-2000	WWR[3]
Ponta Delgada	37.7	25.7	1865-1997	WWR[3]
Gibraltar	36.2	5.4	1821-2000	*Jones et al.* [1997]
Lisbon	38.7	9.1	1855-2000	WWR[3]
Cadiz/San Fernando	36.5	6.3	1786-1998[2]	*Barriendos et al.* [2002]
Bermuda	32.3	64.7	1837-2000	*Jones et al.* [1987]

[1] Potential to extend monthly series back to 1780

[2] Most of the monthly averages for 1786 to 1820 are missing.

[3] World Weather Records (available in decade books published through the World Meteorological Organization; from 1991 onwards much of data is updated in Monthly Climatic Data for the World, MCDW).

Stockholm and Uppsala in northern Europe and Barcelona, Madrid, Padua and Milan in southern Europe [see e.g. *Jones*, 2001; *Jones et al.*, 2002; *Slonosky et al.*, 2001a]. The potential for locating considerably longer records of MSLP is greatest from sites in and around Paris and London. For Paris, near-continuous MSLP measurement began in the late 1660s [*LeGrand and LeGoff*, 1992] and for London slightly later [*Slonosky et al.*, 2001b]. It would take considerable effort and patience to derive corrections for the measurement units, temperature and the non-standard timings of the readings, but the construction of long MSLP series is feasible at both sites. *Slonosky et al.* [2001b] have shown that a Paris-London pressure index is highly significantly correlated with the concurrent winter, spring and autumn season NAO, despite the short distance between the two sites relative to the distance between NAO nodes generally used. To illustrate the potential of such an index we derive a normalized difference series between pressure data at the two sites for 1774 to 2000, and compare it with two commonly used NAO indices and with surface climate data for various European regions.

Calculation of a 'NAO' or a pressure difference series is achieved by normalizing each station pressure series on an individual monthly basis, in the series presented here, using the mean and standard deviation of the 1951-80 period. The 'NAO' is then the difference between the chosen southern and the northern node. Normalization of the data for the two nodes is necessary to avoid the series being dominated by the greater variability of the northern node.

2.2. Temperature and Precipitation

Table 2 lists the five regional temperature and one precipitation series used here to assess changes in the strength of NAO/surface climate relationships. NH20N (the average temperature of NH land areas north of 20°N) was used by *Hurrell* [1996] to illustrate the influence of the NAO on larger-scale temperature averages during the winter season. Here, we also use a large European temperature average (EUR) back to 1851 and local and regional records (Central England temperatures, CET, back to 1659; Fennoscandia based on six locations, FENN, back to 1750, and Central Europe, CEUR, based on five locations, back to 1781). We also use the longest regional precipitation series in Europe, the England and Wales precipitation (EWP) average. We only consider one precipitation series because relationships with the NAO are generally weaker than for temperature. Table 2 gives further details and references to the source of all six series.

3. NAO INDICES

3.1. Comparison of the 'Standard' NAO Indices

There is no universally accepted definition of the NAO, or of an index with which to measure it. Here, we consider only indices based on MSLP data, because this allows us to study variations in the relationship between the NAO and temperature and precipitation (Section 4), which would be unclear if the latter variables had been included in the NAO

Table 2. Temperature and Precipitation Series

Name	Abbr.	Period of Record	Source
NH land temperatures (N of 20°N)	NH20N	1851-2000	*Jones et al.* [1999a]
Central and Northern European temperatures (40-70°N, 10°W-30°E)	EUR	1851-2000	*Jones et al.* [1999a]
Fennoscandian temperatures	FENN	1751-1998	*Jones et al.* [2002]
Central European temperatures	CEUR	1781-1998	*Jones et al.* [2002]
Central England temperatures	CET	1659-2000	*Manley* [1974] *Parker et al.* [1992]
England and Wales precipitation	EWP	1766-2000	*Jones and Conway* [1997] *Alexander and Jones* [2001]

index itself. Normalized pressure differences between various combinations of the northern and southern pairs [e.g. *Rogers*, 1984; *Hurrell*, 1995; *Jones et al.*, 1997] have been used, as has the first principal component (PC) time series of the regional pressure field [*Rogers*, 1990]. PC time series may be better indices, incorporating additional aspects of the variability in the positions of the centers of action [e.g. *Portis et al.*, 2001] not captured by station pairs, but they are only available for parts of the twentieth century. The first PC of the full NH MSLP field yields a different, though related, pattern of interannual variability [the Arctic Oscillation or Northern Annular Mode, *Thompson and Wallace*, 1998, 2000; *Thompson et al.*, this volume].

Osborn et al. [1999] showed that NAO indices defined using various station pairs or the first PC of the Atlantic-sector pressure field show similar time evolutions, with correlations in the range 0.84 to 0.96. They do point out (their Appendix C), however, that the ratio of interdecadal to interannual variance is greater for those indices using southern node information from further west (e.g., Ponta Delgada rather than Gibraltar). Here, we only consider indices based on station pairs.

The large temporal (interannual) variability in Icelandic MSLP, relative to the spatial variability over Iceland, means that there is little to choose between the three possible 'northern node' locations given in Table 1. As Reykjavik is the longest and the best assessed with respect to homogeneity [*Jones et al.*, 1997; *Jónsson and Miles*, 2001] it is chosen as the northern node in the 'traditional' (i.e., those using the pressure difference between locations in the Azores High and Iceland Low regions) indices derived in this paper.

The traditional location used to represent the southern node has tended to be Ponta Delgada in the Azores, which began recording in 1865 (the Portuguese Met. Service has replaced this station by Santa Maria, so it has not been possible to update the series beyond 1997). *Hurrell* [1995], however, showed that a southwest European location as the southern NAO node correlated slightly better with station temperature and precipitation series across Europe, during the winter season. *Hurrell* [1995] used Lisbon back to 1864, but the record can easily be extended back to 1855 [*Hann*, 1887]. *Jones et al.* [1997] extended the Gibraltar record back to 1821. The Gibraltar record used here has been slightly adjusted from that given in Jones et al. [1997]. Corrections due to barometer relocations and/or changes to standard gravity in 1931 and 1971 (assumed to be the same for all months of the year in the 1997 paper) have been made on a monthly basis using comparisons with Tangiers (1951-1998) and Cape Spartel (1894-1920).

Other potential southern nodes are Cadiz/San Fernando where pressure recording began in 1786 (but much of the early data are missing before 1820) and Bermuda in the western Atlantic where a series exists back to 1837. All series were assessed for homogeneity in *Jones et al.* [1987] or more recently in *Jones et al.* [1999b]. The distances between Bermuda and Ponta Delgada and the Iberian stations means that long-term homogeneity can never be assured, particularly during the earliest times.

Figure 2 shows a comparison of the low-frequency MSLP behaviour in the five possible southern node series. Each series has been normalized on a monthly basis over the 1951-80 period and the plot shows smoothed values after applying a 10 year (120 month) filter through each series, using all months of the year (not just winter months). For the southwest European locations the greatest differences

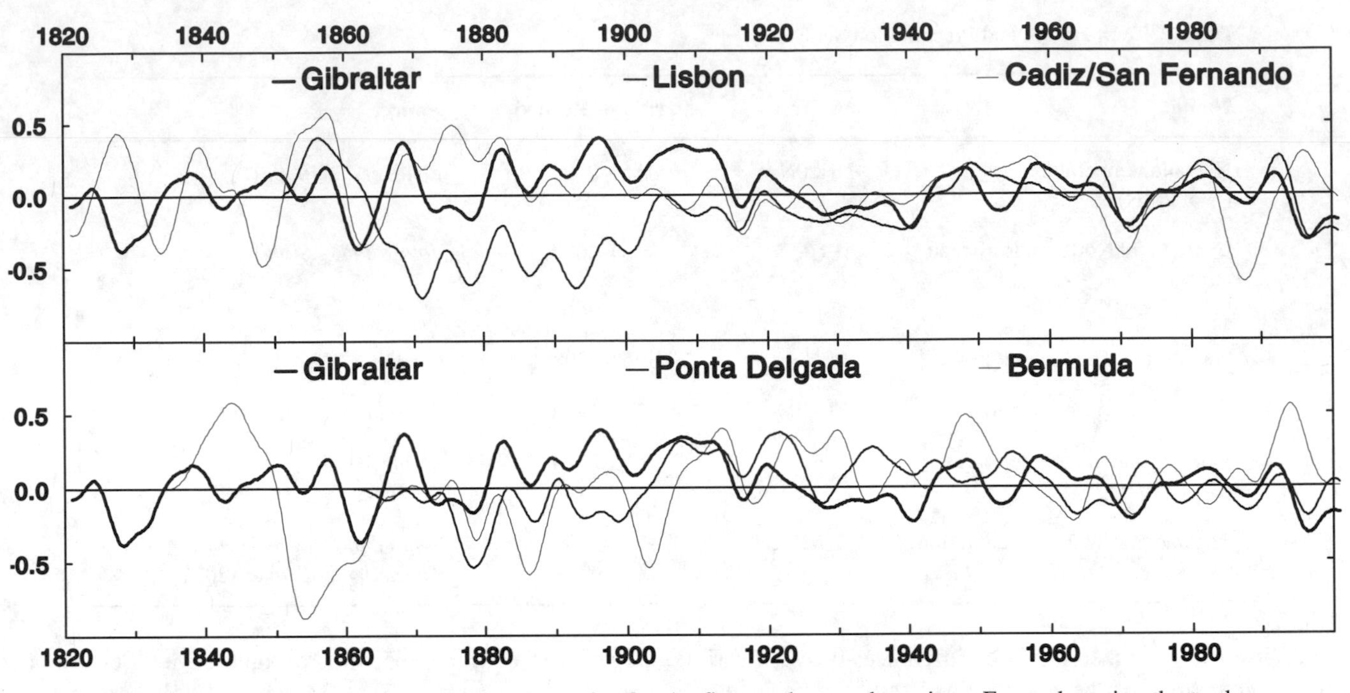

Figure 2. '10-year' low-pass smoothed MSLP series for the five southern node stations. For each station the twelve monthly MSLP series were normalized over the 1951-80 period. The entire monthly series was then smoothed with a 120-term Gaussian filter. The units of the vertical scale therefore represent standard deviation units (~0.6-3.0 hPa depending on month). The top panel shows the three Iberian stations of Gibraltar, Lisbon and Cadiz/San Fernando. The bottom panel compares Gibraltar with the mid-Atlantic locations of Ponta Delgada on the Azores and Bermuda.

occur before about 1920. Between 1870 and 1920 differences reach about 0.5 standard deviation units (0.6-3.0 hPa depending on season). Slightly larger excursions are evident for the Cadiz/San Fernando series after 1980 and before 1860. The agreement between Gibraltar and Lisbon during these times clearly implies uncorrected inhomogeneities in the Cadiz/San Fernando record. The two mid-Atlantic sites (Ponta Delgada and Bermuda) are compared with Gibraltar in the lower half of Figure 2. Agreement is lower between these sites but the distances are considerably greater than for the southern Iberian station comparisons. The Ponta Delgada (Azores) MSLP values are generally closer to Gibraltar than Bermuda. Excursions in the Bermuda record before 1860 suggest probable inhomogeneities.

Table 3 gives correlations for the standard four seasons between each of the five southern nodes and Reykjavik, for four different time periods between 1865 and 1995. Correlations are always highest for the winter season and lowest in summer. Correlations between Bermuda and all the other stations are always poor and mostly insignificant (except during winter, when significant, but weak correlations exist with the other southern nodes during 1865-1900). Gibraltar, Lisbon and Cadiz/San Fernando all exhibit high correlations with each other. These weaken most during the summer season, particularly when Cadiz/San Fernando is involved for some of the analysis periods. Of the three sites, Lisbon always correlates highest with Ponta Delgada. The Azores site correlates most strongly (inversely) with Reykjavik in all seasons except winter. For winter, correlations are slightly larger with the three southern Iberian sites (and are strongest over the full 1865-1995 period, by a small margin, for Gibraltar).

To simplify subsequent analyses, we limit the correlations with surface data to the two alternate NAO series, Gibraltar-Reykjavik (1821-2000) and Ponta Delgada-Reykjavik (1865-1997). These monthly NAO index series are available from the following web site (http://www.cru.uea.ac.uk/cru/data/nao.htm). All comparisons with the surface climate data in Section 4 are made solely over the December-to-March season. These four months exhibit greater common decadal-scale variations than other months [*Osborn et al.*, 1999].

3.2. Comparison With the Paris-London Index

The earliest potential conventional NAO index, based solely on pressure data, can be developed from about 1780 as MSLP data exist (albeit fragmentary up to 1820) for Reykjavik and Cadiz/San Fernando. The longest two European MSLP records, each with data for some parts of the late 17th century, exist for Paris and London. MSLP

Table 3. Correlations between each of the five southern node stations and Reykjavik over four different periods (1865-1900, 1901-1950, 1951-1995 and 1865-1995). There are two panels for each season (top panel, sub-diagonal is 1865-1900 and 1901-50 above diagonal, bottom panel, sub-diagonal is 1951-1995 and 1865-1995 above diagonal). All emboldened values are significant at the 95% level

Winter (DJF)	Gibraltar	Lisbon	Cadiz/SF	Ponta Delgada	Bermuda	Reykjavik
Gibraltar	-	**0.94**	**0.97**	**0.60**	-0.15	**-0.63**
Lisbon	**0.94**	-	**0.95**	**0.73**	-0.13	**-0.53**
Cadiz/SF	**0.92**	**0.91**	-	**0.60**	-0.17	**-0.59**
Ponta Delgada	**0.52**	**0.60**	**0.46**	-	0.02	**-0.50**
Bermuda	**-0.43**	**-0.41**	**-0.39**	**-0.39**	-	-0.14
Reykjavik	**-0.70**	**-0.71**	**-0.66**	**-0.60**	**0.36**	-
	Gibraltar	Lisbon	Cadiz/SF	Ponta Delgada	Bermuda	Reykjavik
Gibraltar	-	**0.93**	**0.92**	**0.60**	-0.08	**-0.68**
Lisbon	**0.95**	-	**0.92**	**0.71**	-0.11	**-0.62**
Cadiz/SF	**0.94**	**0.91**	-	**0.60**	-0.08	**-0.65**
Ponta Delgada	**0.67**	**0.76**	**0.69**	-	-0.01	**-0.59**
Bermuda	0.12	0.08	0.15	0.17	-	-0.11
Reykjavik	**-0.69**	**-0.67**	**-0.71**	**-0.70**	**-0.23**	-
Spring (MAM)	Gibraltar	Lisbon	Cadiz/SF	Ponta Delgada	Bermuda	Reykjavik
Gibraltar	-	**0.95**	**0.94**	**0.44**	**0.30**	**-0.44**
Lisbon	**0.88**	-	**0.94**	**0.51**	**0.27**	**-0.42**
Cadiz/SF	**0.90**	**0.90**	-	**0.39**	**0.24**	**-0.36**
Ponta Delgada	**0.40**	**0.56**	**0.43**	-	0.09	**-0.57**
Bermuda	-0.16	-0.15	-0.12	-0.20	-	-0.10
Reykjavik	**-0.38**	**-0.45**	**-0.37**	**-0.55**	0.04	-
	Gibraltar	Lisbon	Cadiz/SF	Ponta Delgada	Bermuda	Reykjavik
Gibraltar	-	**0.89**	**0.90**	**0.38**	0.13	**-0.35**
Lisbon	**0.84**	-	**0.88**	**0.50**	0.12	**-0.40**
Cadiz/SF	**0.85**	**0.81**	-	**0.36**	0.15	**-0.25**
Ponta Delgada	**0.25**	**0.40**	**0.21**	-	0.03	**-0.57**
Bermuda	0.07	0.09	0.16	0.10	-	0.02
Reykjavik	**-0.23**	**-0.27**	-0.01	**-0.59**	**0.23**	-

records for these sites have been used by *Slonosky et al.* [2001b] as a measure of westerly wind strength over northwestern Europe since 1774. Figure 3 compares running 31-year correlations between the DJFM average for the Gibraltar-Reykjavik NAO, the Ponta Delgada-Reykjavik NAO and the Paris-London normalized pressure difference. It should be remembered that these correlations highlight common interannual variability and not necessarily common trends on multi-decadal timescales. The correlation between the two commonly used NAO series for the DJFM season varies between 0.8 and 0.93. The Paris-London index correlates against the Gibraltar-Reykjavik NAO with almost comparable values, between 0.7 and 0.9. Agreement between the Paris-London index and the Ponta Delgada-Reykjavik NAO is poorer with values between 0.5 and 0.8. Temporal changes in the strength of the correlations are clear in all three curves with the lowest values evident between the 1930s and 1970s. Correlations also weaken between the Paris-London index and the Gibraltar-Reykjavik NAO before 1880. None of these temporal variations, however, are outside the range of statistical sampling error, and are not, therefore, statistically significant.

4. THE INFLUENCE OF THE NAO ON EUROPEAN SURFACE CLIMATE

Several recent studies using running correlations have indicated that the strength of relationships between surface climate and the NAO varies with time [e.g. *Osborn et al.*, 1999; *Slonosky et al.*, 2001a]. This section considers these changes in influence using the longest possible NAO and surface climate records. Figure 4 shows running 31-year correlations between the Gibraltar-Reykjavik NAO for the DJFM season with the six surface climate series listed in

Table 3. (continued) ...

Summer (JJA)	Gibraltar	Lisbon	Cadiz/SF	Ponta Delgada	Bermuda	Reykjavik
Gibraltar	-	**0.74**	**0.75**	**0.28**	-0.11	**0.27**
Lisbon	**0.42**	-	**0.70**	**0.36**	0.03	0.09
Cadiz/SF	0.19	**0.28**	-	**0.36**	-0.11	0.14
Ponta Delgada	**0.32**	**0.32**	**0.30**	-	0.02	**-0.46**
Bermuda	0.10	-0.02	0.17	-0.03	-	-0.17
Reykjavik	0.10	-0.12	**-0.40**	**-0.48**	0.06	-
	Gibraltar	Lisbon	Cadiz/SF	Ponta Delgada	Bermuda	Reykjavik
Gibraltar	-	**0.50**	**0.42**	0.17	-0.04	0.18
Lisbon	**0.56**	-	**0.31**	**0.44**	0.01	0.06
Cadiz/SF	**0.22**	**0.54**	-	0.16	-0.08	-0.02
Ponta Delgada	0.02	**0.48**	**0.25**	-	0.02	**-0.36**
Bermuda	0.01	-0.10	0.00	-0.11	-	0.01
Reykjavik	0.14	0.14	0.08	-0.14	**0.23**	-
Autumn (SON)	Gibraltar	Lisbon	Cadiz/SF	Ponta Delgada	Bermuda	Reykjavik
Gibraltar	-	**0.87**	**0.88**	**0.36**	-0.03	**-0.38**
Lisbon	**0.86**	-	**0.91**	**0.55**	0.01	**-0.45**
Cadiz/SF	**0.92**	**0.94**	-	**0.40**	-0.05	**-0.41**
Ponta Delgada	**0.41**	**0.64**	**0.49**	-	-0.05	**-0.53**
Bermuda	**0.28**	0.06	0.19	-0.09	-	0.05
Reykjavik	**-0.35**	**-0.42**	**-0.36**	**-0.51**	0.17	-
	Gibraltar	Lisbon	Cadiz/SF	Ponta Delgada	Bermuda	Reykjavik
Gibraltar	-	**0.85**	**0.89**	**0.35**	0.02	**-0.31**
Lisbon	**0.83**	-	**0.85**	**0.54**	0.04	**-0.37**
Cadiz/SF	**0.88**	**0.75**	-	**0.37**	-0.02	**-0.30**
Ponta Delgada	**0.42**	**0.54**	**0.41**	-	-0.06	**-0.53**
Bermuda	-0.14	0.04	-0.13	-0.11	-	0.08
Reykjavik	**-0.25**	**-0.25**	-0.19	**-0.59**	0.14	-

Table 2. Figures 5 and 6 show similar curves using the Ponta Delgada-Reykjavik NAO and the Paris-London normalized pressure difference respectively. Again these correlations highlight interannual timescale variability. Against the four European temperature records, the Gibraltar-Reykjavik NAO shows significant positive correlations throughout (Figure 4). Over the entire period, correlations show a gradual increase, peaking at between 0.7 and 0.9 during the 31 years centered on 1960, compared to a range of 0.4 to 0.7 typical of the mid-nineteenth century correlations. Superimposed upon these underlying trends is the expected level of sampling variability, plus some more abrupt or larger magnitude variations, such as those centered at 1921 that are most evident in the correlations with the FENN and EUR temperature series, and the sudden drop in correlations when the sliding window encompasses values from 1824 (i.e. centered on 1839) and earlier, evident in all three temperature records with data back this far.

The overall trends in the correlations indicate either a gradual deterioration in the homogeneity and reliability of the Gibraltar-Reykjavik NAO index, or a gradual change in the character of the NAO and its influence on European temperature (i.e., teleconnection patterns). Similarly, the shorter transient reductions in correlation could result from either a temporary shift in climate or teleconnection pattern, or from a small number of unreliable values. The synchroneity of the changes in correlation between different series pairs does not rule out a real change in climate because the various temperature series are correlated with one another and would, therefore, be expected to exhibit some synchronous features. The synchroneity does imply, however, that if any inhomogeneity exists then it is more likely to be found in the common series (i.e., the Gibraltar-Reykjavik NAO index in the case of Figure 4).

The England and Wales precipitation series (EWP) shows much weaker correlations with the NAO. Over the British Isles, the NAO only seems to influence orographic precipitation as opposed to the vorticity of the circulation that is linked with depressions and frontal troughs [*Osborn and Jones*, 2000]. Given the dominance of vorticity in generat-

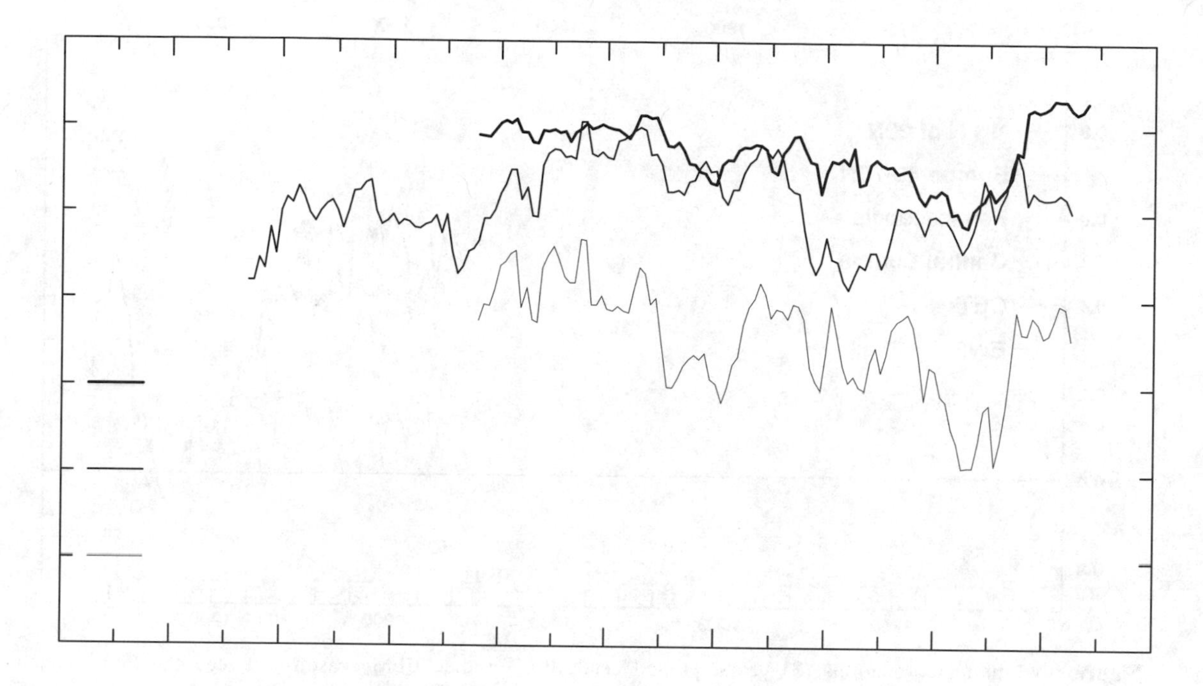

Figure 3. Running correlations (31-years, plotted centrally), for the DJFM season, between the two traditional NAO series of Gibraltar-Reykjavik and Ponta Delgada-Reykjavik and the Paris-London pressure index.

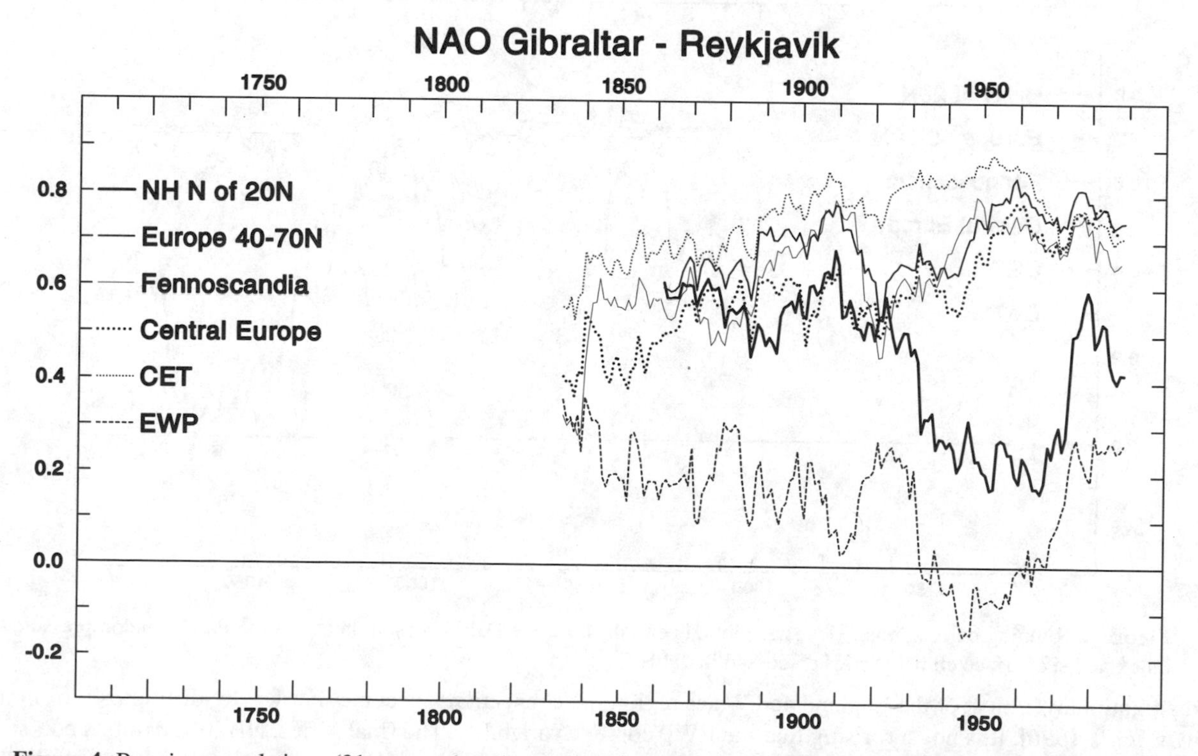

Figure 4. Running correlations (31-years, plotted centrally), for the DJFM season, between the Gibraltar-Reykjavik NAO series and six surface climate series discussed in Table 2.

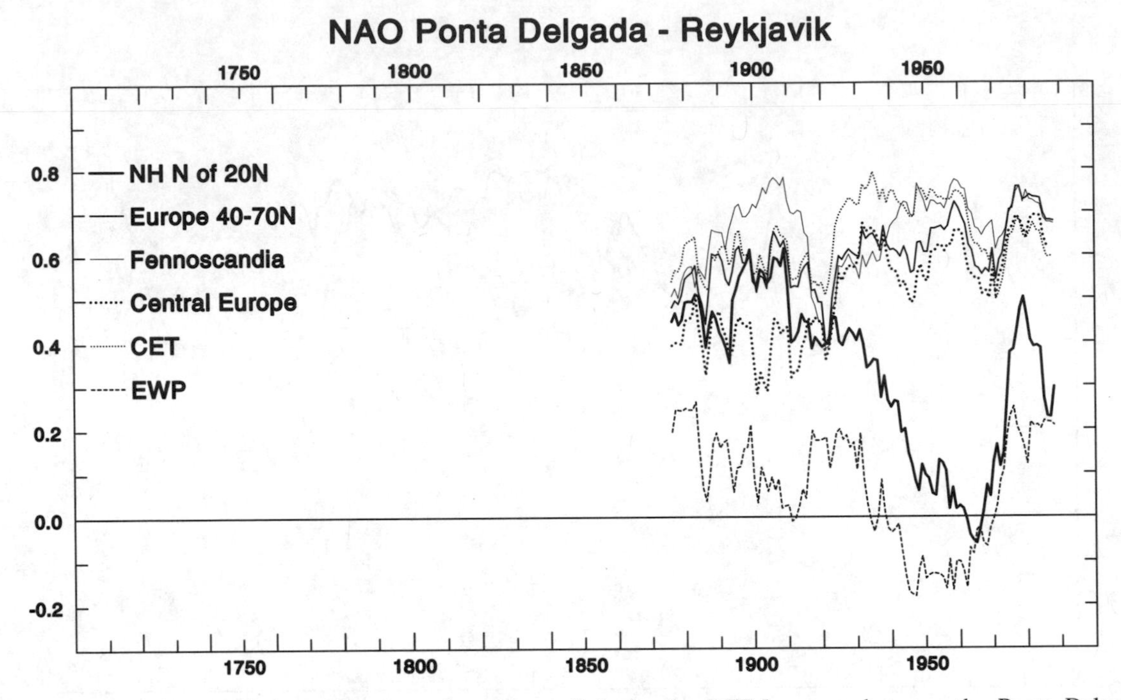

Figure 5. Running correlations (31-years, plotted centrally), for the DJFM season, between the Ponta Delgada-Reykjavik NAO series and six surface climate series discussed in Table 2.

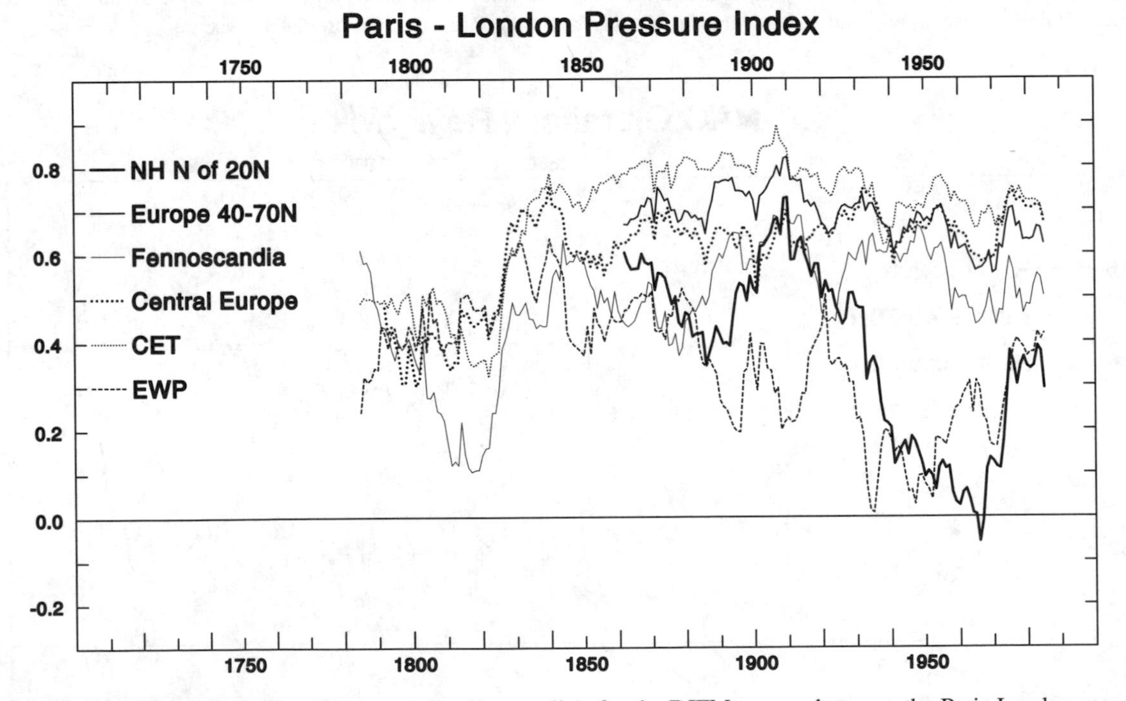

Figure 6. Running correlations (31-years, plotted centrally), for the DJFM season, between the Paris-London pressure index and six surface climate series discussed in Table 2.

ing precipitation over most of the England and Wales region [*Conway et al.*, 1996], it is not surprising that the EWP correlates less strongly with the NAO indices. The lower correlations also exhibit variability, and real variations in teleconnections become harder to distinguish from random variability. The final series, NH20N, exhibits non-stationary conditions [see also Figure 6a of *Osborn et al.*, 1999], with significant correlations for windows centered on years before about 1935 and after 1975, but not during the period 1935-75. Some of these variations may be due, in part, to changes in coverage of the land air temperature data [see also *Broccoli et al.*, 2001]. These will be dominated by European data prior to 1900 (note the clear correspondence with the EUR correlations in Figure 4), and will be affected by increasing coverage over Asia and North America during the first half of the 20th century. Nevertheless, the notable rise in correlations around 1975 occurs at a time when the coverage of NH20N shows little change, perhaps indicating a real shift in climatic behaviour.

The second commonly used NAO index (Ponta Delgada – Reykjavik) shows similar time-dependent correlations (Figure 5). Despite the shorter series, there is some indication of an upward trend in the correlations with the four European temperature series. The transient drop in correlations during the period centered on 1921 is even more distinct than it was in Figure 4. Although the synchroneity of the anomalous correlations still does not prove that there are erroneous values in the climatic series, it does imply that if any erroneous values are the cause, they must be in the Reykjavik MSLP record because that is the only common series. The NH20N series shows similar time-dependent correlations against the Ponta Delgada – Reykjavik NAO as it did in Figure 4, though with a more notable minimum during the 31-year window centered on 1965.

Figure 6, showing running correlations between Paris-London MSLP index and the six climatic series, is similar in many respects to Figures 4 and 5. The fall in correlation with NH20N is still apparent between 1935 and 1970. The link with EWP is now strong enough, however, to identify real changes in correlation: the positive correlations being suppressed during the 1910-70 period. Against the four European temperature series there is no indication of a long-term trend in correlation with Paris-London from 1830 to the present. This is at odds with the other two NAO indices. If a gradual shift in teleconnection structure is responsible, then the results are understandable: the traditional NAO indices are more distant from the temperature records and this might more easily allow a change in the relationship or its strength, while for the Paris-London index, being more local and direct, it is harder to imagine how this pressure gradient can affect central/northern European temperatures differently during different periods.

There is still an indication of the suppressed correlations during the periods centered around 1921, providing support for some real climatic features rather than a data problem with the Reykjavik MSLP. Correlations do not fall so dramatically (cf., Figure 6 with Figure 4) when the running window encompasses values from 1824 and earlier (implying a possible problem with Gibraltar-Reykjavik), though there is a sudden drop in correlations between Paris-London and all four climatic series with data, with windows centered around 1823 and earlier (i.e., those that incorporate data from 1808 and earlier).

5. CONCLUSIONS

The NAO is the most important mode of variability in the NH pressure field. During the winter (DJFM) season the NAO exerts a strong influence on temperature and precipitation throughout much of Europe. A commonly used measure of the NAO is the difference between normalized pressures at Ponta Delgada (Azores) and Reykjavik (Iceland), which can be extended back to 1865. During the winter half of the year, an NAO index using Gibraltar as the southern node performs slightly better than Ponta Delgada and enables an NAO index to be established back to the early 1820s. Earlier pressure data in Iceland (Reykjavik) and southern Spain (Cadiz/San Fernando) gives prospects of an index back to 1780, but both locations contain much missing data.

Running correlations have been used to illustrate the variable nature of the influence of the NAO on six series of surface temperature and precipitation over Europe and the NH. Relationships are weaker in the 19th century (part of the reduction may be due to poorer data quality) and also for some periods in the 20th century for the more distant teleconnections. Some of these reductions can be shown to be real (and statistically significant) and this cautions attempts to reconstruct the NAO from surface climate data and natural proxies (including documentary) as results will include not only changes in the NAO but also changes in its influence on surface climate. The NAO phenomenon would seem to be best reconstructed from pressure data alone and possibilities exist for an index based on Paris and London observations back to the late 17th century, specifically for the winter half of the year.

Acknowledgements. This work has been supported by the U.S. Department of Energy (grant DE-FG02-98ER62601).

REFERENCES

Alexander, L. V., and P. D. Jones, Updated precipitation series for the United Kingdom, *Atmos. Sci. Lett.*, *1*, 142-150, doi:10.1006/asle. 2001.0025, 2001.

Barriendos, M., J. Martin-Vide, J. C. Peña, and R. Rodriguez, Daily meteorological observations in Cadiz-San Fernando: Analysis of the documentary sources and the instrumental data content (1786-1996), *Clim. Change 53*, 151-170, 2002.

Broccoli, A. J., T. L. Delworth, and N-C. Lau, The effect of changes in observational coverage on the association between surface temperature and the Arctic Oscillation, *J. Climate*, *14*, 2481-2485, 2001.

Conway, D., R. L. Wilby, and P. D. Jones, Precipitation and air flow indices over the British Isles, *Climate Research*, *7*, 169-183, 1996.

Hann, J., *Die Vertheilung des Luftdruckes über Mittel und Süd-Europa*, Eduard Hölzel, Wien, 1887.

Hulme, M., Validation of large-scale precipitation fields in GCMs, in *Global Precipitations and Climate Change*, edited by M. Desbois and F. Désalmond, pp. 387-405, Springer-Verlag, Berlin, 1994.

Hurrell, J. W., Decadal trends in the North Atlantic Oscillation: Regional temperatures and precipitation, *Science*, *269*, 676-679, 1995.

Hurrell, J. W., Influence of variations in extratropical wintertime teleconnections on Northern Hemisphere temperature, *Geophys. Res. Lett.*, *23*, 665-668, 1996.

Hurrell, J. W., and H. van Loon, Decadal variations in climate associated with the North Atlantic Oscillation, *Clim. Change*, *36*, 301-326, 1997.

Jones, P. D., Early European Instrumental Records, in *History and Climate: Memories of the Future?*, edited by P. D. Jones, A. E. J. Ogilvie, T. D. Davies and K. R. Briffa, pp. 55-77, Kluwer Academic/Plenum Publishers, New York, 2001.

Jones, P. D., and D. Conway, Precipitation in the British Isles: An analysis of area-average data updated to 1995, Int. *J. Climatol.*, *17*, 427-438, 1997.

Jones, P. D., T. M. L. Wigley, and K. R. Briffa, Monthly Mean Pressure Reconstructions for Europe (back to 1780) and North America (to 1858), *Technical Note TR037*, U.S. Dept. of Energy, Washington, D.C., 99pp, 1987.

Jones, P. D., T. Jónsson, and D. Wheeler, Extensions to the North Atlantic Oscillation using early instrumental pressure observations from Gibraltar and southwest Iceland, *Int. J. Climatol.*, *17*, 1433-1450, 1997.

Jones, P. D., M. New, D. E. Parker, S. Martin, and I. G. Rigor, Surface air temperature and its variations over the last 150 years, *Revs. Geophys.*, *37*, 173-199, 1999a.

Jones, P. D. et al., Monthly mean pressure reconstructions for Europe for the 1780-1995 period, Int. *J. Climatol.*, *19*, 347-364, 1999b.

Jones, P. D., K. R. Briffa, T. J. Osborn, A. Moberg, and H. Bergström, Relationships between circulation strength and the variability of growing season and cold season climate in Northern and Central Europe, *The Holocene, 12*, 643-656, 2002.

Jónsson, T., and M. W. Miles, Anomalies in the seasonal cycle of sea level pressure in Iceland and the North Atlantic Oscillation, *Geophys. Res. Lett.*, *28*, 4231-4234, 2001.

LeGrand, J. P., and M. LeGoff, Les observationes météorologie Nationale, *Monographie Nr. 6*, Météo-France, Trappes, 1992.

Luterbacher, J., C. Schmutz, D. Gyalistras, E. Xoplaki, and H. Wanner, Reconstruction of monthly NAO and EU indices back to AD 1675, *Geophys. Res. Lett.*, *26*, 2745-2748, 1999.

Luterbacher, J. et al., Extending North Atlantic Oscillation reconstructions back to 1500, *Atmos. Sci. Lett.*, *2*, 114-124, 10.1006/asle.2001.0044, 2002.

Manley, G., Central England temperatures: Monthly means 1659 to 1973, Quart. J. *Roy. Met. Soc.*, *100*, 389-405, 1974.

Osborn, T. J., K. R. Briffa, S. F. B. Tett, P. D. Jones, and R. M. Trigo, Evaluation of the North Atlantic Oscillation as simulated by a coupled climate model, *Clim. Dyn.*, *15*, 685-702, 1999.

Osborn, T. J., and P. D. Jones, Air-flow influences on local climate: Observed United Kingdom climate variations, *Atmos. Sci. Lett.*, *1*, 62-74, 10.1006/asle.2000.0017, 2000.

Parker, D. E., T. P. Legg, and C. K. Folland, A new daily Central England temperature series, 1772-1991, *Int. J. Climatol.*, *12*, 317-342, 1992.

Portis, D. H., J. E. Walsh, M. El Hamly, and P. J. Lamb, Seasonality of the North Atlantic Oscillation, *J. Climate*, *14*, 2069-2078, 2001.

Rogers, J. C., The association between the North Atlantic Oscillation and Southern Oscillation in the Northern Hemisphere, *Mon. Wea. Rev.*, *112*, 1999-2015, 1984.

Rogers, J. C., Patterns of low-frequency monthly sea level pressure variability (1899-1986) and associated wave cyclone frequencies, *J. Climate*, *3*, 1364-1379, 1990.

Slonosky, V. C., and P. Yiou, Secular changes in the North Atlantic Oscillation and its influence on 20th century warming, *Geophys. Res. Lett.*, *28*, 807-810, 2001.

Slonosky, V. C., P. D. Jones, and T. D. Davies, Variability of the surface atmospheric circulation over Europe, 1774-1995, *Int. J. Climatol.*, *20*, 1875-1897, 2000.

Slonosky, V .C., P. D. Jones, and T. D. Davies, Atmospheric circulation and surface temperature in Europe from the 18th century to 1995, *Int. J. Climatol.*, *21*, 63-75, 2001a.

Slonosky, V. C., P. D. Jones, and T. D. Davies, Instrumental pressure observations from the seventeenth and eighteenth centuries: London and Paris, *Int. J. Climatol.*, *21*, 285-298 2001b.

Stephenson, D. B., H. Wanner, S. Brönnimann, and J. Luterbacher, The history of scientific research on the North Atlantic Oscillation, this volume.

Thompson, D. W. J., and J. M. Wallace, The Arctic Oscillation signature in the wintertime geopotential height and temperature fields. *Geophys. Res. Lett.*, *25*, 1297-1300, 1998.

Thompson, D. W. J., and J. M. Wallace, Annular modes in the extratropical circulation. Part I: Month-to-month variability, *J. Climate*, *13*, 1000-1016, 2000.

Thompson, D. W. J., S. Lee, and M. P. Baldwin, Atmospheric processes governing the Northern Hemisphere Annular Mode/North Atlantic Oscillation, this volume.

Walker, G. T., Correlations in seasonal variations of weather IX, *Mem. Ind. Meteorol. Dept.*, *24*, 275-332, 1924.

Wanner, H, S. Bronnimann, C. Casty, D. Gyalistras, J. Luterbacher, C. Schmutz, D. B. Stephenson, and E. Xoplaki, North Atlantic Oscillation – concepts and studies, *Survey Geophys.*, *22,* 321-381, 2001.

K.R. Briffa, Climatic Research Unit, University of East Anglia, Norwich, NR4 7TJ, U.K.
k.briffa@uea.ac.uk

P.D. Jones, Climatic Research Unit, University of East Anglia, Norwich, NR4 7TJ, U.K.
p.jones@uea.ac.uk

T.J. Osborn, Climatic Research Unit, University of East Anglia, Norwich, NR4 7TJ, U.K.
t.osborn@uea.ac.uk

Multi-Proxy Reconstructions of the North Atlantic Oscillation (NAO) Index: A Critical Review and a New Well-Verified Winter NAO Index Reconstruction Back to AD 1400

Edward R. Cook

Lamont-Doherty Earth Observatory of Columbia University, Palisades, New York

The reconstruction of the North Atlantic Oscillation (NAO) index from high-resolution paleoclimate records is critically reviewed. In so doing, previously reported problems in most of the published NAO reconstructions are verified. A new multi-proxy reconstruction of the winter NAO index is developed, which largely corrects these reported problems, particularly the lack of reconstruction validity prior to 1850. It covers the period AD 1400-1979 and successfully verifies against independent estimates of the winter NAO index based on European instrumental and non-instrumental data as far back as 1500. This new success is due to a greatly expanded network of proxy records available for reconstruction. It also appears to be related to the use of an extended reconstruction model calibration period that reduced an apparent bias in selected proxies associated with the effect of anomalous 20th century NAO variability on climate teleconnections over circum-North Atlantic land areas.

1. INTRODUCTION

The North Atlantic Oscillation (NAO) is a key pattern of internal climate variability in the extratropical Northern Hemisphere winter, influencing temperature, precipitation, and atmospheric circulation over a wide region [*Hurrell*, 1995; *Hurrell and van Loon*, 1997]. The NAO impacts Arctic sea ice export [*Hilmer and Jung,* 2000], Tigris-Euphrates streamflow [*Cullen and deMenocal*, 2000], plant and animal populations [*D'Arrigo et al.,* 1993; *Post and Stenseth*, 1999], and human activities both present and past [*Cullen et al.*, 2002; *Weiss* 2002]. It has also been linked to widespread warming over the past few decades [*Hurrell*, 1996]. Therefore, it is understandable that a great deal of effort is being spent on observing and modeling the dynamics of the NAO in order to develop a predictive model of its behavior. At the same time, there is a "pressing need" to develop longer records of the NAO to place recent variability in a better long-term context [*Jones et al*., 2001].

The North Atlantic Oscillation:
Climatic Significance and Environmental Impact
Geophysical Monograph 134

10.1029/134GM04

The NAO is defined by a seesaw of atmospheric mass between its northern pole near Iceland (the Icelandic Low) and its southern pole near the Azores (the Azores High) [also see *Hurrell et al.,* this volume; *Jones et al.,* this volume]. Its strongest and most climatologically effective expression occurs during the cold season months [*Rogers*, 1984; *Hurrell*, 1995; *Jones et al.,* 1997] when the north-south pressure gradient is greatest between its quasi-stationary centers of action. As such, a convenient expression of NAO behavior over time is usually obtained by calculating the normalized sea level pressure (SLP) difference between the Azores High and Icelandic Low for the four-month winter (December-March) season [*Hurrell*, 1995]. This index is calculated using long instrumental SLP records from Iceland and either Ponta Delgada in the Azores [*Rogers*, 1984], Lisbon in Portugal [*Hurrell*, 1995], or Gibraltar in Spain [*Jones et al.*, 1997]. See *Jones et al.* [this volume] for detailed comparisons.

The power spectrum of the annual winter NAO index indicates the presence of significant band-limited power at 7-25 year timescales [*Rogers*, 1984; *Cook et al.*, 1998; *Mann*, 2002]. *Schneider and Schonwiese* [1989] also report significant spectral power at 1.7 and 2.2 years in monthly NAO data. In addition, a 65-70 year climate signal has been identified in Northern Hemisphere instrumental records [*Schlesinger and Ramankutty*, 1994], which is centered over the North Atlantic and may be related to the NAO and ocean-atmosphere variability. Similar findings have been reported by *Mann and Park* [1994], *Delworth and Mann* [2000], and *Enfield et al.* [2001]. *Cook et al.* [1998] also noted the existence of a 70-year spectral peak in their NAO reconstruction, but only when the 20th century data were included. However, a signal of this duration is difficult to characterize and test when its period approaches the length of climate record being studied [*Mann*, 2002]. Consequently, much longer records of the NAO are needed to verify the existence of the multi-decadal signal found by *Schlesinger and Ramankutty* [1994]. Using early SLP records from Iceland and Gibraltar, *Jones et al.* [1997] were able to extend the instrumental NAO index record from 1874 [*Rogers*, 1984] and 1864 [*Hurrell*, 1995] back to 1821. Early London/Paris SLP data may eventually extend the winter NAO index back to the 1670s if the original raw pressure measurements can be found [*Jones et al.*, this volume]. These additional years of data are extremely valuable, but still insufficient to rigorously evaluate the multi-decadal behavior of the NAO as suggested by the *Schlesinger and Ramankutty* [1994] results. They also do not provide an adequate "natural" baseline of NAO activity prior to the 20th century impact of greenhouse gases on climate [e.g., *Paeth et al.*, 1999; *Ulbrich and Christoph*, 1999; *Gillett et al.*, this volume].

2. EXTENSIONS OF THE NAO INDEX BACK IN TIME

The need for longer NAO records has led to the development of numerous extensions of monthly and seasonal NAO indices based on early monthly European instrumental climate data (e.g., SLP, temperature, and precipitation, as well as several documentary proxy predictors) and paleoclimatic proxies (e.g., annual tree-ring chronologies and ice core data). Europe is unique in the world for its wealth of exceptionally long and early instrumental climate records [e.g., *Jones et al.*, 1999; *Slonosky et al.*, 2001] that contain information related to NAO variability and forcing [e.g., *Jones et al.*, 1997, 1999; *Luterbacher et al.*, 1999, 2001, 2002]. In addition, the circum-North Atlantic land areas have abundant annually resolved tree ring and ice core records of past climate that are also useful for reconstructing the NAO [e.g., *Cook et al.*, 1998; *Appenzeller et al.*, 1998; *Glueck and Stockton*, 2001].

Three expressions of the extended winter (December-March) NAO index, all based on early European instrumental climate records, are relevant here. Figure 1a shows the Gibraltar-Iceland NAO index estimated from the normalized monthly station SLP data [*Jones et al.*, 1997; referred to hereafter as *J97*] and updated to 2001 (data obtained from http://www.cru.uea.ac.uk/cru/data/nao.htm). It begins in 1824 when the winter season becomes serially complete. Figure 1b shows estimates of the same winter NAO index back to 1781 using the normalized SLP reconstructions for the grid points closest to Iceland and Gibraltar [*Jones et al.*, 1999; hereafter *J99*]. These data were kindly provided by Dr. P. Jones. And, Figure 1c is a recently updated and revised winter NAO index reconstruction [*Luterbacher et al.*, 2001; hereafter *L01*] extending back to 1659 based on early European instrumental SLP, temperature, and precipitation records recovered as part of the European project ADVICE (Annual to Decadal Variability in Climate in Europe). This series was kindly provided by Dr. J. Luterbacher. *L01* is an improved and extended version of an earlier NAO reconstruction back to 1675 described in *Luterbacher et al.* [1999], which was based on many of the same early instrumental European climate records used in constructing *L01*.

It is apparent that these three SLP-based NAO index series are extremely similar over the 1824-1990 common period, in part because they share some common data. The *J97* and *J99* series have a correlation of 0.96, and the correlations of *L01* with *J97* and *J99* are 0.88 and 0.86, respectively. The correlation between *L01* and *J99* is also 0.88 over the earlier 1781-1823 overlap period. Besides sharing some common data, the three NAO index series are similar because they share the same European climate system that is affected by the NAO. This fact suggests that they may unduly emphasize climate impacts over Europe at the expense of those over eastern North America where the NAO also has an influence on climate [*Hurrell and van Loon*, 1997]. It is also apparent that the NAO has experienced periods of consecutive westerly (positive index) years since the late 1980's and in the early 20th century. This may be an expression of the Schlesinger–Ramankutty oscillation described earlier or even a manifestation of greenhouse gas forcing [*Paeth et al.*, 1999; *Ulbrich and Christoph*, 1999; *Shindell et al.*, 1999; *Gillett et al.*, this volume] and increasing tropical sea surface temperatures [*Hoerling et al.*, 2001]. Prior to 1900, this inter-decadal pattern largely disappears in *J97*, *J99*, and *L01*, which suggests that it is not a long-term feature of the NAO.

Alternately, well-dated and highly resolved proxy records of North Atlantic climate variability have been used to reconstruct the winter NAO index. These include annual tree-ring chronologies from circum-North Atlantic land areas [*Cook et al.*, 1998], an ice core accumulation record from Greenland [*Appenzeller et al.*, 1998], and multi-proxy assemblages [*Stockton and Glueck*, 1999; *Cullen et al.*, 2001; *Glueck and Stockton*, 2001; *Mann*, 2002]. Proxy-based reconstructions of this kind all rely on the connection between the NAO and climate (e.g., temperature and precipitation) over the areas where the proxies originate. As such, the proxies (e.g., tree rings) are responding to the influences of the NAO on climate somewhat removed from its actual SLP centers of action. This distinction must be kept in mind because no proxies

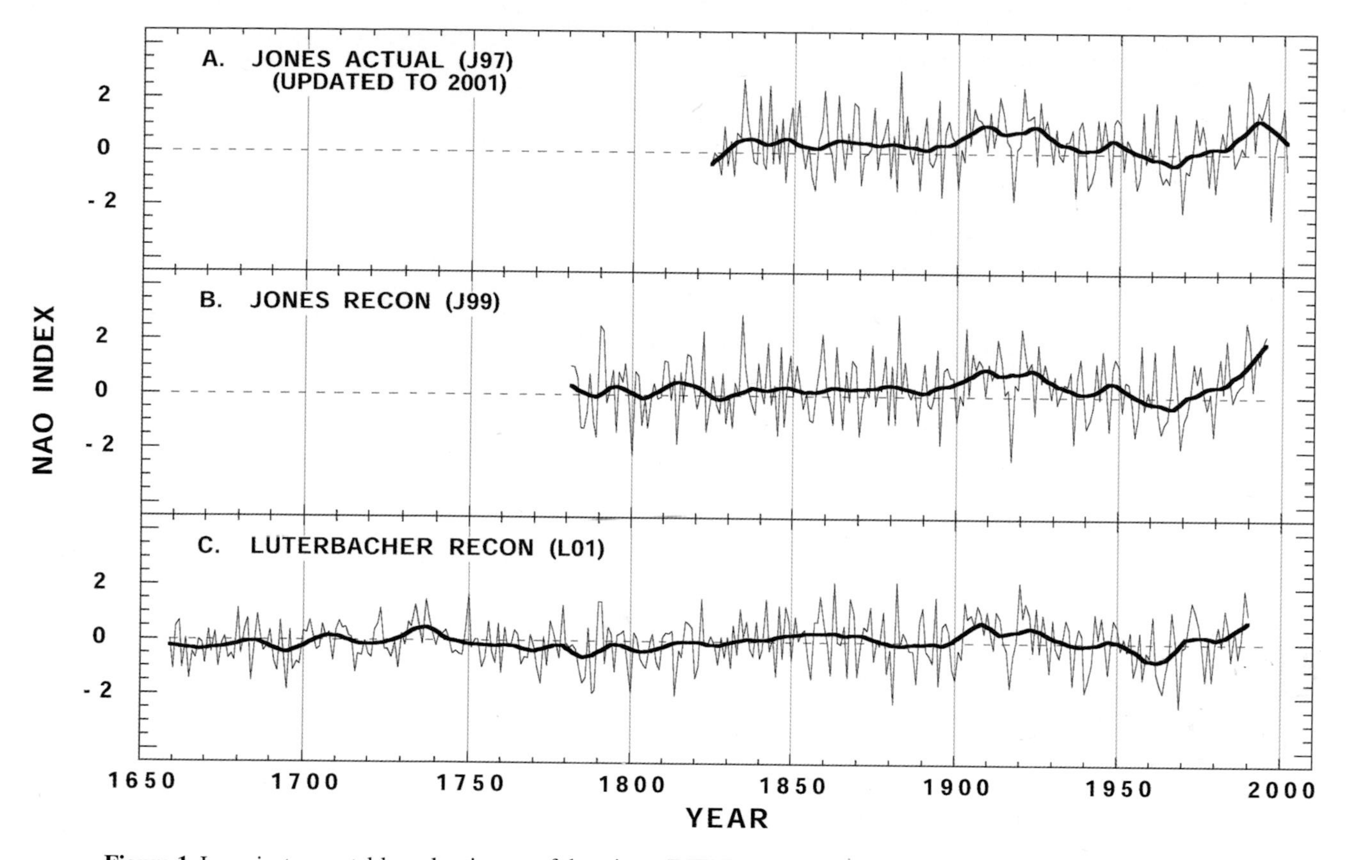

Figure 1. Long instrumental-based estimates of the winter (DJFM) NAO index from Europe data. The two "Jones" series are based strictly on SLP data, while the "Luterbacher" series is based on SLP, temperature, and precipitation data, as well as several documentary proxy predictors. The *J97* series has been updated to 2001 and can be downloaded from http://www.cru.uea.ac.uk/cru/data/. The heavy smooth curve is the 10-year low-pass filtered version of each series.

directly record SLP information. While the winter NAO index reconstructions presented here may contain useful information on the behavior of the NAO itself, they undoubtedly express variations associated with the teleconnected impacts of the NAO on climate as well [e.g., *Jones et al.*, this volume].

Figure 2 shows plots of seven NAO index reconstructions based strictly on natural proxy data. All were originally calibrated with instrumental NAO index data described in *Rogers* [1984], *Hurrell* [1995], *Jones et al.* [1997], and *Cullen et al.* [2002]. However, the NAO indices and seasons used for calibration purposes differ somewhat. Figures 2a-b show winter NAO index reconstructions based on circum-North Atlantic tree-ring series from *Cook et al.* [1998] (hereafter *C98*) and *D'Arrigo* [*D01*] in *Cullen et al.*, [2001]. The latter differs from that described by *D'Arrigo and Cook* [1997] in that no long instrumental data were used in the *D01* reconstruction analyzed here. See *Cullen et al.* [2001] for details. For calibration purposes, *C98* used the *Rogers* [1984] December-February (DJF) NAO index based on Azores/Iceland SLP data, while *D01* used a December-March (DJFM) NAO index of *Cullen et al.* [2002] based on north Atlantic sea surface temperatures from *Kaplan et al.* [1998] that correlated best with the *Hurrell* [1995] DJFM Lisbon/Iceland SLP NAO index. These are followed by a reconstruction of annual (April-March) NAO indices using a smoothed ice core accumulation record from southwest Greenland [*Appenzeller et al.*, 1998] [*A98*; Figure 2c], and a multi-proxy DJF reconstruction using Greenland ice core oxygen isotope ratios and tree rings from Finland and Morocco [*Stockton and Glueck*, 1999] [*S99*; Figure 2d]. These reconstructions also used the *Hurrell* [1995] Lisbon/Iceland monthly NAO indices for their calibration targets. The *Stockton and Glueck* [1999] results have also been updated and described in *Glueck and Stockton* [2001], but any changes made to their reconstruction since the 1999 version were not available in time to be used in this review. The next NAO reconstruction is based on a multi-proxy, spatial reconstruction of gridded sea surface temperatures [*Mann*, 2002] [*M02*; Figure 2e] that were used to estimate an October-March NAO index derived from the monthly Gibraltar/Iceland NAO data of *Jones et al.* [1997]. The version of the *Mann* [2002] reconstruction used here is described in *Cullen et al.* [2001] and does not include any long instru-

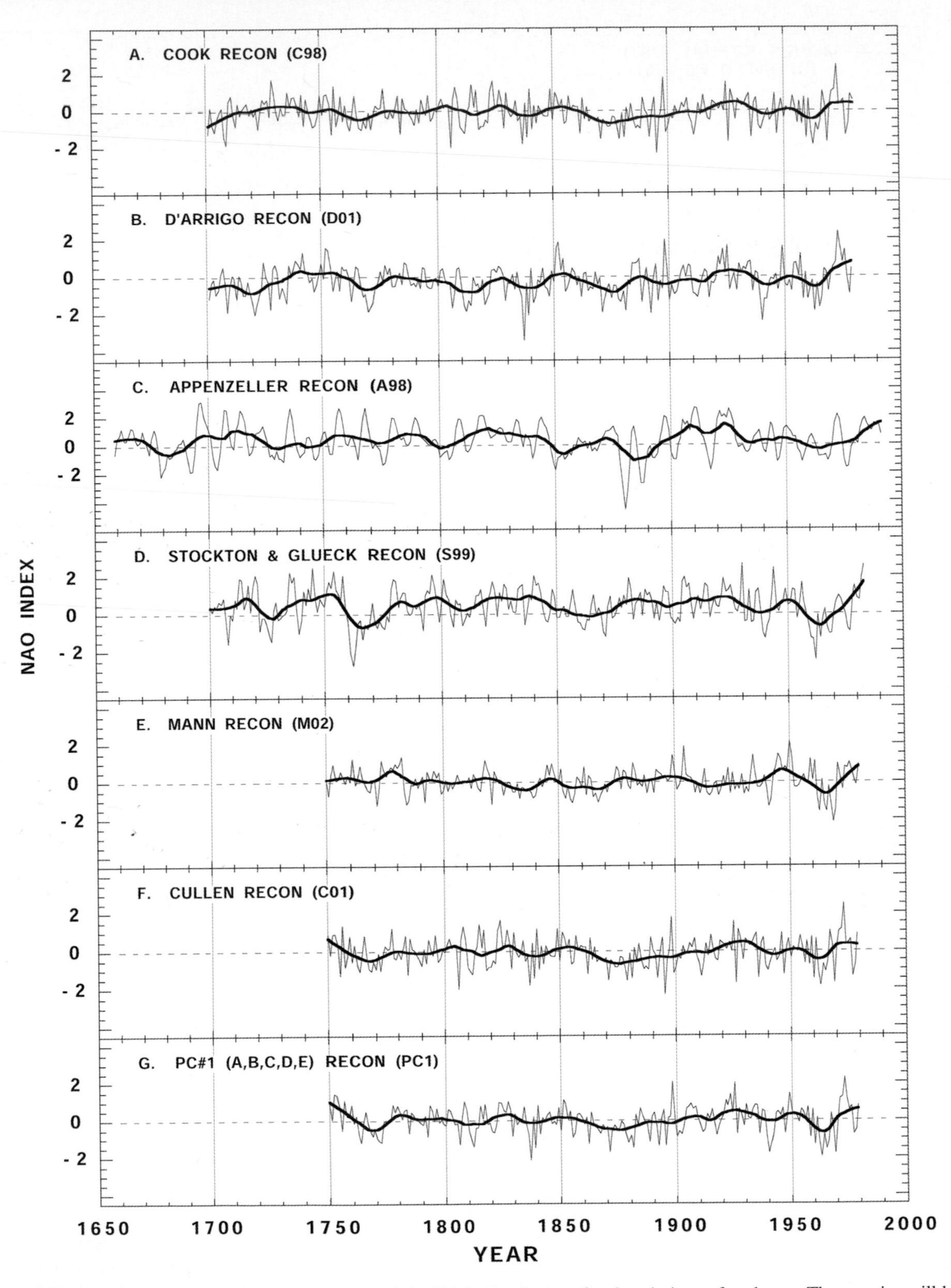

Figure 2. Seven proxy-based reconstructions of the NAO. See the text for descriptions of each one. These series will be tested against the long *L01* reconstruction (Figure 1c) to determine the degree to which the proxy-based reconstructions provide reliable estimates of past NAO variability.

mental climate records in the estimates. The remaining two reconstructions are a regression-weighted average of *C98*, *A98*, *D01*, and *M02* using the Lisbon/Iceland DJFM NAO index as the target [*Cullen et al.*, 2001] [*C01*; Figure 2f], and the first principal component of the *C98*, *D01*, *A98*, *S99*, and *M02* series developed for this study [*PC1*; Figure 2g], which expresses the most common mode of variability among all six series. *C01* and *PC1* are attempts to improve upon the other reconstructions by combining their mutual information in different ways. As is apparent, differences in the NAO seasons chosen for the original reconstructions may be contributing to the uncertainties in these composite estimates. Plate 1 is a map of the locations of the proxies used in the independently derived NAO reconstructions just described. The M02 reconstruction differs most from the others in its much more extensive use of eastern North American tree-ring data.

Each of the proxy-based NAO index reconstructions were subjected to various statistical calibration/verification procedures and all appeared to be valid. However, the validity of the proxy-based reconstructions was called into question by *Schmutz et al.* [2000] when three of them [*C98*, *A98*, *S99*], plus the average of those three series, failed to correlate significantly with the *J97* NAO index (Figure 1a) prior to 1850. In contrast, the *Luterbacher et al.* [1999] reconstruction (with Gibraltar and Iceland SLP data removed from it to avoid any bias in the comparisons) correlated significantly with *J97* back to 1821. Further comparisons of the proxy-based reconstructions with the *Luterbacher et al.* [1999] reconstruction prior to 1821 indicated that their failure to verify extended well back into the 18th century. These results led *Schmutz et al.* [2000] to conclude that the proxy-based NAO index reconstructions could not be trusted prior to 1850.

3. REVISITING THE SCHMUTZ ET AL. [2000] RESULTS

Given the disconcertingly negative conclusion of *Schmutz et al.* [2000], it is worth revisiting that study by comparing the seven proxy-based reconstructions shown in Figure 2 (including three used by Schmutz) with the *L01* reconstruction (Figure 1c). It is possible that the improved *L01* series might be better related to one or more of the reconstructions than that found by *Schumtz et al.* [2000] using the less robust NAO reconstruction described in *Luterbacher et al.* [1999]. However, instead of using the same ad hoc running correlation procedure of *Schmutz et al.* [2000] to compare the reconstructions over time, the Kalman filter will be used here as a dynamic regression modeling procedure [*Harvey*, 1981; *Visser and Molenaar*, 1988; *Van Deusen,* 1990]. In this case, *L01* will act as the predictor (i.e., independent variable) and each proxy-based reconstruction will serve as a predictand (i.e., dependent variable) in a series of bivariate time-dependent regression models.

The Kalman filter method allows for the identification of time-dependence between predictor and predictand variables in a totally objective way using maximum likelihood estimation (MLE). This is accomplished by casting the simple regression model into state-space form with a state equation

$$Y_t = a_t X_t + e_t$$

and a state transition equation

$$a_t = a_{t-1} + n_t$$

In this form, the regression coefficient a_t is allowed to vary as a random walk, with its variance dependent on n_t. Determining if the variance of $n_t > 0$ (i.e., the regression is dynamic) is accomplished by MLE. The MLE solution also provides theoretical standard errors for determining the significance of a_t over time. See *Harvey* [1981] and *Visser and Molenaar* [1988] for details.

The results of the Kalman filter modeling are shown in Figure 3. The time interval analyzed was either 1701-1979 [*C98*, *D01*, *A98*, *S99*] or 1750-1979 [*M02*, *C01*, *PC1*], depending on the overlap with *L01*. Each fitted model is expressed as a time series of dynamic standardized regression coefficients, with 2-standard error limits. As long as the standard error limits do not cross zero, it is considered to be significantly different from zero. This is a conservative test because the standard error limits shown here are 2-tailed, when in fact the only meaningful association between *L01* and each reconstruction is positive (i.e., 1-tailed). This difference is not large enough to change in any meaningful way the interpretation of the results, however. All of the proxy-based reconstructions are significantly related to *L01* over the 20th century and the latter 25-50 years of the 19th century. This is the period from which all of the proxies were calibrated. Prior to 1850, all but *M02* weaken significantly against *L01* and become (mostly) non-significant. These results are generally consistent with those of *Schmutz et al.* [2000] even after using the updated *L01* series as the predictor. Only *M02* shows some degree of long-term stability, which may be due in part to the inclusion of a larger spatial array of proxies in the estimates (see Plate 1). Even so, it has a correlation of only 0.30 with *L01* over the full 1750-1980 overlap period. While statistically significant, this relationship (R^2~10%) may not be strong enough to be of much operational value, nor does it compete in length with *L01*. Therefore, the need still exists for a strongly calibrated, well-verified, and significantly lengthened reconstruction of the winter NAO index from long proxy records, if paleoclimatic information is to contribute meaningfully to our understanding of the NAO.

In the following section, a new multi-proxy reconstruction of the winter NAO index will be described, which possesses these attributes. It calibrates and verifies strongly against all extended instrumental-based NAO indices produced thus far and extends back to AD 1400. This improvement arises from the use of an expanded network of proxies from circum-North Atlantic land areas and an extended calibration period that may be less affected by possible 20th century changes in the teleconnection between the NAO and climate over the regions from where the proxies originate.

4. A NEW MULTI-PROXY WINTER NAO INDEX RECONSTRUCTION

4.1. Introduction

The results of the Kalman filter analyses largely confirmed the negative findings of *Schmutz et al.* [2000] vis-à-vis the trustworthiness of the proxy estimates of the NAO index so far produced. However, they also provide some guidance on how to proceed in developing a new, hopefully improved, multi-proxy reconstruction of the winter NAO index. As noted above, the M02 reconstruction, which included a large spatial array of proxies, proved to be the most stable estimate of an NAO index when compared to *L01*. This finding indicates that the *C98*, *D01*, and *S99* reconstructions might have performed better had a more spatially diverse set of proxies been used. The way in which most of the reconstructions declined rapidly in skill prior to AD 1850 also suggests that the anomalous behavior of the NAO during the 20th century may have biased the selection of the proxies and the calibration procedure. Such might be the case using either the *Rogers* [1984] index back to 1874 or the *Hurrell* [1995] index back to 1864 because the calibration is dominated by 20th century data. Therefore, it is probably not coincidental that the *Cook et al.* [1998], *Appenzeller et al.* [1998], and *Stockton and Glueck* [1999] reconstructions, calibrated using the Rogers and Hurrell records, failed in the tests performed by *Schmutz et al.* [2000] using longer instrumental-based NAO indices. However, the mix of NAO seasons and target series used in creating the reconstructions analyzed by *Schmutz et al.* [2000] and those in this chapter complicate the interpretations of the results reported in each study. Consequently, it is probably wise not to over-interpret and generalize those findings.

4.2. Calibration and Verification Procedures

Given the possible calibration bias associated with anomalous 20th century NAO variability, expanding the calibration time period to include more of the 19th century is desirable. This is possible now because of the long instrumental-based NAO indices shown in Figure 1. Specifically, the *J99* series over the 1826-1974 interval will be used to calibrate the proxies, which have a common end year of 1974. This period covers the serially complete interval of *J97* and gives nearly equal weighting to both the 19th and 20th centuries. The 1781-1825 data in *J99* and the 1659-1825 data from *L01* will be withheld from all calibration procedures to verify the accuracy of the proxy estimates. Based on the NAO studies of *Hurrell* [1995] and *Hurrell and van Loon* [1997], which showed that NAO variability during the extended winter season (DJFM) had strong climatic impacts over circum-North Atlantic land areas, the DJFM seasonal NAO index was chosen as the target for reconstruction.

The teleconnection studies of *Hurrell and van Loon* [1997] (and indirectly the relative stability of the *M02* reconstruction) indicate that the NAO significantly forces climate variability over parts of eastern North America, Europe, North Africa, and the eastern Mediterranean region. Fortunately, these are regions with abundant supplies of centuries-long, climatically-sensitive, annual tree-ring chronologies, some of which have already been used in previous reconstructions of the winter NAO index [*Cook et al.*, 1998; *Stockton and Glueck*, 1999; *Cullen et al.*, 2001; *Glueck and Stockton*, 2001; *Mann,* 2002]. In addition, a number of well-dated and highly resolved ice core records from Greenland (both stable isotope ratios and accumulation rates), covering the past several hundred years, are available for reconstructing the NAO index. As before, some of them have been used previously for that purpose [*Appenzeller et al.*, 1998; *Stockton and Glueck*, 1999; *Glueck and Stockton*, 2001]. Most of these records are in the public domain and can be downloaded anonymously from the NOAA National Geophysical Data Center, World Data Center for Paleoclimatology web site (http://www.ngdc.noaa.gov/paleo/data.html). However, a number of tree-ring series used here, especially from Europe, are not yet in the public domain. These include important pine chronologies from coastal Norway (kindly supplied by T. Melvin) and a number of oak chronologies from the British Isles (kindly supplied by K. Briffa). The full collection of tree-ring chronologies comes from many tree species and all have been processed to remove biological growth trends thought to be non-climatic in origin.

A total of 367 candidate predictors of the winter NAO index were amassed, all covering the common interval 1750-1974. Of those, 329 extend back to 1700, 141 back to 1600, 86 to 1500, and 49 back to 1400. In order to take full advantage of the longer proxies, a series of nested principal component regression (PCR) models were developed in which the common starting year of the candidate predictors was stepped backwards in 25-year increments. This resulted in a total of 15 model runs, each with its own calibration and verification statistics. This approach is similar to that used by *Luterbacher et al.* [1999, 2001, 2002] to reconstruct the winter NAO index using early European instrumental records of varying lengths. However, unlike the European climate data used by *Luterbacher et al.* [1999, 2001], it is difficult to know a priori which proxies are useful records of climate associated with winter NAO variability. Therefore, some sort of predictor variable screening is desirable. In this case, only those proxies that correlated significantly ($p < 0.05$; 2-tailed hypothesis test) with the winter NAO index over the 1826-1974 calibration period were retained. This resulted in a dramatic reduction in the number of proxies actually entered into each nested PCR run, typically about 1/3 of the original number of candidate predictors. The dimensionality of the predictor variable set was further reduced within PCR itself by only retaining those principal components for regression with eigenvalues >1.0 (the Kaiser-Guttman rule). This rule reduced the number of PC candidate predictors to about 1/3 again on average. Finally, the best subset of the retained principal components entered into the actual regression model was determined using the minimum Akaike information criterion [*Akaike*, 1974], with a small-sample bias correction applied [*Hurvich and Tsai*, 1989]. This PCR procedure with prior predictor variable screening is described and tested in detail in *Cook et al.* [1999].

The calibration and verification statistics described here are the same as those provided by *Luterbacher et al.* [1999] for their reconstruction: the coefficient of multiple determination (R^2) for the calibration period, and the square of the Pearson product-moment correlation ($R*R$), and reduction of error (*RE*) for the verification period. These statistics are all measures of variance in common between the actual (*J99*) and estimated (multi-proxy) winter NAO indices. See *Cook et al.* [1999] for more details on these statistics and their comparative differences.

4.3. The New Winter NAO Index Reconstruction

Plate 2 shows the map of proxies retained after the initial screening for correlation with the *J99* winter NAO index (Plate 2a) and the standardized regression coefficients (i.e., beta weights, Plate 2b) of those series used in reconstructing the winter NAO index. All series cover the common interval 1750-1974. However, some of the useful proxies extend out to 1979, which allowed the reconstruction to be extended forward to that year. The distribution of sites is well balanced on both sides of the Atlantic and agrees well with the NAO teleconnection patterns with climate found by *Hurrell* [1995] and *Hurrell and van Loon* [1997] over the North Atlantic sector. This result suggests the additional data from North America may provide a more complete estimate of winter NAO variability than that available from European-only data. The beta weights support this suggestion. The relative explanatory power of the proxies is evenly distributed across the network. The distribution of sites is also consistent with that used in the M02 reconstruction (see Plate 1), which proved to be most stable in the Kalman filter analyses.

Figure 4 shows the complete winter NAO index reconstruction (Figure 4a), based on the statistical calibration procedures described earlier, and the number of proxies available for each 25-year time step. The series has been extended out to 2001, using appropriately scaled and updated *J97* estimates, for full comparison of the 20th century with estimated past NAO behavior. Below Figure 4a are plots of the calibration and verification statistics described earlier (Figure 4b). The calibration R^2 ranges from 0.636 in the post-1700 period to 0.305 for the 1400-1424 period. This decline in R^2 closely tracks the decline in the number of proxies available in each time period, a result that is consistent with expectation [*Rencher and Pun*, 1980]. In contrast, the verification $R*R$ and *RE* statistics remain remarkably constant over time, with ranges of 0.269-0.367 and 0.237-0.350, respectively. With 45 paired observations in the 1781-1825 verification period to compare, these results strongly support the validity of this multi-proxy winter NAO index reconstruction.

Further verification of this reconstruction against *L01* is shown in Figure 5. As before, the Kalman filter is used to allow for time dependence in the relationship. The MLE solution indicates that the relationship between the multi-proxy reconstruction and *L01* is weakly time dependent, but always positive and statistically significant at the 2-standard error limit. Therefore, this new NAO reconstruction verifies against the updated and revised *L01* reconstruction back to 1659, a significant improvement over the negative findings of *Schmutz et al.* [2000] and any of the Kalman filter results shown in Figure 3.

Also shown in Figure 5 are four Pearson correlations that reveal more discretely how the actual correlation between the two reconstructions has changed. These correlations are for the 1659-1780 ($r = 0.53$), 1675-1780 ($r = 0.59$; minus 1714-1721) 1781-1825 ($r = 0.52$), and 1826-1979 ($r = 0.68$) time periods. The 1826-1979 period is effectively the same as the calibration period used here and is provided for comparison with the R^2 results shown in Figure 4. The 1781-1825 period is the same as the verification period used previously and its r agrees favorably (when squared) with that shown in Figure 4. For the 1659-1780, which is independent of the previous verification tests, the verification r remains about the same. However, this result may underestimate the true fidelity of the multi-proxy reconstruction. Only the 1671-1713 and 1722-1780 periods of *L01* used early SLP data, which significantly improved the calibration/verification statistics [see Figure 2 in *Luterbacher et al.*, 2001] over other periods for which SLP data were not available. When only those years of recon-

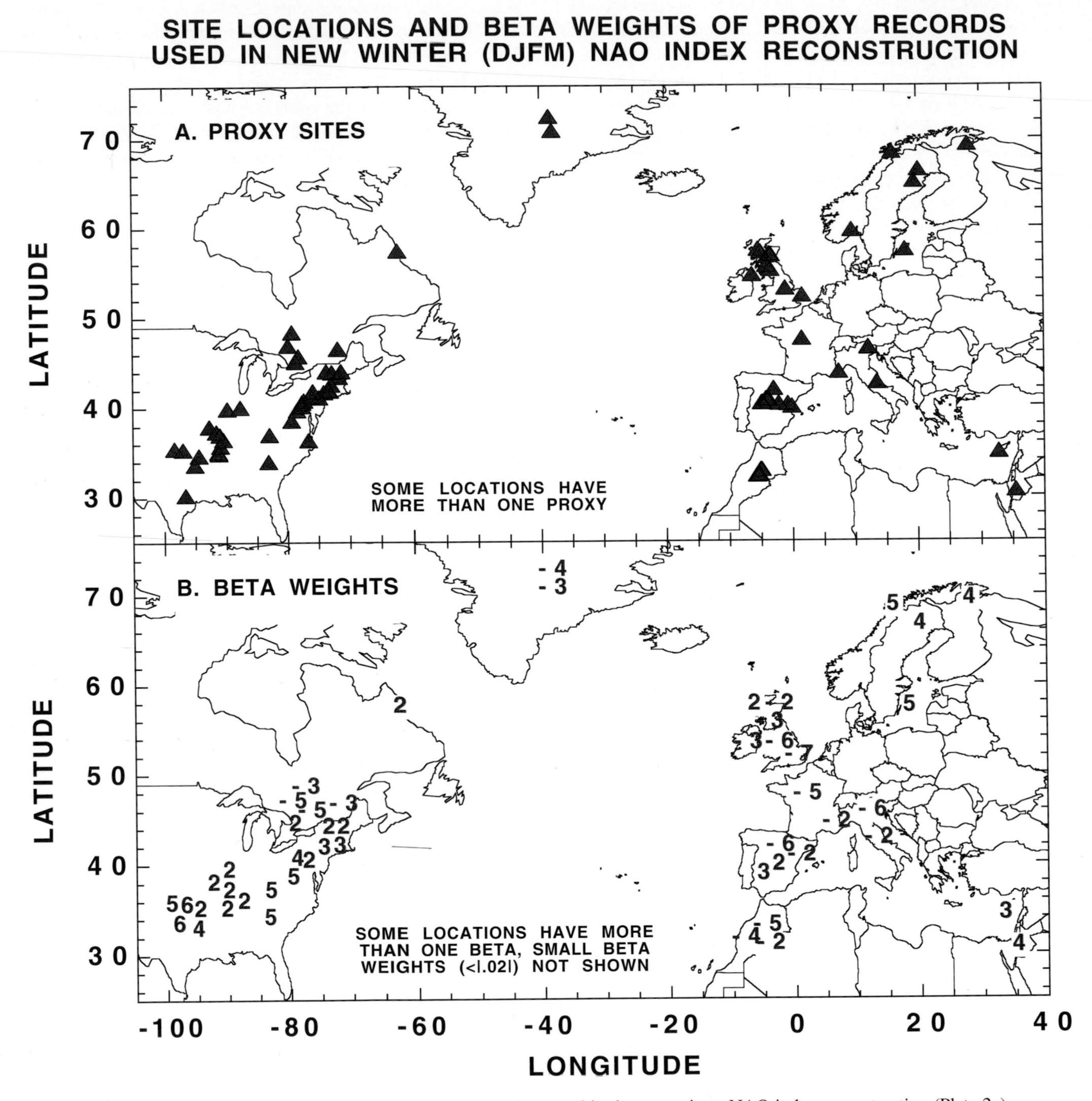

Plate 2. Maps showing the locations of the proxy variables used in the new winter NAO index reconstruction (Plate 2a) and the standardized regression coefficients (beta weights) of the proxies used in the reconstruction (Plate 2b). The spatial distribution of the selected proxies and their beta weights shows that the climatic influences of the NAO on circum-North Atlantic land areas is nearly equal on both sides of the Atlantic.

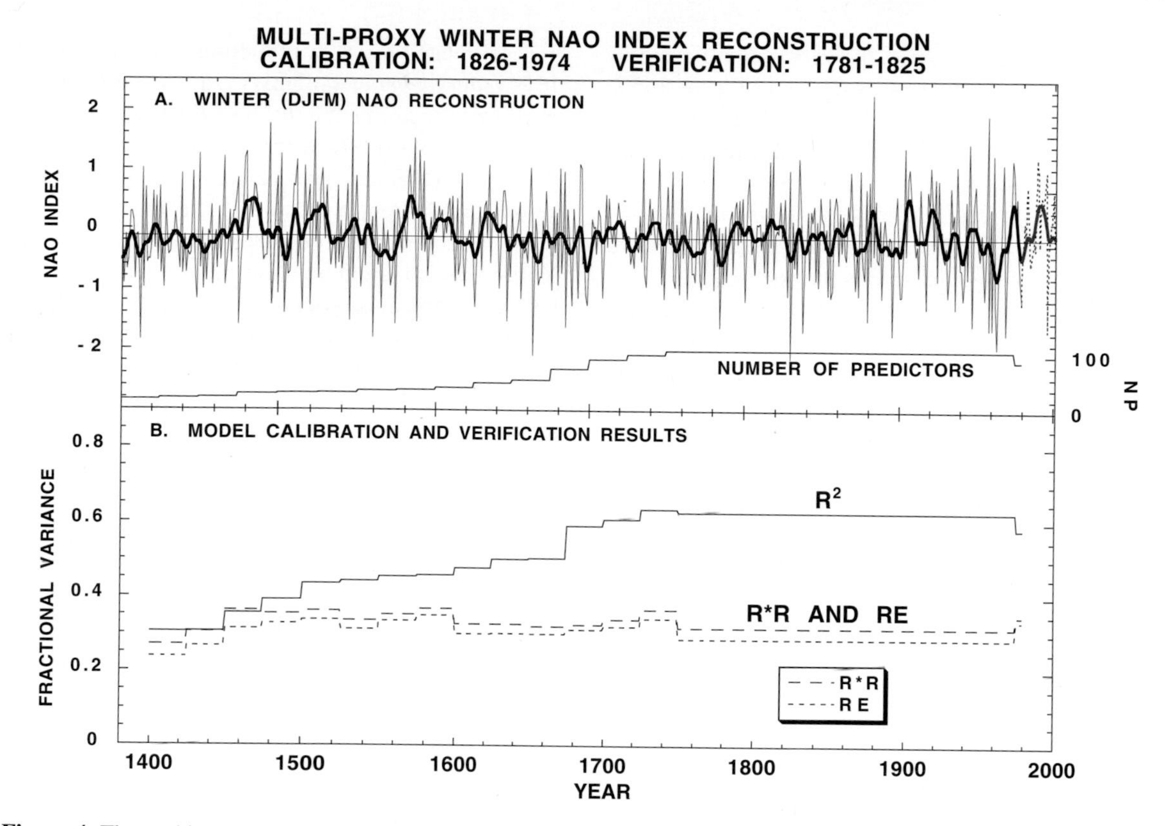

Figure 4. The multi-proxy reconstruction of the winter NAO index and the number of proxies used (4a), with the time varying calibration and verification statistics (4b) over the period 1400-1979. The heavy smooth curve is the 10-year low-pass filtered version of each series. The calibrated variance (R^2) declines with the number of predictors, but the square of the Pearson correlation ($R*R$) and the reduction of error (RE) verification period statistics remain remarkably constant over time. See the text for details.

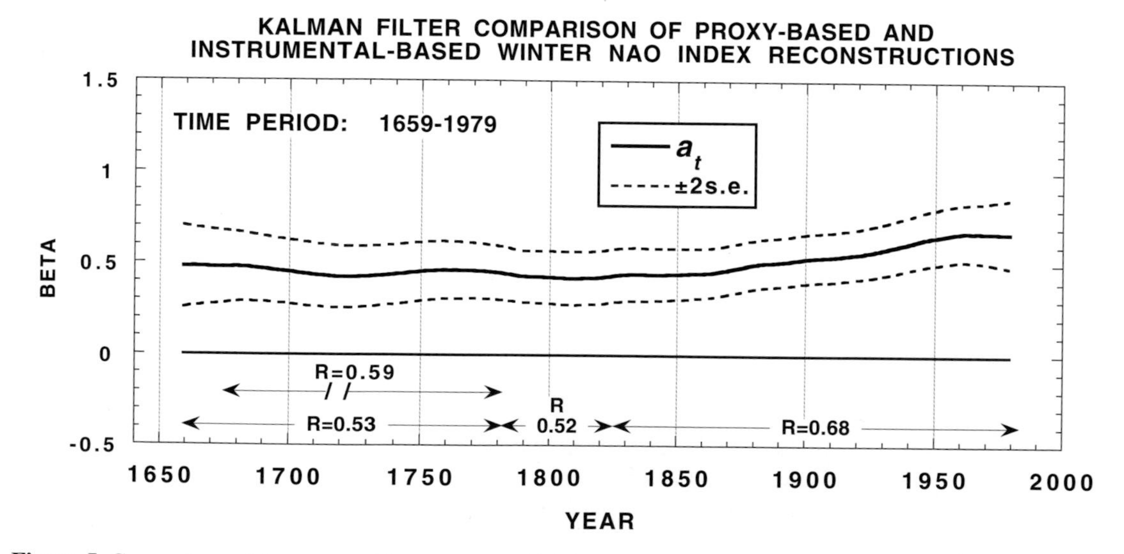

Figure 5. Comparison of the new multi-proxy winter NAO index reconstruction with *L01* (Figure 1c) using the Kalman filter as a dynamic regression modeling procedure. The relationship is time dependent, but always statistically significant back to 1659. The simple correlations show more discretely how the relationship changes. These results are significantly better than those shown in Figure 3 for the other proxy-based reconstructions of the NAO. See the text for more details.

structed NAO index are compared, the verification r increases to 0.59. In contrast, when the reconstructions for the 1659-1670 and 1714-1721 periods based on no SLP data are correlated, the verification r drops to 0.36. Thus, this drop appears to come primarily from the *L01* reconstruction when it is not based on SLP data. Consequently, the other verification results with *L01* that includes SLP data are probably more indicative of the true accuracy of the multi-proxy reconstruction.

One additional comparison was made between the new multi-proxy reconstruction and estimates of the winter (DJF) NAO index produced by *Luterbacher et al.* [2001] for the period 1500-1658 from early seasonal non-instrumental European data. The Pearson r between these two series is 0.22, which is much lower than before, but still significant well above the 95% confidence level. The drop in correlation may be related to some reduction in the quality of the pre-1659 *Luterbacher et al.* [2001] estimates as their primary data shift from instrumental to historical climate sources. The comparison of slightly different seasonal reconstructions here (DJFM vs. DJF) may also be contributing to the lower correlation. In addition, a change in the relative strength of the NAO teleconnections with climate over Europe and eastern North America could be contributing to this change in correlation. For now it is probably wise to regard both NAO reconstructions as equally valid. Thus, assuming that these European estimates of the winter NAO index are reasonably accurate back to 1500, the multi-proxy reconstruction can be regarded as being significantly verified against independently derived estimates covering the past 479 years. This result is quite remarkable given the earlier negative findings of *Schmutz et al.* [2000].

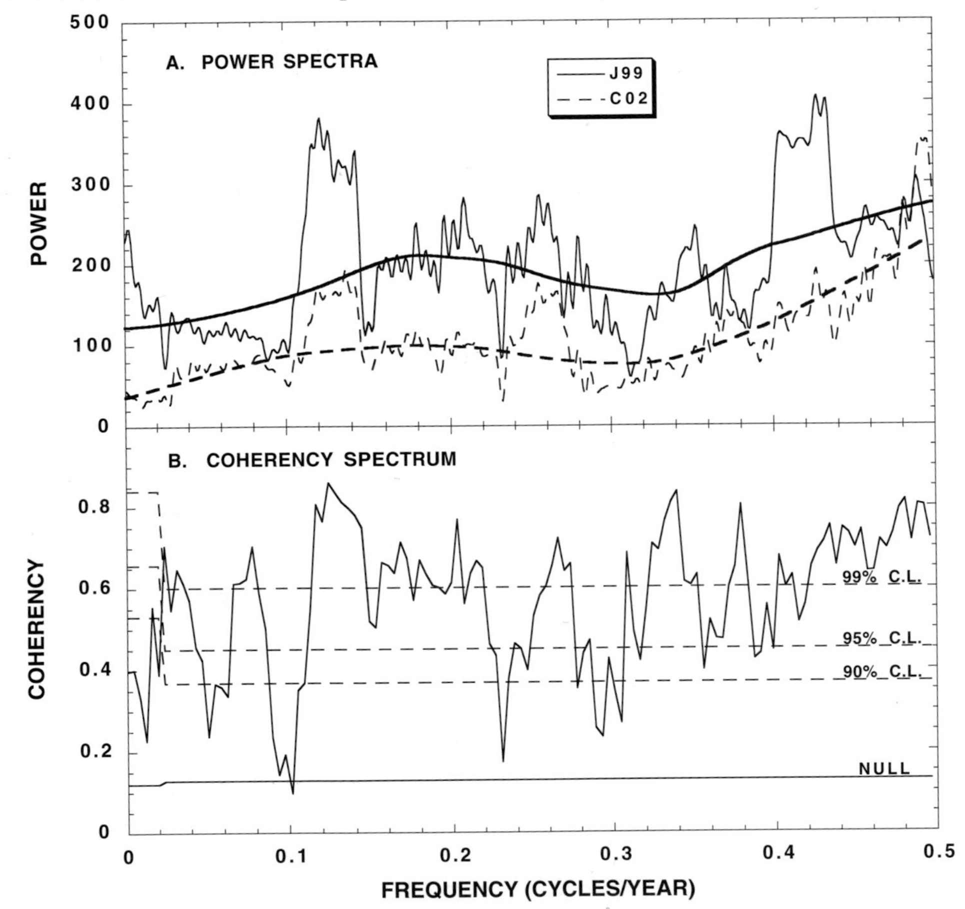

Figure 6. Cross-spectral analysis of the instrumental *J99* data (Figure 1b) and the new reconstruction (*C02*; Figure 4a) using the multi-taper method. The power spectra (6a) are qualitatively similar over much of the frequency range, with *C02* showing consistently less power due to variance not accounted for by the regression model. There is also some suggestion that the reconstruction is missing proportionally more multi-decadal (i.e. >50 years) variance relative to the *J99* series. This may be due to the way in which the tree-ring data were processed prior to their use in the reconstruction. The overall pattern of variance as a function of frequency (smooth solid and dash curves) is also somewhat "blue" in both records, with variance being concentrated more so in the higher frequencies. The coherency spectrum (6b) shows that the two series are broad-band coherent, with a small drop off in the multi-decadal band.

Auto- and cross-spectral analyses of the new reconstruction and *J99* over the common period (1826-1979) are shown in Figure 6, based on the multi-taper method [*Thomson*, 1982; *Mann and Lees*, 1996]. The difference in overall level between the auto-spectra reflects the lost variance due to regression. Both auto-spectra (Figure 6a) have a "blue noise" quality that emphasizes high-frequency variance over low-frequency variance. This characteristic probably reflects the fast response time of the NAO as an atmospheric pressure index and its relatively weak coupling with slower oceanic processes. However, there is some indication of power in the multi-decadal (especially >50 years) band that may reflect a slower coupled atmosphere-ocean response of the NAO, which is reminiscent of the 60-70 year Schlesinger–Ramankutty oscillation. The *J99* spectrum shows this effect more strongly than the spectrum of the new reconstruction, which suggests that the latter is deficient in low-frequency power. The cause of this deficiency is unclear, but the way in which the proxies have been processed may be a contributing factor. It is also possible that both *J99* and the new reconstruction are not fully capturing the low-frequency variations in the NAO. The coherency spectrum (Figure 6b) of the two series indicates that the reconstruction is broad-band coherent in a way that is consistent with the distribution of power in the two series (i.e., periods with more power tend to be more highly coherent). This desirable covariance between power and coherency largely breaks down at periods greater then 10 years, again for reasons that may be related to the way in which the proxies were processed. Resolving this issue would require a very detailed re-evaluation of the proxies, which is beyond the scope of this present study.

Given the successful long-term verification of this reconstruction, its power spectrum was computed to determine the degree to which it has maintained some of the same band-limited power found in the shorter instrumental records [e.g., *Rogers*, 1984; *Cook et al*,. 1998; *Mann*, 2002]. The spectrum (Figure 7) shows that most of the significant band-limited variance is restricted to periods of less than 10 years, with particular concentrations around 7.7 and 4 years and in the quasi-biennial (i.e., 2-3 year) band. This is consistent with previous results, which suggests that these band-limited properties of the winter NAO are organized long-term features of North Atlantic SLP. With regards to the 60-70 year Schlesinger–Ramankutty oscillation, there is no indication of its strong presence in the spectrum, although there is a "lump" of variance with a mean period of about 50-60 years. However, the decrease in calibrated variance at periods greater than about 10 years, and even more significant loss of variance at timescales greater than 50 years seen in Figure 6a, means that conclusions cannot be reached regarding the multidecadal and century-scale variations in the NAO at this stage. Finally, unlike many climatological time series that

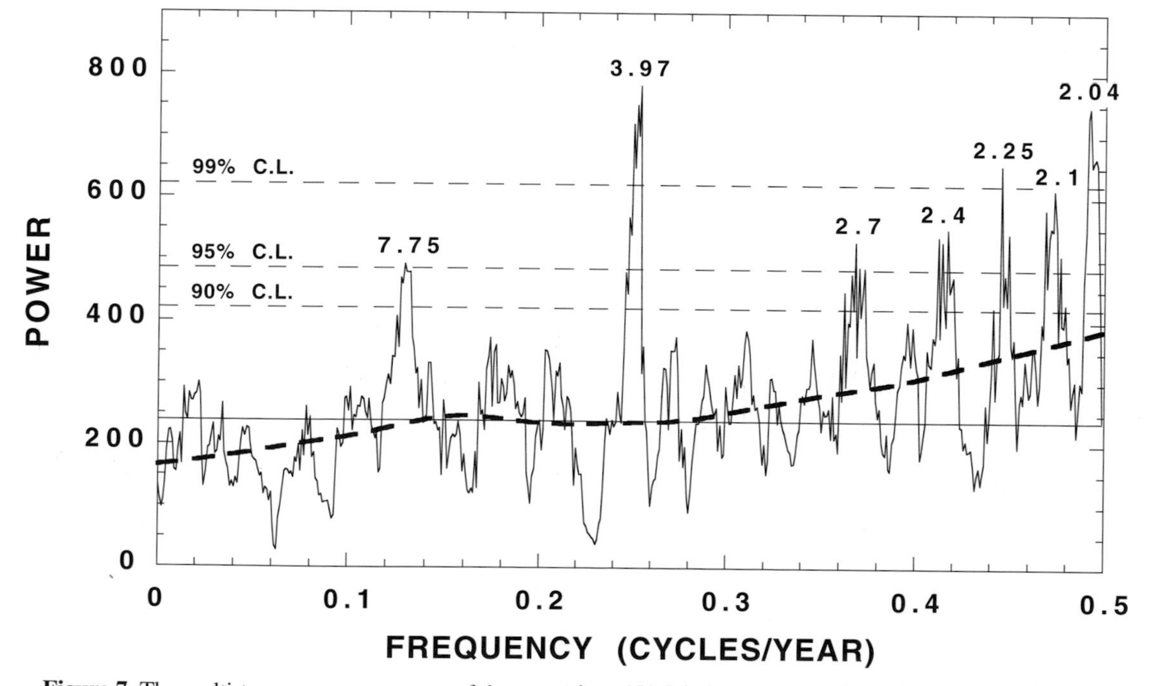

Figure 7. The multi-taper power spectrum of the new winter NAO index reconstruction. The reconstruction is dominated by band-limited power at periods shorter than 10 years and behaves overall as a "blue noise" process with power increasing with frequency (thick dash line). The latter result is consistent with the spectrum of the instrumental SLP-based *J99* record shown in Figure 6a.

behave as "red noise" processes, the winter NAO index reconstruction behaves more like a "blue noise" process. That is, most of its variance is concentrated in the high-frequency end of the spectrum. The lowess robust smooth of the spectrum (the thick dashed line in Figure 7) shows this overall increasing trend in variance. Therefore, the "blueness" of the reconstruction appears to be a real feature of SLP in the North Atlantic as measured by the winter NAO index. Even so, the interpretation of the lowest-frequency variability, as discussed before, is limited by the possible loss of calibrated variance at very low frequencies.

5. WHY DID PREVIOUS PROXY-BASED NAO RECONSTRUCTIONS FAIL?

The successful reconstruction of the winter NAO index begs the question: why did most of the previous reconstructions fail to verify against the instrumental-based NAO indices prior to 1850? The use here of a large number of widely distributed proxies may have contributed to more robust estimates of the NAO index back in time, as suggested by the relative stability of the *M02* reconstruction in the Kalman filter tests. Yet, the verification statistics in Figure 4 remained very stable even with large reductions in the number of proxies used. Therefore, the answer may also be related to the choice of the extended calibration period used here.

As noted earlier, the behavior of the NAO in the 20th century appears to be more persistent and extreme than that seen in the 19th century (see Figure 1). This need not be a problem in selecting the proxies for reconstruction if the teleconnection pattern between the NAO and circum-North Atlantic climate during the 20th century [e.g., *Hurrell*, 1995; *Hurrell and van Loon*, 1997] is similar to that in the past. However, it is possible that some changes in NAO/climate teleconnections have occurred during the 20th century [see *Jones et al.*, this volume]. A documented example of such a teleconnection change has been reported between ENSO and drought over the United States [*Cole and Cook*, 1998], with one of the larger changes occurring since 1940 in the southeastern US where NAO forcing is also operating. Therefore, it is possible that ENSO is interacting with NAO in ways that were not commonly seen prior to the 20th century in the North Atlantic sector. Another possible explanation for anomalous 20th century behavior in NAO teleconnections is that the surface expression of the natural NAO pattern is strongly overprinted with a similar, but distinct, more zonally-symmetric NAO-like trend [the Arctic Oscillation or "AO"; see, e.g., *Thompson and Wallace*, 2000; *Thompson et al.*, 2001; *Thompson et al.*, this volume] in recent decades that could be related to an interaction between greenhouse radiative forcing and stratospheric dynamics [*Shindell et al.*, 1999; *Gillett et al.*, this volume]. If this were the case, the anthropogenic pattern would effectively "contaminate" the 20th century with a related but distinct pattern of variability, and would complicate the accurate calibration of the natural NAO component in a restricted 20th century interval [*Cook et al.*, 2002].

Evidence for change in the teleconnection between the NAO and circum-North Atlantic climate during the 19th and 20th centuries is suggested in maps (Plate 3) of the proxies selected by the same screening procedure described above for two different calibration periods: 1826-1899 and 1900-1974. This split breaks the *J97* series into equal length sub-series that also have distinctly different low-frequency characteristics (see Figure 1a). A comparison of the maps indicates that the 1900-1974 calibration period chooses many more proxies in Fennoscandia and the lower Mississippi Valley than does the 1824-1899 calibration period. This change is unlikely to be related to changing quality of the proxies themselves. In particular, the tree-ring chronologies are highly replicated in the time periods being tested. Rather, the change in the geographical distribution of selected proxies is indirect evidence for a change in the teleconnection of the winter NAO with circum-North Atlantic climate [see *Jones et al.*, this volume]. Consequently, the proxies used in previous reconstructions [e.g., those tested by *Schmutz et al.*, 2000] may have been geographically biased towards regions that were strongly affected by the anomalous 20th century behavior of the NAO. Prior to the 20th century, and especially before 1850, this bias led to a significant loss of reconstruction skill. By expanding the calibration period to equally weight the 19th and 20th centuries, it was possible to identify a less geographically biased set of proxies to reconstruct the winter NAO index. Other factors may be playing lesser roles, but this is probably a significant cause of the greatly improved verification success reported here.

6. CONCLUSIONS AND RECOMMENDATIONS

A new multi-proxy reconstruction of the winter NAO index has been developed, which verifies well against previous estimates from early European instrumental [*Jones et al.*, 1999; *Luterbacher et al.*, 1999] and non-instrumental [*Luterbacher et al.*, 2001] data. The verification is statistically significant ($p < 0.05$) back to 1500, making the reconstruction one of the most strongly validated proxy-based records of past climate yet produced. A comparison of 20th century NAO behavior with that estimated back to 1400 (Figure 4a) indicates that NAO variability over the most recent 100 years is unusual, but not unique. In particular, the 15th and 16th centuries appear to have experienced episodes of persistent positive-phase NAO behavior that are comparable to those seen in the 20th century [see also *Luterbacher et al.*, 2001, 2002]. In contrast, during 1640-1880 interval the NAO appears to have

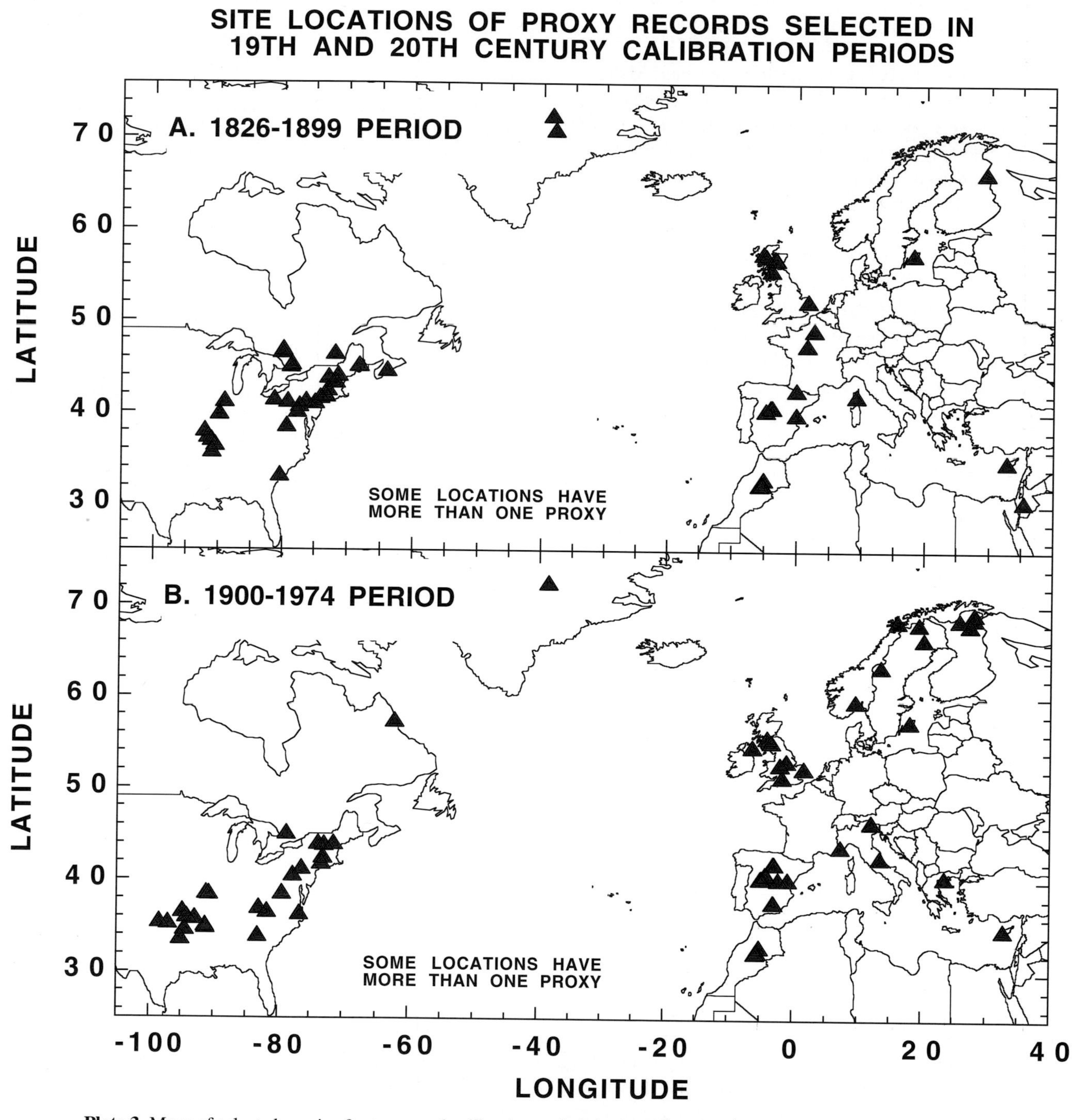

Plate 3. Maps of selected proxies for two equal calibration periods in the 19th and 20th centuries. In each case, the selected proxies are significantly correlated ($p < 0.05$) with the *J99* index. The change in the distribution of proxies, especially in Fennoscandia and the lower Mississippi Valley, indicates that their selection based mainly on 20th century data is geographically biased. This finding helps explain why previous proxy-based reconstructions of the NAO index failed to verify prior to 1850 in most cases, because all were based on calibrations that were strongly weighted towards the 20th century.

been less vigorous. This may be an expression of the impact of Little Ice Age cooling on climate in the North Atlantic sector via a reduction in the vigor in the NAO. Alternately, *Shindell et al.* [2001] suggest that the NAO may have been in a persistent low-index phase during the late Maunder Minimum (1680-1780) due to reduced solar irradiance, a result supported somewhat by the *Luterbacher et al.* [2001] reconstruction (Figure 1c). The lack of a conspicuous low-index period in the new multi-proxy NAO reconstruction during the late 17th century (Figure 4a) may therefore reflect some missing multi-decadal variability, as suggested in the power spectra comparison shown in Figure 6. Clearly, more effort is needed to better reconstruct multi-decadal to centennial timescale variability in the multi-proxy estimates.

The reconstruction's improved verification is most likely due to the use of an extended calibration period that reduced an apparent geographic bias in the selection of proxies used for reconstruction in previous studies. This bias appears to be related to the anomalous nature of NAO teleconnections over circum-North Atlantic land areas during the 20th century, which may relate to anthropogenically-forced non-stationarity of the climate during the 20th century. Changing teleconnections between ENSO and drought over the United States during the late-19th and 20th centuries have also been documented [*Cole and Cook*, 1998]. So, the possible effect of such changes on the subsequent quality of climate reconstructions must be kept in mind when selecting and calibrating proxies with (mostly) 20th century climate data using indirect, teleconnected relationships.

The apparent loss of multi-decadal variance in the reconstruction is cause for concern and probably represents its greatest deficiency. Although the exact cause of this problem is unclear now, a careful re-evaluation of the proxies and the statistical modeling procedures used here is highly recommended. In so doing, it may be possible to reconstruct more completely the multi-decadal variance that is clearly important to our understanding of the NAO and its sources of variability.

The use of this reconstruction for identifying and characterizing long-term teleconnections between the NAO and other climate reconstructions must be done carefully. Because of the extended nature of the North Atlantic proxy network used here, long proxy reconstructions of associated climate variables, like precipitation over Morocco or the lower Mississippi Valley, can not be compared to our multi-proxy reconstruction without the danger of circularity. However, it should be useful for examining potential proxy-based teleconnections in more distant regions of the world potentially affected by the NAO. And, providing that multi-proxy reconstructions of other ocean-atmosphere processes like ENSO are based on completely independent data [e.g., *Stahle et al.*, 1998], it will be possible to look for relationships and interactions between these important internal global forcings.

Acknowledgements. The research is supported by the Cooperative Institute for Arctic Research (CIFAR) through a partnership between NOAA and the University of Alaska, Grant UAF 02-0033. I thank Phil Jones and Juerg Luterbacher for providing the long European NAO records for calibrating and testing our multi-proxy reconstruction, and for stimulating communications on various aspects of this paper. I would also like to thank Keith Briffa and one anonymous referee for their useful comments on an earlier version of this paper. Lamont-Doherty Earth Observatory Contribution No. 6339.

REFERENCES

Akaike, H., A new look at the statistical model identification, *IEEE Trans. Auto. Con.*, *AC-19*, 716-723, 1974.

Appenzeller, C., T. F. Stocker, and M. Anklin, North Atlantic oscillation dynamics recorded in Greenland ice cores, *Science*, *282*, 446-449, 1998.

Cole, J. E., and E. R. Cook, The changing relationship between ENSO variability and moisture balance in the continental United States, *Geophy. Res. Lett.*, *25*, 4529-4532, 1998.

Cook, E. R., Proxy reconstructions of the NAO: a critical review and comparison with NAO indices estimated from long European instrumental records. Invited paper presented at Chapman Conference On The North Atlantic Oscillation, University of Vigo, Galicia, Spain, November 28-December 1, 2000.

Cook, E. R., R. D. D'Arrigo, and K. R. Briffa, A reconstruction of the North Atlantic Oscillation using tree-ring chronologies from North America and Europe, *The Holocene*, *8*, 9-17, 1998.

Cook, E. R., D. M. Meko, D. W. Stahle, and M. K. Cleaveland, Drought reconstructions for the continental United States, *J. Climate*, *12*, 1145-1162, 1999.

Cook, E. R., R. D. D'Arrigo, and M. E. Mann, A well-verified, multi-proxy reconstruction of the winter North Atlantic oscillation index since AD 1400, *J. Climate*, *15*, 1754-1764, 2002.

Cullen, H. M., A. Kaplan, P. A. Arkin, and P. B. deMenocal, Impact of the North Atlantic Oscillation on Middle Eastern climate and streamflow, *Clim. Change*, *55*, 315-338, 2002.

Cullen, H., and P. B. deMenocal, North Atlantic influence on Tigris-Euphrates streamflow, *Int. J. Climatol.*, *20*, 853-863, 2000.

Cullen, H., R. D'Arrigo, E. Cook, and M. E. Mann, Multiproxy-based reconstructions of the North Atlantic Oscillation over the past three centuries, *Paleoceanography*, *15*, 27-39, 2001.

D'Arrigo, R. D., and E. R. Cook, North Atlantic sector tree-ring records and SST variability. Paper presented at *Meeting on Atlantic Climate Variability*, Lamont-Doherty Earth Observatory, Palisades, NY, 1997.

D'Arrigo, R. D., E. R. Cook, G. C. Jacoby, and K. R. Briffa, NAO and sea surface temperature signatures in tree-ring records from the North Atlantic sector, *Quat. Sci. Rev.*, *12*, 431-440, 1993.

Delworth, T. D., and M. E. Mann, Observed and simulated multidecadal variability in the North Atlantic, *Clim. Dyn.*, *16*, 661-676, 2000.

Enfield, D. B., A. M. Mestes-Nuñez, and P. J. Trimble, The Atlantic multidecadal oscillation and its relation to rainfall and river flows in the continental U.S., *Geophy. Res. Lett.*, *28*, 2077-2080, 2001.

Gillett, N. P., H. F. Graf, and T. J. Osborn, Climate Change and the North Atlantic Oscillation, this volume.

Glueck, M. F., and C. W. Stockton, Reconstruction of the North Atlantic Oscillation, 1429-1983, *Int. J. Climatol.*, *21*, 1453-1465, 2001.

Harvey, A. C., *Time series models*, P. Allen Publ. Ltd., Oxford, 1981.

Hilmer, M., and T. Jung, Evidence for a recent change in the link between the North Atlantic Oscillation and Arctic sea ice export, *Geophy. Res. Lett.*, *27*, 989-992, 2000.

Hoerling, M. P., J. W. Hurrell, and T. Xu, Tropical origins for recent North Atlantic climate change, *Science*, *292*, 90-92, 2001.

Hurrell, J. W., Decadal trends in the North Atlantic Oscillation regional temperatures and precipitation, *Science*, *269*, 676-679, 1995.

Hurrell, J. W., Influence of variations in extratropical wintertime teleconnections on Northern Hemisphere temperatures, *Geophys. Res. Lett.*, *23*, 665-668, 1996.

Hurrell, J. W., and H. van Loon, Decadal variations in climate associated with the North Atlantic oscillation, *Clim. Change*, *36*, 301-326, 1997.

Hurrell, J. W., Y. Kushnir, G. Ottersen, and M. Visbeck, An Overview of the North Atlantic Oscillation, this volume.

Hurvich, C. M., and C. Tsai, Regression and time series model selection in small samples, *Biometrika*, *76*, 297-307, 1989.

Jones, P. D., T. Jónsson, and D. Wheeler, Extension of the North Atlantic Oscillation using early instrumental pressure observations from Gibraltar and south-west Iceland, *Int. J. Climatol.*, *17*, 1433-1450, 1997.

Jones, P. D., et al., Monthly mean pressure reconstructions for Europe for the 1780-1995 period, *Int. J. Climatol.*, *19*, 347-364, 1999.

Jones, P. D., T. J. Osborn, and K. R. Briffa, The evolution of climate over the last millennium, *Science*, *292*, 662-667, 2001.

Jones, P. D., T. J. Osborn, and K. R. Briffa, Pressure-based measures of the North Atlantic Oscillation (NAO): A comparison and an assessment of changes in the strength of the NAO and in its influence on surface climate parameters, this volume.

Kaplan A., M. A. Cane, Y. Kushnir, A. C. Clement, M. B. Blumenthal, and B. Rajagopalan, Analyses of global sea surface temperature 1856-1991, *J. Geophys. Res.*, *103(C9)*, 18,567-18,589, 1998.

Luterbacher, J., C. Schmutz, D. Gyalistras, E. Xoplaki, and H. Wanner, Reconstruction of monthly NAO and EU indices back to AD 1675, *Geophy. Res. Lett.*, *26*, 2745-2748, 1999.

Luterbacher, J., et al., Extending North Atlantic Oscillation reconstructions back to 1500, *Atmos. Sci. Lett.*, *2*, 114-124, doi:10.1006/asle.2001.0044, 2002.

Mann, M. E., Large-scale climate variability and connections with the Middle East during the past few centuries, *Clim. Change*, *55*, 287-314, 2002.

Mann, M. E., and J. Park, Global-scale modes of surface temperature variability on interannual to century timescales, *J. Geophy. Res.*, *99*, 25,819-25,833, 1994.

Mann, M. E. and J. Lees, Robust estimation of background noise and signal detection in climatic time series, *Clim. Change*, *33*, 409-445, 1996.

Paeth, H., A. Hense, R. Glowienka-Hense, R. Vose, and U. Cubasch, The North Atlantic Oscillation as an indicator for greenhouse-gas induced regional climate change, *Clim. Dyn.*, *15*, 953-960, 1999.

Post, E., and N. C. Stenseth, Climate change, plant phenology, and northern ungulates, *Ecology*, *80*, 1322-1339, 1999.

Rencher, A. C., and F. C. Pun, Inflation of R^2 in best subset regression, *Technometrics*, *22*, 49-53, 1980.

Rogers, J. C., The association between the North Atlantic Oscillation and the Southern Oscillation in the Northern Hemisphere, *Mon. Wea. Rev.*, *112*, 1999-2015, 1984.

Schmutz, C., J. Luterbacher, D. Gyalistras, E. Xoplaki, and H. Wanner, Can we trust proxy-based NAO index reconstructions?, *Geophy. Res. Lett.*, 27, 1135-1138, 2000.

Schlesinger, M. E., and N. Ramankutty, An oscillation in the global climate system of period 65-70 years, *Nature*, *367*, 723-726, 1994.

Schneider, U., and C.-D. Schonwiese, Some statistical characteristics of the El Niño/Southern Oscillation and North Atlantic Oscillation indices, *Atmosfera*, *2*, 167-180, 1989.

Shindell, D. T., R. L. Miller, G. A. Schmidt, and L. Pandolfo, Simulation of recent northern winter climate trends by greenhouse-gas forcing, *Nature*, *399*, 452-455, 1999.

Shindell, D. T., G. A. Schmidt, M. E. Mann, D. Rind, and A. Waple, Solor forcing of regional climate change during the Maunder Minimum, *Science*, *294*, 2149-2152, 2001.

Slonosky, V. C., P. D. Jones, and T. D. Davies, Atmospheric circulation and surface temperature in Europe from the 18th century to 1995, *Int. J. Climatol.*, *21*, 63-75, 2001.

Stahle, D. W., et al., Experimental dendroclimatic reconstruction of the Southern Oscillation, *Bull. Am. Met. Soc.*, *79*, 2137–2152, 1998.

Stockton, C. W., and M. F. Glueck, Long-term variability of the North Atlantic oscillation (NAO), *Proceedings of the American Meteorological Society, Tenth Symposium on Global Change Studies*, January 11-15, 1999, Dallas, TX, pp. 290-293, 1999.

Thompson, D. W. J., and J. M. Wallace, Annular modes in the extratropical circulation. Part 1: Month-to-month variability, *J. Climate*, *13*, 1000-1016, 2000.

Thompson, D. W. J., J. M. Wallace, and G. C. Hegerl, Annular modes in the extratropical circulation. Part II: Trends, *J. Climate*, *13*, 1018-1036, 2000.

Thompson, D. W. J., S. Lee, and M. P. Baldwin, Atmospheric processes governing the Northern Hemisphere Annular Mode/North Atlantic Oscillation, this volume.

Thomson, D. J., Spectrum estimation and harmonic analysis, *Proceedings of the IEEE, 70*, 1055-96, 1982.

Ulbrich, U., and M. Christoph, A shift of the NAO and increasing storm tract activity over Europe due to anthropogenic greenhouse gas forcing, *Clim. Dyn.*, *15*, 551-559, 1999.

Van Deusen, P. C., Evaluating time-dependent tree ring and climate relationships, *J. Envir. Qual.*, *19*, 481-488, 1990.

Visser, H., and Molenaar, J., Kalman filter analysis in dendroclimatology, *Biometrics*, *44*, 929-940, 1988.

Weiss, H., Social responses to abrupt climate changes in the Middle East, *Clim. Change*, in press, 2002.

E. R. Cook, Lamont-Doherty Earth Observatory, Tree-Ring Laboratory, P.O. Box 1000, Palisades, NY, 10964-8000
drdendro@ldeo.columbia.edu

Atmospheric Processes Governing the Northern Hemisphere Annular Mode/North Atlantic Oscillation

David W. J. Thompson

Colorado State University, Ft. Collins, Colorado

Sukyoung Lee

The Pennsylvania State University, University Park, Pennsylvania

Mark P. Baldwin

Northwest Research Associates, Bellevue, Washington

The North Atlantic Oscillation, referred to herein as the Northern Hemisphere annular mode (NAM), owes its existence entirely to atmospheric processes. In this chapter, we review the structure of the NAM in the atmospheric general circulation, discuss opposing perspectives regarding its physical identity, examine tropospheric processes thought to give-rise to NAM-like variability, and review the role of the stratosphere in driving variability in the NAM. The NAM is characterized by a deep, nearly barotropic structure, with zonal wind perturbations of opposing sign along ~55° and ~35° latitude. It has a pronounced zonally symmetric component, but exhibits largest variance in the North Atlantic sector. During the Northern Hemisphere (NH) winter, the NAM is strongly coupled to the circulation of the NH stratosphere. The NAM also affects tropical regions, where it perturbs the temperature and wind fields of both the tropical troposphere and stratosphere. The structure of the NAM is remarkably similar to the structure of the leading mode of variability in the Southern Hemisphere circulation. The processes that give rise to annular variability are discussed. In the troposphere, the NAM fluctuates on timescales of ~10 days and is associated with anomalous fluxes of zonal momentum of baroclinic waves across ~45°N. It is argued that the tropospheric component of the NAM exhibits largest variance in the Atlantic sector where the relatively weak thermally driven subtropical flow and the relatively warm lower boundary conditions at subpolar latitudes permit marked meridional excursions by baroclinic waves. In the stratosphere, fluctuations in the NAM evolve on timescales of several weeks. Evidence is presented that long-lived anomalies in the stratospheric NAM frequently precede similarly persistent anomalies in the tropospheric NAM. It is argued that variability in the lower stratospheric polar vortex yields a useful level of predictive skill for NH wintertime weather on both intraseasonal and seasonal timescales. The possible dynamics of these linkages are outlined. The recasting of the North Atlantic Oscillation as an expression of an annular mode has generated a debate over the physical identity of the mode in question. This debate attests to the absence of a unique theory for the existence of annular modes in the first place. Our current understanding of the fundamental processes to which the NAM owes its existence is discussed.

The North Atlantic Oscillation:
Climatic Significance and Environmental Impact
Geophysical Monograph 134

10.1029/134GM05

1. INTRODUCTION

The North Atlantic Oscillation (NAO) owes its existence, not to coupled ocean-atmosphere interactions, but to dynamics intrinsic to the extratropical atmosphere. Numerical simulations of the climate system have repeatedly demonstrated that atmospheric processes alone are sufficient to generate an NAO-like pattern of variability [*Hurrell et al.,* this volume]. Results of both theoretical and modeling studies suggest that midlatitude sea-surface temperature anomalies have only a weak impact on the NAO, at least on month-to-month and year-to-year timescales [*Kushnir et al.,* 2002; *Czaja et al.*, this volume].

In this chapter, we review ongoing research efforts aimed at improving our understanding of the atmospheric processes thought to give rise to NAO-like variability. In Section 2, we document the structure of the NAO in the atmospheric circulation. In Section 3, we review the debate over the physical interpretation of the NAO. Section 4 explores the tropospheric dynamics of the NAO in greater detail, with emphasis on two current research questions: 1) what tropospheric processes give rise to NAO-like variability? and 2) why is variability in the NAO largest over the North Atlantic sector? In Section 5 we outline our current understanding of troposphere/stratosphere coupling in the context of the NAO. Section 6 offers a summary of the chapter and discusses priorities for future research.

2. THE STRUCTURE OF THE NAO IN THE ATMOSPHERIC CIRCULATION

Since its discovery, the NAO has been widely viewed as a zonally asymmetric pattern restricted primarily to the North Atlantic sector [see *Stephenson et al.*, this volume, for a history of NAO research]. Its structure has been defined on the basis of one point correlation maps [*Wallace and Gutzler,* 1981] or rotated empirical orthogonal functions [*Barnston and Livezey,* 1987]. Its temporal variability has been characterized by differences in sea level pressure between stations located near Iceland and the Azores/Portugal [*Rogers*, 1984; *Hurrell*, 1995a; *Hurrell and van Loon*, 1997; *Jones et al.,* this volume]. Not surprisingly, variability in the NAO has often been interpreted as a reflection of coupled ocean-atmosphere processes [e.g., *Grotzner et al.*, 1998; *Rodwell et al.* 1999].

Recently, it has been proposed that the dynamics of the NAO are analogous to those that drive the leading mode of variability in the Southern Hemisphere (SH) extratropical circulation [*Thompson and Wallace,* 2000], a pattern characterized by a zonally-symmetric seesaw in geopotential height between the polar cap and the surrounding zonal ring along ~45°S [*Rogers and van Loon*, 1982; *Szeredi and Karoly*, 1987; *Yoden et al.*, 1987; *Kidson*, 1988a, b; *Karoly*, 1990; *Hartmann and Lo*, 1998; *Gong and Wang*, 1999] (Note that throughout this chapter, the term "mode" is used to describe a dominant pattern of variability. It does not necessarily refer to the normal modes of a physical system). The analogy between the leading mode of variability in the SH and the NAO suggests that the NAO reflects internal atmospheric dynamics that transcend the striking differences between the land-sea geometry of the two hemispheres. It has hence been suggested that the structure of the NAO is best captured, not by station based indices located over the North Atlantic, but by the leading empirical orthogonal function (EOF) of the Northern Hemisphere (NH) sea level pressure (SLP) anomaly field [*Thompson and Wallace*, 1998; 2000]. (The leading EOF is the state vector that explains the largest fraction of the total variance in the data. It is the eigenvector corresponding to the largest eigenvalue of the covariance matrix. Anomalies are defined as departures from climatology.) In contrast to the structure of the NAO that emerges from one-point correlation maps, the leading EOF of NH SLP has a strong zonally symmetric component [see also *Kutzbach*, 1970; *Trenberth and Paolino*, 1981; *Wallace and Gutzler*, 1981]. Its structure is characterized by a meridional dipole in SLP not only between centers of action located near Iceland and the Azores/Portugal, but by fluctuations in atmospheric mass between the entire Arctic basin and the surrounding zonal ring.

In this section, we examine the structure of the NAO in the atmospheric circulation, as defined on the basis of the leading EOF of SLP and its associated expansion coefficient (referred to as the principal component, or PC) time series (as noted later in this chapter, the analysis is not sensitive to the choice of SLP as a base level). Following *Thompson and Wallace* [2000], we present evidence of the similarity between the resulting structure and the leading mode of variability in the SH circulation, defined here as the leading EOF of the SH (90°S-20°S) 850-hPa geopotential height field. We use 850- hPa geopotential height in the SH to partially alleviate the ambiguities introduced by the reduction to sea-level over the Antarctic Plateau. In both hemispheres, the analysis is based on data from the NCEP/ NCAR Reanalysis [*Kalnay et al.,* 1996].

Through the rest of this chapter, we will refer to structures derived from the leading EOFs of the NH and SH geopotential height fields as the NH and SH annular modes (NAM and SAM), respectively, and the corresponding expansion coefficient time series of these patterns as the NAM and SAM indices (the NH annular mode is also known as the Arctic Oscillation, e.g., *Thompson and Wallace* 1998). As noted later in this chapter, the NAM and SAM indices can be found as a function of level by projecting geopotential height anomalies at a given level onto the structure of the corresponding leading EOF. The annular mode nomenclature is used for three reasons: 1) it suggests that the NAO reflects dynamics that would occur in the absence of the North Atlantic ocean; 2) it emphasizes the analogy between the two hemispheres; and 3) it acknowledges the fact that the polar center of action of the NAO has a high degree of zonal symmetry. In practice, the NAM-index time series used in this section is highly correlated with the leading PC time series calculated for SLP data over the North Atlantic half of the hemisphere (the leading PC of NH SLP is correlated with the leading PC of SLP over the North Atlantic half of the hemisphere at a level of r=0.95; *Deser*, 2000; *Hurrell et al.,* this volume). Hence, the results presented here are not strongly dependent on the specific definition of the NAO/NAM index. The implications of recasting the NAO as the NAM are discussed further in Section 3.

While the results presented in this section are based on monthly-mean data, it should be born in mind that the NAM and the SAM also exhibit variability on submonthly timescales: the *e*-folding timescale of the annular modes in the troposphere is ~10 days [*Hartmann and Lo*, 1998; *Feldstein*, 2000; *Lorenz and Hartmann*, 2001; 2002], and somewhat longer in the NH during winter. The intramonthly variability in the NAM does not impact the structures derived in this section, but it is further evidence that the annular modes would exist in the absence of coupled-ocean atmosphere interactions. The processes that give rise to the ~10 day timescale of the NAM are explored in Section 4.

2.1. Structure

Following *Thompson and Wallace* [2000], the structure of the NAM in the SLP field is found by regressing monthly mean SLP anomalies onto standardized values of the NAM index (positive values of the annular mode indices are defined as low SLP over the polar regions, and vice versa). As evidenced in Figure 1 (middle), the resulting pattern is characterized not only by a meridional seesaw in SLP over the North Atlantic sector, but by a hemispheric scale seesaw of atmospheric mass between polar latitudes and centers of action located over both the North Pacific and the North Atlantic. The structure of the NAM in lower tropospheric geopotential height bears a strong resemblance to that of the SAM, which is found by regressing lower tropospheric geopotential height in the SH onto the SAM-index time series (Figure 1, left). The patterns in the left and middle panels of Figure 1 are both predominantly zonally symmetric (the SH pattern somewhat more so), and both exhibit similar amplitudes and meridional scales. As noted earlier, the pattern shown in the middle panel of Figure 1 is virtually identical to the leading EOF of the Euro-Atlantic sector

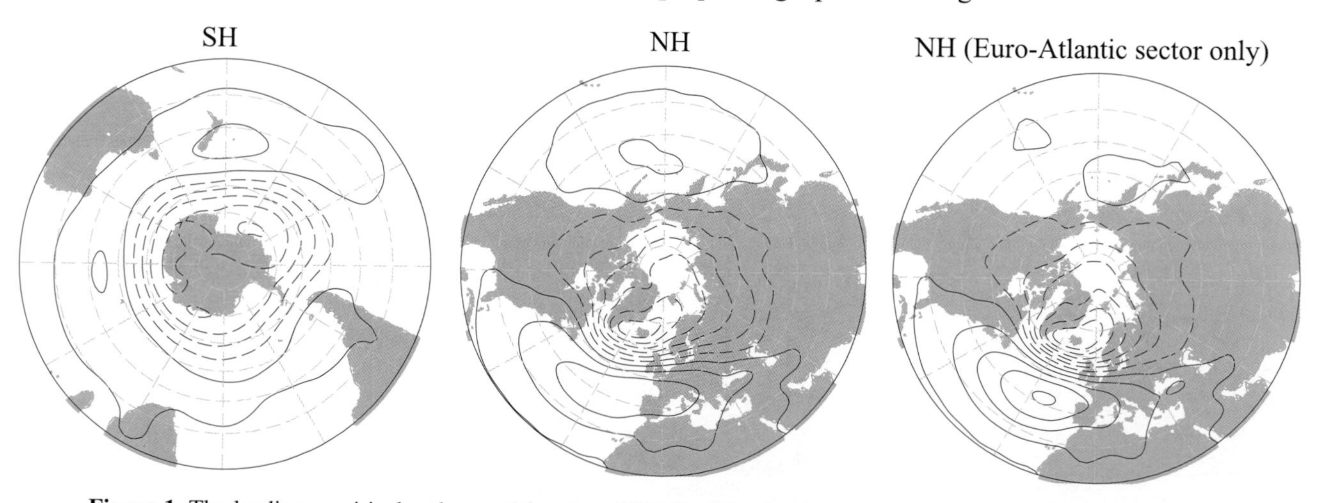

Figure 1. The leading empirical orthogonal function (EOF 1) of the Southern Hemisphere (SH; 90°S-20°S) monthly-mean 850-hPa height field (*left panel*); the Northern Hemisphere (NH; 20°N-90°N) monthly-mean SLP field (*middle panel*); and the monthly-mean SLP field in the Euro-Atlantic sector (20°N-90°N, 60°W- 30°E) (*right panel*). The pattern in the right panel has been extended to include the entire hemisphere by regressing the monthly-mean SLP field upon the corresponding principal component time series. SLP is displayed in units of geopotential height at 1000-hPa. Contour interval 10 m (-5, 5, 15...); negative contours are dashed. Results are based on monthly-mean fields of the NCEP/NCAR Reanalyses (January-March for the NH; all months for the SH) for the period 1958-99. Figures duplicated from *Thompson and Wallace* [2000] and *Wallace and Thompson* [2002].

(Figure 1, right panel; the EOF of the Euro-Atlantic sector is extended to the hemispheric domain by regressing hemispheric SLP anomalies onto the respective PC time series). Both patterns have a strong zonally symmetric component, but the hemispheric EOF has larger amplitude over the Pacific sector.

The similarity between the NAM and SAM is even more apparent when one compares their vertical structures. The top panels in Figure 2 show the zonal-mean zonal wind and mean meridional circulations regressed onto "active" season segments of the NAM and SAM indices for the domains extending from pole to pole and from 1000-hPa to 30-hPa. The active seasons are defined as times of year when troposphere/stratosphere coupling is most vigorous, and correspond to periods when the zonal flow in the lower stratosphere is disturbed by waves dispersing upwards from the troposphere. Theory predicts that these interactions should occur during seasons when the zonal flow in the lower stratosphere is westerly, but less than a threshold value [*Charney and Drazin*, 1961]. Observations reveal that these interactions occur throughout the winter in the NH, but only during late spring (November) in the SH [*Thompson and Wallace*, 2000].

During the active seasons, both the NAM and the SAM are characterized by equivalent barotropic, meridional dipoles in the extratropical circulation of their respective hemispheres. The annular modes are also marked by weak

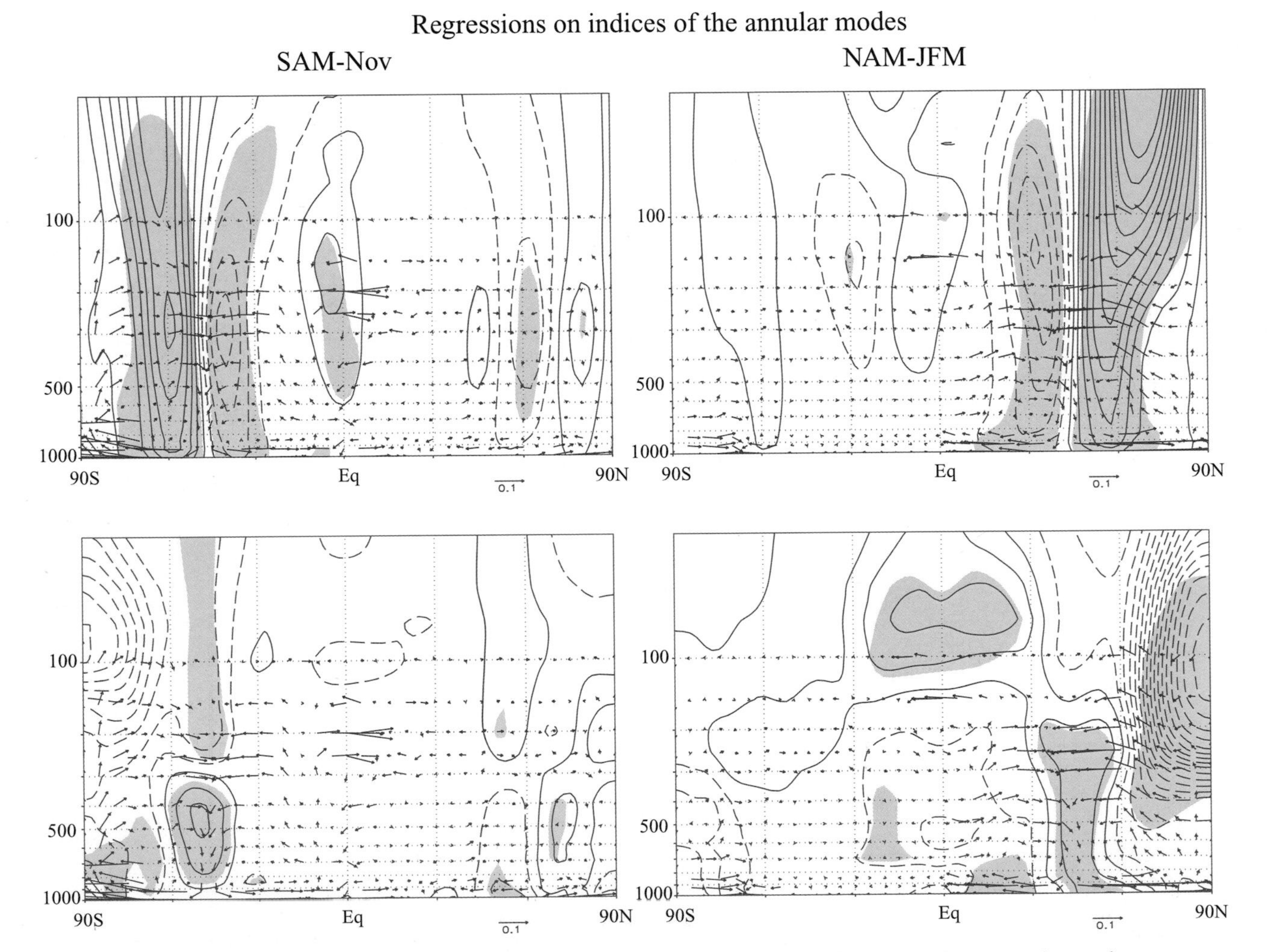

Figure 2. Top: Zonal-mean zonal flow (contours) and mean meridional circulation (vectors) regressed onto the standardized monthly time series of the annular modes for the "active seasons" of stratosphere/troposphere coupling based on monthly data, January-March (NH) and November (SH), from 1979-1999. Bottom: As in the top panel but contours are for zonal-mean temperature. Contour intervals are 0.5 m s^{-1} (-0.75, -0.25, 0.25) for zonal wind and 0.2 K (-0.3, -0.1, 0.1) for temperature. Vectors are in units of m s^{-1} for the meridional wind component; cm s^{-1} for the vertical component (scale at bottom). Shading indicates correlations of $r > 0.4$. The top of the diagram corresponds to 50-hPa. Figure adapted from *Thompson and Wallace* [2000].

zonal wind anomalies in tropical latitudes (the anomalies are somewhat larger in association with the SAM), which are consistent with global profiles of SLP observed in association with the annular modes [*Baldwin*, 2001]. The anomalies in the poleward center amplify with height from the surface upward into the lower stratosphere, the axis tilting slightly poleward and the meridional scale broadening from the troposphere to the lower stratosphere [see also *Black*, 2002]. The zonal wind anomalies located at ~35° N extend equatorward into the subtropics near the surface (particularly in the NH), but exhibit a narrow maximum at the 200-hPa level. A more detailed treatment of the linkage between the NAM and the tropical zonal wind anomalies is presented in the next subsection.

The patterns derived by regressing the mean meridional circulation onto the NAM and SAM indices are shown by the vectors in Figure 2. Both the NAM and the SAM are marked by paired subtropical and high latitude circulation cells that rotate in the opposite sense. The high index polarity of the annular modes is accompanied by anomalous rising motion over subpolar latitudes and subsidence between ~40 and 50° latitude. The high latitude cells extend into the lower stratosphere, while the subtropical cells are confined to the troposphere. As noted in *Thompson and Wallace*, [2000], the Coriolis force acting on the upper branch of the circulation cells acts to weaken the corresponding zonal wind anomalies. Hence, consistent with findings reported in *Yoden et al.* [1987], *Shiotani* [1990], *Karoly* [1990], *Kidson and Sinclair* [1995], *Hartmann and Lo* [1998], and *Limpasuvan and Hartmann* [1999; 2000], the wind anomalies associated with the high index polarity of the annular modes must be maintained by anomalous poleward eddy fluxes of westerly momentum in the upper troposphere centered near 45° latitude. The role of eddy fluxes in driving the NAM is discussed further in Section 4. The mean meridional circulation anomalies in the tropics are noisy and difficult to interpret.

The lower panels in Figure 2 show zonal-mean temperature regressed onto the same indices used as a basis for the analysis in the top panels. The positive polarity of the annular modes is marked by anomalously low temperatures over the polar cap that amplify with height, and by anomalously high temperatures in a band centered near 45°. That the temperature anomalies are adiabatically driven is suggested by the fact that cool anomalies tend to coincide with anomalous rising motion, and vice versa. Note that rising motion in regions of cooling is consistent with mechanical driving by anomalous eddy fluxes of zonal momentum. The shallow warm anomalies near the surface between 55°N and 75°N are consistent with anomalous temperature advection over the NH landmasses during the positive polarity of the NH annular mode [*Hurrell*, 1995a; 1996; *Thompson and Wallace*, 1998; 2000; *Xie et al.*, 1999].

The positive polarity of the NAM is also characterized by positive temperature anomalies at the tropical tropopause. Since water vapor is sparse in the lower stratosphere, the out-of-phase relationship between lower stratospheric temperature anomalies at tropical and polar latitudes presumably reflects adiabatic temperature changes induced by a weakening of the mean-meridional circulation of the lower stratosphere during the high index polarity of the NAM. The SAM has a much weaker signature in tropical tropopause temperature, which is consistent with the fact that stratosphere/troposphere coupling occurs during a much shorter season in the SH. As discussed in the next section, the NAM is also associated with weak temperature anomalies throughout the tropical troposphere.

2.2. *Global-scale Features of the NAM*

Several studies have noted a linkage between the NAM and variability in the tropical Atlantic region. For example: *Meehl and van Loon* [1979] have shown that the NAO is significantly correlated with the strength of the trade-winds over the North Atlantic and with the position of the ITCZ over Africa; *Lamb and Peppler* [1987] have noted that the NAO has a significant impact on Moroccan rainfall; *Moulin et al.* [1997] have demonstrated that the NAO impacts dust transport from the Sahara desert; *Malmgren et al.* [1998] have shown that the NAO is reflected in Puerto Rican rainfall; and *McHugh and Rogers* [2001] have suggested that the NAO impacts eastern African rainfall. *Baldwin* [2001] has shown that fluctuations in the NAM are characterized by significant interhemispheric exchanges of atmospheric mass. Here we offer evidence that the NAM has a distinct signature in the temperature and zonal wind field of the tropics that transcends the tropical Atlantic sector.

Plate 1 shows tropospheric temperature anomalies from the Microwave Sounding Unit Channel 2LT data [*Spencer et al.*, 1990; *Spencer and Christy*, 1992; the weighting function for MSU2LT temperature data is centered near 600 hPa] and lower stratospheric temperature anomalies from the MSU4 data [*Spencer and Christy*, 1993; the weighting function for MSU4 data is centered near 70 hPa] regressed onto JFM monthly values of the NAM index from 1979-1999. Consistent with results presented in the previous section, the high index polarity of the NAM (low SLP over the pole) is characterized by pronounced cooling over the NH polar cap that amplifies with height from the troposphere to the lower stratosphere, and warming in the midlatitude troposphere. However, the signature of the NAM is clearly not restricted to the extratropics of the NH. The regressions also

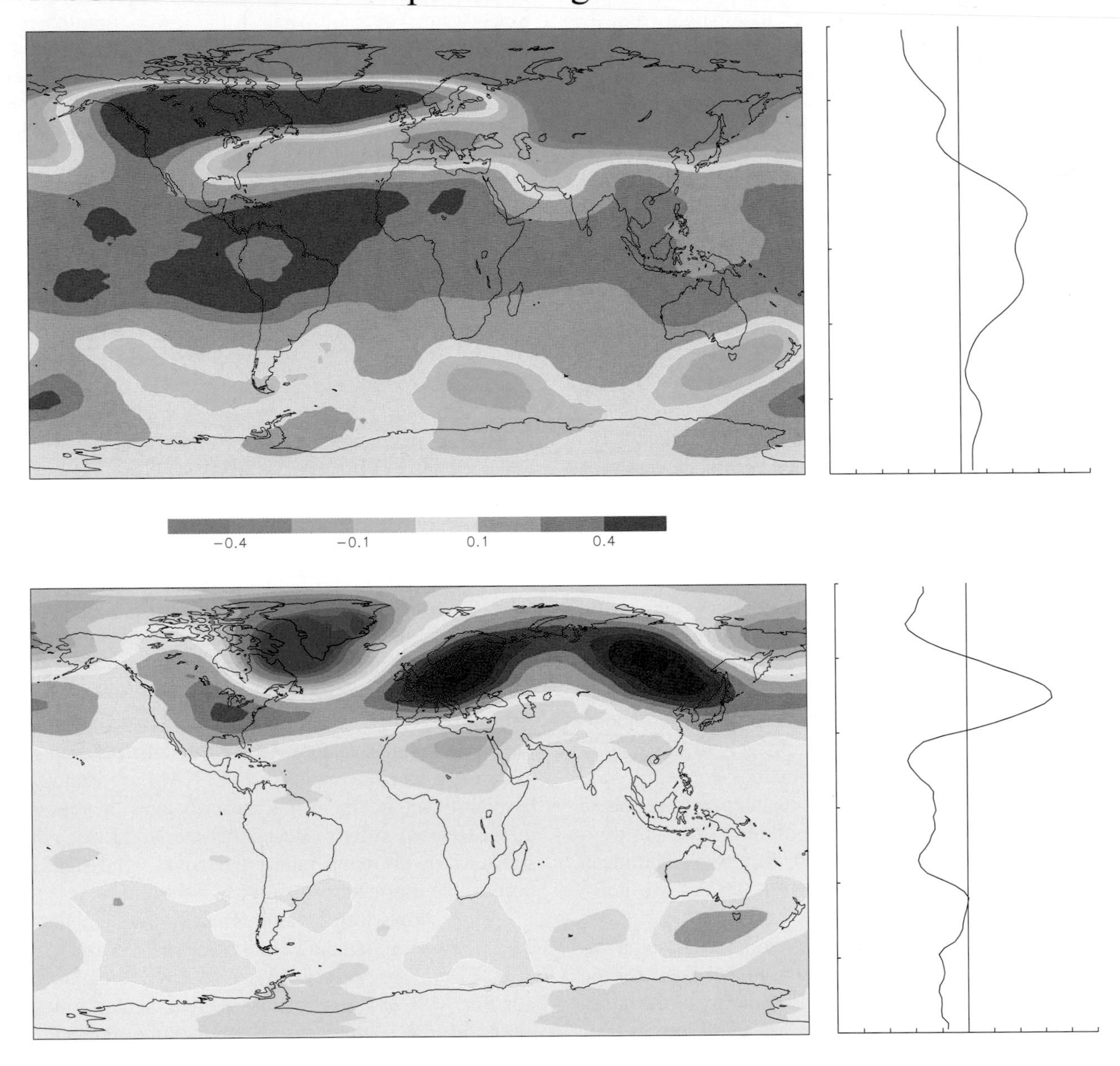

Plate 1. Left panels: MSU4 (top) and MSU2LT (bottom) temperature anomalies (K) regressed upon January-March (JFM) standardized monthly values of the NAM-index. Right panels: The correlation coefficients (*r*) between JFM values of the NAM index and zonal mean MSU4 (top) and MSU2LT (bottom) temperature anomalies. Tickmarks are at $r = 0.2$. Figure from research in progress in collaboration with D. J. Lorenz (University of Washington).

reveal a distinct pattern of temperature anomalies throughout the tropics, with warm anomalies found at the tropical tropopause region, and cool anomalies found throughout much of the tropical troposphere. Tropical-mean (20°S-20°N) tropopause temperatures are 0.36 K warmer and tropical-mean tropospheric temperatures are 0.12 K cooler per one standard deviation increase in the NAM index (the correlation coefficient between the NAM index and tropical-mean tropopause temperature is r=+0.49, and the correlation between the NAM index and tropical mean tropospheric temperature is r=-0.32. Both correlations are significant at the 95% level based on the t-statistic). In both the troposphere and stratosphere, the largest temperature anomalies in the tropics are found, not over the equator, but over the subtropical regions of both hemispheres, as clearly evidenced in the correlations between zonal mean temperature and the annular mode indices (Plate 1, right panels).

The cooling of the tropical troposphere observed in the MSU2LT data is more pronounced than that observed in surface temperature data (not shown), which hints that the mid-tropospheric anomalies are dynamically induced through eddy-driven vertical motions. For example, tropical mean surface temperature decreases by only 0.06 K per standard deviation increase in the NAM index [based on data described in *Jones*, 1994] in contrast to the corresponding 0.12 K decrease observed in the MSU2LT data. To what extent the NAM has contributed to the widely publicized discrepancy between trends in temperatures at the surface and in the free troposphere remains to be assessed [see the National Research Council Report on Reconciling Temperature Observations for an overview of the discrepancies between recent tropospheric and surface temperature trends].

Regressions based on daily data reveal that the linkage between the NAM and the circulation of the tropics is most pronounced when the tropical circulation lags variability in the NAM by several weeks (the daily NAM index used in the regressions was generated by projecting daily-mean fields of SLP from the NCEP/NCAR Reanalysis onto the structure in the middle panel of Figure 1). Figure 3 shows regression coefficients between the daily-mean zonal-mean zonal wind anomalies at 200-hPa and daily values of the NAM index for lags ranging from -35 to +35 days, where positive lags indicate the zonal flow lags the NAM index, and vice versa. The zonal wind anomalies in the tropics and in the SH are weak in the simultaneous regressions (i.e., at lag 0), but they exhibit a marked structure roughly ~2-3 weeks following the development of the largest wind anomalies in the extratropical NH. Hence, the NAM does not project strongly onto the tropics when the daily-mean zonal-mean circulation is regressed on contemporaneous daily values of the NAM index (Figure 4, top). However, when the zonal-mean circulation in Figure 4 is regressed onto daily values of the NAM index lagged by two weeks (bottom), the resulting patterns exhibit pronounced features in both the tropical troposphere and subtropical SH. The zonal wind anomalies in the bottom panel of Figure 4 reflect a banded structure that extends all the way from the polar regions of the NH deep into the subtropics of the SH. The distinct off-equatorial cooling maxima evident in the monthly regressions shown in Plate 1 are also clearly evident in these lagged daily regressions.

As discussed earlier, the warming of the tropical tropopause during the high index phase of the NAM is consistent with a weakening of the wave-driven Brewer-Dobson circulation in the lower stratosphere [*Thompson and Wallace*, 2000]. The processes that drive the tropical tropospheric features evident in Plate 1 and Figs. 3-4 remain to be determined, but the lag between the extratropics and tropics suggest that they may reflect dynamics associated with waves originating in the extratropical NH.

3. THE NAO AS AN ANNULAR MODE

As noted at the beginning of the previous section and in *Stephenson et al.* [this volume], the NAM has been widely viewed as a fundamentally zonally asymmetric structure, i.e., a pattern that owes its existence to regional sources of heat and momentum. Month-to-month variability in the NAM has been interpreted as a reflection of dynamics unique to the North Atlantic stormtrack. Decadal variability

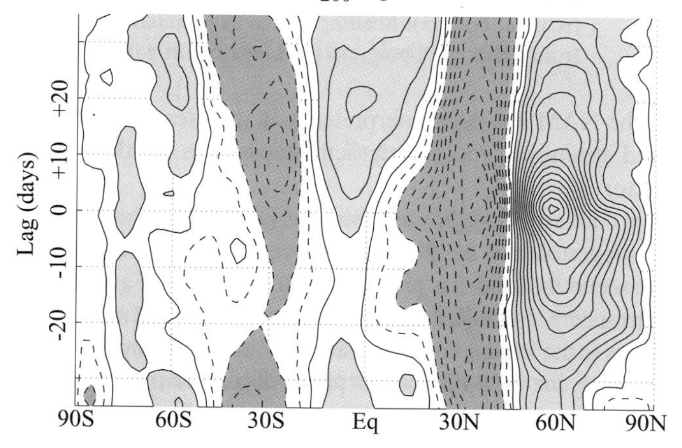

Figure 3. Zonal-mean zonal flow at 200-hPa regressed onto the standardized daily time series of the NAM for lags as indicated. Results based on JFM data from 1979-1999. Positive lags are for the zonal flow lagging the NAM index, and vice versa. Contour intervals are 0.2 m s^{-1} (-0.3, -0.1, 0.1). Values exceeding +/-0.2 m s^{-1} are shaded. Figure from research in progress in collaboration with D. J. Lorenz (University of Washington).

agating along the gradient are trapped meridionally [*Swanson et al.*, 1997]. As such, the pronounced PV gradient of the subtropical jet acts to hold eddies that grow along the jet, limiting their movement in the meridional direction. Hence, it follows that strong thermally driven subtropical jets, although turbulent, are characterized by relatively small meridional meanders in the zonal flow.

In contrast to thermally driven jets, eddy-driven jets are driven entirely by eddy fluxes of momentum. That eddies are capable of maintaining a jet in the absence of thermal forcing is evidenced in Eq. 1, and has been extensively documented in the literature. For example, *Panetta and Held* [1988], *Panetta* [1993], and *Lee* [1997] examined jets that spontaneously organize in a baroclinically unstable flow. In all three studies, the models neither support a subtropical jet nor force a prescribed jet. Instead, the model flow is driven by a broad zone of uniform baroclinicity that is maintained against mixing by the eddies either by a fixed meridional temperature gradient of the basic flow [*Panetta and Held*, 1988; *Panetta*, 1993] or by relaxation of the model temperature towards radiative equilibrium [*Lee*, 1997]. Hence, the jets in these models are driven entirely by eddy momentum flux convergences: when the flow is perturbed, baroclinic waves spontaneously grow and drive a westerly jet through the associated meridional convergence of the westerly eddy momentum flux.

Eddy-driven jets are generally accompanied by marked jet meanders because their associated meridional PV gradients are not externally imposed: eddy-driven jets can be interpreted as the signature of decaying eddies. However, the extent to which jet meanders dominate the variability of eddy-driven jets also depends on the meridional scale of the midlatitude eddies and their associated baroclinic zone. Since the role of the baroclinic eddies (i.e., the synoptic-scale waves in the atmosphere) is to stabilize the flow, jet pulsation and jet meanders can be viewed as a competition between the removal of the baroclinicity by the eddy fluxes and the restoration of the baroclinicity by radiative relaxation. Results from an idealized two-layer beta-plane channel model suggest that jet meanders occur when the meridional size of the eddies is insufficiently large to fully remove the negative lower layer PV gradient at any given time. In this case, the eddies and their attendant momentum fluxes meander towards regions where the lower layer PV gradient is large and negative in their perpetual attempt to stabilize the flow [*Lee and Feldstein*, 1996]. The same set of calculations suggest that jet pulsation occurs when the baroclinic zone is comparable to the meridional scale of the eddies. Except for very limited values of the baroclinic zone width, the most dominant form of variability of the eddy-driven jet in this case is found to be the north-south meander of the zonal flow about the latitude corresponding to its climatological mean [*Lee and Feldstein*, 1996].

The above discussion suggests that the variability of eddy-driven jets is generally larger than that of subtropical jets. It also suggests that the variability of eddy-driven jets is largest when the size of the baroclinic zone exceeds the size of the eddies, hence permitting meanders in the extratropical zonal flow. Subsequently, to the extent that the NAM is driven by variability in eddy momentum fluxes, one expects the amplitude of the NAM pattern to be largest in regions where the eddy-driven jet is most prominent and the subtropical jet is weakest, and vice versa. In the following section, we draw on this argument to provide an explanation for the zonally varying structure of the NAM.

4.2. Why is Variability in the NAM Most Pronounced Over the North Atlantic Sector?

Idealized model experiments suggest that, for a given sector of the Northern Hemisphere, the strength and variability of the local eddy-driven jet is intrinsically linked to the strength and latitude of the corresponding subtropical jet [*Lee and Kim*, 2002]. Lee and Kim's results suggest that when the subtropical jet is sufficiently strong, it tends to organize a large fraction of the baroclinic eddies, hence inhibiting the growth of eddy-driven flow at higher latitudes. Conversely, when the subtropical jet is weak, the baroclinic eddies tend to organize themselves in the poleward baroclinic zone, hence driving a purely eddy-driven jet.

While the results of such idealized model experiments should be interpreted with caution, the findings in *Lee and Kim* [2002] are consistent with several key features of the climatological flow. Plate 3 shows the climatological December-March mean 300-mb zonal flow (top) and the divergent component of the 300-hPa meridional wind (bottom) averaged over the 1958-97 period, and Figure 9 shows the zonal wind and high-pass (periods less than 10 days) eddy momentum flux convergence (i.e., analogous to the first term on the rhs of Eq. 1) averaged over longitude bands corresponding to the Atlantic, Asian, and Pacific sectors. The climatology of the Atlantic sector is clearly dominated by two jets (Plate 3 (top); Figure 9a). The jet along 20°N corresponds to the thermally driven subtropical jet: it is driven by a thermally direct meridional overturning cell (as revealed by the poleward flow on its equatorward side (Plate 3 (bottom)), and is largely restricted to the upper troposphere (Figure 9a). In contrast, the jet along 55°N corresponds to the eddy-driven jet: it is driven by eddy momentum flux convergence (Figure 9a) and extends throughout the depth of the troposphere. (That eddy driven jets extend throughout the depth of the troposphere is illustrated in Eq. 1. Since the lhs is zero in the steady-state and the second term on the rhs is zero when integrated vertically, it follows

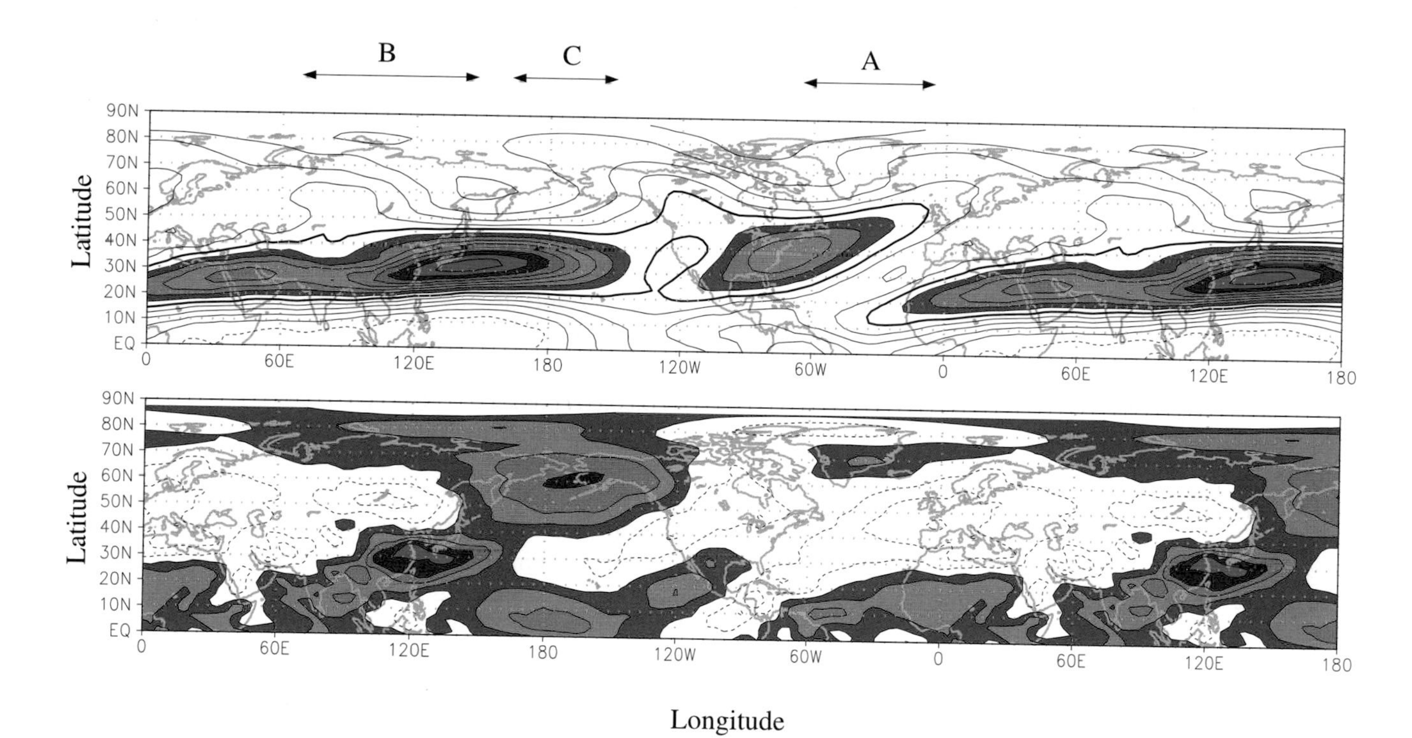

Plate 3. The NH 300-mb (a) zonal and (b) divergent component of the meridional wind averaged over December-January-February, 1958-1997. Contour interval is 5 m s^{-1} for (a) and 2 m s^{-1} for (b). Shading is above 20 m s^{-1} for (a) and 0 m s^{-1} for (b).

from 1000 to 10-hPa (in practice, correlations between the "multi-level NAM index" and zonal-mean wind and temperature yield structures virtually identical to those presented in Section 2). Figure 10 illustrates the lag correlations between the 90-day low-pass NAM index time series at 10 hPa on January 1 (the key date for the calculation) with the NAM index at all other levels during November–March. The choice of January 1 is not critical; similar results are obtained throughout the winter. The dominant feature in Figure 10 is the downward propagation of the signature of the NAM through the lower stratosphere into the troposphere, with lag correlations exceeding 0.65 at 1000 hPa –3 weeks later. The thick line illustrates the peak correlation at each level. The lag correlations in Figure 10 should not be interpreted as a precise downward propagation speed; rather, they indicate an average tendency for downward propagation on a timescale of a few weeks.

The results in Figure 10 suggest that at low frequencies, the annular mode is strongly coupled in the vertical and that the phase of the patterns tends to move downward with time, often reaching Earth's surface. *Baldwin and Dunkerton* [2001] extended the analysis to 26 pressure levels from 1000 to 0.316 hPa. They defined the annular mode separately at each level as the first EOF of 90-day low-pass filtered November–April geopotential anomalies north of 20°N. In this case, time series for each level were found by projecting daily geopotential anomalies onto each level's EOF. In practice, the resulting time series are virtually identical to those used in *Baldwin and Dunkerton* [1999]. In the stratosphere, the local NAM time series are a measure of the strength of the polar vortex, similar to the zonal-mean zonal wind at 60°N; at the surface, they are identical to those used in Section 2 of this Chapter.

Plate 2 illustrates the time-height development of the NAM at daily resolution during the northern winter of 1998–1999. In the stratosphere, the time scale is relatively long: the polar vortex is warm and weak (indicated by the red shading) during middle December and late February, and cold and strong (indicated by the blue shading) through all of January and early February. Consistent with the discussion in Section 5.1, the largest anomalies in the stratosphere originate above 1-hPa (~50km), and descend through the stratosphere over a period of ~1-2 weeks. As stated in Section 4, the relatively high frequency variations in the NAM in the troposphere during 1998-1999 appear for the most part unrelated to the stratosphere. However, most winters over the past several decades (1958–1999) have descending positive and negative anomalies that appear as similarly signed anomalies in the troposphere [*Baldwin and Dunkerton*, 1999]. In general, only the strongest anomalies of either sign appear to connect to the surface, while weaker anomalies typically remain within the stratosphere. In some cases, tropospheric anomalies appear to precede stratospheric anomalies.

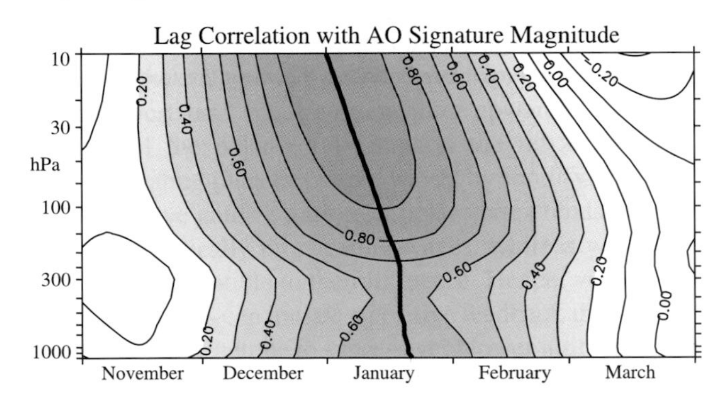

Figure 10. Correlations between the 90-day low-pass annular mode values at 10 hPa on January 1 with the annular mode values at all levels during November-March. From *Baldwin and Dunkerton* [1999].

Following *Baldwin and Dunkerton* [2001], the average behavior of extreme events in the stratosphere can be seen by forming composites based on large negative and positive anomalies in the NAM index at 10-hPa. Since the NAM index at 10-hPa is highly correlated with the corresponding zonal-mean zonal wind along 60°N (r=0.95), it follows that large positive values in the NAM index at 10-hPa correspond to a strong, well-organized vortex, while large negative values correspond to a weak, disorganized vortex (i.e., the most extreme negative values correspond to major stratospheric warmings; *Gillett et al.*, 2001). Weak and strong vortex "events" are defined here as dates on which the 10-hPa annular mode index dips below -3.0 or rises above +1.5 thresholds, respectively. The thresholds yield 18 weak vortex events and 30 strong vortex events over the period 1958-99, with the highest concentration of events occurring during December– February.

Composites based on the extreme events defined above reveal that, on average, large circulation anomalies at 10 hPa descend to the lower stratosphere over a period of ~10 days, where they tend to persist for ~60 days [Plate 4]. That the anomalies in the lower stratosphere persist longer than those at 10-hPa presumably reflects the longer radiative time scale in the lower stratosphere [*Shine*, 1987]. The composites also reveal that the persistent anomalies do not stop at the tropopause level, but that they are reflected as a shift in the mean of the relatively high frequency variability that occurs at tropospheric levels. Hence, the composites in Plate 4 reveal that, on average, large amplitude anomalies at stratospheric levels are generally followed by a bias in the mean of the tropospheric variability that persists for ~60 days.

The average surface circulation anomalies during the 60-day period following the onset of weak and strong vortex events bear a remarkable resemblance to the signature of the NAM, with the largest effect on pressure gradients in the North Atlantic and Northern Europe (Figure 11). Hence, the 60-day periods following the onset of opposite signed anomalies in the stratosphere are marked by shifts in the mean value of the NAM-index (the projection of the pattern in the middle panel of Figure 1) and by definition, the mean value of the NAO-index [the daily NAO-index shown in Plate 5 is defined in *Baldwin and Dunkerton,* 2001]. The probability density functions (PDFs) of these indices for the contrasting weak and strong vortex conditions are compared in Plate 5. Large amplitude anomalies in the stratosphere are followed not only by shifts in the mean of the PDFs, but also by substantial changes in the shapes of the PDFs. Values of the NAM (or NAO) indices greater than 1.0 are 3-4 times as likely following strong vortex conditions than they are following weak vortex conditions. Similarly, index values less than –1.0 are 3-4 times more likely following weak vortex conditions than they are following strong vortex conditions. Since large swings in the NAM (and NAO) indices are associated with significant changes in the probabilities of weather extremes such as cold air outbreaks, snow, and high winds across Europe, Asia, and North America [*Thompson and Wallace*, 2001], it follows that the observed circulation changes following weak and strong vortex conditions in the stratosphere have substantial implications for prediction of Northern Hemisphere wintertime weather up to two months in advance. In the next section, we assess the direct linkage between the stratospheric polar vortex and the climate impacts of the NAM in greater detail.

5.3. Connection Between the Stratospheric Polar Vortex and Surface Temperature Anomalies

Following *Thompson et al.* [2002], in this section we present results that examine the direct linkage between stratospheric circulation anomalies and surface weather. The analysis technique is very similar to that used in *Baldwin and Dunkerton* [2001], but in this case variability in the lower stratospheric polar vortex is defined as the time series of 10-hPa geopotential height anomalies averaged 60°–90° N, and the onset dates of weak and strong vortex conditions are defined as da ys when this index dips below -1 standard deviations and exceeds +1 standard deviations, respectively. Note that the onset dates of weak and strong vortex conditions can be assessed in real-time; they do not depend on how long the stratospheric anomalies persist.

The differences between surface temperature anomalies averaged over the 60-day periods following the onset of weak and strong stratospheric polar vortex conditions are shown in Plate 6 [left panel; see *Thompson et al.,* 2002 for details of the analysis]. The pattern of surface temperature anomalies in the left panel of Plate 6 is largely consistent with the pattern of surface temperature anomalies associated with the surface signature of the NAM [*Hurrell*, 1995a;

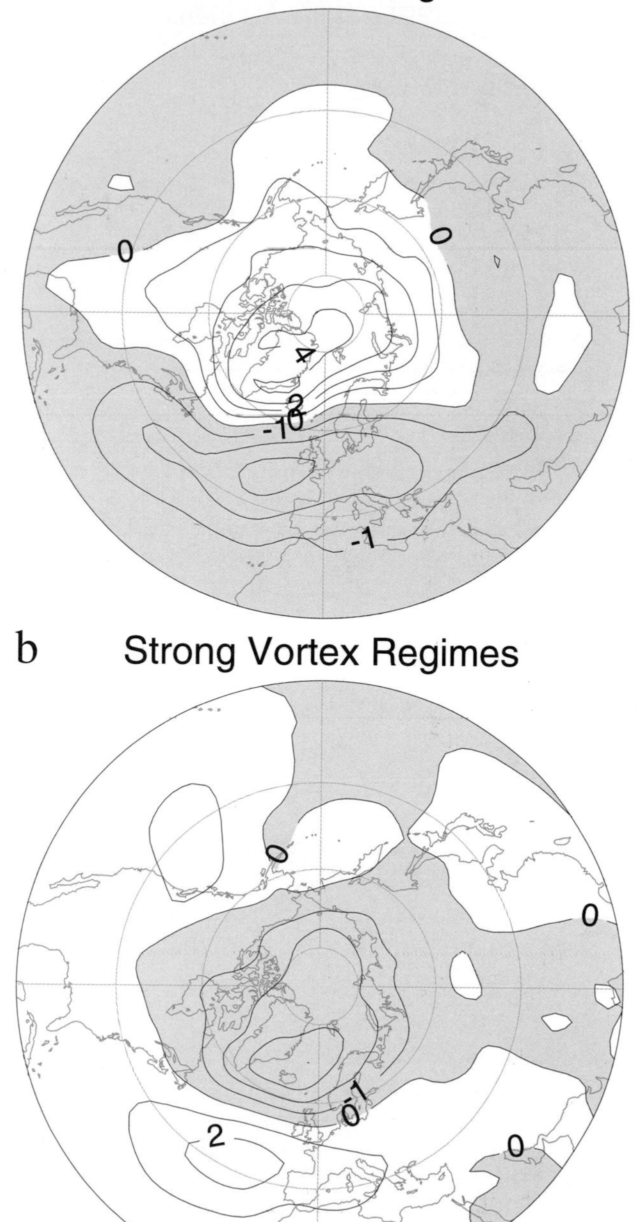

Figure 11. Average sea level pressure anomalies (hPa) for (a) the 1080 days during weak vortex conditions and (b) the 1800 days during strong vortex conditions. From *Baldwin and Dunkerton* [2002].

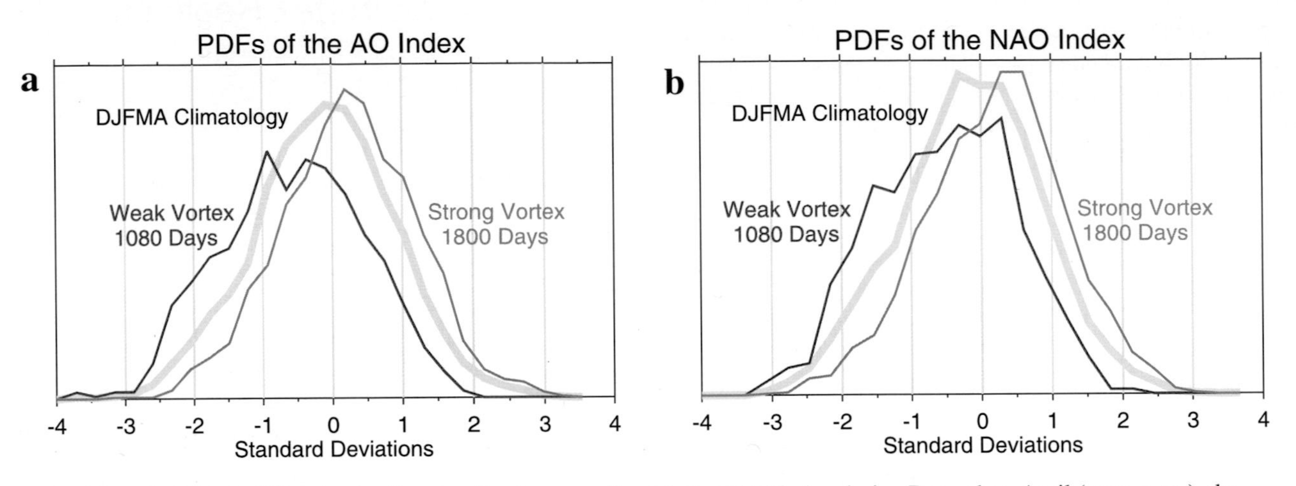

Plate 5. (a) Probability density function for the normalized daily NAM index during December–April (gray curve), the 1080 days during weak vortex conditions (red curve), and the 1800 days during strong vortex conditions (blue curve). (b) As in (a), but for the index of the NAM based solely on North Atlantic data. From *Baldwin and Dunkerton* [2002].

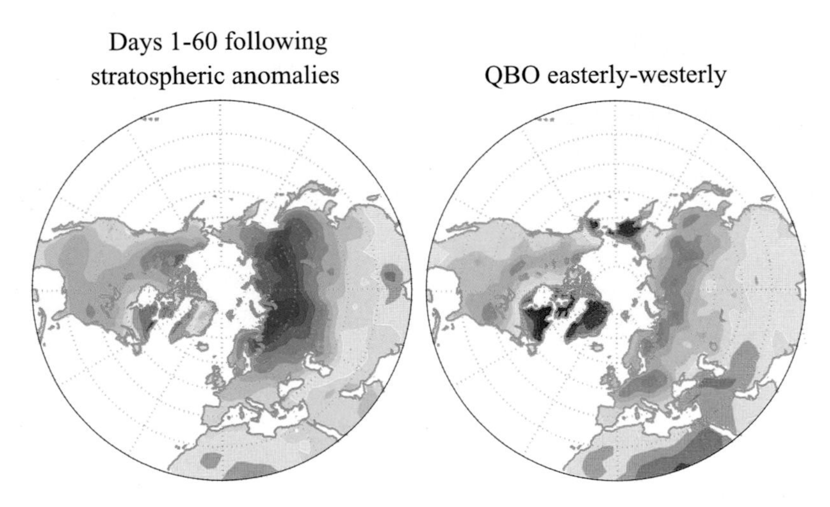

Plate 6. The difference in daily mean surface temperature anomalies between the 60-day interval following the onset of weak and strong vortex conditions at 10-hPa (left) and between Januarys when the QBO is easterly and westerly (right panel). The samples used in the analysis are documented in *Thompson et al.* [2002]. Contour levels are at 0.5 C. From *Thompson et al.* [2002].

Thompson and Wallace, 2000]: most of the mid-high latitude land masses tend to be anomalously cold following the onset of weak stratospheric polar vortex conditions while extreme eastern Canada and North Africa are anomalously warm (the pattern would be entirely consistent with the surface signature of the NAM if eastern Siberia and Alaska were of the opposite sign). Densely populated regions such as eastern North America, northern Europe, and eastern Asia are ~1-2 K colder following the onset of weak vortex conditions. As noted in *Thompson et al.* [2002], the surface temperature anomalies shown in the left panel of Plate 6 are roughly comparable to those associated with the contrasting phases of the El-Niño/Southern Oscillation cycle.

The 60-day interval following the onset of weak vortex conditions is also characterized by an enhanced frequency of occurrence of extreme low temperatures in most large cities that lie in NH midlatitudes [*Thompson et al.*, 2002]. Extreme cold events are roughly two times more likely during the 60-day period following weak vortex conditions than they are during the 60-day period following strong vortex conditions throughout much of North America to the east of the Rocky Mountains, northern Europe and Asia [*Thompson et al.*, 2002].

5.4. Connection Between the Quasi-Biennial Oscillation and the NAM

The equatorial quasi-biennial oscillation (QBO) is a downward propagating quasi-periodic reversal in the direction of the zonal flow in the equatorial stratosphere with a mean period of ~27 months [*Reed et al.*, 1961; *Baldwin et al.*, 2001]. While the QBO is a tropical stratospheric phenomenon, it also impacts the strength and stability of the NH wintertime stratospheric polar vortex [*Holton and Tan*, 1980]: the easterly phase of the QBO favors a weaker stratospheric polar vortex, and vice versa. That the impact of the QBO on the extratropical circulation may extend to the surface is suggested by the fact that anomalies in the NH polar vortex frequently precede similarly signed anomalies at Earth's surface, as demonstrated in the previous section. Hence, the easterly phase of the QBO should not only favor a weaker stratospheric polar vortex, but through the linkages observed in Plates 3-4, the low index polarity of the NAM at the surface as well.

That the QBO does in fact impact the NAM is evidenced in SLP composites calculated for the opposing phases of the QBO [*Holton and Tan*, 1980; *Baldwin et al.*, 2001]. It is also evidenced by the recent finding that time series of the NAM and the QBO exhibit statistically significant coherence on ~27 month timescales [*Coughlin and Tung*, 2001]. Hence, the dynamical coupling between the troposphere and stratosphere not only has implications for the predictability of NH wintertime weather on month-to-month timescales, but on seasonal timescales as well.

The right panel in Plate 6 shows the differences in daily mean temperature between Januarys when the QBO is easterly and westerly [see *Thompson et al.*, 2002 for details of the analysis]. The amplitudes of the SAT anomalies are weaker than those obtained for the 60-day period following stratospheric anomalies [Plate 6 left panel], but the structure of the anomalies clearly bears a striking resemblance to the signature of the NAM in surface temperature. During the easterly phase of the QBO, midwinter temperatures are lower over much of North America and northern Eurasia, and most large cities that lie in NH midlatitudes experience a greater frequency of occurrence of extreme cold events [*Thompson et al.*, 2002]. Since the phase of the QBO can be predicted several months in advance, the results in Plate 6 strongly suggest that the contrasting phases of the QBO provide a useful level of predictive skill for NH wintertime weather on seasonal timescales.

5.5. Stratospheric Contribution to Predictability of the NAM

The results reviewed in this section suggest that stratospheric processes yield a useful level of predictability for the climate impacts of the NAM on timescales longer than the ~10 day limit of deterministic weather prediction. This predictability derives from three key observations: 1) NAM anomalies tend to propagate downward, as evidenced in the ~10 day time lag between stratospheric and tropospheric anomalies; 2) the timescale of the attendant surface anomalies is ~60 days, considerably longer than the timescale of internal tropospheric dynamics; and 3) the QBO impacts the strength of the extratropical zonal flow, not only in the stratosphere, but in the troposphere as well. The results imply that high frequency variability in the NAM in the troposphere is sometimes "nudged" by low frequency variability in the lower stratosphere.

The dynamics of the apparent impact of the stratosphere on the tropospheric circulation are currently under investigation. The impact could occur directly through momentum forcing of the extratropical circulation: stratospheric anomalies should induce a deep, thermally indirect mean meridional circulation below the level of the forcing that acts to transport momentum downwards [*Haynes et al.*, 1991; *Hartley et al.*, 1998; *Black*, 2002] and they should similarly induce anomalies in tropopause height that act to alter PV at troposphere levels [*Ambaum and Hoskins*, 2002]. It may also occur indirectly through the effect of stratospheric circulation anomalies on the refraction of planetary waves dispersing upwards from the troposphere: westerly flow in the extratropical stratosphere favors increased equatorward propagation and anomalous poleward flux of westerly momentum in the upper troposphere/lower stratosphere, and vice versa [*Hartmann et al*., 2000; *Shindell et al*., 2001].

Despite the skill evidenced in this section, stratosphere/tropospheric coupling has yet to be applied in operational numerical weather prediction. Operational forecast models at the European Centre for Medium-range Weather Forecasts (ECMWF) and at the Meteorological Research Institute in Japan both have well-resolved stratospheres and include adequate representations of the relevant stratospheric dynamics. As such, these models presumably capture the dynamics of downward-propagating zonal wind anomalies. In principle, an ensemble of forecasts run out to 60-90 days with slightly perturbed initial conditions should yield results similar to those in the composites in Plate 4. If the forecast model is capable of simulating the observed downward propagation of NAM anomalies, the model surface NAM should be nudged towards the same sign as those in the stratosphere.

Forecasts which include stratospheric information have the potential of benefiting society in a manner similar to the benefits derived from ENSO forecasts. However, forecasts based on stratospheric information will differ from those based on ENSO in three principal ways. First, since NH stratosphere/troposphere coupling is most vigorous during the winter months, subseasonal stratospheric forecasts only apply to the NH winter season. Second, while the QBO offers some hope for predictability on seasonal timescales, it only appears to impact the surface during late December-January. Third, since the stratospheric flow changes more rapidly than ENSO, forecasts may be updated daily throughout the winter season.

In light of the research emphasis placed on forecasting the NAM via midlatitude sea-surface temperature anomalies [see *Rodwell*, this volume], we feel that the evidence outlined in this section argues for increased emphasis on the skill that derives from the dynamical coupling between the tropospheric and stratospheric circulations.

6. SUMMARY AND CONCLUSIONS

6.1. Summary

This chapter provides an overview of the state of the art of our understanding of the atmospheric processes that underlie NAO-like variability.

Section 2 documents the structure of the NAO when defined on the basis of the leading EOF of the NH SLP field. The results suggest that the NAO can be interpreted as the NH analogue to the leading mode of variability in the SH circulation: both patterns are characterized by vacillations in the strength of the zonal flow with centers of action located ~55 and 35° latitude, and both are marked by polar centers of action in the geopotential height field with a high degree of zonal symmetry. As such, the observations presented in Section 2 motivate recasting the NAO as the Northern Hemisphere annular mode.

The most compelling argument in favor of abandoning the perspective that the dynamics of the NAM are restricted to the North Atlantic sector is its striking similarity to the leading mode of variability in the SH, the so-called SAM. Nevertheless, whether the mode in question is viewed as a statistical artifact of locally occurring stormtrack dynamics (the regional perspective) or as a physical phenomenon that organizes climate variability on a hemispheric scale (the annular mode perspective) remains open to debate.

Regardless of which perspective one subscribes to, the mode in question clearly has a pronounced signature in climate variability throughout much of the NH. For example, fluctuations in the NAM are strongly coupled to variability in the strength of the wintertime stratospheric polar vortex: a colder and stronger stratospheric polar vortex is associated with anomalously strong tropospheric westerlies along ~55°N, and vice versa. Fluctuations in the NAM are also coupled with the circulation of the tropics: the high index polarity of the NAM is marked by stronger than normal trade winds over both the Atlantic and Pacific sectors, low temperatures throughout the tropical free troposphere, and weak westerly anomalies along the equator centered at ~200-hPa. The high index polarity of the NAM also favors positive temperature anomalies, and hence anomalously weak upwelling, in the tropical tropopause region.

Section 4 examines the tropospheric dynamics that give rise to NAM-like variability in greater detail. The section focuses on two key questions related to variability in the tropospheric component of the NAM: 1) what are the dynamical processes that determine the structure of the NAM?, and 2) why is variability in the NAM largest over the North Atlantic sector? Section 4 begins with a review of the basic physical processes that drive variability in localized maxima in the extratropical zonal flow.

Variability in localized maxima in the extratropical zonal flow (referred to as jets) can be divided into two general classes: jet meandering, and jet pulsation. Jet meandering occurs when the zonal flow exhibits marked meridional excursions about its climatological mean latitude; jet pulsation occurs when the strength of the jet varies in strength at a fixed latitude. Whether jet pulsation or jet meandering dominates the variability of the extratropical zonal flow depends on both the strength and the meridional scale of the subtropical jet, and the meridional scale of the midlatitude baroclinic zone. In general, a strong jet and/or a small baroclinic zone acts to restrict the meridional excursions of the midlatitude eddies, and hence favors jet pulsation, and vice versa.

Extratropical jets are either thermally driven (e.g., subtropical jets) or driven by the convergence of eddy momentum flux (eddy-driven jets). Eddy-driven jets are generally weaker than their thermally driven counterparts and are found in broader baroclinic zones. Hence, variability in eddy driven jets is generally characterized by meanders in the zonal flow. Since the forcing mechanism that drives eddy-driven jets (transient waves) exhibits more variability than the mechanism that drives subtropical jets (meridional gradients in heating), one expects that eddy driven jets are also generally marked by greater variability than their subtropical counterparts.

In the troposphere, the NAM is characterized by variability in the eddy-driven jet. Hence, the amplitude of the NAM is largest in regions where the eddy-driven jet is most prominent and the subtropical jet is weakest. Observations and results from idealized model studies suggest that, over a given sector of the hemisphere, the strength of the local eddy-driven zonal flow is an inverse function of the strength of the local subtropical zonal flow. Over the Pacific sector, the thermally driven zonal flow is very strong and the eddy-driven zonal flow is relatively weak; over the Atlantic sector, the subtropical zonal flow is weaker than its Pacific counterpart, while the eddy-driven zonal-flow in the North Atlantic is relatively strong. Hence, the results in Section 4 suggest that the observed distortion of the NAM from zonal symmetry reflects the zonally varying climatological strength of the subtropical and eddy-driven jets. It is suggested that variability in the NAM is most pronounced over the North Atlantic sector because the subtropical zonal flow is weakest and the eddy- driven zonal flow is strongest in that region. It is also noted that the presence of warm lower boundary conditions at subpolar latitudes in the North Atlantic sector should permit eddy activity over a relatively broad range of latitudes there.

In the last section in this chapter we examined the relationship between stratosphere/troposphere coupling and temporal variability in the NAM. While most studies of predictability of the NAM emphasize atmosphere/ocean coupling on decadal timescales [e.g., see *Rodwell*, this volume], the results presented in this section suggest that the dynamical coupling with the stratosphere yields a significant level of predictability on both subseasonal and winter-to-winter timescales.

Variability in the circulation of the NH lower stratosphere is driven by waves dispersing upwards from the troposphere. Since only ~25% of the mass of the extratropical atmosphere lies above the tropopause, it has generally been assumed that wave-induced variability in the strength of the stratospheric polar vortex has little impact on the circulation of the troposphere. At least two key pieces of observational evidence outlined in Section 5 suggest otherwise: 1) large amplitude anomalies in the strength of the zonal flow along ~60°N frequently originate in the middle stratosphere and descend into the troposphere. Lag correlations between the circulation at ~10-hPa and the surface reveal that variability in the strength of the lower stratospheric polar vortex leads similar signed variability in the troposphere by ~1-2 weeks; 2) the downward propagating stratospheric circulation anomalies appear to modulate relatively high frequency tropospheric variability for periods up to ~60 days following the initiation of the stratospheric signal. Since the ~60 day timescale far exceeds the ~10 day timescale of extratropical tropospheric variability, the results in Section 5 may be interpreted as reflecting the impact of the lower stratosphere on the tropospheric circulation.

The coupling between the stratosphere and troposphere also has implications for predictability of the NAM on winter-to-winter timescales. In this case, the predictability derives from the impact of the QBO in the equatorial stratosphere on the strength and stability of the extratropical polar vortex. For example, the easterly phase of the QBO is associated with a weaker and more disturbed extratropical polar vortex and, through the linkages described above, weaker zonal flow in the troposphere consistent with the low index polarity of the NAM.

The observed linkages between long-lived anomalies in the stratospheric circulation and the surface signature of the NAM are not only of theoretical interest, but are of practical interest as well. For example, the 60 day period following weakenings in the strength of the stratospheric polar vortex are marked by lower temperatures and substantial increases in the frequency of extreme cold events throughout much of the NH. The connection between stratospheric and tropospheric circulations yields a level of predictability for NH weather that is comparable to that observed in association with the El-Niño/Southern Oscillation phenomenon.

6.2. Theoretical Considerations

The debate over whether the mode in question is more accurately described as a "Northern Hemisphere annular mode" or as the "North Atlantic Oscillation" attests to a key shortcoming in our understanding of the NAM, namely the absence of a unique theory for its existence in the first place. The discussion in this chapter highlights what we do understand in this regard.

In the troposphere, NAM-like variability is driven primarily by the meridional convergence of zonal momentum of baroclinic waves; in the stratosphere, it is driven by the meridional convergence of zonal momentum of waves with zonal wavenumbers 1-2. That the NAM is driven by wave-mean flow interactions is suggested by the fact that its centers of action in zonal wind are located on the poleward and equatorward flanks of the latitude band where the climatological mean eddy fluxes are largest. In Section 4, we argued that this latitude band corresponds to the region where the midlatitude eddies exhibit pronounced meridional meanders. On the basis of results from an idealized model experiment, we further suggested that tropospheric eddies exhibit pronounced meridional meanders in regions where their meridional scale is less than the meridional scale of the baroclinic zone. Hence, we conclude that the characteristics of annular-mode like variability are determined in large part by the meridional scale of the eddies.

What determines the relevant length scale of the eddies? In the context of linear baroclinic instability, the relevant length scale is the Rossby radius of deformation. However, since the energy associated with individual eddies cascades to larger spatial scales in a quasi-two dimensional circulation [*Kraichnan*, 1967], a more appropriate length scale is given by *Rhines* [1975]. This spatial scale, commonly referred to as the Rhines scale, corresponds to the length scale at which the upscale energy cascade is balanced by Rossby wave radiation, and is determined by the meridional gradient of planetary vorticity and the root-mean-square velocity of the flow. But the Rhines scale itself may also not be entirely appropriate, as it only holds for an inviscid flow contained in a sufficiently large domain [*Held*, 1999]. In the presence of friction, the inverse energy cascade may be halted before it ever reaches the Rhines scale. In the case where the Rhines scale is greater than the size of Earth, the length scale of the energy containing eddies will be that of Earth itself.

While our understanding of the fundamental dynamics of the NAM is incomplete, the above discussion highlights two important conclusions: 1) annular modes are constrained by the eddy length scale and hence are governed by fundamental quantities such as the radius of Earth, its rotation rate, and stratification; and 2) the physical process of the NAM are rooted in the dynamics of large-scale turbulence. Still, the above discussion does not necessarily provide a theory for the possible existence of coordinated variability on a hemispheric scale. That is, the above discussion does not provide proof that the NAM is a coherent physical mode that organizes climate variability throughout the hemisphere rather than a statistical artifact of locally occurring wave-mean flow interactions. It is possible that the NAM reflects the organization of two- dimensional turbulence into zonal jets, which occurs in cases where β is large [*Rhines*, 1975; *Williams*, 1978; *Pedlosky*, 1987]. It is also possible that the NAM reflects the organization of eddy activity by anomalies in the zonal flow [e.g., *Lorenz and Hartmann*, 2002]. However, whether the radiative forcing that continuously energizes the atmosphere allows enough time for these processes to take place remains unclear.

6.3. Concluding Remarks

As noted above, the absence of a unique theory for the existence of the NAM constitutes a key shortcoming in our understanding of extratropical climate variability. Another key shortcoming regards the dynamical coupling between the stratospheric component of the NAM and the circulation of the troposphere. The results in this chapter clearly suggest that an improved understanding of this coupling is of practical use for weather prediction. Several theories have been proposed to explain how stratospheric anomalies can impact the circulation of the troposphere, and several models are capable of simulating the observed linkages. Nevertheless, our understanding of the dynamics of the coupling remains incomplete.

The NAM has played an important role in recent climate change [*Hurrell*, 1995a; 1996; *Thompson et al.*, 2001; *Gillett et al.*, this volume], and similar trends have been observed in the Southern Hemisphere [*Hurrell and van Loon*, 1994; *Meehl et al.*, 1998; *Thompson and Solomon*, 2002]. Recent research suggests that both annular modes are sensitive to a wide array of forcing mechanisms, including increasing greenhouse gases [*Shindell et al.*, 1999; 2001; *Fyfe et al.*, 1999; *Kushner et al.*, 2001], feedbacks between greenhouse gases and ozone depletion [*Hartmann et al.*, 2000], increases in tropical sea-surface temperatures [*Hoerling et al.*, 2001], and variations in solar forcing [*Shindell et al.*, 2001b]. Nevertheless, it is unlikely that the source(s) of the observed trends in the annular modes can be unequivocally isolated in the absence of a consensus regarding the atmospheric processes that give rise to annular variability in the first place. In our view, establishing a theory for the existence of annular variability is of paramount importance for future research.

Acknowledgments. Thanks to W. J. Randel, W. A. Robinson, J. W. Hurrell, and one anonymous reviewer for their helpful comments and suggestions. Thanks also to T. J. Dunkerton, S. Feldstein, H.-K. Kim, D. J. Lorenz, and J. M. Wallace for their assistance and insight at various stages of this research, and to S.-W. Son for generating Plate 3. DWJT is supported by the National Science Foundation under grant CAREER: ATM-0132190. SL is supported by NSF grant ATM-0001473. MPB is supported by NSF grant ATM-0002485, NASA's SR&T Program for Geospace Science, contract NASW-00018, and NOAA's OGP CLIVAR Atlantic grant NOAA-0572.

REFERENCES

Ambaum, M. H. P. and B. J. Hoskins, The NAO troposphere-stratosphere connection, *J. Climate, 15*, 1969-1978, 2002.

Ambaum, M. H. P., B. J. Hoskins, and D. B. Stephenson, Arctic Oscillation or North Atlantic Oscillation?, *J. Climate, 14*, 3495-3507, 2001.

Baldwin, M. P., X. Cheng, and T. J. Dunkerton, Observed correlations between winter-mean tropospheric and stratospheric circulation anomalies, *Geophys. Res. Lett., 21*, 1141-1144, 1994.

Baldwin, M. P., and T. J. Dunkerton, Propagation of the Arctic Oscillation from the stratosphere to the troposphere, *J. Geophys. Res., 104*, 30937-30946, 1999.

Baldwin, M. P., Annular Modes in global daily surface pressure, *Geophys. Res. Lett., 28*, 4115-4118, 2001.

Baldwin, M. P., and T. J. Dunkerton, Stratospheric harbingers of anomalous weather regimes, *Science, 294*, 581-584, 2001.

Baldwin, M. P. et al., The Quasi-Biennial Oscillation, *Rev. Geophys., 39*, 179-229, 2001.

Barnston, A., and R. E. Livezey, Classification, seasonality and persistence of low-frequency circulation patterns, *Mon. Wea. Rev., 115*, 1083-1126, 1987.

Black, R. X., Stratospheric forcing of surface climate in the Arctic Oscillation, *J. Climate, 15*, 268-277, 2002.

Branstator, G., Circumglobal teleconnections, the jetstream waveguide, and the North Atlantic Oscillation, *J. Climate, 15*, 1903-1910, 2002.

Chang, E. K. M., Downstream development of baroclinic waves as inferred from regression analysis, *J. Atmos. Sci., 50*, 2038-2053, 1993.

Chang, K. M. E., S. Lee, and K. L. Swanson, Storm track dynamics, *J. Climate, 15*, 2163-2183, 2002.

Charney, J. G., and P. G. Drazin, Propagation of planetary- scale disturbances from the lower into the upper atmosphere, *J. Geophys. Res., 66*, 83-109, 1961.

Christiansen, B., Downward propagation of zonal mean zonal wind anomalies from the stratosphere to the troposphere: Model and reanalysis, *J. Geophys. Res., 106*, 27307-27322, 2001.

Coughlin, K., and K. -K. Tung, QBO Signal found at the extratropical surface through Northern Annular Modes, *Geophys. Res. Lett., 28*, 4563-4566, 2001.

Czaja, A., A. W. Robertson, and T. Huck, The role of Atlantic Ocean-Atmosphere coupling in affecting North Atlantic Oscillation variability, this volume.

Deser C, On the teleconnectivity of the "Arctic Oscillation," *Geophys. Res. Lett., 27*, 779-782, 2000.

DeWeaver, E., and S. Nigam, Do stationary waves drive the zonal-mean jet anomalies of the Northern winter?, *J. Climate, 13*, 2160-2176, 2000a.

DeWeaver, E., and S. Nigam, Zonal-eddy dynamics of the North Atlantic Oscillation, *J. Climate, 13*, 3893-3914, 2000b.

Dommenget D., and M. Latif, A cautionary note on the interpretation of EOFs, *J. Climate,* 15, 216-225, 2002.

Dunkerton, T. J., C. P. F. Esu, and M. E. McIntyre, Some Eulerian and Lagrangian diagnostics for a model stratospheric warming, *J. Atmos. Sci., 38*, 819-843, 1981.

Dunkerton, T. J., Midwinter deceleration of the subtropical mesospheric jet and interannual variability of the high-latitude flow in UKMO analyses, *J. Atmos. Sci., 57*, 3838-3855, 2000.

Feldstein, S. B., and S. Lee, Mechanisms of zonal index variability in an aquaplanet GCM, *J. Atmos. Sci.*, 53, 3541- 3555, 1996.

Feldstein, S. B., and S. Lee, Is the atmospheric zonal index driven by an eddy feedback?, *J. Atmos. Sci., 55*, 2077-3086, 1998.

Feldstein, S. B., Teleconnections and ENSO: The timescale, power spectra, and climate noise properties, *J. Climate, 13*, 4430-4440, 2000.

Feldstein, S. B., On the recent trend and variance increase of the Annular Mode, *J. Climate, 15*, 88-94, 2002.

Fyfe, J. C., G. J. Boer, and G. M. Flato, The Arctic and Antarctic Oscillations and their projected changes under global warming, *Geophys. Res. Lett., 26*, 1601-1604, 1999.

Gillett, N. P., M. P. Baldwin, and M. R. Allen, Nonlinearity in the stratospheric response to external forcing, *J. Geophys. Res., 106*, 7891-7901, 2001.

Gillett, N. P., H. F. Graf, T. J. Osborn, Climate change and the North Atlantic Oscillation, this volume.

Gong, D., and S. Wang, Definition of Antarctic Oscillation index, *Geophys. Res. Lett., 26*, 459-462, 1999.

Grotzner, A., M. Latif, and T. P. Barnett, A decadal climate cycle in the North Atlantic Ocean as simulated by the ECHO coupled GCM, *J. Climate, 11*, 831-847, 1998.

Hartmann, D. L., A PV view of zonal flow vacillation, *J. Atmos. Sci., 52*, 2561-2576, 1995.

Hartmann, D. L., and F. Lo, Wave-driven zonal flow vacillation in the Southern Hemisphere, *J. Atmos. Sci., 55*, 1303-1315, 1998.

Hartley, D. E., J. Villarin, R. X. Black, and C. A. Davis, A new perspective on the dynamical link between the stratosphere and troposphere, *Nature, 391*, 471-474, 1998.

Hartmann, D. L., J. M. Wallace, V. Limpasuvan, D. W. J. Thompson, and J. R. Holton, Can ozone depletion and greenhouse warming interact to produce rapid climate change?, *Proc. Nat. Acad. Sci., 97*, 1412-1417, 2000.

Haynes, P. H., C. J. Marks, M. E. McIntyre, T. G. Shepherd, and K. P. Shine, On the "downward control" of extratropical diabatic circulations by eddy-induced mean zonal forces, *J. Atmos. Sci., 48*, 651-678, 1991.

Held, I. M., The Macroturbulence of the troposphere, *Tellus, 51A-B*, 59-70, 1999.

Hines, C. O., *Upper Atmosphere in Motion,* AGU Geophysical Monograph, No. 18, 1974.

Hoerling, M. P., J. W. Hurrell, T. Xu, Tropical origins for recent North Atlantic climate change, *Science*, *292*, 90-92, 2001.

Holton, J. R., and C. Mass, Stratospheric vacillation cycles, *J. Atmos. Sci., 33*, 2218-2225, 1976.

Holton, J. R., and H. -C. Tan, The influence of the equatorial quasi-biennial oscillation on the global circulation at 50 mb. *J. Atmos. Sci., 37*, 2200-2208, 1980.

Holton, J. R., *An Introduction to Dynamic Meteorology*, 511 pp., third edition, Academic Press, USA., 1992.

Hoskins, B. J., and D. J. Karoly, Steady linear response of a spherical atmosphere to thermal and orographic forcing, *J. Atmos. Sci., 38*, 1179-1196, 1981.

Hurrell, J. W., and H. van Loon, A modulation of the atmospheric annual cycle in the Southern Hemisphere, *Tellus*, *46A*, 325-338, 1994.

Hurrell, J. W., Decadal trends in the North Atlantic Oscillation region temperatures and precipitation, *Science*, *269*, 676-679, 1995a.

Hurrell, J. W., An evaluation of the transient eddy forced vorticity balance during northern winter, *J. Atmos. Sci., 52*, 2286-2301, 1995b.

Hurrell, J. W., Influence of variations in extratropical wintertime teleconnections on Northern Hemisphere temperature, *Geophys. Res. Lett, 23*, 665-668, 1996.

Hurrell, J. W., and H. van Loon, Decadal variations in climate associated with the North Atlantic Oscillation, *Climatic Change, 36*, 301-326, 1997.

Hurrell, J. W., Y. Kushnir, G. Ottersen, and M. Visbeck, An overview of the North Atlantic Oscillation, this volume.

Intergovernmental Panel on Climate Change, *Climate Change 2001*: *The Science of Climate Change*, Cambridge University Press, Cambridge, UK, 2001.

James I. N., Suppression of baroclinic instability in horizontally sheared flows. *J. Atmos. Sci., 44*, 3710-3720, 1987.

James, I. N., and L. J. Gray, Concerning the effect of surface drag on the circulation of a baroclinic planetary atmosphere, *Quart. J. Roy. Meteor. Soc., 112*, 1231-1250, 1986.

James, I. N., and P. M. James, Spatial structure of ultra-low frequency variability of the flow in a simple atmospheric circulation model, *Quart. J. Roy. Meteor. Soc*., *118*, 1211-1233, 1992.

Jones, P. D., Hemispheric surface air temperature variations: A reanalysis and update to 1993, *J. Climate*, *7*, 1794-1802, 1994.

Jones, P. D., T. J. Osborn, and K. R. Briffa, Pressure-based measures of the North Atlantic Oscillation (NAO): A comparison and an assessment of changes in the strength of the NAO and in its influence on surface climate parameters, this volume.

Kalnay, M. E., et al., The NCEP/NCAR Reanalysis Project, *Bull. Amer. Meteor. Soc.*, *77*, 437-471, 1996.

Karoly, D. J., The role of transient eddies in low-frequency zonal variations of the Southern Hemisphere circulation, *Tellus*, *42A*, 41-50, 1990.

Kerr, R. A., A new force in high-latitude climate *Science*, *284*, 241-242, 1999.

Kidson, J. W., Indices of the Southern Hemisphere zonal wind, *J. Climate*, *1*, 183-194, 1988a.

Kidson, J. W., Interannual variations in the Southern Hemisphere circulation, *J. Climate*, *1*, 1177-1198, 1988b.

Kidson, J. W., and M. R. Sinclair, The influence of persistent anomalies on Southern Hemisphere storm tracks, *J. Climate*, *8*, 1938-1950, 1995.

Kidson, J. W., and I. G. Watterson, The structure and predictability of the "High-Latitude Mode" in the CSIRO9 general circulation model, *J. Atmos. Sci*., *56*, 3859-3873, 1999.

Kodera, K., K. Yamazaki, M. Chiba, and K. Shibata, Downward propagation of upper stratospheric mean zonal wind perturbation to the troposphere, *Geophys. Res. Lett*., *17*, 1263-1266, 1990.

Kraichnan, R. H., Inertial ranges in two-dimensional turbulence, *The Physics of Fluids, Supplement II*, pp.233-239, 1967.

Kushner, P. J., I. M. Held, and T. L. Delworth, Southern Hemisphere atmospheric circulation response to global warming, *J. Climate*, *14*, 2238-2249, 2001.

Kushnir, Y., and J. M. Wallace, Low frequency variability in the Northern Hemisphere winter: Geographical distribution, structure and time dependence, *J. Atmos. Sci.*, *46*, 3122-3142, 1987.

Kushnir, Y., W. A. Robinson, I. Blade, N. M. J. Hall, S. Peng, R. Sutton, Atmospheric GCM response to extratropical SST anomalies: Synthesis and evaluation, *J. Climate, 15,* 2233-2256, 2002.

Kutzbach, J. E., Large-scale features of monthly mean Northern Hemisphere anomaly maps of sea-level pressure *Mon. Wea. Rev., 98*, 708-716, 1970.

Lamb, P. J., and R. A. Peppler, North Atlantic Oscillation: Concept and application, *Bulletin of the Amer. Met. Soc., 68*, 1218-1225, 1987.

Lau, N. -C., Variability of the observed midlatitude storm tracks in relation to low-frequency changes in the circulation pattern, *J. Atmos. Sci., 45,* 2718-2743, 1988.

Lee, S., and I. M. Held, Baroclinic wave packets in models and observations, *J. Atmos. Sci., 50*, 1413-1428, 1993.

Lee, S., and S. Feldstein, Mechanism of zonal index evolution in a two-layer model, *J. Atmos. Sci*., *53*, 2232-2246, 1996.

Lee, S., Maintenance of multiple jets in a baroclinic flow, *J. Atmos. Sci*., *54*, 1726-1738, 1997.

Leith, C. E., The standard error of time-averaged estimates of climatic means, *J. Appl. Meteor.*, *12*, 1066-1069, 1973.

Limpasuvan, V., and D. L. Hartmann, Eddies and the annular modes of climate variability, *Geophys. Res. Lett.*, *26*, 3133-3136, 1999.

Limpasuvan, V., and Hartmann, D. L., Wave-maintained annular modes of climate variability, *J. Climate*, *13*, 4414- 4429, 2000.

Lorenz, E. N., Seasonal and irregular variations of the northern hemisphere sea-level pressure profile, *J. Meteorol.*, *8*, 52-59, 1951.

Lorenz, D. J., and D. L. Hartmann, Eddy-zonal flow feedback in the Southern Hemisphere, *J. Atmos. Sci.*, *58*, 3312-3327, 2001.

Lorenz, D. J., and D. L. Hartmann, Eddy-zonal flow feedback in the Northern Hemisphere winter, *J.Climate*, in press, 2002.

Madden, R. A., Estimates of the natural variability of time-averaged sea-level pressure, *Mon. Wea. Rev.*, *104*, 942-952, 1976.

Malmgren, B. A., A. Winter, and D. Chen, El Niño-Southern Oscillation and North Atlantic Oscillation control of climate in Puerto Rico, *J. Climate, 11*, 2713-2718, 1998.

McHugh, M. J., and J. C. Rogers, North Atlantic Oscillation influence on precipitation variability around the Southeast African Convergence Zone, *J. Climate*, *14*, 3631- 3642, 2001.

Meehl, G. A., and H. van Loon, The seesaw in winter temperatures between Greenland and Northern Europe, Part II: Teleconnections with lower latitudes, *Mon. Wea. Rev.*, *107*, 1095-1106, 1979.

Meehl, G. A., J. W. Hurrell, and H. van Loon, A modulation of the mechanism of the semiannual oscillation in the Southern Hemisphere, *Tellus*, *50A*, 442-450, 1998.

Moulin, C., C. E. Lambert, F. Dulac, and U. Dayan, Control of atmospheric export of dust from North Africa by the North Atlantic Oscillation, *Nature*, *387*, 691-694, 1997.

Namias, J., The index cycle and its role in the general circulation, *J. Meteor.*, *7*, 130-139, 1950.

National Research Council Panel on Reconciling Temperature Observations, *Reconciling Observations of Global Temperature Change*, 85 pp., National Academy Press, Washington, DC, 2000.

Panetta, R. L., Zonal jets in wide baroclinically unstable regions: Persistence and scale selection, *J. Atmos. Sci.*, *50*, 2073-2106, 1993.

Panetta, R. L., and I. M. Held, Baroclinic eddy fluxes in a one-dimensional model of quasi-geostrophic turbulence, *J. Atmos. Sci.*, *45*, 3354-3365, 1988.

Pedlosky, J., *Geophysical Fluid Dynamics*, 710 pp., Springer-Verlag, 1987.

Perlwitz, J., and H. -F. Graf, The statistical connection between tropospheric and stratospheric circulation of the Northern Hemisphere in winter, *J. Climate, 8*, 2281-2295, 1995.

Randel, W. J., and J. L. Stanford, Observational study of medium-scale wave dynamics in the Southern Hemisphere summer, pt. 2, Stationary transient wave interference, *J. Atmos. Sci., 42*, 1364-1373, 1985.

Randel, W. J., Coherent wave-zonal mean flow interactions in the troposphere, *J. Atmos. Sci., 47*, 439-456, 1990.

Reed, R. J., W. J. Campbell, L. A. Rasmussen, and R. G. Rogers, Evidence of downward propagating annual wind reversal in the equatorial stratosphere, *J. Geophys. Res., 66*, 813-818, 1961.

Rhines, P., Waves and turbulence on a beta-plane, *J. Fluid. Mech., 69,* 417-443, 1975.

Robinson, W. A., The dynamics of the zonal index in a simple model of the atmosphere, *Tellus, 43A*, 295-305, 1991.

Robinson, W., Predictability of the zonal index in a global model, *Tellus, 44A*, 331-338, 1992.

Robinson, W. A., Eddy feedbacks on the zonal index and eddy-zonal flow interactions induced by zonal flow transience, *J. Atmos. Sci., 51*, 2553-2562, 1994.

Robinson, W. A., Does eddy feedback sustain variability in the zonal index?, *J. Atmos. Sci.*, *53*, 3556-3569, 1996.

Robinson, W. A., A baroclinic mechanism for the eddy feedback on the zonal index, *J. Atmos. Sci., 57*, 415-422, 2000.

Rodwell, M. J., On the predictability of North Atlantic climate, this volume.

Rodwell, M. J., D. P. Rowell, and C. K. Folland, Oceanic forcing of the wintertime North Atlantic Oscillation and European climate, *Nature*, *398*, 320-323, 1999.

Rogers, J. C., and H. van Loon, Spatial variability of sea level pressure and 500-mb height anomalies over the Southern Hemisphere, *Mon. Wea., Rev., 110*, 1375-1392, 1982.

Rogers, J. C., Association between the North Atlantic Oscillation and the Southern Oscillation in the Northern Hemisphere, *Mon. Wea. Rev.*, *112*, 1999-2015, 1984.

Rossby, C. -G., Relations between variations in the intensity of the zonal circulation of the atmosphere and the displacements of the semipermanent centers of action, *J. Mar. Res.*, *3*, 38-55, 1939.

Rossby, C. -G., and H. C. Willett, The circulation of the upper troposphere and lower stratosphere, *Science*, *108*, 643- 652, 1948.

Schubert, S. D., and C.-K. Park, Low-frequency intraseasonal tropical-extratropical interactions, *J. Atmos. Sci.*, *48*, 629-650, 1991.

Shindell, D. T., R. L. Miller, G. Schmidt, and L. Pandolfo, Simulation of recent northern winter climate trends by greenhouse-gas forcing, *Nature*, *399*, 452-455, 1999.

Shindell, D. T., G. A. Schmidt, R. L. Miller, and D. Rind, Northern Hemisphere winter climate response to greenhouse gas, ozone, solar, and volcanic forcing, *J. Geophys. Res., 106*, 7193-7210, 2001.

Shindell, D. T., G. A. Schmidt, M. E. Mann, D. Rind, A. Waple, Solar forcing of regional climate change during the maunder minimum, *Science*, *294*, 2149-2152, 2001b.

Shine, K. P., Middle atmosphere in the absence of dynamical heat fluxes, *Q. J. Roy. Met. Soc.*, *113*, 603-633, 1987.

Shiotani, M., Low-frequency variations of the zonal mean state of the Southern Hemisphere troposphere, *J. Met. Soc. Japan*, *68*, 461-471, 1990.

Spencer, R. W., J. R. Christy and N. C. Grody, Global Atmospheric temperature monitoring with satellite microwave measurements: Method and results 1979-84, *J. Climate*, *3*, 1111-1128, 1990.

Spencer, R. W. and J. R. Christy, Precision and radiosonde validation of satellite grid point temperature anomalies, Part I: MSU Channel 2, *J. Climate*, *5*, 847-857, 1992.

Spencer, R. W., and J. R. Christy, Precision lower stratospheric temperature monitoring with the MSU: technique, validation, and results 1979-1991, *J. Climate, 6*, 1194-1204, 1993.

Stephenson, D. B., V. Pavan, and Ro. Bojariu, Is the North Atlantic Oscillation a random walk?, *Int. J. Climtol, 20*, 1-18, 2000.

Stephenson, D. B., H. Wanner, S. Bronnimann, and J. Luterbacher, The history of scientific research on the North Atlantic Oscillation, this volume.

Szeredi, I., and D. J. Karoly, Horizontal structure of monthly fluctuations of the Southern Hemisphere troposphere from station data, *Australian Met. Mag.*, *35*, 119-129, 1987.

Swanson, K. L., P. J. Kushner, and I. M. Held, Dynamics of barotropic storm tracks, *J. Atmos. Sci.*, *54*, 791-810, 1997.

Thompson, D. W. J, and J. M. Wallace, The Arctic Oscillation signature in the wintertime geopotential height and temperature fields, *Geophys. Res. Lett.*, *25*, 1297-1300, 1998.

Thompson, D. W. J., and J. M. Wallace, Annular modes in the extratropical circulation, Part I: Month-to-month variability, *J. Climate*, *13*, 1000-1016, 2000.

Thompson, D. W. J., J. M. Wallace, and G. C. Hegerl, Annular modes in the extratropical circulation, Part II: Trends, *J. Climate*, *13*, 1018-1036, 2000.

Thompson, D. W. J., Annular modes in the atmospheric general circulation. Ph. D. Thesis, Department of Atmospheric Sciences, University of Washington, (J. M. Wallace, thesis advisor), 179 pp., 2000.

Thompson, D. W. J., and J. M. Wallace, regional climate impacts of the Northern Hemisphere annular mode, *Science*, *293*, 85-89, 2001.

Thompson, D. W. J., M. P. Baldwin, and J. M. Wallace, Stratospheric connection to Northern Hemisphere wintertime weather: implications for prediction, *J. Climate*, in press, 2002.

Thompson, D. W. J., and S. Solomon, Interpretation of recent Southern Hemisphere climate change, *Science*, *296*, 895-899, 2002.

Trenberth, K. E., and D. A. Paolino, The Northern Hemisphere sea level pressure data set: Trends, errors and discontinuities, *Mon. Wea. Rev.*, *108*, 855-872, 1980.

Trenberth, K. E., and D. A. Paolino, Characteristic patterns of variability of sea level pressure in the Northern Hemisphere, *Mon. Wea. Rev., 109*, 1169-1189, 1981.

van Loon, H., and J. Rogers, The seesaw in winter temperatures between Greenland and Northern Europe, Part I: General description, *Mon. Wea. Rev.*, 106, 296-310, 1978.

von Storch, J. -S., On the reddest atmospheric modes and the forcings of the spectra of these modes, *J. Atmos. Sci., 56*, 1614-1626, 1999.

Wallace, J. M., and D. S. Gutzler, Teleconnections in the geopotential height field during the Northern Hemisphere winter, *Mon. Wea. Rev., 109*, 784-812, 1981.

Wallace, J. M., and H.-H. Hsu, Another look at the index cycle, Tellus 37A, 478-486, 1985.

Wallace, J. M., North Atlantic Oscillation / Annular Mode: Two paradigms - One Phenomenon, *Q. J. Royal Met. Soc., 126*, 791-805, 2000.

Wallace, J. M., and D. W. J. Thompson, The Pacific center of action of the Northern Hemisphere Annular Mode: Real or artifact?, *J. Climate, 15*, 1987-1991, 2002.

Walsh, J. E., W. L. Chapman and T. L. Shy, Recent decrease of sea level pressure in the central Arctic, *J. Climate, 9*, 480-486, 1996.

Willett, H. C., Patterns of world weather changes, *Trans. Amer. Geophys. Union, 29*, 803-809, 1948.

Williams, G. P., Planetary circulations: 1. Barotropic representation of Jovian and terrestrial turbulence, *J. Atmos. Sci., 35*, 1399-1426, 1978.

Xie, S.-P., H. Noguchi, and S. Matsumura, A hemispheric- scale quasi-decadal oscillation and its signature in Northern Japan, *J. Met. Soc. Japan, 77*, 573-582, 1999.

Yamazaki, K., Y. Shinya, Analysis of the Arctic Oscillation simulated by AGCM, *J. Met. Soc. Japan, 77*, 1287-1298, 2000.

Yoden, S., M. Shiotani, and I. Hirota, Multiple planetary flow regimes in the Southern Hemisphere, *J. Met. Soc. Japan, 65*, 571-586, 1987.

Yu, J.-Y., and D. L. Hartmann, Zonal flow vacillation and eddy forcing in a simple GCM of the atmosphere, *J. Atmos. Sci., 50*, 3244-3259, 1993.

Mark P. Baldwin, Northwest Research Associates, 14508 NE 20th Street, Bellevue, WA 98007-3713, USA
mark@nwra.com

Sukyoung Lee, Department of Meteorology, 510 Walker Building, University Park, PA 16802, USA
sl@essc.psu.edu

David W. J. Thompson,Department of Atmospheric Science, Colorado State University, Fort Collins, CO 80523, USA
davet@atmos.colostate.ed

The Ocean's Response to North Atlantic Oscillation Variability

Martin Visbeck[1], Eric P. Chassignet[2], Ruth G. Curry[3], Thomas L. Delworth[4],
Robert R. Dickson[5], Gerd Krahmann[1]

The North Atlantic Oscillation (NAO) is the dominant mode of atmospheric variability in the North Atlantic Sector. Basin scale changes in the atmospheric forcing significantly affect properties and circulation of the ocean. Part of the response is local and rapid (surface temperature, mixed-layer depth, upper ocean heat content, surface Ekman transport, sea ice cover). However, the geostrophically balanced large-scale horizontal and overturning circulation can take several years to adjust to changes in the forcing. The delayed response is non-local in the sense that waves and the mean circulation communicate perturbations at the air-sea interface to other parts of the Atlantic basin. A delayed and non-local response can potentially give rise to oscillatory behavior if there is significant feedback from the ocean to the atmosphere. We conjecture that, on decadal and longer time scales, changes in the ocean's heat storage and transport should have an increasingly important impact on the climate. Finally, changes in the ocean circulation and distribution of heat and freshwater will also alter ventilation rates and pathways. Thus we expect a change in the net uptake of gases (e.g., O_2, CO_2), altered nutrient balance, and changes in the dispersion of marine life. We review what is known about the oceanic response to changes in NAO-induced forcing from combined theoretical, numerical experimentation and observational perspectives.

1. INTRODUCTION

More than two centuries ago, missionaries noticed that interannual fluctuations in wintertime air temperature were out of phase between Greenland and Denmark [see *Stephenson et al.*, this volume; *Hurrell et al.*, this volume].

[1]Department of Earth and Environmental Sciences, Lamont-Doherty Earth Observatory, Columbia University, Palisades, New York, U.S.A.

[2]RSMAS/MPO, University of Miami, Miami, Florida, U.S.A.

[3]Physical Oceanography Department, Woods Hole Oceanographic Institution, Woods Hole, Massachusetts, U.S.A.

[4]Geophysical Fluid Dynamics Laboratory, National Oceanic and Atmospheric Administration, Princeton, New Jersey, U.S.A.

[5]Center for Environment, Fisheries and Aquaculture Science, Lowestoft Laboratory, United Kingdom

The North Atlantic Oscillation:
Climatic Significance and Environmental Impact
Geophysical Monograph 134

10.1029/134GM06

In the early part of the twentieth century, Walker analyzed the spatial correlation patterns of seasonal weather and noticed a surface pressure correlation pattern within the Atlantic sector that he referred to as the North Atlantic Oscillation (NAO) [*Walker and Bliss*, 1932]. In a seminal paper several decades later, Bjerknes discussed his views of air-sea interactions in the Atlantic, addressing "... causes of the variations in the surface temperature of the Atlantic Ocean from year to year and over longer periods" [*Bjerknes*, 1964]. He interpreted the changes in ocean temperatures as partly due to radiative transfer and heat exchanges at the interface between the ocean and the atmosphere, and partly due to advective heat transport divergence as a result of varying ocean currents. Most of the heat flux changes depend directly on the strength of the wind. Bjerknes refers to the large-scale wind changes in terms of "low-index" and "high-index" conditions, the associated pattern in sea level pressure (SLP) of which is the NAO [*Wallace and Gutzler*, 1981; *Stephenson et al.*, this volume; *Jones et al.*, this volume]. The advent of global data sets and advances in theory and computer model simulations over the last decade have revitalized interest in the NAO and its interaction with the ocean, sea ice and other parts of the climate system. Many of the early concepts have survived. However, the increased observational database, models and a more complete theory of oceanic and atmospheric circulation have helped to sharpen hypotheses and understanding. The NAO is now widely recognized as the most significant pattern of climate variability in the North Atlantic Sector, and it is a strong competitor to the El Niño-Southern Oscillation (ENSO) phenomenon in terms of global significance [e.g., *Marshall et al.*, 2001a; *Hurrell et al.*, 2001; *Visbeck et al.*, 2001].

Here we examine the oceanic response to changes in the atmospheric forcing associated with the NAO. The ocean's large heat capacity (2.5 m of water contains as much thermal energy as the entire atmospheric column) makes it the 'flywheel' of the climate system. If the ocean and atmosphere were tightly coupled, one might question our consideration of only the ocean's response. However, outside of the deep tropics, most studies find that the ocean largely reacts to the high frequency changes of the atmospheric forcing and that its influence back to the atmosphere is weak on time scales shorter than a decade [*Kushnir et al.*, 2002]. Thus the emerging null hypothesis is one where the ocean merely responds to and integrates in time the atmospheric forcing anomalies in the spirit of the stochastic climate model of *Hasselman* [1976] and *Frankignoul and Hasselman* [1977]. Is it really true? Or does the ocean have more in stock than one could explain by a local response? We will give evidence for both local and remote responses, but refer to *Czaja et al.* [this volume] for a comprehensive review of possible ocean-atmosphere coupling mechanisms in the context of the NAO [see also *Czaja and Marshall*, 2001].

In most of the following, we will focus on the winter season as the time of NAO forcing. As shown by *Hurrell et al.* [this volume], the NAO forcing is most active between November and April, at a time when the ocean mixed layers are deep and much of the ocean uptake of gases takes place. One can identify an NAO signature in the summer season, but it is weaker, less persistent and explains only a small fraction of the overall variance. We will define a 'typical' NAO perturbation using regression techniques and study the oceanic response to it. Such NAO forcing patterns are quite robust between different methods, data sets, and decades. However, we caution that any attempts to identify 'normal' or 'standard' patterns of NAO behavior and the associated ocean response will necessarily be oversimplified. For example, a slight eastward shift of the NAO dipole pattern in certain winters of the late 1990s was enough to reverse a cooling and freshening trend in Labrador Sea Water (LSW) that had persisted for decades despite a positive NAO index. *Hilmer and Jung* [2000] suggest that the centers of maximum interannual variability in SLP associated with the NAO have been located further to the east since the late 1970s and dramatically changed the sea-ice response. Those observations (i.e., that NAO-positive index conditions can locally drive quite different ocean responses depending on the detailed configuration of the associated SLP pattern) offer a timely reminder of the limitations of using a simple 2-point pressure difference as our index of NAO behavior [see also *Hurrell et al.*, this volume]. It remains to be seen whether those shifts in the 'standard' NAO pattern are further evidence of the chaotic nature of the atmospheric circulation or whether they are part of a more concerted trend in NAO behavior. In other words, both the detailed configuration of the NAO and the ocean's response to it can be expected to change.

First we revisit the observational evidence of NAO-related changes in the sea surface temperatures (SST), and then discuss what is known about the air-sea fluxes that force them. The next two sections apply ocean circulation theory to provide an idea about NAO-induced ocean circulation changes. Those circulation changes can directly impact the oceanic heat transport, which potentially could introduce non-local atmospheric feedback. The advection of the anomalies themselves by the mean circulation is of interest for the same reason. In subsection 5 and 6, we review the observational evidence of the modulation of ocean deep convection and changes in water mass composition with special emphasis on recent trends. The last section reviews the sea-ice response to changes in NAO forcing, and we close with a brief summary and discussion.

2. OBSERVED SST RESPONSE PATTERN

The surface temperature response of the North Atlantic Ocean to changes in the NAO index has been described carefully by the early work of *Bjerknes* [1962]. He contrasted individual winter seasons with a "high" and "low" index and found warmer temperatures between 30°N and 45°N in the western North Atlantic and cooler temperatures in the subpolar gyre region (north of 45°N). More recent studies [e.g., *Cayan*, 1992a; *Visbeck et al.*, 1998; *Seager et al.*, 2000; *Marshall et al.*, 2001b] correlated SST anomalies from the NCEP/NCAR reanalysis with the NAO index and reproduced a similar response pattern. However, the amplitudes were somewhat lower (~0.5°C for a strong NAO event) when compared to Bjerknes' maps (~2°C, but for individual years and smaller regions). As an example, we present, covariance and correlation maps between a 100-year long SST anomaly data set [*Kaplan et al.*, 1997; 1998] and the NAO index [*Hurrell*, 1995] (Figure 1). In addition to Bjerknes' North Atlantic SST dipole, a large third lobe in the northern tropical Atlantic is apparent. The latter plays an important role in tropical Atlantic variability [e.g., *Marshall et al.*, 2001a] and can displace the location of tropical precipitation within the intertropical convergence zone. This three lobe SST response pattern is often referred to as the "NAO SST tripole" pattern. It also appears as the second EOF when global SST anomalies are analyzed. Maximum correlation is 0.4 in the centers of action, and the maximum SST covariance with the NAO index is about 0.3°C at zero lag. Thus the NAO explains only 20–40% of the local winter season SST variance.

Lag correlation between SST and the NAO index is interesting since it contains information about the coupling between the atmosphere and ocean. If a coherent SST pattern leads the NAO index and SST itself varies either slowly or predictably, one would be able to construct a statistical NAO prediction system [see *Czaja et al.*, this volume; *Rodwell*, this volume]. However, if the ocean is merely responding to NAO-related atmospheric forcing, one would then expect a strong asymmetry in the lag correlation, with weak correlation prior to an NAO event and more persistent correlation when the ocean lags the atmosphere. Lag correlation between the Kaplan SST data set and the NAO index from 1901–2000 shows that the largest correlation is at zero lag, with some limited ocean memory (lags of up to 3 years) in the Gulf Stream extension region (Figure 2). *Watanabe and Kimoto* [2000] remark that this persistence is longer than what would be expected from local damping due to air-sea interaction, which yields a decay scale of about 3 months [*Frankignoul et al.*, 1998]. This is consistent with the so-called reemergence mechanism in which a shallow summer thermocline shields deeper temperature anomalies from the atmosphere, which then are reentrained in the following winter season [*Alexander and Deser*, 1995]. However, the displacement of SST anomalies along the path of the mean ocean currents might suggest some role for ocean advection [*Sutton and Allan*, 1997]. Only small regions of significant correlation are found when SST anomalies lead NAO events (Figure 2). Thus the observations support a strong and immediate response of the surface ocean, with limited evidence for multi-season persistence and little support for a strong atmospheric response [*Kushnir et al.*, 2002; see also *Czaja et al.*, this volume].

A similar three-lobe mode of SST variability within the Atlantic sector has been found in a large number of observational, modeling, and theoretical studies of North Atlantic climate variability [e.g., *Deser and Blackmon*, 1993; *Kushnir*, 1994; *Battisti et al.*, 1995; *Luksch*, 1996; *Delworth*, 1996; *Halliwell*, 1998; *Seager et al.*, 2000; *Grötzner et al.*, 1998; *Selten et al.*, 1999; *Xie and Tanimoto*, 1998]. In all cases, the ocean is responding to variable atmospheric forcing with a preferred spatial structure. However, there are considerable differences in opinion, and possible model sensitivity, with regard to mechanisms that cause changes in SST, in particular the relative role of air-sea heat fluxes versus momentum flux-induced changes in the ocean circulation, and the role of ocean circulation in general.

For example, *Battisti et al.* [1995] and *Seager et al.* [2000] show that SST variability on interannual time scales over the Atlantic can be understood in terms of one-dimensional mixed-layer processes, except in the region of the separated Gulf Stream/North Atlantic current where advection by ocean currents is hypothesized to be important.

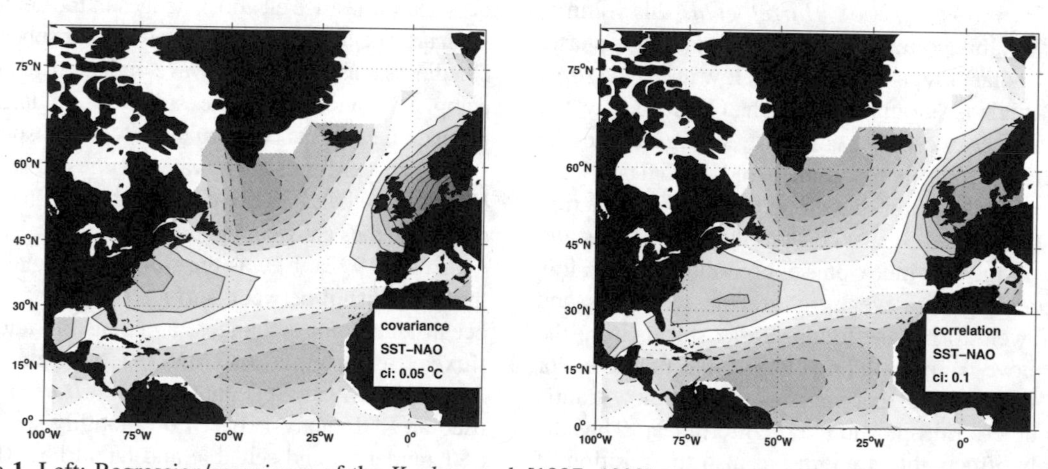

Figure 1. Left: Regression/covariance of the *Kaplan et al.* [1997; 1998] reconstructed SST data set with the *Hurrell* [1995] normalized NAO index, both averaged over the winter (December–March) season from 1900–2000. Positive values are shown with solid contours, negative values with dashed contours, and zero regression is represented by a dotted line. Note that a strong NAO index will yield about twice the temperature anomaly given in the plot. When shorter SST records from other data sets have been used, the SST covariance tends to be somewhat larger. However, pattern and correlation magnitudes are similar. Right: same as left but correlation is shown. Maximum values are on the order of 0.3–0.5, which means that the NAO explains only 8–25% of the total winter SST variance.

While there is consensus on the fast (seasonal to interannual) response of the ocean to NAO forcing, the decadal and longer-term response is controversial. *Kushnir* [1994] shows that the decadal mode of SST variability is more a hemispheric one sign response, and is thus different from the interannual tripolar SST pattern. *Visbeck et al.* [1998] and *Krahmann et al.* [2001] argue that this is the expected response for low frequency NAO forcing and, that on multi-decadal time scales, the NAO-induced SST response switches from a dipolar pattern between the subpolar and subtropical gyre to a one-sign monopole response. They propose that this is due to advection of anomalous temperatures by the Gulf Stream/North Atlantic Current system. *Delworth and Greatbatch* [2000] and *Eden and Willebrand* [2001] find a similar low frequency basin scale response. However, they attribute it to NAO-induced modulation in the upper ocean circulation as part of the Atlantic Ocean meridional overturning.

Thus care has to be taken when attributing a particular SST response pattern to the NAO since the ocean's dynamical response, which takes several years to be fully established, is non-local and is able to substantially alter the air-sea heat flux driven SST response pattern. In the following sections, we discuss in more detail several of the proposed mechanisms that cause the observed NAO-induced SST response.

3. AIR-SEA FLUX

The NAO involves a shift in atmospheric mass between the subtropics and the polar regions [e.g., *Hurrell and van Loon*, 1997; *Hurrell et al.*, 2001, *Hurrell et al.*, this volume; *Thompson et al.*, this volume]. During the high index phase, the Icelandic (polar) low is anomalously low and the Azores (subtropical) high is anomalously high. Consequently the midlatitude surface westerly winds are strong, as are the easterly surface winds in the trade wind belt. The North Atlantic storm track is well developed and has a signature that extends from the U.S. east coast to the British Isles and Scandinavia. In the low index phase, both the Icelandic low and the Azores high are weak, as are the westerlies and trades, and fewer storms are found. Some storms have the tendency to move from the United States into the Labrador Sea region, while those that make it across the Atlantic move into southern Europe and the Mediterranean. Thus the NAO not only affects the strength but also the position of the maximum westerlies and storm frequency and intensity. Variations in the NAO index and associated wind fields imply strong changes in surface air-sea flux fields of heat, momentum, and water. These changes impact both the local thermodynamic response of the mixed layer and the large-scale circulation field. We continue to examine how each of these fluxes varies with the NAO, and their implications for the North Atlantic Ocean.

3.1. Momentum Flux

Figure 3 shows the correlation and covariance between the NAO index and the winter season averaged wind stress. High correlation is found in two bands, one centered at ~55°N and the other at ~30°N, each of which is ~15° latitude wide. The NAO-induced changes in the wind stress are largest near 55°N; however, their relative magnitude compared to the mean wind stress is largest at the northern and southern boundaries of the west wind regime (65°N and 35°N). The overall pattern shows enhanced and northerly displaced westerlies north of 45°N, as well as slightly enhanced trade winds between 10°N and 30°N. Notice also the increased advection of cold Arctic air masses within the Labrador Sea and the western part of the Greenland Sea, as inferred from the covariance of wind stress. We agree with *Marshall et al.* [2001b] that the associated changes in the wind stress curl can be best understood as a meridional shift in the mean pattern rather than a modulation of its strength.

3.2. Heat Flux

Changes in the local air-sea heat fluxes are a likely cause for the observed SST anomaly pattern. The heat flux can be divided into four components, the net short wave and long wave radiation and the sensible and latent heat flux anomalies. Variability in the net short wave radiation will depend on changes in cloudiness and the sea-ice albedo. Changes in the net long wave radiation are due to changes in the lower atmospheric temperature, cloudiness, or SST. Long wave radiation anomalies tend to damp SST anomalies. The sensible and latent heat fluxes depend on gradients between the lower atmosphere and the sea surface in temperature or water vapor pressure respectively. However, both heat fluxes depend strongly on the surface wind speed and thus are well correlated.

Cayan [1992a, b] was the first to systematically examine the relationship between surface pressure anomalies, upper ocean temperature changes, and sensible and latent air-sea fluxes using the COADS (Comprehensive Ocean-Atmosphere Data Set) data set [*Woodruff et al.*, 1987]. He used an EOF analysis based on monthly sea level pressure, SST tendency, and sensible and latent heat flux anomalies. In comparison, we show heat flux anomalies regressed on the NAO index, using the NCEP/NCAR winter season averaged (DJFM) anomalies (Figure 4). In the subpolar gyre, the largest flux anomalies are due to the sensible heat loss closely followed by evaporative heat loss. In the subtropical gyre, the flux anomalies are weaker and are dominated by

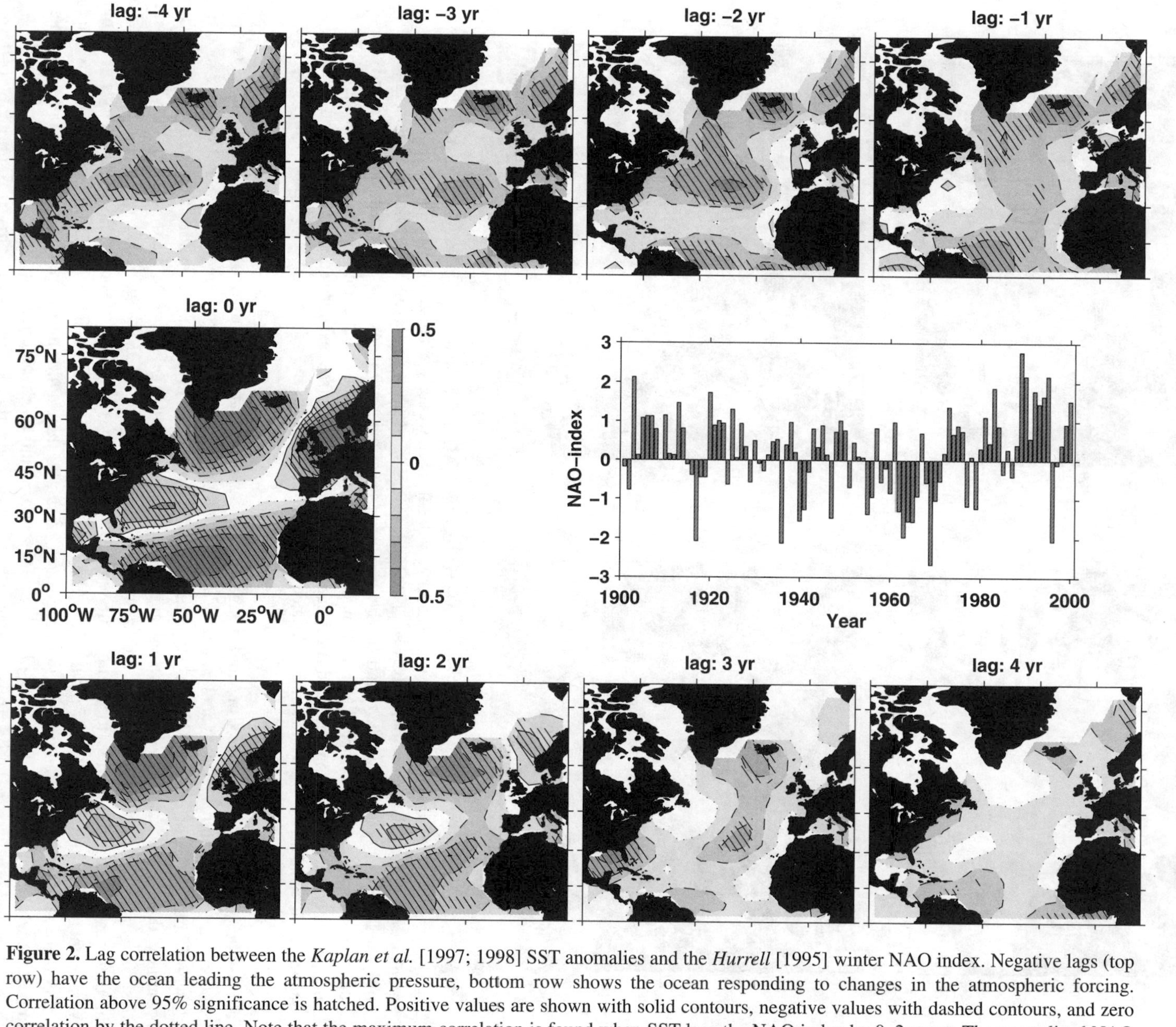

Figure 2. Lag correlation between the *Kaplan et al.* [1997; 1998] SST anomalies and the *Hurrell* [1995] winter NAO index. Negative lags (top row) have the ocean leading the atmospheric pressure, bottom row shows the ocean responding to changes in the atmospheric forcing. Correlation above 95% significance is hatched. Positive values are shown with solid contours, negative values with dashed contours, and zero correlation by the dotted line. Note that the maximum correlation is found when SST lags the NAO index by 0–2 years. The normalized NAO index is given in the middle for reference.

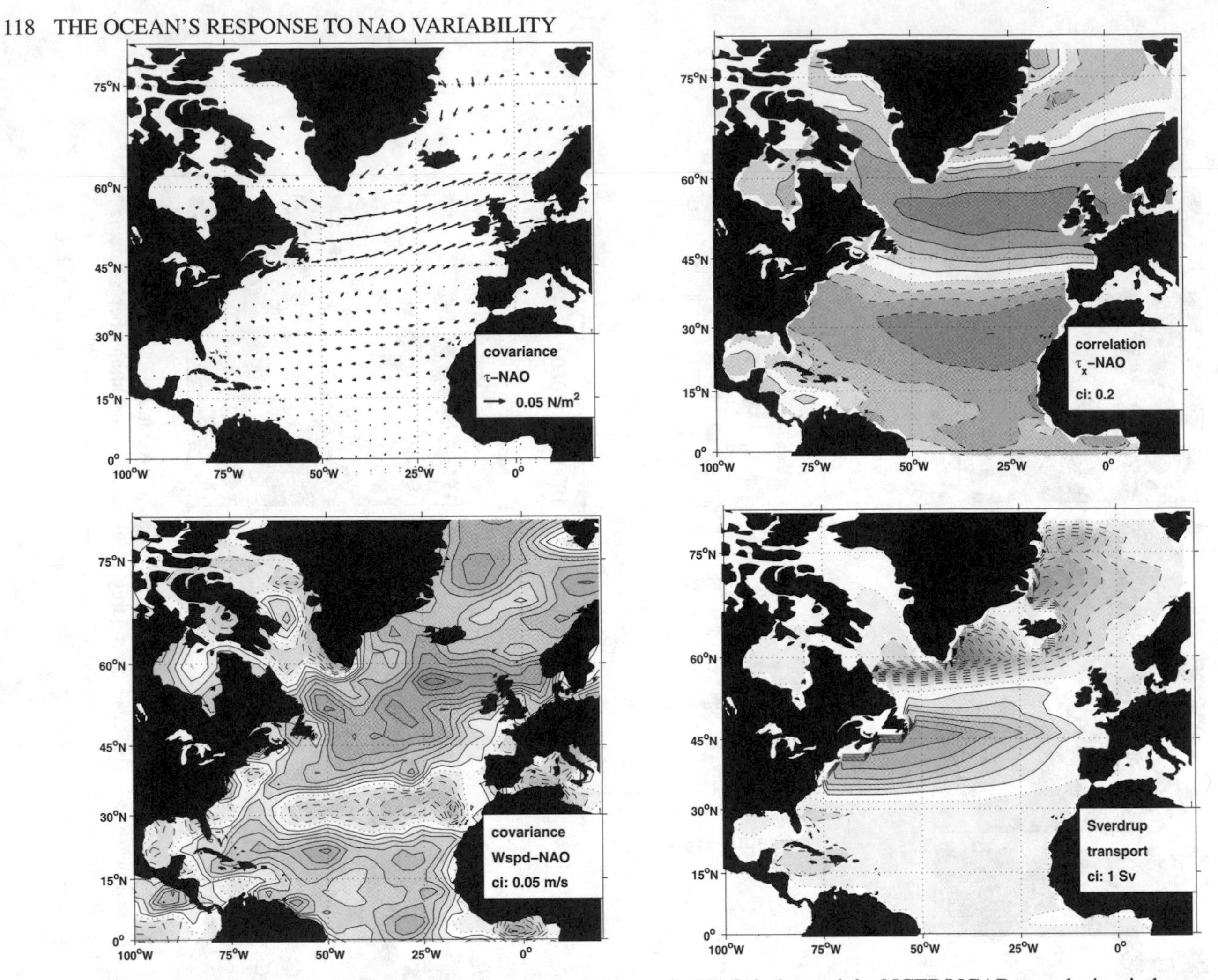

Figure 3. The upper left graph shows the covariance between the NAO-index and the NCEP/NCAR reanalysis wind stress on the ocean. The maximum response is found at 50°N with enhanced westerly wind stress by 0.1 N m^{-2}. The upper right graph shows the correlation coefficient between NAO-index zonal wind stress and NAO-index with maximum values of 0.8. The lower left graph shows the covariance between wind speed and NAO-index showing enhanced wind speed over the subpolar gyre and in the tropical North Atlantic. The lower right graph shows the barotropic equilibrium stream function anomaly as predicted from the wind stress curl using Sverdrup theory (see text for details). We expect an enhanced anticyclonic wind driven circulation of 6 Sv located between the subtropical and subpolar gyre (the 'inter-gyre gyre').

evaporation. The sum is about 20–40 W m^{-2} and is significantly lower than the monthly based analysis of *Cayan* [1992a], who found values of up to 100 W m^{-2}. The difference is largely due to the use of winter season average anomalies. *Seager et al.* [2000] discuss the relative importance of the local air-sea fluxes due to changes in wind speed versus changes in the advection of air masses. They find that equatorward of 40°N changes in wind speed are most important while poleward of 40°N both play an equal role. Variations in short wave radiative fluxes induced by changes in the cloud cover are approximately an order of magnitude smaller and are not shown. The response of the net long wave radiation is also small in the NCEP/NCAR reanalysis data set, but one might question how realistic cloud variability is represented in the assimilating model.

Changes in the Ekman transport induced upper ocean heat transport divergences (see next section) are significant. Following *Marshall et al.* [2001b] one can compute from the changes in wind stress a field of upper ocean Ekman transport anomalies. This NAO-induced Ekman transport can be then multiplied by the mean upper ocean temperature, and its divergence can be expressed as a pseudo surface heat flux (Figure 4c). This flux anomaly pattern is roughly similar to the sum of sensible and latent heat fluxes, with values of up to 20 W m^{-2} near the Gulf Stream/ North Atlantic Current region east of 50°W. We have also

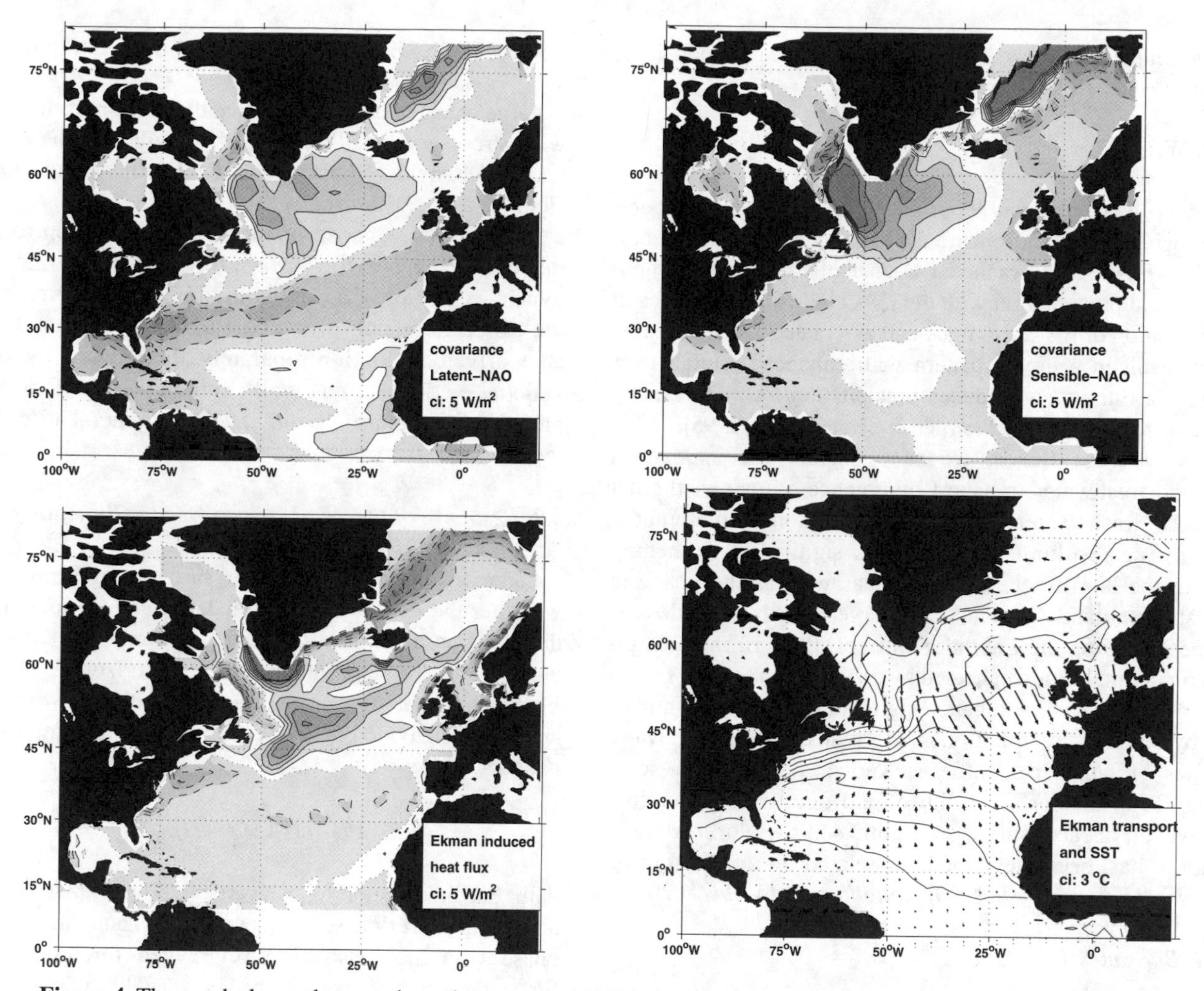

Figure 4. The graph shows the covariance between the NAO-index and the NCEP/NCAR reanalysis (1958–2000) latent heat flux (top left) and the sensible heat flux (top right). Changes in the Ekman transport (bottom right) alter the upper ocean heat transport and its divergence is expressed as a surface heat flux (bottom left). The solid lines in the lower right graph show the climatological winter SSTs and the arrows represent the NAO-induced surface Ekman transport.

estimated the contribution due to the mean surface Ekman transport divergence acting on the anomalous temperature gradient and found this term to be quite a bit smaller compared to other flux anomalies.

Can the air sea flux anomalies explain the observed SST anomalies? A 50 W m^{-2} heat flux anomaly would warm a 100 m deep mixed layer by 1°C in about 100 days, which is of the same order as the observed interannual variations in SST. A more precise answer to this question would allow us to accept or refute the "null hypothesis" of the oceanic response being merely a one dimensional integrator of air-sea fluxes (section 1). Unfortunately, one would need both good observations of the upper ocean heat content (SST and mixed-layer depth, which is quite variable in the mean and interannually) and air-sea flux anomalies, neither of which exists with accuracy high enough to rule out a significant role of ocean dynamics. Model aided studies [e.g., *Seager et al.*, 2000] suffer from uncertainties in the air-sea flux formulation and ignore the contribution of a variable depth mixed layer; thus, they cannot provide final proof. However, let us consider a sustained NAO induced heat flux anomaly winter after winter. One would expect slowly growing SST anomalies, even if the heat flux anomalies were reduced on longer time scales. Thus one might suspect that, on long time scales, changes in the ocean circulation and, thus, changes in ocean heat transport divergence will balance the remaining long-term air sea flux anomaly. On interannual time scales, however, the good spatial correspondence of the air-sea flux anomalies with the SST anomalies, as well as their corresponding magnitude, explains why the short term upper ocean temperature response can be rationalized in terms of a simple local mixed-layer heat budget calculation [e.g., *Battisti et al.*, 1995; *Seager et al.*, 2000].

3.3. Water Flux

Much less is known about the balance between evaporation, which is proportional to the latent heat flux, and precipitation. Figure 5 shows the covariance of winter mean evaporation and precipitation with the NAO index. The changes in the position of the storm track and associated moisture transport result in a dipole pattern with enhanced rainfall over northern Europe and reduced precipitation from the Canary Islands toward the Mediterranean Sea [*Hurrell*, 1995].

As for heat fluxes, we can consider the anomalous Ekman freshwater transport divergence as part of the net surface fresh water flux [Mignot, personal communication]. The resulting pattern shows significant freshening along the North Atlantic Current as well as the East and West Greenland Currents. The largest fresh water loss is expected along the Labrador Current. Parts of the precipitation anomalies are canceled by evaporation, however, and the net fresh water exchange sign and pattern are dominated by both the precipitation and Ekman induced fresh water fluxes (Figure 5d). Little is known about basin scale changes in the surface salinity field due to NAO forcing. However, decadal salinity variability in the North Atlantic region has been described as related to NAO forcing [*Reverdin et al.*, 1997; 1999; *Houghton and Visbeck*, 2002].

3.4. Buoyancy Flux

Changes in the large-scale surface density gradient of the ocean can alter the strength and character of the basin-scale meridional ocean overturning. The surface density flux depends on both the air-sea heat flux and the net fresh water flux, which can be combined into a buoyancy flux anomaly by multiplying them by the thermal and haline expansion coefficients, respectively (Figure 6). We find that the zonally-averaged response is dominated by the heat flux contribution with maximum values of 1×10^{-8} m^2 s^{-3}. One of the important impacts of enhanced or reduced surface buoyancy flux is on the maximum late winter mixed-layer depth. In general one would expect an increase in late winter mixed-layer depth in the subpolar gyre and a reduction in mixed-layer depth in the subtropical gyre for a positive NAO index season. *Khatiwala et al.* [2002] and *Marsh* [2000] have calculated the implied variability in water mass transformation rates due to interannual changes in the surface buoyancy flux. In the Labrador Sea, the production of Labrador Sea Water was expected to have increased from its mean value of 2.7 Sv (1 Sv = 10^6 m^3 s^{-1}) to 4 Sv during the high NAO index period of the early 1990s.

4. RESPONSE OF THE OCEAN CIRCULATION

Large changes in the strength and direction of the surface wind stress associated with the NAO will alter the momentum and buoyancy balance of the North Atlantic Ocean. Here we ask:

What is the response of the ocean circulation to NAO induced changes in the surface forcing? Simplified ocean dynamics can provide us with some guidance regarding the expected response. We first treat the wind and buoyancy driven circulation separately, then review the observational evidence, and finally summarize the more complex NAO response found in ocean general circulation models.

4.1. Response of the Wind Driven Ocean Circulation

Large-scale anomalies of the surface wind stress field will alter the surface frictional balance and cause an immediate response of the upper wind driven circulation. The upper layer vertical integral of the horizontal flow, the Ekman transport (M_{ek}), is proportional to the wind stress anomaly (τ') divided by a reference density (ρ_0) and Coriolis parameter (f):

$$M_{ek} = (- k \times \tau') / (\rho_0 f)$$

with k a unit vector in the vertical. The Ekman transport adjusts on short time scales (several days) and thus will be in phase with the seasonal NAO related forcing anomaly. Since the wind stress anomalies vary little in the zonal direction (Figure 3), we limit our discussion to the zonally-averaged response. During a positive NAO index phase, there is a poleward Ekman transport south of 40°N and an equatorward transport north of 40°N (Figure 7). Note, as mentioned in the previous section, a changing Ekman transport also induces an upper ocean heat transport divergence [*Marshall et al.*, 2001b] that consequently acts in a manner similar to that of heat and freshwater air-sea fluxes.

The zonally-integrated Ekman transport shows maximum southward advection of 3 Sv at 58°N and maximum northward surface flow at 30°N. Between those latitudes, the zonally averaged Ekman transport is convergent and must be balanced by downwelling below the surface layer. If we assume no net meridional transport, which is true in the Atlantic to about 1 Sv uncertainty, the zonally-averaged mean surface transport (mean and anomaly) must be balanced by an equal and opposite flow below. Thus, the zonally-averaged surface Ekman transport contributes to the meridional overturning circulation (MOC) stream function with a maximum below the Ekman layer at a depth h_{ek} ~300m.

This Ekman part of the ocean's meridional circulation response to changes in NAO forcing can be expressed as:

$$\Phi_{ek,max} = \int M_{ek} dx$$

For a homogeneous ocean with small internal friction but significant bottom friction, the return flow will occur in the frictional bottom boundary layer. On the other hand, a strongly stratified ocean basin will allow for a geostrophically balanced shallower return flow. Note that in both scenarios, we expect rapid communication with the deeper ocean circulation. Figure 7 shows the $\Phi_{ek,\max}$ expected for a positive NAO index phase. At 50°N, we expect the meridional overturning to be reduced by 1–2 Sv while at 30°N, the poleward surface transport should be enhanced by 1 Sv.

Changes in the Ekman transport divergence (Figure 7c,d) cause up/downwelling and thus perturb the large-scale potential vorticity balance of the ocean gyres. The adjustment towards a new balanced state is established by coastally trapped (boundary) waves, equatorial Kelvin waves, and long Rossby waves. While boundary and Kelvin waves have relatively fast phase speeds, the westward propagating long Rossby are much slower with decreasing phase speeds in higher latitudes. The Rossby wave phase speed is also a function of the baroclinic mode, with higher vertical modes taking much longer to adjust. The basin adjustment time scale is governed by the slowest component of the most

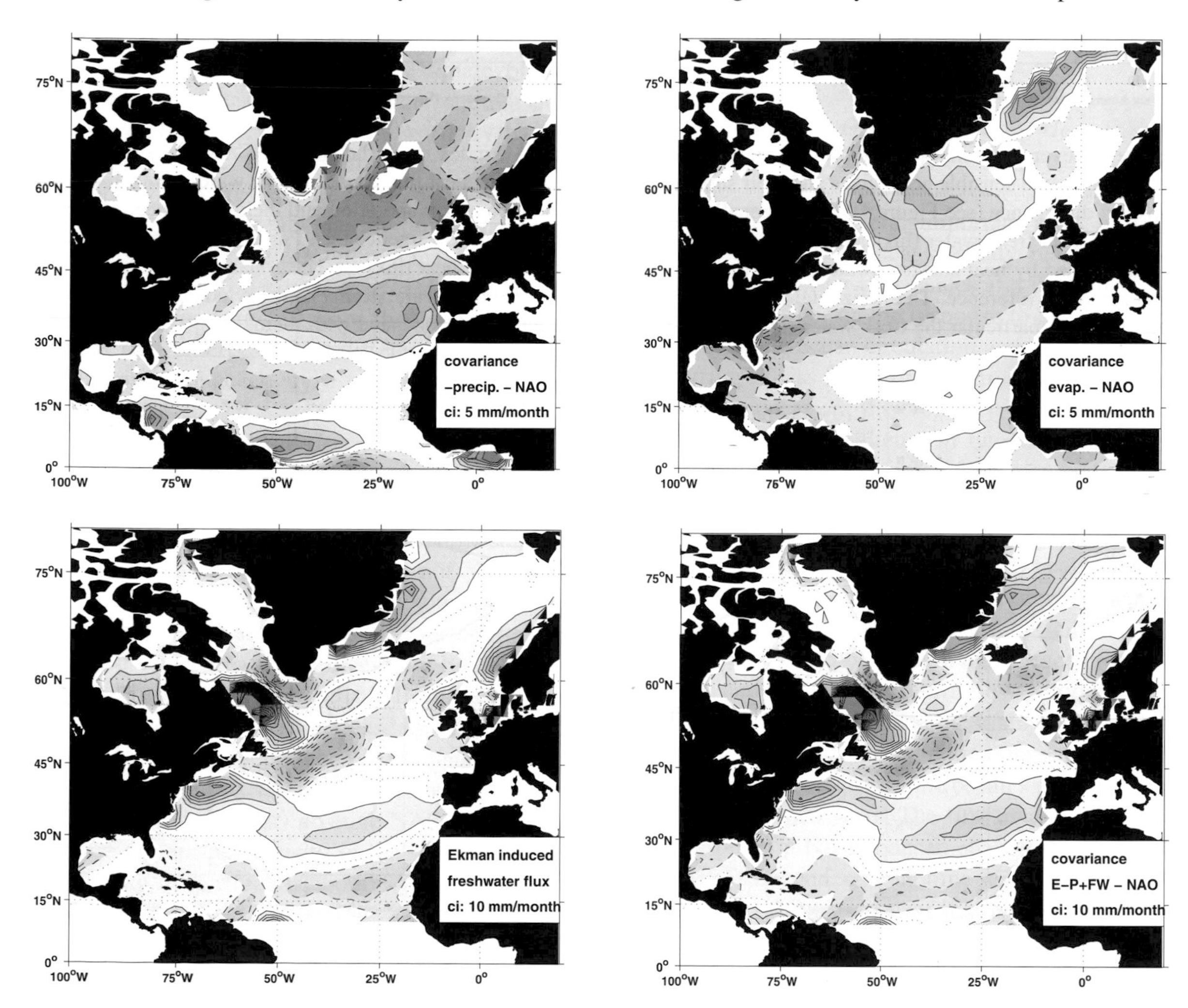

Figure 5. The top left graph shows the covariance between the NAO index and the NCEP/NCAR reanalysis precipitation (multiplied by –1). The top right graph shows the covariance between the NAO index and NCEP/NCAR reanalysis evaporation. The bottom left graph shows how changes in the surface Ekman transport acting on the mean salinity gradient will affect the surface fresh water balance (expressed here as a fresh water flux anomaly). The bottom right graph shows the sum of all fresh water fluxes. In most regions, precipitation, and in particular the Ekman induced fresh water flux anomalies, dominate the net surface fresh water flux.

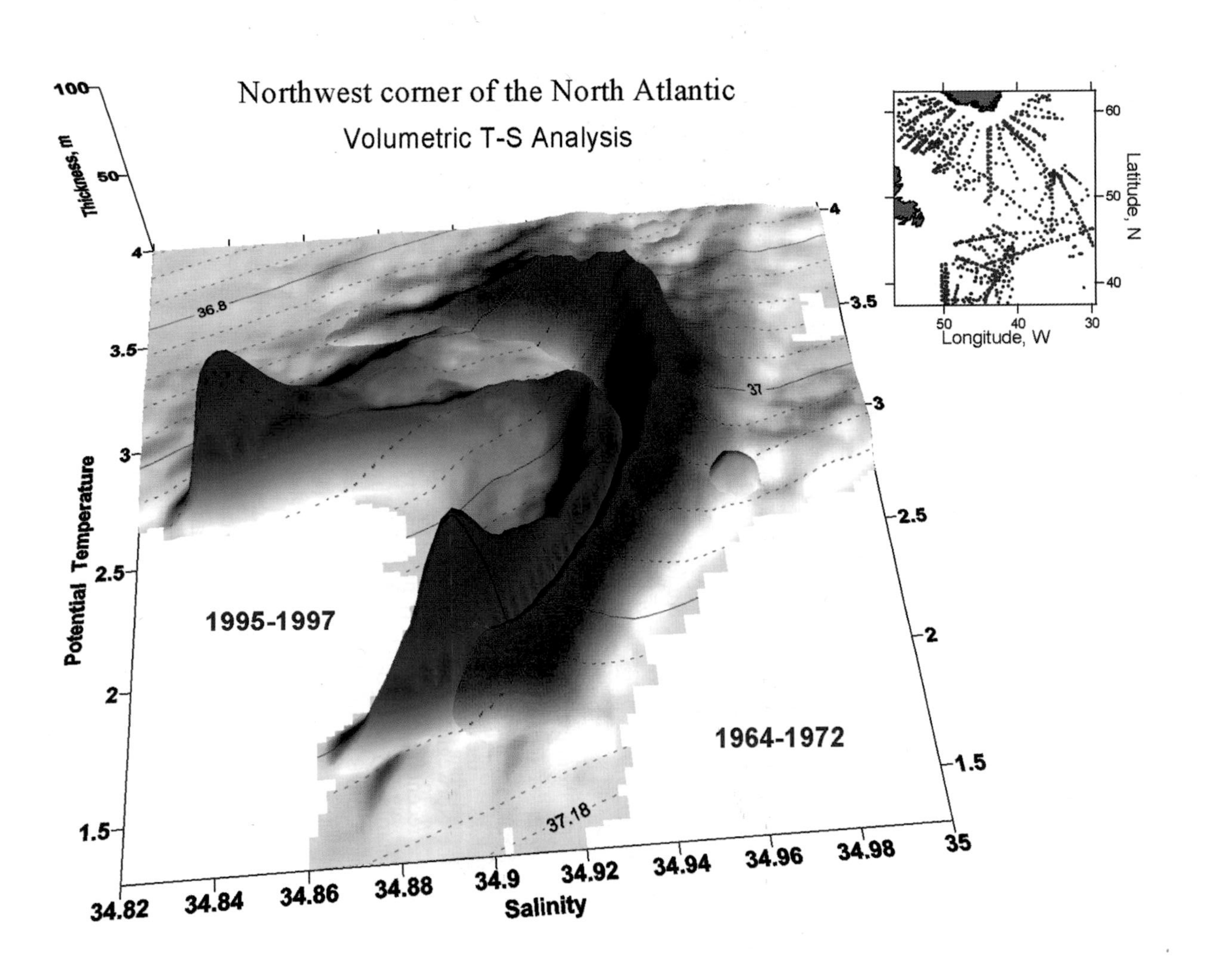

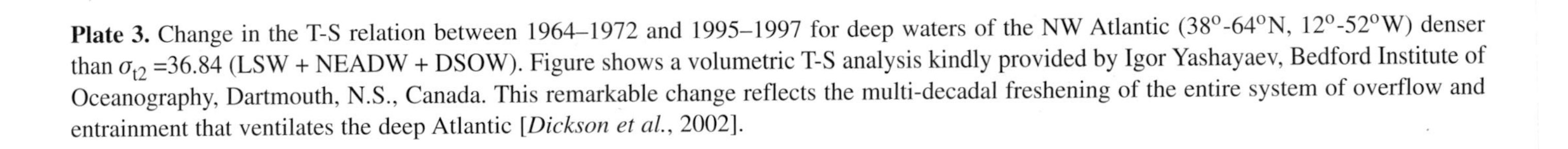

Plate 3. Change in the T-S relation between 1964–1972 and 1995–1997 for deep waters of the NW Atlantic (38°-64°N, 12°-52°W) denser than σ_{t2} =36.84 (LSW + NEADW + DSOW). Figure shows a volumetric T-S analysis kindly provided by Igor Yashayaev, Bedford Institute of Oceanography, Dartmouth, N.S., Canada. This remarkable change reflects the multi-decadal freshening of the entire system of overflow and entrainment that ventilates the deep Atlantic [*Dickson et al.*, 2002].

sea ice cover. This proximity and NAO related variability in the northward extension of the North Atlantic Current into the Greenland-Iceland-Norwegian (Nordic) Seas suggests that the Arctic sea ice cover and ice export is likely to be modified by changes in the phase of the NAO [*Dickson et al.*, 2000]. Observations indeed show a strong correlation between winter sea ice extent and the NAO index (see Figure 9). The NAO can influence Arctic sea ice through several different processes [*Deser et al.*, 2000]. These include changes in the air-ice flux of momentum and heat and changes in the divergence of the oceanic heat transport. Other mechanisms, such as variations in cloud cover that influence the Arctic radiation budget, might also play a significant role. Enhanced wind stress associated with a positive NAO index generally forces the sea ice edge southward in the Labrador Sea and further to the northeast in the Barents Sea. The effects due to variations in air-ice-ocean heat fluxes are more complex. During a positive NAO index phase, strong winds bring more warm air masses towards the Nordic Seas and Arctic Ocean thus reducing the winter sea ice production. The associated changes in the wind driven ocean circulation result in enhanced ocean advection of warm water into the Nordic Seas, in particular if the forcing persists over several winter seasons. Both mechanisms reduce sea ice cover, similar to the direct response to changes in wind stress.

Unfortunately, reliable long-term data of Arctic sea ice are mostly restricted to the position of the sea ice edge during the pre-satellite era and sea ice concentration thereafter. In addition, there are a few decades of satellite-tracked sea ice motion, but only the Fram Strait (between Greenland and Spitzbergen) has a continuous record of ice thickness over several years [*Vinje et al.*, 1997].

Recent studies of changes in Arctic sea ice coverage [e.g., *Chapman and Walsh*, 1993; *Cavalieri et al.*, 1997; *Deser et al.*, 2000; *Vinje*, 2001] have found a long-term decrease of ~3% per decade, with the strongest signal occurring during the summer months. This warrants a short discussion of the role of the seasonal cycle.

Modeling studies of *Zhang et al.* [2000] suggest that the NAO impacts the wintertime ice thickness in the Arctic, which may then precondition the summer ice concentrations even in the absence of additional anomalous atmospheric forcing during summer. The long-term decline of summer sea ice cover coincides with the period of increasing NAO index values since the mid 1960s. The associated pattern is shown in Figure 9, calculated by regressing the sea ice concentration onto the NAO index. The pattern is very similar to the first empirical orthogonal function (EOF) of the sea ice variability [*Deser et al.*, 2000] and shows a slight increase in sea ice extent in the Labrador Sea and a reduction in the Greenland and Barents Seas. Though the correlation between the principal component (PC) of the first EOF of sea ice concentration and the NAO index is high (r=0.63), *Deser et al.* [2000] note that "individual winters can be radically different." The pattern in Figure 9 is thus mostly associated with the decadal variations in the NAO index.

Figure 10 shows the evolution of winter sea ice concentration for the three main subpolar regions: the Labrador, Greenland, and Barents Seas. All time series show considerable variability on interannual to decadal time scales. Trends are weak in the Labrador and Barents Seas, while the Greenland Sea sea ice concentration is decreasing in response to the positive trend of the NAO index. Decadal variability of the NAO index is well mirrored in the respective sea ice concentration time series, with short or negligible lags (lagged correlations peak at 0-1 year lags).

The export of sea ice from the Arctic into the Nordic seas through the Fram Strait exhibits pronounced variability of the ice volume transport [*Vinje et al.*, 1997; *Harder et al.*, 1998]. Note, that increased sea ice export has been proposed as the cause of the "Great Salinity Anomaly" in the 1970s [*Dickson et al.*, 1988]. *Harder et al.* [1998] show that the ice transport is mainly a function of the southward wind component in the Fram Strait. A positive NAO index is generally associated with increased northerly winds in this region, though occasionally a zonal shift in the NAO pressure pattern leads to significant changes in the local wind direction. Such anomalous wind patterns are the cause of the "radically different winters" mentioned by *Deser et al.* [2000]. *Dickson et al.* [2000] and *Hilmer and Jung* [2000], in particular, show how the east-west shift of the Icelandic Low (which is part of the NAO index) alters Fram Strait ice export by both changes in the local atmospheric pressure gradient as well as thermodynamically induced variations in sea ice concentration.

Thus we have two competing mechanisms that affect the sea ice concentration in the Greenland Sea. For a positive NAO index winter, the large scale EOF pattern indicates decreased ice cover, while increased ice cover is suggested by changes in Fram Strait ice export (the main source of Greenland Sea sea ice). This apparent paradox might be related to the different time scales on which the two processes are important. The ice transport shows high correlation with the NAO index on short time scales, while the larger scale ice concentration response pattern (first EOF) shows better correlation with decadal and longer-term NAO variations. This difference is also apparent in the very high winter-to-winter autocorrelation (0.69) of the first PC [*Deser et al.*, 2000] and the lower autocorrelation of the NAO index for the same time period of only 0.43. In addition to east-west shifts of the Icelandic low pressure center, one might speculate that an alternative explanation is given by the slow response of ocean dynamics to NAO forcing which result in increased advection of warm waters from the North Atlantic into the Nordic seas. More research is needed to substantiate both arguments.

8. SUMMARY AND DISCUSSION

The response of the Atlantic Ocean to changes in the phase of the NAO has been of interest since the discovery of the phenomenon. Early investigators were convinced that changes in sea surface temperature and associated air-sea heat fluxes would have a significant impact on the atmospheric state and, thus, might reinforce or control the phase of the NAO. In recent years, extensive numerical experimentation with a large number of atmospheric and climate models has attempted to quantify how strong this feedback might be. Although there is still disagreement, the consensus seems to be a modest feedback from the ocean back to the atmosphere exists [*Kushnir et al.*, 2002; *Czaja et al.*, this volume].

We have argued, however, that the ocean's role is likely more important on longer time scales. To date, because of the expense to run such experiments, only very few modeling studies have addressed coupling on decadal and longer time scales [e.g., *Delworth*, 1996, *Delworth and Dixon*, 2000]. While this problem is interesting, we have restricted our discussion to the oceanic response to atmospheric NAO-like forcing through insights gained from observations, theory, and general circulation ocean model results.

Ocean and climate observations have become more abundant since the late 1960s. This period has been characterized by a trend in the boreal winter NAO index punctuated with several strong interannual events, resulting in spectacular records of climate variability and change. However, the NAO can only partially explain the observed variability. This is mostly due to the fact that theory and models indicate both rapid and slow responses of the ocean to a change in the NAO index, and also that other modes of forcing induce variability in the climate system.

On interannual time scales, NAO-induced changes in air-sea heat fluxes dominate the SST response and thus produce a well-known response pattern (Figure 1; SST tripole). This has led several investigators to conclude that, to first order, the response is essentially that of a one dimensional mixed layer in good agreement with *Hasselman's* [1976] stochastic climate model theory. However, we have shown that the anomalous Ekman transport induced heat and fresh water flux divergences are significant, as are the mean and variable geostrophic circulations in modifying the SST response. The lack of high quality basin scale air-sea flux

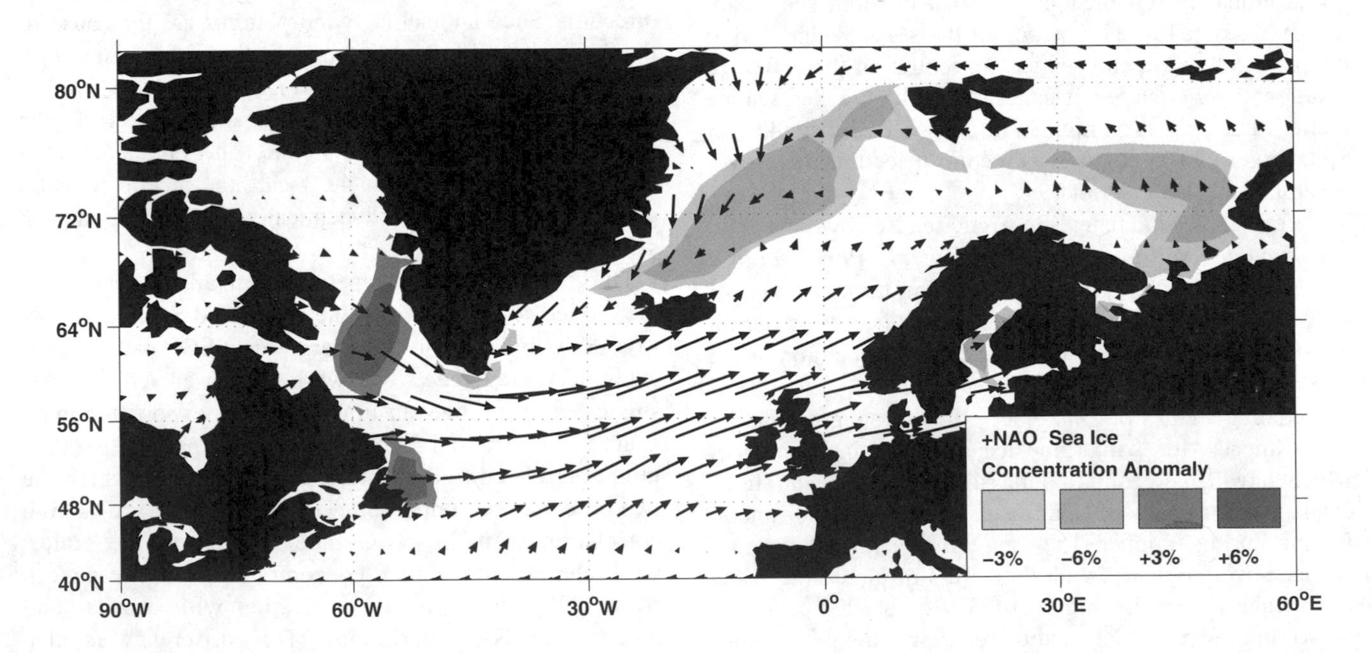

Figure 9. Winter (JFM) Arctic sea ice concentration from 1950 to 1995 [*Chapman and Walsh*, 1993] regressed onto the NAO index. Contoured are 3 and 6% changes in ice concentration. The arrows are the winter wind anomaly regressed onto the NAO index (see also Figure 3). The darker patches show increased ice concentration during a positive NAO index winter while the lighter patches show the areas where ice concentration is reduced. The sea ice concentration response to the NAO shows a pronounced seesaw pattern between the Labrador and Nordic Seas.

measurements make it difficult to give precise ratios of the changes in flux due to local atmospheric forcing versus those due to slowly adjusting ocean currents, which are often remotely forced.

We have also shown that the ocean response to NAO forcing is strongly dependant upon the frequency of the forcing. Figure 11 shows in a somewhat schematic way how one can summarize the high frequency (interannual, left) and low frequency (decadal and longer, right) response of the upper ocean. If the NAO index persists for several winter seasons, a more complex response emerges due to changes in ocean circulation. The strength and position of the boundary currents respond with a delay between 0–3 years, while the ocean meridional overturning takes up to a decade to adjust. Preferential dispersion of temperature (property) anomalies along the pathway of mean currents

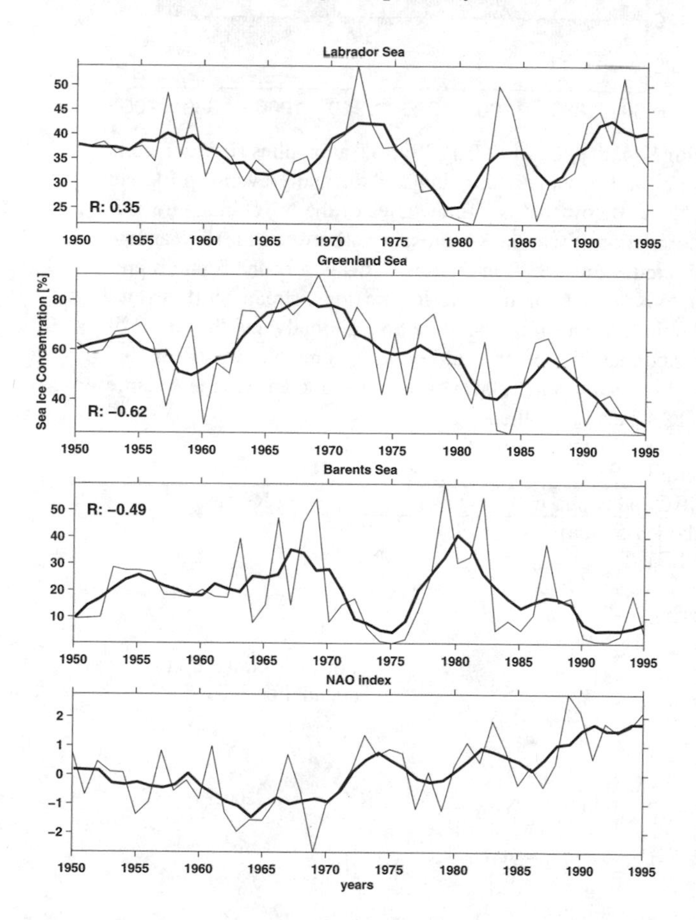

Figure 10. Time series of winter (JFM) sea ice concentration [*Chapman and Walsh*, 1993] area averaged for the Labrador Sea, Greenland Sea, and Barents Seas as well as the NAO index (lower panel). The heavy lines are obtained with a 5-year running mean filter. For each region the correlation coefficient with the NAO index is given and ranges from r=0.35 in the Labrador Sea to r= -0.62 in the Greenland Sea region.

also takes several years to show significant effects. All of these slow changes, mostly non-local, are the result of adjusting ocean dynamics that modify the response to the NAO. From theory, one would anticipate that it would take several decades to obtain a converged equilibrium response of the ocean. This time scale is set by the time it takes to ventilate the upper ocean water masses and for long planetary waves to communicate changes in pressure gradients across the ocean basin.

While the Atlantic is one of the best observed ocean basins, direct measurements of ocean circulation changes at depth are lacking, especially of the evolution of the three-dimensional salinity field. Fresh water exchanges with the Arctic Ocean are believed to be a key process and are not faithfully represented in the current generation of ocean and climate models. At this point in time, we can only speculate on what might have caused the observed decrease of salinity in the Labrador Sea that is equivalent to an input of 6 meters of fresh water. Our best estimate of local precipitation versus evaporation cannot explain a sizeable fraction of this signal.

There is mounting evidence that slowly adjusting ocean dynamics and advection of temperature anomalies can yield a change of sign of the subpolar gyre temperature response to NAO forcing. Ocean model experiments [*Visbeck et al.*, 1998; *Krahmann et al.*, 2001; *Eden and Willebrand*, 2000; *Eden and Jung*, 2001; *Delworth and Dixon*, 2000] have shown basin wide warming of upper ocean temperatures for sustained (low frequency) NAO forcing (see also Figure 11). The switch from a cold subpolar gyre to a warmer than normal state seems to occur if the positive NAO index phase persists for more than 6–8 years. The emerging consensus is that both advection of thermal anomalies along the GS/NAC and an increase in the Atlantic MOC are responsible for the subpolar gyre warming equally [Eden, personal communication]. Such insights are important when ocean properties are used to reconstruct climate changes of the past.

Table 1 lists changes in observed ocean properties that have been identified with the North Atlantic Oscillation. Table 1 of *Marshall et al.* [2001a] provides a brief summary of how those changes in the physical climate system yield to societal impacts of the NAO [see also *Drinkwater et al.*, this volume; *Mysterud et al.*, this volume; and *Straile et al.*, this volume]. In particular, *Drinkwater et al.* [this volume] review the impact of the NAO on the marine ecosystem where changes in the physical properties of the ocean are felt most directly. The marine ecosystem also plays a role in the cycling of carbon in the North Atlantic Ocean. Changes in the NAO index, for example, have been shown to directly effect primary productivity and the biological export of carbon [e.g., *Dutkiewicz et al.*, 2001; *Oschlies*, 2001]. Changes in the ocean mixed-layer depths and surface buoy-

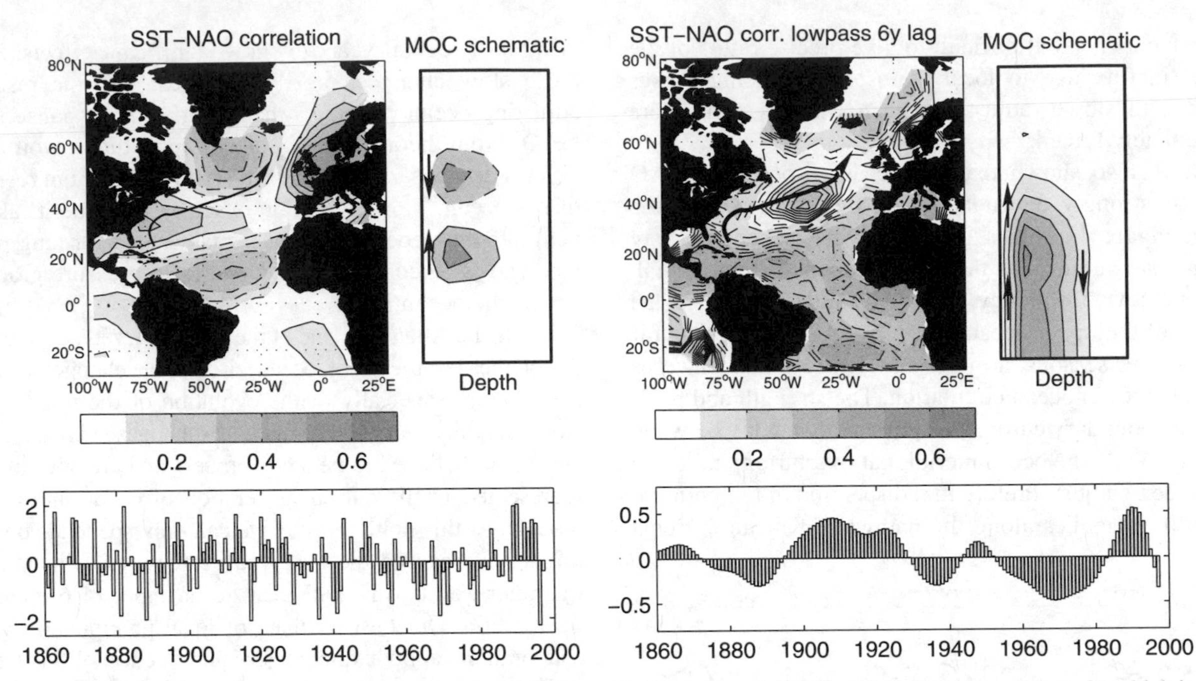

Figure 11. Correlation of the *Hurrell* [1995] NAO index with *Kaplan et al.* [1997; 1998] SST anomalies (left) which is dominated by the interannual variability. Immediately to the right, a schematic drawing of the zonally-averaged Ekman induced meridional overturning circulation (MOC) in the ocean. Below that is a time series of the NAO index for reference. The right part of the figure shows the 6 year lag correlation (ocean lags atmosphere) between the15-year low pass filtered NAO index (shown below) and the low pass filtered Kaplan SST anomalies. This part of the figure represents the decadal and lower frequency response of the ocean to NAO like forcing. Notice the down stream shift (arrows indicate position and strength of the Gulf Stream/North Atlantic Current) of the positive SST anomaly and the "loss" of the subpolar gyre cooling region. Model results have suggested that this switch could either be due to downstream dispersion of the warm temperature anomaly by the mean flow [*Krahmann et al.*, 2001] or due to an increase in the Atlantic MOC [*Eden and Willebrand*, 2001], as indicated by the MOC schematic to the left.

Table 1.

Property	High NAO index phase	Remarks
SLP subtropical High	Stronger subtropical High (+ 3-5 hPa)	
SLP Icelandic Low	Deeper polar Low (- 7-9 hPa)	
Storm tracks	More northeasterly tilt and extended tracks	
Heat flux over subpolar gyre	Enhanced ocean heat loss by 20-50 W m^{-2}	
Heat flux over subtropical gyre	Reduced ocean heat loss by 15-35 W m^{-2}	
SST within subpolar gyre	0.5-1.0 °C colder	For interannual up to decadal periods.
SST western subtropical gyre	0.3-0.7°C warmer	
SST northeastern tropical Atlantic	0.4-0.8°C colder	
Gulf Stream position	20-50 km north of its mean position (~39°N) between 70-60°W	[*Joyce et al.* 2000]
Baroclinic transport (Labrador Sea - Bermuda)	Enhanced eastward transport between the gyres by 5-9 Sv	[*Curry and McCartney* 2001]
Thickness change of Labrador Sea Water	50-100 m increase per year of forcing	[*Curry et al.* 1998]
Transport of Florida Current	Reduced by 1-2 Sv out of a mean of 32 Sv	[*Baringer and Larsen*, 2001]
Ice cover in Labrador Sea	Enhanced	[*Deser et al.* 2000]
Ice cover in Greenland Sea	Reduced	[*Deser et al.* 2000]
Deep mixing (convection) in the Labrador Sea	Enhanced up to 2500m maximum depth	[*Dickson et al.* 1996; *Lazier* 1995]

ancy fluxes directly effect the subduction rate [*Marshall et al.*, 1993; *Joyce et al.*, 2000; *Marsh*, 2000] and thus the ventilation pattern and strength of the thermocline. All of those directly affect the ocean uptake of CO_2 and other gases.

The combination of theory, numerical models, and observations has allowed us to review the basic response of the ocean circulation to NAO-induced forcing. However, several of the mechanisms await final observational and/or theoretical proofs. Competing hypotheses are being offered to explain the same phenomena. We expect to gain new insights from the next generation of data assimilating ocean models that will allow a more complete, three dimensional, description of the variability of the present and past. The oceanic response to the NAO is also a key ingredient in the development of conceptual and numerical predictive models of the coupled climate system [see also *Czaja et al.*, this volume; *Rodwell*, this volume] to tackle problems such as global climate change [*Gillett et al.*, this volume; *IPCC*, 2001].

Acknowledgements. We acknowledge conversations with countless colleagues at numerous meetings over the past years as well as insightful comments from the four reviewers. This work would have not been possible without the generous support of our funding agencies in the US and Europe.

REFERENCES

Aagaard K., L. A. Barrie, E. C. Carmack, C. Garrity, E. P. Jones, D. Lubin, R. W. Macdonald, J. H. Swift, W. B. Tucker, P. A. Wheeler and R. H. Whritner, U.S., Canadian researchers explore Arctic Ocean, *Eos*, *77*, 209 and 213, 1996.

Adlandsvik, B., Wind-driven variations in the Atlantic Inflow to the Barents Sea, ICES CM 1989/C:*18* 13 pp (mimeo), 1989.

Adlandsvik, B., and H. Loeng, A study of the climate system in the Barents Sea, *Polar Res.*, *10*, 45–49, 1991.

Alexander, M. A., and C. Deser, A mechanism for the recurrence of winter time midlatitude SST anomalies. *J. Phys. Oceanogr.*, *25*, 122–137, 1995.

Alexandersson, H., T. Schmith, K. Iden, and H. Tuomenvirta, Long-term variations of the storm climate over NW Europe, *The Global Ocean Atmosphere System*, *6*, 97–120, 1998.

Baringer, M. O., and J. C. Larsen, Sixteen years of Florida Current transport at 27N, *Geophys. Res. Lett.*, *28*, 3179–3182, 2001.

Battisti, D. S., U. S. Bhatt, and M. A. Alexander, A modeling study of the interannual variability in the wintertime North Atlantic Ocean, *J. Climate*, *8*, 3067–3083, 1995.

Belkin, I. M., S. Levitus, J. I. Antonov, and S.-A. Malmberg, "Great Salinity Anomalies" in the North Atlantic, *Prog. Oceanogr.*, *41*, 1–68, 1998.

Bersch, M., J. Meincke, and A. Sy, Interannual thermocline changes in the northern North Atlantic 1991–1996, *Deep-Sea Res. II*, *46*, 55–75, 1999.

Bjerknes, J., Atlantic air-sea interactions, *Advances in Geophysics*, *10*, 1-82, 1964.

Blindheim, J., V. Borovkov, B. Hansen, S. -A. Malmberg, W. R. Turrell, and S. Osterhus, Upper layer cooling and freshening in the Norwegian Sea in relation to atmospheric forcing, *Deep-Sea Res. I*, *47*, 655–680, 2000.

Boenisch, G., J. Blindheim, J. L Bullister, P. Schlosser, and D. W. R. Wallace, Long-term trends of temperature, salinity, density and transient tracers in the central Greenland Sea. *J. Geophys Res.*, *102*, 18553–18571, 1997.

Carmack, E. C., R. W. Macdonald, R. G. Perkin, F. A. McLaughlin, and R. Pearson, Evidence for warming of Atlantic water in the southern Canadian Basin of the Arctic Ocean: Results from the Larsen-93 Expedition, *Geophys. Res. Lett.*, *22*, 1061–1064, 1995.

Carmack, E. C., K. Aagaard, J. H. Swift, R. W. Macdonald, F. A. McLaughlin, E. P. Jones, R. G. Perkin, J. N. Smith, K. M. Ellis and L. R. Kilius, Changes in temperature and tracer distributions within the Arctic Ocean: Results from the 1994 Arctic Ocean section, *Deep-Sea Res. II, 44*, 1487–1502, 1997.

Cavalieri, D. J., P. Gloersen, C. L. Parkinson, J. C. Comiso, and H. J. Zwally, Observed hemispheric asymmetry in global sea ice changes, *Science*, *278*, 1104–1106, 1997.

Cayan, D. R., Latent and sensible heat flux anomalies over the northern oceans: Driving the sea surface temperature, *J. Phys. Oceanogr.*, *22*, 859–881, 1992a.

Cayan, D. R., Latent and sensible heat flux anomalies over the northern oceans: The connection to monthly atmospheric circulation, *J. Climate*, *5*, 354–369, 1992b.

Chapman, W. L., and J. E. Walsh, Recent variations of sea ice and air temperature in high latitudes. *Bull. Am. Meteorol. Soc.*, *74*, 33–47, 1993.

Clarke, R. A., J. H. Swift, J. A. Reid, and K. P. Koltermann, The formation of Greenland Sea Deep Water: Double diffusion or deep convection? *Deep-Sea Res.*, *37*, 1385–1424, 1990.

Cessi, P., Thermal feedback on wind stress as a contributing cause of climate variability, *J. Climate*, *13*, 232–244, 2000.

Curry, R. G., M. S. McCartney, and T. M. Joyce, Oceanic transport of subpolar climate signals to mid-depth subtropical waters, *Nature*, *391*, 575–577, 1998.

Curry, R. G., and M. S. McCartney, Ocean gyre circulation changes associated with the North Atlantic Oscillation, *J. Phys. Oceanog.*, *31*, 3374–3400, 2001.

Czaja A., and J. Marshall, Observations of atmosphere-ocean coupling in the North Atlantic, *Quart. J. Roy. Meteor. Soc.*, *127*, 1893–1916, 2001.

Czaja A., A. W. Robertson, and T. Huck, The role of Atlantic ocean-atmosphere coupling in affecting North Atlantic Oscillation variability, this volume.

Delworth, T. L., North Atlantic interannual variability in a coupled ocean-atmosphere model, *J. Climate*, *9*, 2356–2375, 1996.

Delworth, T. L. and R. J. Greatbatch, Multidecadal thermohaline circulation variability driven by atmospheric surface flux forcing, *J. Climate*, *13*, 1481–1495, 2000.

Delworth, T. L., and K.W. Dixon, Implications of the recent trend in the Arctic/North Atlantic Oscillation for the North Atlantic thermohaline circulation, *J. Climate*, *13*, 3721–3727, 2000.

Deser, C., and M. L. Blackmon, Surface climate variations over the North Atlantic Ocean during winter: 1900–1993, *J. Climate*, *6*, 1743–1753, 1993.

Deser, C., J. E. Walsh, and M. S. Timlin, Arctic sea ice variability in the context of recent atmospheric circulation trends, *J. Climate*, *13*, 617–633, 2000.

Dickson, R. R., J. Meincke, S. A. Malmberg, and A. J. Lee, The "Great Salinity Anomaly" in the northern North Atlantic 1968–1982, *Prog. Oceanogr.*, *20*, 103–151, 1988.

Dickson, R. R., From the Labrador Sea to global change, *Nature*, *386*, 649–650, 1997.

Dickson, R. R., All change in the Arctic, *Nature*, *397*, 389–391, 1999.

Dickson, B., J. Meincke, I. Vassie, J. Jungclaus, and S. Østerhus, Possible predictability in overflow from the Denmark Strait, *Nature*, *397*, 243–246, 1999.

Dickson, R. R., J. Lazier, J. Meincke, P. Rhines, and J. Swift, Long-term co-ordinated changes in the convective activity of the North Atlantic, *Prog. Oceanogr.*, *38*, 241–295, 1996.

Dickson, R. R., T. J. Osborn, J. W. Hurrell, J. Meincke, J. Blindheim, B. Adlandsvik, T. Vigne, G. Alekseev, and W. Maslowski, The Arctic Ocean response to the North Atlantic Oscillation, *J. Climate*, *13*, 2671–2696, 2000.

Dickson, R. R., I. Yashayaev, J. Meincke, W. Turrell, S. Dye, and J. Holfort, Rapid freshening of the deep North Atlantic over the past four decades, *Nature*, *416*, 832–837, 2002.

Dutkiewicz, S., Follows, M. J., J. C. Marshall and W. W. Gregg, 2001: Interannual variability of phytoplankton abundance in the North Atlantic. *Deep-Sea Res. II.*, *48*, 2323–2344, 2001.

Drinkwater, K. F., A. Belgrano, Á. Borja, A. Conversi, M. Edwards, C. H. Greene, G. Ottersen, A. J. Pershing, and H. A. Walker, The response of marine ecosystems to climate variability associated with the North Atlantic Oscillation, this volume.

Dye S., A century of variability of flow through the Faroe-Shetland Channel, PhD thesis, University of East Anglia, 189 pp, 1999.

Eden, C., and T. Jung, North Atlantic interdecadal variability: Oceanic response to the North Atlantic Oscillation (1865–1997), *J. Climate*, *14*, 676–691, 2001.

Eden, C., and J. Willebrand, Mechanism of interannual to decadal variability of the North Atlantic circulation, *J. Climate*, *14*, 2266–2280, 2001.

Esselborn, S., and C. Eden, Sea surface height changes in the North Atlantic Ocean related to the North Atlantic Oscillation, *Geophys. Res. Lett.*, *28*, 3473–3476, 2001.

Frankignoul, C., A. Czaja, and B. L'Hevender, Air-sea feedback in the North Atlantic and surface boundary conditions for ocean models, *J. Climate*, *11*, 2310–2324, 1998.

Frankignoul, C., and K. Hasselmann, Stochastic climate models. Part 2. Application to sea-surface temperature variability and thermocline variability, *Tellus*, *29*, 284–305, 1977.

Frankignoul, C., G. de Cotlogon, T. M. Joyce, and S. Dong, Gulf Stream variability and ocean-atmosphere interactions, *J. Phys. Oceanogr.*, *31*, 3516–3529, 2001.

Gangopadhyay, A., P. Cornillon, and R. D. Watts, A test of the Parsons-Veronis hypothesis on the seperation of the Gulf Stream, *J. Phys. Oceanogr.*, *22*, 1286–1301, 1992.

Gillett, N. P., H. F. Graf, and T. J. Osborn Climate change and the North Atlantic Oscillation, this volume.

Girton, J. B., T. B. Sanford, and R. H. Kase, Synoptic sections of the Denmark Strait Overflow, *Geophys. Res. Lett.*, *28*, 1619–1622, 2001.

Gordon A., Weddell deep water variability, *J. Mar. Res.*, *40*, 199–217, 1982

Greatbatch, R. J., A. F. Fanning, A. G. Goulding, and S. Levitus, A diagnosis of interpentadal circulation changes in the North Atlantic, *J. Geophys. Res.*, *96*, 22,009–22,023, 1991.

Grotefendt K., K. Logemann, D. Quadfasel, and S. Ronski, Is the Arctic Ocean warming? *J. Geophys. Res.*, *103*, 27,679-27,687, 1998.

Grötzner, A., M. Latif, and T. P. Barnett, A decadal climate cycle in the North Atlantic Ocean as simulated by the ECHO coupled GCM, *J. Climate*, *11*, 831–847, 1998.

Hansen, D. V., and H. F. Bezdek, On the nature of decadal anomalies in North Atlantic sea surface temperature, *J. Geophys. Res.*, *101*, 9749–9758, 1996.

Hansen, B., and S. Osterhus, North Atlantic-Nordic Seas exchanges, *Prog. Oceanogr*, *45*, 109–208, 2000.

Hansen, B., W. R. Turrell, and S. Østerhus, Decreasing overflow from the Nordic seas into the Atlantic Ocean through the Faroe-Shetland Channel since 1950, *Nature*, *411*, 927–930, 2001.

Häkkinen, S., Variability of the simulated meridional heat transport in the North Atlantic for the periods 1951–1993, *J. Geophys. Res.*, *104*, 10,991–11,007, 1999.

Halliwell, G., Simulation of North Atlantic decadal/multidecadal winter SST anomalies driven by basin-scale atmospheric circulation anomalies, *J. Phys. Oceanogr.*, *28*, 5–21, 1998.

Hansen, D. V., and H. F. Bezdek, On the nature of decadal anomalies in North Atlantic sea surface temperature, *J. Geophys. Res.*, *101*, 8749–8758, 1996.

Harder, M., P. Lemke, and M. Hilmer, Simulation of sea ice transport through Fram Strait: natural variability and sensitivity to forcing, *J. Geophys. Res.*, *103*, 5595–5606, 1998.

Hasselman, L., Stochastic climate models: Part I: theory, *Tellus*, *28*, 289–305, 1976.

Hilmer, M., and T. Jung, Evidence for a recent change in the link between the North Atlantic Oscillation and Arctic sea ice export, *Geophys. Res. Lett.*, *27*, 989–992, 2000.

Houghton, R. W., Subsurface quasi-decadal fluctuations in the North Atlantic, *J. Climate*, *9*, 1363–1373, 1996.

Houghton, B., and M. Visbeck, Quasi-decadal salinity fluctuations in the Labrador Sea, *J. Phys. Oceanogr.*, *32*, 687–701, 2002.

Hurrell, J. W., Decadal trends in the North Atlantic Oscillation: Regional temperatures and precipitation, *Science*, *269*, 676–679, 1995.

Hurrell, J. W., and H. van Loon, Decadal variations in climate associated with the North Atlantic Oscillation, *Clim. Change*, *36*, 301–326, 1997.

Hurrell, J. W., Y. Kushnir, and M. Visbeck, The North Atlantic Oscillation, *Science*, *291*, 603–605, 2001.

Hurrell J. W., and R R. Dickson, Climate variability over the North Atlantic, *Ecological effects of climate variations in the North Atlantic*, N. C. Stenseth, G. Ottersen, J. W. Hurrell, and A. Belgrano, Eds., Oxford University Press, in press, 2002.

Hurrell, J. W., Y. Kushnir, G. Otterson, and M. Visbeck, An overview of the North Atlantic Oscillation, this volume.

IPCC, *Climate Change 2001: The Scientific Basis*, Contribution of Working Group I to the Third Assessment Report of the Intergovernmental Panel on Climate Change (IPCC), J. T. Houghton, Y. Ding, D. J. Griggs, M. Noguer, P. J. van der Linden and D. Xiaosu, Eds., Cambridge University Press, 944 pp, 2001.

Jenkins, W. J., On the climate of a subtropical ocean gyre: decade time scale variations in water mass renewal in the Sargasso Sea, *J. Mar. Res., 42*, 265–290, 1982.

Jones, P. D., T. J. Osborn, and K. R. Briffa, Pressure-based measures of the North Atlantic Oscillation (NAO): A comparison and an assessment of changes in the strength of the NAO and in its influence on surface climate parameters, this volume.

Jónsson, S., Seasonal and interannual variability of wind stress curl over the Nordic Seas, *J. Geophys. Res*., *96*, 2649–2659, 1991.

Joyce, T. M., and P. Robbins, The long-term hydrographic record at Bermuda, *J. Climate*, *9*, 3121–3131, 1996.

Joyce, T. M., C. Deser, and M. A. Spall, The relationship between decadal variability of subtropical mode water and the North Atlantic Oscillation, *J. Climate*, *13*, 2550–2569, 2000.

Kaplan, A., M. Cane, Y. Kushnir, A. Clement, B. Blumenthal, B. Rajagopalan, Analyses of global sea surface temperature 1856–1991, *J. Geophys. Res.*, *103*, 18,567–18,589, 1998.

Kaplan, A., Y. Kushnir, M. Cane, and M. B. Blumenthal, Reduced space optimal analysis for historical data sets: 136 years of Atlantic sea surface temperatures, *J. Geophys. Res.*, *102*, 27,835–27,860, 1997.

Khatiwala, S., and M. Visbeck, An estimate of the eddy-induced circulation in the Labrador Sea, *Geophys. Res. Lett.*, *27*, 2277–2280, 2000.

Khatiwala, S., P. Schlosser, and M. Visbeck, Tracer observations in the Labrador Sea, *J. Phys. Oceanogr.*, *32*, 666–686, 2002.

Kolatschek, J., H. Eicken, V. Yu. Alexandrov, and M. Kreyscher, The sea-ice cover of the Arctic Ocean and the Eurasian marginal seas: A brief overview of present day patterns and variability, R. Stein, G. I. Ivanov, M. A. Levitan and K. Fahl, Eds., Berichte zur Polarforschung, Alfred-Wegener Instute fur Polar und Meeresforschung, Bremerhaven, Germany, *212*, 2–18, 1996.

Koltermann, K. P., A. V. Sokov, V. P. Tereschenkov, S. A. Dobroliubov, K. Lorbacher, A. Sy, Decadal changes in the thermohaline circulation of the North Atlantic, *Deep-Sea Res.*, *46*, 109–138, 1999.

Kushnir, Y., Interdecadal variations in North Atlantic sea surface temperature and associated atmospheric conditions, *J. Climate*, *7*, 142–157, 1994.

Kushnir, Y., W. A. Robinson, I. Blade, N. M. J. Hall, S. Peng, and R. Sutton, Atmospheric GCM response to extratropical SST anomalies: Synthesis and evaluation, *J. Climate,* in press, 2002.

Krahmann, G., M. Visbeck, and G. Reverdin, Formation and propagation of temperature anomalies along the North Atlantic Current, *J. Phys. Oceanogr.*, *31*, 1287–1303, 2001.

Kwok, R., and D. A. Rothrock, Variability of Fram Strait ice flux and the North Atlantic Oscillation, *J. Geophys. Res.*, *104*, 5177–5189, 1999.

Lascaratos, A., W. Roether, K. Nittis, and B. Klein, Recent changes in deep water formation and spreading in the eastern Mediterranean Sea, *Prog. Oceanogr.*, *44*, 5–36, 1999.

Lazier, J. R. N., Oceanographic conditions at O.W.S. Bravo, *Atmos.-Ocean*, *18*, 227–238, 1981.

Lazier, J. R. N., Temperature and salinity changes in the deep Labrador Sea, 1962–1986, *Deep Sea Res.*, *35*, 1247–1253, 1988.

Lazier, J. R. N., The salinity decrease in the Labrador Sea over the past thirty years, *Natural Climate Variability on Decade-to-Century Time Scales,* D. G. Martinson, K. Bryan, M. Ghil, M. M. Hall, T. M. Karl, E. S. Sarachik, S. Sorooshian, and L. D. Talley, Eds., National Academy Press, 295-304, 1995.

Levitus, S., Interpentadal variability of temperature and salinity at intermediate depths of the North Atlantic ocean, 1970–1974 versus 1955–1959, *J. Geophys. Res.*, *94*, 6091-6131, 1989.

Levitus, S., Interpentadal variability of Steric Sea level and geopotential thickness of the North Atlantic Ocean, 1970–1974 versus 1955–1959, *J. Geophys. Res.*, *95*, 5233–5238, 1990.

Loeng, H., V. Ozhigin, and B. Adlandsvik, 1997. Water fluxes through the Barents Sea, *ICES J. Mar. Sci., 54*, 310-317, 1997.

Lorbacher, K., Niederfrequente variabilitaet meridionaler transporte in der divergenzzone des nordatlantischen subtropen- und subpolarwirbels – Der WOCE-Schnitt A2. Berichte des BSH, *22*, 156 pp., 2000

Luksch, U., Simulation of North Atlantic low-frequency SST variability, *J. Climate*, *9*, 2083–2092, 1996.

Marotzke, J., Boundary mixing and the dynamics of three-dimensional thermohaline circulations, *J. Phys. Oceanogr.*, *27*, 1713–1728, 1997.

Marsh, R., Recent variability of the North Atlantic thermohaline circulation inferred from surface heat and freshwater fluxes, *J. Climate*, *13*, 3239–3260, 2000.

Marshall, J., A. J. G. Nurser, and R. G. Williams, Inferring the subduction rate and period over the North Atlantic, *J. Phys. Oceanogr.*, *23*, 1315–1329, 1993.

Marshall, J., Y. Kushnir, D. Battisti, P. Chang, A. Czaja, R. Dickson, J. Hurrell, M. McCartney, R. Saravanan, and M. Visbeck, North Atlantic climate variability: phenomena, impacts and mechanisms, *Int. J. Climatol.*, *21*, 1863–1898, 2001a.

Marshall, J., H. Johnson, and J. Goodman, A study of the interaction of the North Atlantic Oscillation with the ocean circulation, *J. Climate*, *14*, 1399–1421, 2001b.

McLaughlin, F. A., E. C. Carmack, R. W. Macdonald, and J. K. B. Bishop, Physical and geochemical properties across the Atlantic/Pacific water mass front in the southern Canadian Basin, *J. Geophys. Res.*, *101*, 1183–1197, 1996.

Meincke, J., S. Jonsson, and J. H. Swift, Variability of convective conditions in the Greenland Sea, *ICES Mar. Sci. Symp. 195*, 32–39, 1992.

Meincke, J., and B. Rudels,. Greenland Sea deep water: A balance between convection and advection. *Extended Abstracts, Nordic Seas Symposium*. Hamburg, Germany, 143–148, 1995.

Meincke, J., B. Rudels, and H. J. Friedrich, The Arctic Ocean-Nordic Seas thermohaline system, *J. Mar Sci.*, *54*, 283–299, 1997.

Molinari, R. L., D. Mayer, J. F. Festa, and H. F. Bezdek, Multiyear variability in the near-surface temperature structure of the mid-latitude western North Atlantic Ocean, *J. Geophys. Res.*, *102*, 3267–3278, 1997.

Molinari, R. L., R. A. Fine, W. D. Wilson, R. G. Curry, J. Abell, and M. S. McCartney, The arrival of recently formed Labrador Sea Water in the Deep Western Boundary Current at 26.5N, *Geophys. Res. Lett.*, *25*: 2249–2252, 1998.

Morison, J., M. Steele, and R. Anderson, Hydrography of the upper Arctic Ocean measured from the Nuclear Submarine, USS PARGO, *Deep-Sea Res., I, 45*, 15–38, 1998a.

Morison, J, K. Aagaard, and M. Steele, Report on the Study of the Arctic Change Workshop, November 10–12, 1997, University of Washington, ARCSS Report No. 8., 63 pp, 1998b.

Morison, J, K. Aagaard, and M. Steele, Recent changes in the Arctic: a review, *Arctic*, in press, 2002.

Mork, K. A., and J. Blindheim, Variations in the Atlantic inflow to the Nordic Seas, 1955–1996, *Deep-Sea Res., I, 47,* 1035–1057, 2000.

Mysterud, A., N. C. Stenseth, N. G. Yoccoz, G. Ottersen, R. Langvatn, The response of terrestrial ecosystems to climate variability associated with the North Atlantic Oscillation, this volume.

Orvik, K. A., O. Skagseth, and M. Mork, Atlantic inflow to the Nordic Seas: current structure and volume fluxes from moored current meters, VM-ADCP and SeaSoar-CTD observations, 1995–1999. *Deep-Sea Res., I, 48*, 937–957, 2001.

Østerhus. S., and T. Gammelsrod, The abyss of the Nordic Seas is warming, *J. Climate*, *12*, 3297–3304, 1999.

Oschlies, A., NAO-induced long-term changes in nutrient supply to the surface waters of the North Atlantic, *Geophys. Res. Lett.*, *28*, 1751–1754, 2001.

Paiva, A. M., and E. P. Chassignet, North Atlantic modeling of low-frequency variability in mode water formation, *J. Phys. Oceanogr.*, *32,* 2666–2680.

Parilla, G., A. Lavin, H. Bryden, M. Garcia, and R. Millard, Rising temperatures in the subtropical North Atlantic Ocean over the past 35 years, *Nature*, *369,* 48–51, 1994.

Parsons, A. T., Two layer model of Gulf Stream separation, *J. Fluid Mech.*, *39*, 511–528, 1969.

Petrie, B., and K. Drinkwater, Temperature and salinity variability on the Scotian Shelf and in the Gulf of Maine 1945–1990, *J. Geophys. Res.*, *98,* 20,079-20,089, 1993.

Quadfasel, D., A. Sy, D. Wells, and A. Tunik, Warming in the Arctic, *Nature*, *350*, 385, 1991.

Rahmstorf, S., and A.Ganopolski, Long-term global warming scenarios computed with an efficient coupled climate model, *Climatic Change*, *43*, 353-367, 1999.

Rahmstorf, S., A simple model of seasonal open-ocean convection. Part I: Theory, *Ocean Dyn.*, *52*, 26–35, 2001.

Reverdin, G., D. Cayan, and Y. Kushnir, Decadal variability of hydrography in the upper northern North Atlantic in 1948–1990, *J. Geophys. Res.*, *102*, 8505–8531, 1997.

Reverdin, G., N. Verbrugge, H. Valdimarsson, Upper ocean variability between Iceland and Newfoundland 1993–1998, *J. Geophys. Res.*, *104*, 29,599-29,611, 1999.

Rodwell, M. J., On the predictability of the North Atlantic climate, this volume.

Rogers, J. C., Patterns of low-frequency monthly sea-level pressure variability (1899–1986) and associated wave cyclonic frequencies, *J. Climate*, *3*, 1364–1379, 1990.

Rudels, B., and D. Quadfasel, Convection and deep water formation in the Arctic Ocean-Greenland Sea system, *J. Mar. Sys.*, *2*, 435–450, 1991.

Saravanan, R., and J. C. McWilliams, Advective ocean-atmosphere interaction: an analytical stochastic model with implications for decadal variability, *J. Climate*, *11*, 165–188, 1998.

Schlosser, P., G. Bonisch, M. Rhein, and R. Bayer, Reduction of deepwater formation in the Greenland Sea during the 1980's: Evidence from tracer data, *Science*, *251*, 1054–1056, 1991.

Seager, R., Y. Kushnir, M. Visbeck, N. Naik, J. Miller, G. Krahmann, and H. Cullen, Causes of Atlantic Ocean climate variability between 1958 and 1998, *J. Climate*, *13*, 2845–2862, 2000.

Selten, F. M., R. J. Haarsma, and J. D. Opsteegh, On the mechanism of North Atlantic decadal variability. *J. Climate*, *12*, 1956–1973, 1999.

Smethie, W. M., R. A. Fine, A. Putzka, and E. P. Jones, Tracing the flow of North Atlantic Deep Water using chlorofluorocarbons, *J. Geophys. Res.*, *105*, 14,297–14323, 2000.

Stephenson, D. B., H. Wanner, S. Brönnimann, J. Luterbacher, The history of scientific research on the North Atlantic Oscillation, this volume.

Sturges, W., and B. G. Hong, Wind forcing of the Atlantic thermocline along 32N at low frequencies, *J. Phys. Oceanogr.*, *25*, 1706–1715, 1995.

Straile, D., D. M. Livingstone, G. A. Weyhenmeyer, and D. G. George, The response of freshwater ecosystems to climate variability associated with the North Atlantic Oscillation, this volume.

Sutton, R. T., and M. R. Allen, Decadal predictability of North Atlantic sea surface temperature and climate. *Nature*, *388*, 563–567, 1997.

Swift, J. H., E. P. Jones, K. Aagaard, E. C. Carmack, M. Hingston, R. W. Macdonald, F. A. McLaughlin, and R. G. Perkin, Waters of the Makarov and Canada Basins, *Deep-Sea Res.*, *44*, 1503–1529, 1997.

Sy, A,. M. Rhein, J. R. N. Lazier, K. P. Koltermann, J. Meincke, A. Putzka, and M. Bersch, Surprisingly rapid spreading of newly formed intermediate waters across the North Atlantic Ocean, *Nature, 386*, 675–679, 1997.

Talley, L. D., and M. E. Raymer, Eighteen degree water variability, *J. Mar. Res.*, *40*, 757–775, 1982.

Talley, L. E., North Atlantic circulation and variability, reviewed for the CNLS conference, *Physica D*, *98*, 625–646, 1996.

Taylor, A. H., and A. Gangopadhyay, A simple model of interannual shifts of the Gulf Stream, *J. Geophys. Res.*, *106*, 13,849–13,860, 2001.

Taylor, A. H., and J. A. Stephens, The North Atlantic Oscillation and the latitude of the Gulf Stream, *Tellus*, *50*,134–142, 1998.

Tereshchenko, V. V., Seasonal and year-to-year variations of temperature and salinity along the Kola meridian transect. *ICES CM 1996/C:11* 24 pp (mimeo), 1996.

Thompson, D. W. J., S. Lee, and M. P. Baldwin, Atmospheric processes governing the Northern Hemisphere annular mode/North Atlantic Oscillation, this volume.

Turrell, W. R., G. Slesser, R. D. Adams, R. Payne, and P. A Gillibrand, Decadal variability in the composition of Faroe-Shetland Channel bottom water, *Deep-Sea Res.*, *46*, 1–25, 1999.

Verduin, J., and D. Quadfasel Long-term temperature and salinity trends in the central Greenland Sea, European Sub-Polar Ocean Programme II, Final Scientific Report, E. Jansen, Ed., 1999.

Veronis, G., Model of World Ocean Circulation: I. Wind-driven, two-layer, *J. Mar. Res.*, *31*, 228–288, 1973.

Vinje, T., N. Nordlund, and A. Kvambekk, Monitoring ice thickness in Fram Strait, *J. Geophys. Res.*, *103*, 10,437-10,449, 1998.

Vinje, T., Anomalies and Trends of Sea-Ice Extent and Atmospheric Circulation in the Nordic Seas during the Period 1864-1998. *J. Climate*, *14*, 255-267, 2001a.

Vinje, T., Fram Strait ice fluxes and atmospheric circulation, 1950–2000, *J. Climate*, *14*, 3508–3517, 2001b.

Visbeck, M., H. Cullen, G. Krahmann, and N. Naik, An oceans model's response to North Atlantic Oscillation like wind forcing. *Geophys. Res. Lett.*, *25*, 4521–4524, 1998.

Visbeck, M., J. Hurrell, L. Polvani, and H. Cullen, The North Atlantic Oscillation, present, past and future, *PNAS*, *98*, 12,876–12,877, 2001.

Visbeck M., and M. Rhein, Is bottom boundary layer mixing slowly ventilating Greenland Sea Deep Water? *J. Phys. Oceanogr.*, *30*, 215–224, 2000.

Yang, J., A linkage between decadal climate variations in the Labrador Sea and the tropical Atlantic Ocean, *Geophys. Res. Lett.*, *26*,1023–1026, 1999.

Wallace, J. M., and D. S. Gutzler, Teleconnections in the geopotential height field during the Northern Hemisphere winter, *Mon. Weather. Rev.*, *109*, 784–812, 1981.

Walker, G. T., and E. W. Bliss, World Weather V. *Mem. R. Meteorol. Soc,*. *44*, 53–83, 1932.

Watanabe, M., and M. Kimoto, On the persistence of decadal SST anomalies in the North Atlantic, *J. Climate*, *13*, 3017–3028, 2000.

Weng, W., and J. D. Neelin, On the role of ocean-atmosphere interaction in midlatitude interdecadal variability, *Geophys. Res. Lett.*, *25*, 167–170, 1998.

Woodruff, S. D., R. J. Slutz, R. L. Jenne, and P. M. Streurer, A comprehensive ocean-atmosphere data set, *Bull. Am. Meteorol. Soc.*, *68*, 1239–1250, 1987.

Xie, S.-P., and Y. Tanimoto, A pan-Atlantic decadal climate oscillation, *Geophys. Res. Lett.*, *25*, 2185–2188, 1998.

Zhang, J., D. Rothrock, and M. Steele, Recent changes in Arctic sea ice: The interplay between ice dynamics and thermodynamics, *J. Climate*, *13*, 3099–3114, 2000.

Martin Visbeck, Lamont-Doherty Earth Observatory of Columbia University, Department of Earth and Environmental Sciences, 206D Oceanography, 61 Route 9W, Palisades, New York, 10964, USA.
visbeck@ldeo.columbia.edu

Eric P. Chassignet, Rosenstiel School of Marine & Atmospheric Science/MPO, University of Miami, 4600 Rickenbacker Causeway, Florida, 33149, USA.
echassignet@rsmas.miami.edu

Ruth Curry, Department of Physical Oceanography, Woods Hole Oceanographic Institution, Woods Hole, Massachusetts, 02543 USA.
rcurry@whoi.edu

Tom Delworth, Geophysical Fluid Dynamics Laboratory/National Oceanic and Atmospheric Administration, Princeton, New Jersey, 08542, USA.
td@gfdl.gov

Robert R. Dickson, Center for Environment, Fisheries and Aquaculture Science, Lowestoft Laboratory, Pakefield Road, Lowestoft, Suffolk NR33 0HT United Kingdom
r.r.dickson@cefas.co.uk

Gerd Krahmann, Lamont-Doherty Earth Observatory of Columbia University, Oceanography Division, 206C Oceanography, 61 Route 9W, Palisades, New York 10964, USA.
krahmann@ldeo.columbia.edu

The Role of Atlantic Ocean-Atmosphere Coupling in Affecting North Atlantic Oscillation Variability

Arnaud Czaja

Massachusetts Institute of Technology, Cambridge, Massachusetts, USA

Andrew W. Robertson

International Research Institute for Climate Prediction, Palisades, New York, USA

Thierry Huck

Laboratoire de Physique des Océans, Brest, France

We review the role of ocean-atmosphere interactions over the Atlantic sector in North Atlantic Oscillation (NAO) variability. The emphasis is on physical mechanisms, which are illustrated in simple models and analyzed in observations and numerical models. Some directions of research are proposed to better assess the relevance of Atlantic air-sea interactions to observed and simulated NAO variability.

INTRODUCTION

In *Thompson et al.* [this volume], it was suggested that extratropical atmospheric dynamics alone might set the horizontal and vertical scales of North Atlantic Oscillation (NAO) variability, as well as its background 'white' temporal spectrum. This is a notable feature of extra-tropical climate variability since, unlike the tropical Pacific where both oceanic and atmospheric dynamics are needed to create a strong interannual signal like El Niño-Southern Oscillation (ENSO), NAO variability on timescales of months to decades might primarily reflect intrinsic atmospheric processes alone. The atmospheric cap north of the equator is, however, not an isolated system.It exchanges heat, moisture and momentum with the ocean, the land, the biosphere and the cryosphere below, as well as with the tropics at its southern boundary. All these interactions could possibly influence NAO variability on certain timescales. To motivate this statement, we show (Figure 1) a comparison between the observed boreal winter (DJF) NAO index and that simulated by an ensemble of atmospheric general circulation model (GCM) experiments forced by the observed, global, time-varying sea surface temperature (SST) and sea ice anomalies over 1947–1997. One clearly observes some skill in reproducing the low-frequency time evolution of the NAO index over the last decades. Although the interpretation of these experiments is subtle [*Bretherton and Battisti*, 2000; *Czaja and Marshall*, 2000], this quite realistic NAO simulation would be impossible if SST anomalies and/or sea ice anomalies had no impact on the NAO. In this chapter, we will focus on the potential role of the Atlantic Ocean in providing such a modulation of intrinsic NAO variability. We have chosen to decompose NAO Atlantic Ocean interactions into two distinct physical frameworks, according to whether or not they involve ocean dynamics. Each framework is now introduced in turn.

It is firmly established from both observational [e.g., *Cayan*, 1992; *Halliwell and Mayer*, 1996; *Deser and Timlin*, 1997] and modeling studies [e.g., *Battisti et al.*, 1995; *Halliwell*, 1998] that NAO variability intrinsic to the atmosphere drives large-scale changes in SST over the North Atlantic. Changes in windspeed of the westerlies and the trade winds, associated with the NAO, lead to a modulation of turbulent heat loss at the ocean surface and entrainment at the base of the ocean mixed layer. In addition, changes in surface windstress drive anomalous Ekman currents that add constructively to these forcings north of 30°N [e.g., *Marshall et al.*, 2001a; *Visbeck et al.*, this volume]. Thus, even in the absence of further dynamical effects associated with geostrophic ocean circulation (see below), intrinsic NAO variability drives a tripolar anomaly pattern in SST (Figure 2).

The North Atlantic Oscillation:
Climatic Significance and Environmental Impact
Geophysical Monograph 134

10.1029/134GM07

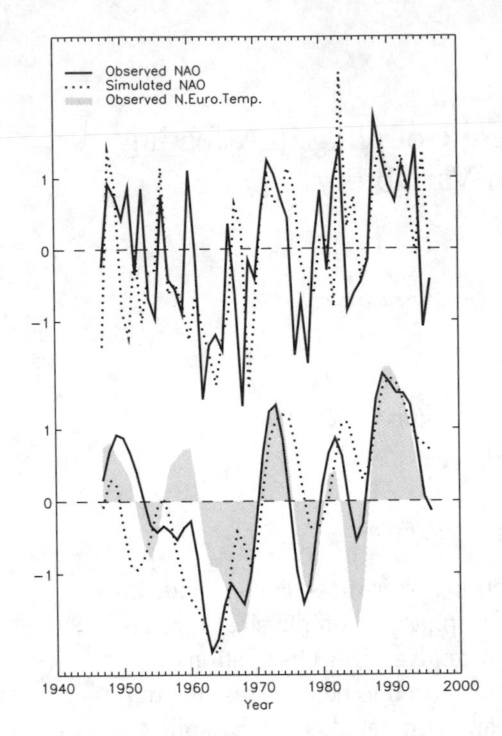

Figure 1. Observed (solid line) and modeled ensemble average (dotted line) of the *normalized winter* (DJF) NAO index. The lower graph shows the normalized NAO index time series after they have been filtered to pass variations with periods greater than 6.5 years. Shading in lower graph is the normalized filtered time series of observed North European surface temperature (averaged over 5–50°E, 50–70°N). The year corresponds to December for each DJF season. From *Rodwell et al.* [1999].

A more open question is the extent to which NAO variability is substantially modified by the presence of the SST tripole.

If air-sea interactions were purely local in space, one would expect thermal equilibration between the upper ocean and the troposphere through surface heat flux to become important in setting the amplitude of temperature anomalies in both fluids. Inspection of temperature anomalies in the free troposphere associated with NAO (Figure 2, contours) indeed suggests a certain degree of thermal equilibration, with warm air above warm SST and cold air above cold SST, and a similar amplitude of perturbation ~0.3 K). The feedback of the SST tripole onto the NAO is, however, expected to be more complex because of nonlocal dynamical effects. In particular, since the temperature anomalies in Figure 2 extend meridionally from the equator to the North Pole, very different mechanisms are involved. The two northern centers of action of the SST tripole modulate the SST gradient in middle latitudes, which could impact the storm track and subsequently the large-scale atmospheric flow. Changes in SST associated with the southern lobe of the SST tripole modulate the cross-equatorial SST gradient and diabatic heating associated with the Intertropical Convergence Zone (ITCZ) and South American monsoon, which could feedback onto the NAO through forced Rossby wavetrains and/or changes in the sectorial Hadley circulation. Below, the relative roles of local thermal equilibration and nonlocal SST feedbacks will be discussed in the light of a linear stochastic climate model. In addition, we will review the physical mechanisms as well as the observational and modeling support for a feedback of the SST tripole on the NAO.

The fact that the Atlantic Ocean is not just a heat reservoir providing sources and sinks of energy for NAO variability, but also carries energy horizontally, introduces new features to Atlantic air-sea interactions. In particular, it makes NAO/Atlantic Ocean interactions not merely restricted to NAO/SST tripole interactions. It is firmly established from observations and numerical simulations that NAO variability drives large-scale anomalies in currents, density and water masses of the North Atlantic Ocean. As reviewed in *Visbeck et al.* [this volume], the dynamical ocean response can occur on a broad range of timescales with distinct spatial patterns. Various observational studies have shown a fast response of the Gulf Stream/recirculation system to NAO variability (about a 1-year lag), leading to localized SST anomalies north of the Gulf Stream extension [e.g., *Taylor and Stephens*, 1998; *Frankignoul et al.*, 2001b]. On longer (decadal and interdecadal) timescales, the ocean response involves basin-scale currents and SST anomalies, as hinted at in various numerical studies [e.g., *Visbeck et al.*, 1998; *Krahman et al.*, 2001; *Eden and Jung*, 2001; *Eden and Willebrandt*, 2001]. In addition, like the atmosphere, the ocean circulation displays intrinsic variability on a broad range of timescales (e.g., interannual to decadal associated with the wind-driven gyres, interdecadal timescales associated with the meridional overturning circulation (MOC)). The possibility then, that ocean circulation changes forced by the NAO, or intrinsic to the Atlantic Ocean, might influence the NAO through their impact on the SST field allows coupled atmosphere-ocean variability on timescales set by ocean dynamics.

In a series of papers, *Bjerknes* [1958; 1962; 1964] was the first to raise this issue, and argued that interdecadal fluctuations of the zonal index [a possible NAO index, see *Hurrell et al.*, this volume] might reflect compensation between changes in oceanic and atmospheric energy transport. In his own words,

"The [above] hypothesis concerning quasi constant total meridional heat flux and opposite fluctuations of its oceanic and atmospheric parts, does explain the possibility of relatively big variations in climate without having recourse to primary solar changes." [*Bjerknes*, 1964].

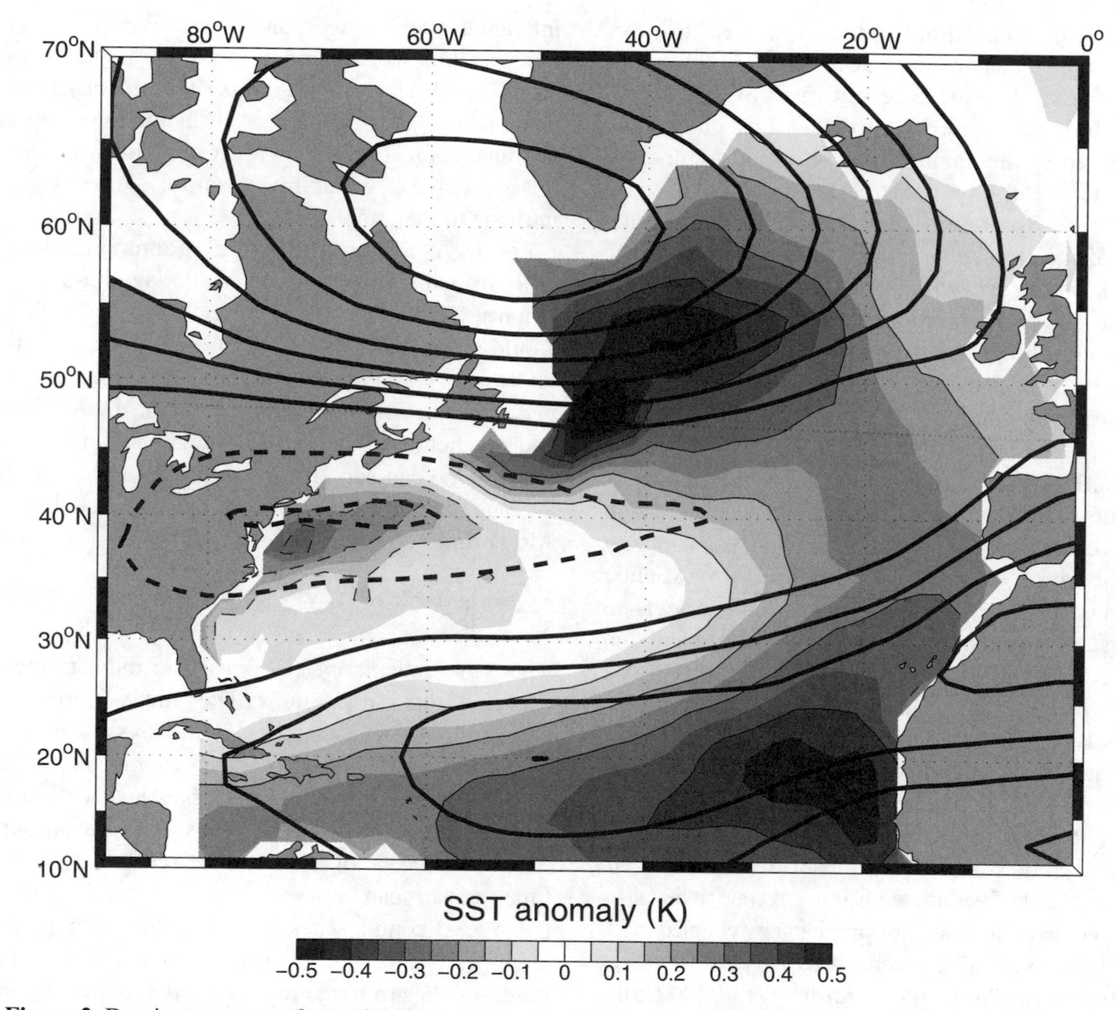

Figure 2. Dominant pattern of covariability between monthly temperature at 500 mb (contoured every 0.1 K, dashed when negative) and SST (shaded with dashed contours when negative). Anomalies from the NCEP/NCAR reanalysis data (1958–1999), as found in a maximum covariance analysis [also called SVD in the literature, see *Bretherton et al.*, 1992]. The mode explains 55% of the square covariance between the temperature fields. All months were considered in the analysis, not only wintertime.

Since Bjerknes' works the observational record has just gained a few decades. It is still too limited in its oceanic coverage (both horizontally and vertically) and its description of the radiative and diabatic processes in the atmosphere to allow a direct investigation of the fluctuations in atmosphere-ocean energy budget at timescales longer than a few years. Nevertheless, the analysis of more conventional surface climate variables like SST, surface winds, and surface pressure collected over the past century has provided some support for, and has refined, Bjerknes earlier findings. Two dominant climate signals involving the NAO and perhaps the ocean circulation have now been identified in the observational record, at decadal and interdecadal timescales. In this review, the limited description of these signals that can be gained from the observational record will be complemented by a parallel discussion of results obtained from a hierarchy of coupled ocean-atmosphere models. The latter show a wide range of behavior, from middle latitude coupled modes on decadal timescales to intrinsic oceanic variability imprinting weakly on the NAO, while some indicate a purely passive oceanic response to an essentially stochastic NAO. This diversity of results reflects the richness of the problems posed by Atlantic air-sea interactions, but also severely limits our ability to draw firm conclusions concerning the true impact of the Atlantic Ocean on NAO variability.

The chapter is structured as follows. We begin in section 2 with an illustration of the null hypothesis for NAO variability, referred to as the climate noise scenario in the following. It proposes that NAO variability on interannual and longer timescales simply reflects the year to year, or decades to decades, changes of its short timescale (weeks to months) statistics, i.e. intrinsic to the atmosphere. This defines a reference for the subsequent scenarios discussed in the chapter. In section 3, we discuss the role of the ocean as a heat reservoir for NAO variability. We begin by presenting a stochastic energy balance model of the atmosphere and the ocean mixed layer (section 3.1), which we will use to analyze the role of local thermal interactions and nonlocal dynamical SST feedbacks in providing departures from the climate noise scenario. The physical mechanisms of SST feedback are further discussed in section 3.2 for the middle latitudes, and in section 3.3 for a possible remote influence from the tropical Atlantic. Observational evidence of the impact of SST anomalies on the NAO is discussed in section 3.4. In section 4, we discuss the modifications to troposphere/ ocean mixed layer interactions introduced by Atlantic Ocean currents. Section 4.1 illustrates advective effects on air-sea interactions at the boundary of the subtropical and subpolar gyres of the North Atlantic, while section 4.2 focuses on basin-scale ocean-atmosphere interactions, possibly involving the ocean's meridional overturning circulation. In both, a parallel discussion of observations and coupled models will be given. Section 5 offers a concluding assessment of the relevance of coupled processes to NAO variability, and some suggestions for future research.

This review clearly overlaps with other recent reviews by *Latif* [1998; interdecadal variability simulated by coupled models], by *Marshall et al.* [2001b; mechanisms and observations of Atlantic climate variability], and by *Kushnir et al.* [2002; the response of atmospheric GCMs to extratropical SST anomalies]. Several new aspects, however, are worth noting. The climate noise paradigm discussed in section 2, as well as the dynamics governing remote forcing of extra-tropical climate variability from the tropics (section 3.3), are topics which, to our knowledge, have not been discussed previously in the context of NAO variability. The thermal interaction of the troposphere and the upper ocean (section 3) is described in detail and compared to some original and unpublished research on the subject. An effort is made throughout the review to first illustrate basic ideas with simplified models, and then test these ideas against the observations and more complex models. The aim is to bring together theories, statistics, observations and model outputs.

2. THE CLIMATE NOISE PARADIGM

Prior to investigating the oceanic influence on NAO variability, it is useful to consider a simpler interpretation, or null hypothesis, for the existence of NAO variability on interannual and longer timescales. We will refer to this interpretation as the climate noise paradigm [*Leith*, 1973; *Madden*, 1981]. Briefly, it argues that the observed interannual and longer timescale NAO fluctuations are essentially a remnant of its energetic weekly to monthly fluctuations. In this view, NAO variability is entirely driven by processes intrinsic to the atmosphere.

Let us consider a first-order autoregressive process (AR(1)), with daily variance σ^2 and lag one-day autocorrelation γ [*Box et al.*, 1993]. A realization of such a process would show fluctuations on all timescales, with the visual impression of larger fluctuations on timescales close to its persistence time τ (τ = -1/ln γ days for the AR(1) process), and weaker amplitude fluctuations on timescales longer than τ. One can actually show that the variance σ^2_{CN} of the average time series (over N days) constructed from the daily AR(1) time series is [*Madden*, 1981]

$$\sigma^2_{CN} = \frac{\sigma^2}{N}[1 + 2(1 - 1/N)\gamma + 2(1 - 2/N)\gamma^2 + \ldots + 2/N\gamma^{N-1}] \quad (1)$$

where the underscript *CN* is used to indicate the variance expected from climate noise. The ratio $\sigma^2{}_{CN}/\sigma^2$ is plotted as a function of γ for monthly (N = 30 days), seasonal (N = 90 days), and yearly (N = 365 days) averages in Figure 3. As expected, when the AR(1) process is nearly decorrelated at one day ($\gamma \sim 0$), the ratio becomes inversely proportional to N, with only 3% of the variance retained by the monthly time series. It can be seen in Figure 3 that the persistence of the process compensates for this decline with, in the limit of large persistence ($\gamma \rightarrow 1$), a ratio of unity. Interestingly, one observes a sharp transition for $\gamma > 0.8$, indicating that a significant fraction of the variance is retained in the averaged

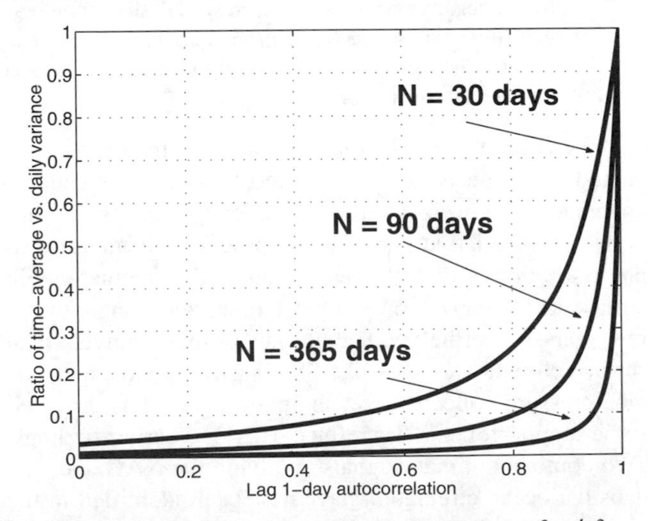

Figure 3. Ratio of time average over daily variance σ^2_{CN}/σ^2 as a function of the 1-day autocorrelation γ. Several choices of time average N are shown, as indicated on the plot.

time series when the process is sufficiently persistent (γ = 0.8 corresponds to a persistence time of 4.5 days).

Feldstein [2000] recently investigated the relevance of the AR(1) model to the observed NAO. Using geopotential height anomalies at 300 mb from the NCEP/NCAR reanalysis, he produced a daily NAO index, deduced from a rotated principal component analysis. Feldstein found a good fit of the observed NAO intraseasonal spectrum to that of an AR(1) with a persistence time of 9.5 days (Figure 4, right curves), clearly governed by atmospheric dynamics [*Thompson et al.*, this volume]. This timescale corresponds to a value γ = 0.9, for which 45% of the variance is retained in the monthly time series, a number decreasing to 20% and 5% for the seasonal and annual time series (Figure 3). Actual comparison of the interannual NAO index spectrum (Figure 4, left thick curve) with the energy level expected from climate noise (lowest thin curve on the left) indicates that the climate noise prediction is about a factor of 2 to 3 too low. The related increase in interannual variance of the observed NAO index compared to σ^2_{CN} is about 60% (from Table 2 of *Feldstein* [2000], $\chi^2/N - 1$), suggesting significant interactions of the troposphere with other components of the climate system to shape the net NAO variability.

In the following sections, we will show that the finite heat capacity of the upper ocean, SST feedbacks and the Atlantic Ocean circulation can be expected to redden and modulate NAO variability in various frequency bands. Nevertheless, it must be kept in mind that the redness of dynamical NAO indices is weak and controversial. *Wunsch* [1999] found that the spectrum of Hurrell's index [*Hurrell*, 1995], based on a longer time series than that used by *Feldstein* [2000], is essentially white [but see also *Stephenson et al.*, 2000], and it can be pointed out that the factor 2 to 3 discrepancy between climate noise and observed spectra is within the estimated error bars (medium and upper thin lines in Figure 4). This suggests that an analysis of the impact of the ocean on NAO variability should not be limited to the sole time series analysis of dynamical NAO indices, like those based on pressure or temperature. Other variables show clearer signatures of an alternative to the climate noise scenario. This will be emphasized in section 5.

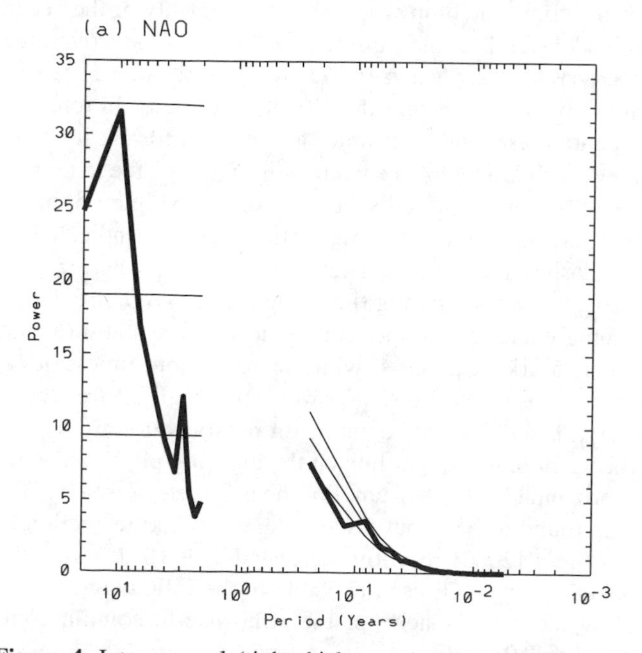

Figure 4. Intraseasonal (right thick curve) and interannual (left thick curve) spectrum of an observed NAO index. The red noise or AR(1) spectrum (lowest thin line), and its a priori (middle thin line) and a posteriori (upper thin line) 95% confidence levels are also shown. From *Feldstein* [2000; his Figure 4].

3. THE NORTH ATLANTIC OCEAN AS A HEAT RESERVOIR

In this section we take a first step in understanding the impact of the Atlantic Ocean on NAO variability, by studying Atlantic air-sea interactions in the absence of ocean currents. By developing a stochastic model of the atmosphere and the ocean mixed layer, it will be seen that local thermal coupling can redden dynamical NAO spectra but that this effect is substantially modulated by any nonlocal dynamical feedbacks of SST anomalies onto the atmosphere (section 3.1). The latter are discussed further in sections 3.2 and 3.3, and searched for in the observations in section 3.4. Section 3.5 summarizes the main results.

3.1. The Reduced Thermal Damping Argument

In a purely one-dimensional (vertical) framework, there should be a local adjustment of atmospheric and oceanic temperature anomalies on a timescale set by the finite heat capacity of the ocean mixed layer (a few months). *Barsugli and Battisti* [1998] noticed further that on longer timescales the anomalous heat exchange between the atmosphere and the ocean mixed layer would vanish, thereby reducing the damping of oceanic and atmospheric thermal anomalies. This could introduce departures from climate noise for the spectra of SST and atmospheric temperature anomalies, by reddening the latter on monthly to interannual timescales. To illustrate this idea, referred to in the following as the reduced thermal damping argument, and take into account (although still crudely) further complications associated with nonlocal dynamical effects, we analyze below the one-dimensional energy balance model developed by *Barsugli and Battisti* [1998; hereafter BB].

The atmosphere is represented as a slab at temperature T_a that exchanges energy with a slab ocean mixed layer at temperature T_o (the SST). The fluctuations of T_a and T_o can be thought of as that of the NAO and the SST tripole, respectively. For small deviations with respect to the stationary state (denoted by a prime), the conservation of energy for each component can be written as

$$\gamma_a \frac{dT'_a}{dt} = -\lambda_{sa}(T'_s - T'_o) - \lambda_a T'_a + F'_a \quad (2)$$

$$\gamma_o \frac{dT'_o}{dt} = \lambda_{so}(T'_s - T'_o) - \lambda_o T'_o \quad (3)$$

where $\gamma_{a,o}$ denote the heat capacity for the column-averaged atmosphere and the ocean mixed layer respectively, and $\lambda_{a,o}$ represent the corresponding radiative damping coefficients (see BB for a derivation). The turbulent heat exchange at the air-sea interface is assumed proportional to the air-sea temperature difference $T_S' - T_O'$, with $T_S' = cT_a'$ and c a model parameter setting the vertical scale of atmospheric temperature anomalies. All dynamical processes are encapsulated in the anomalous convergence of atmospheric energy transport F_a'. It is simply parameterized as

$$F'_a = \lambda_{sa}(N + (b-1)T'_o) \quad (4)$$

where N is a stochastic component representing the forcing of temperature anomalies by transient and stationary eddies associated with the turbulent flow of the atmosphere in middle latitudes. The second term on the r.h.s of (4) represents a sensitivity of atmospheric energy transport convergence to SST, where b is a non-dimensional parameter. The latter controls the net feedback of a SST anomaly onto the atmospheric dynamics, i.e. the sum of 'thermal' $\lambda_{sa}T_o'$ and 'dynamical' $\lambda_{sa}(b-1)\,T_o'$ effects in (2). The linear decomposition (4) thus introduces a random component, N, assumed to be solely determined by intrinsic atmospheric dynamics, and a slowly evolving part associated with SST forcing. Other decompositions could be envisioned with, for instance, a dependence of N on SST [non-deterministic SST feedback - see an illustration in *Neelin and Weng*, 1999]. It will be seen below that b is the crucial parameter of the model. Standard values for the parameters are taken from BB, namely: $c = 1$, $b = 0.5$, $\lambda_{sa} = 23.9$ W m^{-2} K^{-1}, $\lambda_{so} = 23.4$ W m^{-2} K^{-1}, $\lambda_a = 2.8$ W m^{-2} K^{-1}, $\lambda_o = 1.9$ W m^{-2} K^{-1}, $\gamma_a = 10^7$ J m^{-2} K^{-1}, $\gamma_o = 2.10^8$ J m^{-2} K^{-1} (50 m mixed layer depth). Note that both atmospheric and oceanic heat capacities will be increased by a factor 2 in the following.

A comment is warranted on the omission of entrainment and Ekman advection in the above model of the ocean mixed layer. Various studies suggest that the observed winter-to-winter memory of the SST tripole might result from the entrainment of temperature anomalies formed the previous winter [e.g., *Alexander and Deser*, 1995; *Watanabe and Kimoto*, 2000a; *Deser et al.*, submitted; *de Coëtlogon and Frankignoul*, submitted]. This effect can be roughly captured in the model by increasing the depth of the ocean mixed layer to its wintertime value. Accordingly, we set $\gamma_o = 4.10^8$ J m^{-2} K^{-1} (100 m mixed layer depth). As discussed by *Marshall et al.* [2001a] [see also *Visbeck et al.*, this volume], anomalous Ekman advection further enhances the stochastic forcing of SST anomalies driven by NAO, since anomalies in Ekman and surface turbulent heat fluxes have the same signs north of 30°N. This could easily be implemented by a term eT_a in (3), introducing another parameter e, but this does not fundamentally change the results presented below. Accordingly, anomalous Ekman advection will be neglected here.

As a reference system, let us consider a special case of the model (2) through (4) when no SST anomalies are present. This is a climate noise model for the NAO (section 2), as would be found in an atmospheric GCM (AGCM) simulation forced by a climatological mean SST distribution (the limit where the heat capacity of the ocean mixed layer becomes infinite, $\gamma_o \rightarrow \infty$). The governing equations for this system are $\gamma_a dT_a^{CN}/dt = -(c\lambda_{sa}+\lambda_a)\,T_a^{CN} + F_a^{CN}$ and $F_a^{CN} = \lambda_{sa}N^{CN}$ where the CN superscript again refers to climate noise, and the primes are omitted for clarity from now on. Taking N^{CN} as a white noise process, the latter two equations predict a red spectrum for T_a^{CN} (Figure 5, black thin line) for frequencies higher than $(2\pi\tau_a)^{-1}$ (indicated by the dashed black vertical line in Figure 5), where $\tau_a = \gamma_a/(c\lambda_{sa}+\lambda_a)$ is a damping timescale for an NAO circulation anomaly, and a flat spectrum at lower frequency (a first order Markov process with decorrelation timescale τ_a shows an elbow in its power spectrum at frequency $(2\pi\tau_a)^{-1}$, marking the beginning of the transition from a red (i.e., a sloping straight line on the log - log plot) to a white (horizontal line) spectrum. For the parameters used by BB, τ_a is found to be about 5 days. As discussed in section 2, *Feldstein* [2000] recently suggested $\tau_a \sim 10$ days for the NAO, and we will use this value in the following by doubling the heat capacity of the atmospheric column compared to BB.

Comparing the climate noise spectrum (Figure 5, thin black line) to the model prediction for the power spectrum of T_a when SST anomalies are present (Figure 5, thick black line), the latter shows a slightly red spectrum from interannual to decadal timescales. Both T_a and T_o (Figure 5, thick grey line) spectra become white at frequencies lower than $(2\pi\tau_o)^{-1}$, where

$$\tau_o = \frac{\gamma_o}{\lambda_o + \lambda_{so} - b c \lambda_{so}/(c + \lambda_a/\lambda_{sa})} \tag{5}$$

is an oceanic damping timescale ($\tau_o \simeq 10$ months for standard parameters). Including the interaction with the upper ocean makes little difference to the total variance of T_a (less than 10% increase), because most of it lies in the high frequency tail of the spectrum. Nevertheless, it substantially increases the T_a variance at interannual and longer timescales. In Figure 5, the interannual variance of T_a is more than 75% larger than that of T_a^{CN}, as the energy levels of the climate noise and model T_a spectra differ by a factor 2 to 3 on timescales longer than a few years. This order of magnitude is consistent with that discussed in section 2.

The red spectrum prediction for SST is commonly found in simple models of middle latitude climate variability [*Frankignoul and Hasselmann*, 1977] and in observations [e.g., *Frankignoul*, 1985]. The redness of the atmospheric spectrum at interannual timescales is more surprising and deserves further explanation. Let us write the equation governing the equilibrium or low-frequency (denoted by *LF*) amplitude of T_a, easily obtained by setting time derivatives in (2)-(3) to zero:

$$(\lambda_{sa} c[1 - b\lambda_{so}/(\lambda_{so} + \lambda_o)] + \lambda_a) T_a^{LF} = \lambda_{sa} N^{LF} \tag{6}$$

Similarly, for the climate noise model,

$$(\lambda_{sa} c + \lambda_a) T_a^{CN,LF} = \lambda_{sa} N^{CN,LF} \tag{7}$$

Equations (6)-(7) indicate that the amplitude of an atmospheric temperature anomaly at low frequency is proportional to the strength of the stochastic forcing, and inversely proportional to that of the net damping. For the standard values used in BB (c = 1, b = 0.5), a net damping rate of 15.6 W m^{-2} K^{-1} is found in (6), compared with 26.7 W m^{-2} K^{-1} in (7). For similar levels of stochastic forcing N and N^{CN}, the higher energy level found in the spectrum of T_a compared to T_a^{CN} is thus related to the weaker damping of temperature anomalies found at low frequency. This reduced damping reflects the mutual adjustment of T_o and T_a which is artificially suppressed in the climate noise model by considering the upper ocean as an infinite heat reservoir ($\gamma_o \rightarrow \infty$). The degree of adjustment is controlled by the net feedback of SST anomalies onto atmospheric dynamics, the parameter b (see also section 3.2 for further discussion of this issue).

Figure 6 illustrates the sensitivity of BB's model to b, by showing T_a and T_o spectra when b = -0.5 instead of b = 0.5 (Figure 5). Clearly, if the shape of the SST spectrum is essentially unchanged (thick grey line), that of the atmospheric temperature (thick black line) goes from a red to a blue spectrum, with less interannual variance than expected from climate noise (thin black line). The conclusion from Figure 6 is that the reduced thermal damping of oceanic and atmospheric temperature anomalies at low frequency drives the redness of the model NAO spectrum, but it is modified by the strength of the SST feedback. Note that inclusion of ocean dynamics would also modulate the redness of the T_a and T_o spectra (see section 4).

Modeling studies have given some support for the notion of reduced thermal damping of atmospheric and oceanic anomalies. In a purely atmospheric context, *Hendon and Hartman* [1982] showed how inclusion of a flow-dependent diabatic heating parameterization could modulate the amplitude of large-scale baroclinic anomalies in a GCM. *Manabe and Stouffer* [1996], *Delworth* [1996], and *Bhatt et al.* [1998] have clearly shown how the variance of surface air temperature was increased when an AGCM was allowed to interact with the upper ocean. Focusing on the NAO, *Delworth* [1996] and subsequently *Bladé* [1997] found an increase in NAO variability simulated with the GFDL model when coupled to a slab mixed layer compared to that found when the model is forced by climatological mean SST. Nevertheless, it is not a general rule that an AGCM forced with a climatological mean SST distribution will show a less energetic NAO than when the model atmosphere is free to interact with an ocean model.

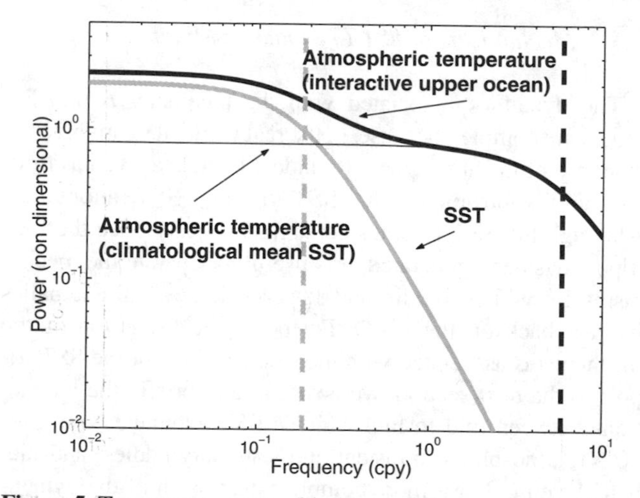

Figure 5. Temperature spectra (black, atmosphere; grey, ocean) predicted by the model of *Barsugli and Battisti* [1998]. Also indicated is the atmospheric temperature spectrum when no SST anomalies are allowed (thin black line), assuming the same energy level for the stochastic forcings N and N^{CN}. The power is non dimensional and the frequency is expressed in cycles per yr. The dashed grey and black vertical lines indicate the frequency $(2\pi\tau_o)^{-1}$ and $(2\pi\tau_a)^{-1}$ respectively.

For instance, *Saravanan* [1998] found a larger NAO variance in the NCAR Community Climate System Model when forced by SST climatology than when coupled to an oceanic GCM. *Robertson* [2001] confirms this finding by reporting similar results with the Hamburg climate model. This might again illustrate the importance of the dynamical SST feedback (possibly negative in these cases) in setting the energy level of NAO dynamical spectra. It might also reflect the different mean states and the impact of ocean currents on the upper ocean heat budget (see section 4).

To our knowledge no study has yet focused on testing the reduced thermal damping argument against observations. Some support might nevertheless be found by inspection of observed NAO spectra for geopotential height (Figure 4) or temperature fields (Figure 7, black curve for temperature at 500 mb, grey curve for SST) which show a qualitative agreement with those predicted above when a weak positive feedback of SST on the NAO is included (Figure 5 – note that the time series used in Figure 7 were linearly detrended. The redness of the spectra can thus not be attributed to such linear trend). A recent analysis of the persistence time of a low-level vorticity anomaly in the atmosphere by *Peña et al.* [2001] is also consistent with longer persistence when cold cyclonic anomalies are associated locally with cold SST anomalies (or warm anticyclonic anomalies associated locally with warm SST anomalies), i.e. thermally locked anomalies, but it is not known yet if similar results apply to large-scale features like NAO.

3.2. Mechanisms of SST Feedback in Middle Latitudes

The dynamics associated with the parameter b in BB's model are complex because, in the real world, they involve the response of the atmosphere to middle latitude as well as tropical and subtropical (the SST tripole) SST anomalies. Although the SST forcings act simultaneously once the SST tripole has been generated, it is useful to separate the mechanisms by which middle and lower latitude SST anomalies interact back on the NAO. Postponing a discussion of the mechanisms associated with the southern lobe of the SST tripole to the next section, we now consider briefly the interactions between mid to high latitude SST anomalies (north of 30ºN), atmospheric transient and stationary eddies, and diabatic heating. For a more complete discussion of the dynamics, the reader is referred to *Frankignoul* [1985] and *Kushnir et al.* [2002].

At the simplest level, linear theory for an atmosphere with zonal flow having constant vertical shear (or through thermal wind, constant meridional temperature gradient) and static stability N_0 predicts that the growth rate of the most unstable perturbation is related to the meridional potential temperature gradient $\partial\theta/\partial y$ through

$$\sigma \simeq 0.31 \frac{\mathrm{g}}{N_o \theta_o} \left| \frac{\partial \theta}{\partial y} \right| \tag{8}$$

where g is gravity and θ_0 is a reference potential temperature profile [see *Gill*, 1982; *Hoskins*, 1983]. As discussed by *Hoskins and Valdes* [1990], (8) captures the spatial variations of the Northern Hemisphere storm track when the low-level (~780 mb) potential temperature gradient is used. Thus, to the extent that SST and air temperature just above the planetary boundary layer are strongly coupled through turbulent air-sea interactions, an enhanced meridional SST gradient is expected to enhance baroclinic wave activity. The SST tripole in a positive NAO index phase [see *Visbeck et al.*, this volume] enhances the meridional temperature gradient close to the U.S. east coast along 45ºN and reduces it along 30ºN (Figure 2). It should thus favor more cyclogenesis north of the separated Gulf Stream. Equation (8) also indicates an inverse dependence upon static stability. As has been observed across the Gulf Stream temperature front, warmer waters to the south of the front are associated with more turbulence and lower static stability compared to the conditions found at low levels north of the front [e.g., *Sweet et al.*, 1981]. The influence of SST on the development of unstable baroclinic waves may depend not just on SST gradient, but also on the SST itself. Note that the direct interaction between the SST anomaly and the storm track occurs on short spatial scales, as the anomalous turbulent heat exchange between the ocean and atmosphere will be deter-

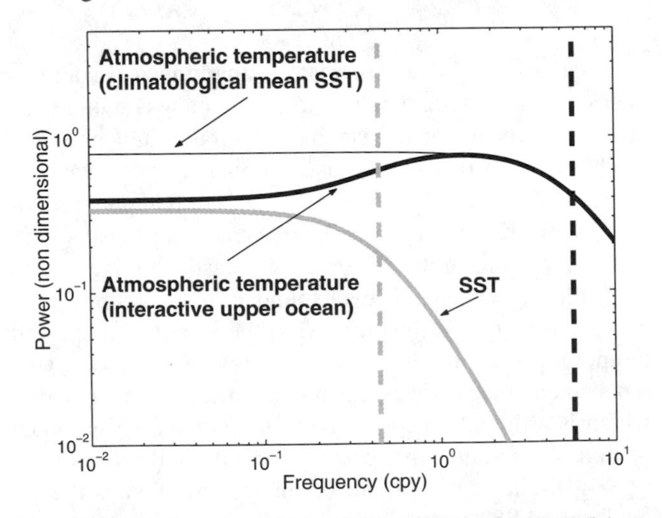

Figure 6. Same as Figure 5 but with b = -0.5 instead of 0.5 (negative SST feedback or further away from thermal equilibration). The energy levels can be compared to Figure 5. Note that the timescale τ_o (indicated by the dashed grey vertical line) has shortened compared to Figure 5, in agreement with (5).

mined by features such as the position of the cold sector and the low-level jet ahead of the cold front [*Bresch*, 1998]. To resolve this likely requires high resolution models.

The subsequent interaction of the modified storm track with its environment determines if the SST anomaly induces an equivalent barotropic structure like that displayed by the NAO. The modified storm track determines the vertical scale of the heating and cooling associated with positive and negative SST anomalies, by transporting heat and moisture upward [note that changes in radiative forcing associated with anomalous cloud cover, and changes in static stability, might also play a role in determining the vertical structure of the heating. See *Kushnir and Held*, 1996]. The associated condensational and transient eddy heating may then create a large-scale baroclinic perturbation which can feedback on the storm track, possibly leading to the forcing of equivalent barotropic structure by eddy vorticity fluxes [e.g., *Palmer and Sun*, 1985; *Ting and Peng*, 1995; *Peng and Whitaker*, 1999; *Watanabe and Kimoto*, 2000b]. It is worth emphasizing that aquaplanet GCMs typically lack such interactions [e.g., *Ting*, 1991], and even GCMs with zonal asymmetries do not systematically produce equivalent barotropic structures in response to middle latitude SST anomalies [*Kushnir and Held*, 1996]. Accordingly, predicted atmospheric responses to an anomalous meridional SST gradient along 45°N can vary between AGCMs, although a

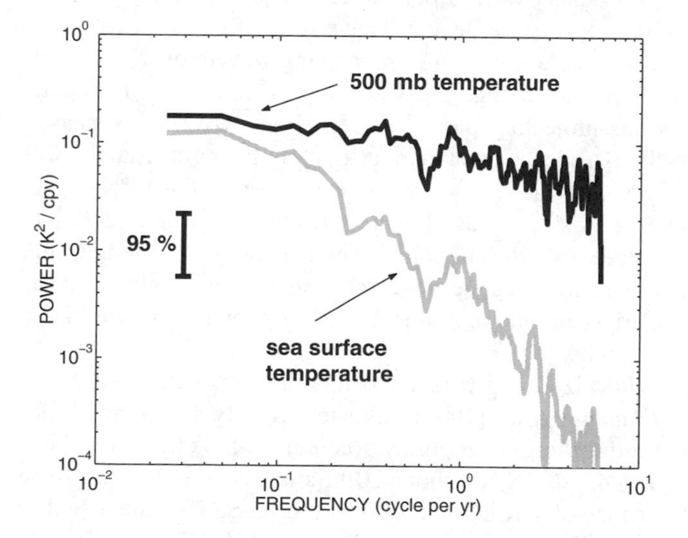

Figure 7. Power spectrum of atmospheric temperature at 500 mb (black) and SST (grey) time series associated with the spatial patterns in Figure 2. The monthly time series were scaled to represent a basin-average amplitude. The frequency is expressed in cycles per yr (*cpy*), and the power in K^2/cpy. The spectra of the linearly detrended time series were estimated using the multitaper method [*Percival and Walden*, 1993] with 9 tapers. The vertical bar indicates the 95% confidence level.

strengthening of the westerlies in response to a strengthening of the meridional SST gradient along 45°N appears to be robust [e.g., *Watanabe and Kimoto*, 200b; *Sutton et al.*, 2001; *Peng et al.*, 2002]. This is consistent with a positive feedback on the NAO from the extra-tropical part of the SST tripole, but model studies suggest the feedback amplitude is weak compared to intrinsic variability. A dipolar height anomaly of 20 m at 500 mb (roughly Icelandic Low minus Azores High) is commonly found for a 1 K difference across 45°N [*Robinson et al.*, 2000; *Kushnir et al.*, 2002].

The above suggests that diabatic heating induced by SST anomalies is sensitive to the state of the atmosphere, and since SST anomalies themselves are to first order created by the atmospheric circulation, there is the possibility of a 'thermally equilibrated' state of the atmosphere and the upper ocean with vanishing diabatic heating. This way of thinking about thermal forcing of the atmosphere was developed by *Döos* [1962] and subsequently elaborated by *Shutts* [1987] and *Marshall and So* [1990; see also the review by *Molteni*, 1992] in a purely atmospheric context. It is not known from observations if one or both phases of the NAO actually reflect such a thermally equilibrated state, but several theoretical studies have built upon this concept to investigate middle latitude ocean-atmosphere interactions [e.g., *Goodman and Marshall*, 1999; *Goodman*, 2001; *Ferreira et al.*, 2001; *Nilsson*, 2001].

The parameter b in BB's model can be related to thermal equilibration. Neglecting energy storage in the atmosphere (which is a very good approximation on timescales longer than seasonal) and using (4), (2) can be rewritten as

$$aT_a = bT_o + N \qquad (9)$$

while the net surface heat flux anomaly (the diabatic heating in BB) can be expressed as

$$F_s' \equiv \lambda_{so}[cT_a - (1 + \lambda_o/\lambda_{so})T_o] = -\lambda_{so}(d - bc/a)T_o + \lambda_{so}N\,c/a$$

where we have introduced the non dimensional parameters $a = c + \lambda_a / \lambda_{sa}$ (=1.12) and $d = 1 + \lambda_o / \lambda_{so}$ (= 1.08). At a critical value of $b = ad/c$ (= 1.2), the sensitivity of F_s to T_o is zero and the model is in a thermally equilibrated regime in the absence of stochastic forcing. For a smaller value of b however, the sensitivity of heat flux to SST (or heat flux feedback) is negative (damping of SST anomalies) although the dynamical feedback can still be positive ($0 < b < ad/c$). It has been suggested that the heat flux feedback could become positive in middle latitudes, implying a strong dynamical SST feedback $b > ad/c$ [e.g., *Latif and Barnett*, 1994; *Grötzner et al.*, 1998]. This has, however, not been supported by the analysis of the observations, at least over the North Atlantic, where a negative heat flux feedback of 20 Wm^{-2} K^{-1} is commonly found [*Frankignoul et al.*, 1998; *Czaja and Frankignoul*, 1999].

3.3. Mechanisms of SST Feedback in the Tropics

Our focus in the previous section was on the forcing of the NAO by the SST gradient in the mid-to-high latitude North Atlantic, which is itself a product of intrinsic NAO variability and the ocean circulation. This SST forcing is mediated by the North Atlantic storm track, which distributes the SST induced heating/cooling in the vertical. Here, we consider remote coupling with the tropical Atlantic within an analogous framework. The NAO has a strong impact on the strength and position of convection in the Atlantic intertropical convergence zone [see the review by *Marshall et al.*, 2001b] and even the South American (austral) summer Monsoon system by forcing the southern lobe of the SST tripole [e.g., *Robertson and Mechoso*, 1998; *Nogues-Paegle et al.*, 2000], but also potentially through ocean circulation changes (section 4). At the same time, diabatic heating anomalies over the tropical Atlantic sector can force atmospheric circulation anomalies in the extratropics through changes in the sectorial Hadley circulation and Rossby wave propagation, potentially closing a feedback loop between the NAO and the tropical Atlantic Ocean.

In the zonally-averaged picture, tropical diabatic heating produces a meridional overturning Hadley cell which, for most of the calendar year, extends from deep within the summer hemisphere to the subtropical jet of the winter hemisphere, maintaining the latter through poleward transport of angular momentum. The hypothesis that an intensified Hadley circulation accelerates the subtropical jet and subsequently impacts the middle latitude climate was addressed in idealized experiments by *Hou* [1993], *Hou and Molod* [1995], *Chang* [1995], and again more recently by *Hou* [1998]. Using an idealized GCM with prescribed anomalous diabatic heating in the tropics (zonally symmetric), *Hou* [1998] argues that a shift of the latter towards the summer hemisphere (dipole in diabatic heating) intensifies the winter hemisphere Hadley cell, and extends the baroclinicity further from the subtropics into middle latitudes. The increased poleward heat transport by more-vigorous low-frequency planetary scale transients was found to lead to middle latitude cooling and high-latitude warming in the winter hemisphere, thus reminiscent of the negative NAO index phase signature on the tropospheric temperature field. For a 6° southward latitude shift of the diabatic forcing (inducing cooling and warming north and south of the equator, respectively, at a rate of about 0.8 K/day, i.e., about a 40% change), the zonally averaged middle latitude response is a ~0.5 K temperature change at 600 mb and a ~1 - 2 ms^{-1} zonal wind change at 250 mb. These changes are comparable to the Northern Hemisphere annular mode fluctuations on the monthly timescale [*Thompson and Wallace*, 2000]. Note however that the relationship between enhanced Hadley circulation and reduced mid to high latitude temperature gradient is dependent upon the latitudinal scale of the convection anomaly that drives the modulation of the Hadley cell. In an experiment where the diabatic heating anomaly was more concentrated in latitude, *Hou* [1998] found a deceleration of the subtropical jet and enhanced mid to high latitude temperature gradient, projecting partly on positive NAO index conditions.

The above paradigm of a Hadley cell regulating the extratropical climate through tropical axisymmetric diabatic forcing may be more applicable to the Atlantic than the Pacific sector, despite its smaller zonal extent, because, unlike during ENSO events over the Pacific, diabatic heating anomalies tend to be associated with meridional shifts in the Atlantic ITCZ [e.g., *Nobre and Shukla*, 1996]. Nonetheless, the heating anomalies are zonally localized so that it may be more appropriate to consider their impact on the extratropical rotational flow in terms of forced Rossby wavetrains [*Hoskins and Karoly*, 1981]. Anomalous descent over a cold SST anomaly in the subtropical North Atlantic is associated with a convergence and anomalous cyclonic vorticity at upper levels. This positive Rossby wave source can excite a northward propagating wavetrain with expected anomalous high pressure over the central north Atlantic (thus projecting partly on a positive NAO index phase). Similar mechanisms may also apply to south Atlantic SST forcing, where a warm anomaly over 40°S–10°N, in an AGCM study, was shown by *Robertson et al.* [2000] to induce a southward shift of the mean regional Hadley cell, producing an upper-level convergence anomaly over the Caribbean and subsequent ridging over the central North Atlantic.

Idealized experiments with a low resolution GCM by *Okumura et al.* [2001] indicate typically a tripolar geopotential height anomaly (centers of action at about 15°–45°–65°N) of about 10 m amplitude at 500 mb for a change of 1 K in the cross-equatorial SST gradient (either induced by north or south off-equatorial SST anomalies, or both in the form of a SST dipole), with a hint at a stronger impact on the 'Azores High' than on the 'Icelandic Low' region. The response is robust and also seen in higher resolution GCMs [e.g., *Terray and Cassou*, submitted]. The NAO sensitivity to the southern lobe of the tripole appears thus of similar strength and sign (positive feedback) to that found in GCMs forced with its middle latitude part (mid to high latitude SST gradient change - see section 3.2).

Two comments are warranted on the above forced Rossby wave mechanism. First, as discussed by *Sardeshmukh and Hoskins* [1988], the relationship between the Rossby wave source and the anomalous diabatic heating is not simple. Due to the advection of vorticity by the divergent component of the flow, the region that acts as a source of Rossby waves in the presence of tropical heating does not generally coincide with the region where the heating takes place. In fact, the largest Rossby wave sources are typically in the subtropics, rather than the Tropics, because of the much stronger vorticity gradients in the vicinity of the subtropical jet. The local Hadley circulation and Rossby wave sources are thus intimately connected. Second, the interaction of the propagating Rossby wavetrain with middle latitude transient eddies ultimately shape the net response of the Northern Hemisphere to the tropical forcing [*Hoskins and Sardeshmukh*, 1987; *Held et al.*, 1989; *Hoerling and Ting*, 1994]. An illustration of this in the context of NAO/tropical Atlantic interactions can be found in, for instance, *Watanabe and Kimoto* [1999], *Sutton et al.* [2001] and *Terray and Cassou* [submitted].

3.4. SST Impact on NAO in Observations

In support of the mechanisms proposed in sections 3.2 and 3.3, many AGCM studies suggest a positive feedback between the SST tripole as a whole and the NAO [*Venzke et al.*, 1999; *Rodwell et al.*, 1999; *Watanabe and Kimoto*, 2000b; *Cassou and Terray*, 2001a; *Sutton et al.*, 2001; *Terray and Cassou*, submitted; *Peng et al.*, 2002], although not all [*Robertson*, 2001]. Analysis of the observational record should thus at least provide a hint for such a feedback.

A major problem, however, arises when considering observational studies. Indeed, the fact that the atmosphere predominantly forces middle latitude SST anomalies makes it difficult to estimate directly the SST feedback on the atmosphere (for a review of the literature on the empirical and theoretical estimates of the SST feedback prior to the 1990s, see *Frankignoul* [1985]). Since the work of *Davis* [1976] and *Frankignoul and Hasselmann* [1977], the study of the temporal covariance or correlation between SST anomalies and various tropospheric variables in lead and lag conditions has nevertheless appeared as a useful tool to bypass this problem [*Frankignoul*, 1985; *Deser and Timlin*, 1997; *Czaja and Frankignoul*, 1999; 2002]. The rationale for this statistical approach can be understood from the model developed in section 3.1. Consider the linear decomposition of NAO variability as in (9) at time t, and multiply by $T_o\,(t-\tau)$, i.e. by the SST anomaly at a previous time $t-\tau$. Taking an ensemble average (denoted by a bracket) yields

$$a\langle T_o(t-\tau)T_a(t)\rangle = b\langle T_o(t-\tau)T_o(t)\rangle + \langle T_o(t-\tau)N(t)\rangle$$

so that if τ is larger than the decorrelation time of N (typically a couple of weeks), then $\langle T_o(t-\tau)N(t)\rangle \simeq 0$ and we get the simple relation

$$a\langle T_o(t-\tau)T_a(t)\rangle \simeq b\langle T_o(t-\tau)T_o(t)\rangle \text{ for } \tau > 1 \text{ month}$$

The latter equation indicates that the strength of the covariance between SST and an atmospheric variable when SST leads is proportional to the dynamical feedback b, with the stronger the feedback, the larger the covariance when SST leads. It is important to emphasize that (10) was obtained by assuming an instantaneous impact of SST onto the atmosphere (storage was neglected in (9)). A non-null covariance when SST leads by, say, τ months should thus not be interpreted as a delayed response of the atmosphere to SST by τ months. It simply reflects that atmospheric variability has an SST induced component that has the persistence of the SST anomaly (τ_o in the model of section 3.1). This persistence ultimately determines the lag over which (10) is useful, since we expect the covariance $\langle T_o(t-\tau)T_o(t)\rangle$ to become negligible when is larger than the SST persistence time. Note also that in practice, we always expect non-null covariance $\langle T_o(t-\tau)T_a(t)\rangle$because of the finite length of the observational record. Statistical significance tests are thus needed to estimate the level of covariance that could arise by chance in the observations [Monte Carlo simulations are traditionally used, e.g., *von Storch and Zwiers*, 1999].

Figure 8 from *Czaja and Frankignoul* [2002] illustrates this technique, and shows the dominant patterns of covariability between monthly SST (shaded) and 500 mb geopotential height (contoured every 5 m) anomalies over the North Atlantic sector in the NCEP/NCAR reanalysis [*Kalnay et al.*, 1996], as a function of time lag. The analysis was applied for several lags, ranging from SST lags by one month (upper left, L = 1 on the figure) to SST leads by 7 months (lower right corner, L = -7 on the figure). One observes a robust dipolar pattern in the height field at all lags, which is reminiscent of the NAO signature on pressure. On the other hand, the SST pattern evolves somewhat from SST lags to SST leads. At zero lag and when SST lags the height field we recover the familiar SST tripole pattern (Figure 2), as found in many studies [*Wallace et al.*, 1990; *Deser and Timlin*, 1997]. When SST leads the height field by up to seven months, we observe a large-scale SST pattern with opposite signed anomalies southeast of

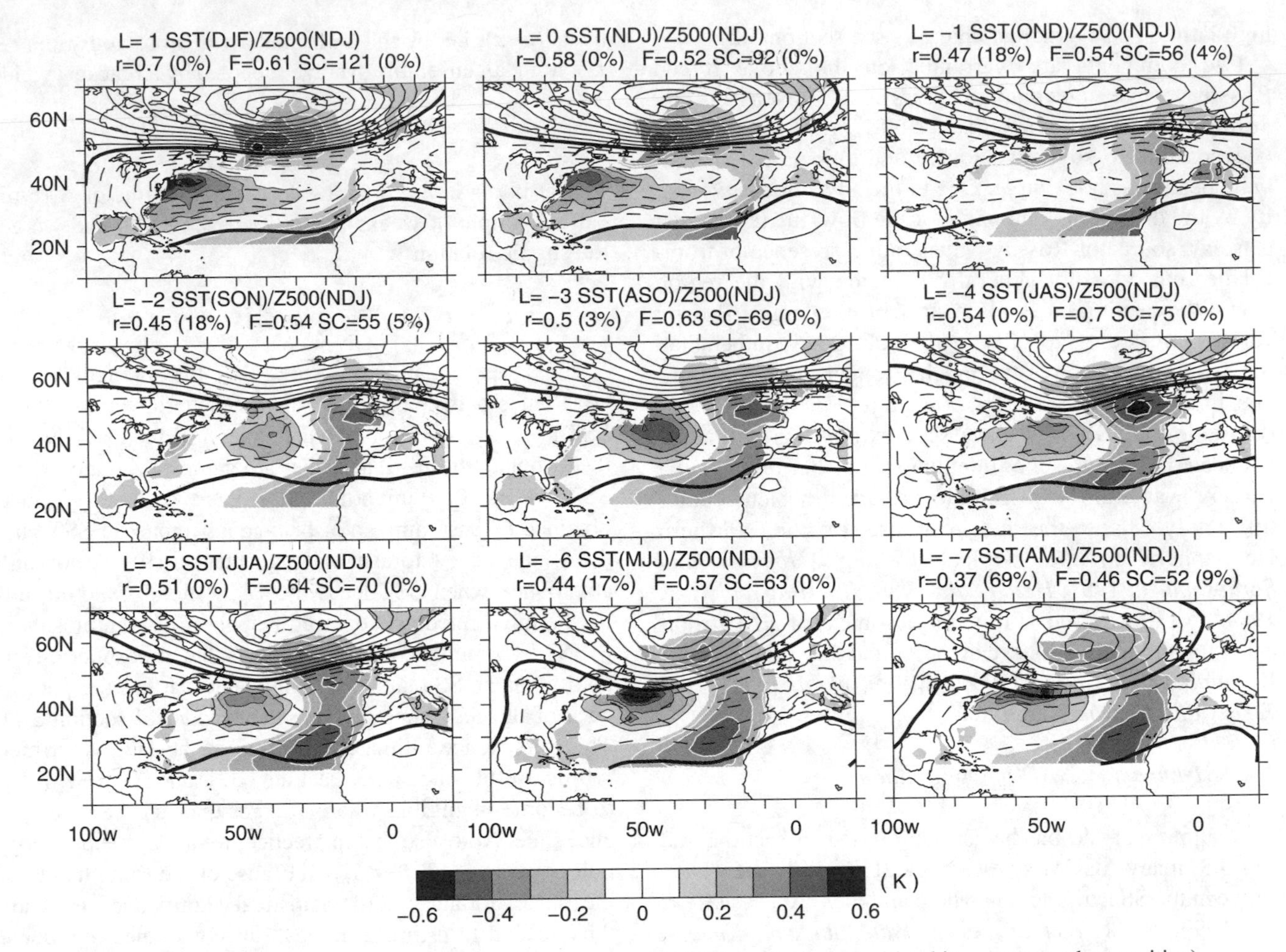

Figure 8. Dominant patterns of covariability between monthly SST (in *K*, shaded with white contours when positive) and 500 mb height (contoured every 5 m, continuous when positive) anomalies. The height field is fixed to November through January and the SST field lagged accordingly by *L* months, as indicated on the plot (SST leads when $L < 0$). Also indicated are the temporal correlation coefficient *r*, the fraction of squared covariance *F* explained by the patterns, and their squared covariance *SC*. The number in parenthesis indicates a significance level (percentage of chance that the observed covariance arise by chance), as deduced from Monte Carlo simulations. From *Czaja and Frankignoul* [2002].

Newfoundland and along the eastern North Atlantic, in the form of a horseshoe pattern (hereafter NAH, for North Atlantic horseshoe). The signs are such that anomalously cold SST southeast of Newfoundland and anomalously warm SST along the eastern boundary of the Atlantic precede a negative NAO index phase by several months [*Czaja and Frankignoul*, 1999; *Drévillon et al.*, 2001]. Although the projection between the NAH and the tripole SST patterns is not perfect (~0.6), the two patterns share common features (reduced meridional SST gradient along 40°N and enhanced meridional SST gradient along 30°N). Figure 8 is thus consistent with (10) and is suggestive of a non zero feedback of the SST tripole onto the NAO. Since the NAO generates the same polarity of the tripole as that found when the tripole leads the NAO, the feedback is positive. It must be emphasized that the interpretation of Figure 8 is delicate. As suggested by one reviewer, one could also argue that Figure 8 reflects a winter-to-winter memory of the NAO (yet to be explained) in addition to a winter to summer memory in the SST tripole. Further discussion of the interpretation of Figure 8 (e.g., impact of tropical SST anomalies, seasonal effects) can be found in *Czaja and Frankignoul* [2002]. As shown in *Drévillon et al.* [2001], changes in baroclinicity and transient eddy activity are consistent with changes in meridional SST gradient associated with the NAH and the SST tripole patterns. This suggests that a direct interaction between anomalous SST and the storm track, as discussed in section 3.2, could be responsible for the oceanic influence suggested by Figure 8.

The fraction of *monthly* NAO variance explained by the positive feedback is, however, weak – no more than 15% [*Czaja and Frankignoul*, 2002] – consistent with the results gained from the simple model in section 3.1. The fraction of *interannual* variance that might be explained by this feedback is related to the deviation from white noise displayed by the NAO indices spectra (e.g., Figure 7). As discussed in sections 2 and 3.1, it might reach 60%.

3.5. Summary

We have discussed the relevance of Atlantic air-sea interactions as a source of enhanced NAO variability in the absence of ocean dynamics in a hierarchy of models and in observations. It was shown how local reduced thermal damping of atmospheric and oceanic temperature anomalies on timescales longer than that set by the thermal inertia of the mixed layer (a few months) will tend to redden dynamical NAO spectra compared to those of climate noise (section 2). This effect is strongly dependent upon a nonlocal dynamical impact of the SST tripole onto the NAO, which seems to be supported (for both its northern centers of action and its southern lobe) by several observational and modeling studies.

There is still however a large spread in modeling studies of the SST tripole-NAO interactions. This spread probably reflects the complexity of the mechanisms involved both in middle latitudes and the subtropics. Quantitatively, the impact of air-sea interactions on the NAO appears weak compared to the level of NAO variability intrinsically generated by the atmosphere. It is also dependent upon timescales, the impact significantly increasing with longer timescales. An indication that ocean dynamics might introduce further modifications to the climate noise scenario is studied in the next section.

4. THE NORTH ATLANTIC OCEAN AS A HEAT CARRIER

Based on the analysis of the historical record (roughly 1900 to present) of surface variables like SST, atmospheric pressure and winds [the COADS dataset, *Woodruff et al.*, 1987] two climate signals potentially involving an active role for ocean circulation have been identified in the observations. Both show NAO-like anomalous surface circulation in the atmosphere, but have distinct timescales: decadal and interdecadal. Encouragingly, the variability described in some coupled atmosphere-ocean models bears some resemblance to that seen in observations and provides more readily available diagnostics and analysis to infer potential mechanisms.

To review the observational and model results, we have chosen to first discuss the impact of ocean heat transport changes at the intergyre boundary of the North Atlantic on the NAO. The idea of interacting ocean gyres and the NAO is indeed a dominant paradigm in the field, especially at decadal timescales (section 4.1). We will then examine NAO variability involving basin scale changes in ocean heat transport associated with the MOC and, in general, longer (interdecadal) timescales (section 4.2). We conclude with a summary in section 4.3.

4.1. NAO/Ocean Circulation Interaction at the Intergyre Boundary

A major climatological feature of the North Atlantic ocean is the region of sharp SST gradient marking the separation of the subtropical and subpolar gyres and the path of the separated Gulf Stream (or simply the North Atlantic current). It was emphasized in section 3 that the NAO may be sensitive to such a meridional SST gradient, through storm track dynamics, but the SST gradient at the intergyre boundary itself is also modulated by the ocean circulation, especially at timescales longer than a few years [e.g., *Bjerknes*, 1964; *Battisti et al.*, 1995; *Halliwell*, 1998]. This yields the possibility of coupled ocean-atmosphere dynamics driven in the vicinity of the intergyre boundary.

It has been suggested in many studies that the enhanced variability displayed by the SST tripole in the decadal band [*Deser and Blackmon*, 1993; *Sutton and Allen*, 1997; *Moron et al.*, 1998; *Tourre et al.*, 1999; *Czaja and Marshall*, 2001; *Costa and Colin de Verdière*, 2002] might reflect such oceanic impact. As a simple illustration, we show in Figure 9 a composite map for SST anomalies based on a long time series of the SST difference (ΔT) across the separated Gulf Stream ([60°–40°W/40°–55°N] minus [80°–60°W/25°–35°N]). This index captures the SST tripole, with anomalously warm SST north of the separated Gulf Stream and in the eastern subtropics, but anomalously cold conditions south of the stream ($\Delta T > 0$). The composite maps reveal that the tripole persists from one year to the next (not shown), but essentially disappears after three years. This is expected if the thermal inertia of the upper ocean is the only mechanism introducing a timescale for the tripole (see section 3.1). Strikingly however, the tripole reappears after six years with opposite signs, indicating a damped oscillation at decadal timescale rather than a simple exponential decay.

Before addressing which ocean processes could be responsible for this behavior of the SST tripole, it is worth investigating if similar features are present in atmospheric variables. The various dynamical NAO indices presented in this chapter exhibit little indication of such a preferred decadal timescale, partly because of the poor frequency resolution associated with the short record used (40 years in Figures 4, 7). Analyses of longer records give a weak indication that a decadal enhancement of power is present in atmospheric variables [e.g., surface winds and sea level pressure for *Deser and Blackmon*, 1993; sea level pressure for *Tourre et al.*, 1999; *Robertson*, 2001]. Figure 10 is an illustration using the reconstructed sea level pressure (SLP) field from *Kaplan et al.* [2000]. It displays the power spectrum of SLP anomalies averaged over the Greenland-Icelandic Low region (black), the northern center of action of the NAO signature in SLP. It shows enhanced variance in the 10–20 year band and reduced power at longer timescales. It is striking that similar spectral signatures are found independently in the ΔT index (Figure 10, grey) with both indices keeping significant coherence at decadal and longer timescales [*Czaja and Marshall*, 2001]. This might reflect a modulation of the Greenland-Icelandic Low by the ocean circulation at decadal and longer timescales, through changes in heat transport across the mean position of the separated Gulf Stream (see below). Alternatively, one can argue that the observed deviations from white SLP spectra are not significant and that the coherence between Greenland-Iceland SLP and SST gradient anomalies at low frequencies just reflects the driving of the latter by the former on much shorter timescales (the climate noise interpretation).

Both mean and anomalous ocean currents could be instrumental in setting a preferred decadal timescale to NAO-SST tripole interactions, through their impact on the meridional SST gradient at the intergyre boundary. Consider first the advection of the anomalous middle latitude SST dipole (the two northern centers of action of the SST tripole) by the mean meridional circulation (Figure 11). At a timescale determined by the meridional scale of the NAO forcing L and the mean current velocity V, advection and surface forcing can reinforce each other, leading (if damping effects are not too strong) to enhanced power in the SST dipole and possibly NAO-circulation anomalies for sufficiently strong SST feedback [*Saravanan and McWilliams*, 1998; see also *Visbeck et al.*, 1998; *Krahman et al.*, 2001; *Drijfhout et al.*, 2001; *Visbeck et al.*, this volume]. For L = 5000 km and V = 1-2 cm s^{-1}, this "resonant timescale" is of order of 10 years, roughly consistent with the damped oscillation timescale seen in Figure 9. Note that it would not exist independently in the ocean and the atmosphere but arises only when the two fluids are considered together.

Another possible explanation involves changes in ocean circulation rather than mean currents. Indeed, the subtropical-subpolar gyres of the North Atlantic and the Gulf Stream-recirculation system strongly respond to NAO variability [*Taylor and Stephens*, 1998; *Joyce et al.*, 2000; *Curry and McCartney*, 2001; *Frankignoul et al.*, 2001b; *Visbeck et al.*, this volume], but also show significant intrinsic variability of their own on interannual and longer timescales [e.g., *Jiang et al.*, 1995; *Meacham*, 2000; *Cessi and Primeau*, 2001; *Cessi and Paparella*, 2001; *Dewar*, 2001]. Both intrinsic and forced dynamics lead to changes in heat transport across the mean path of the separated Gulf Stream through expansion – contraction of the gyres, or equivalently, through large-scale anomalous currents acting on mean temperature gradients. The forced response of the thermocline–ocean mixed layer to large-scale wind forcing orchestrated by the atmosphere and its subsequent feedback on the atmospheric flow has been studied in a hierarchy of idealized coupled models [e.g., *Jin*, 1997; *Münnich et al.*, 1998; *Neelin and Weng*, 1999; *Cessi*, 2000; *Primeau and Cessi*, 2001; *Marshall et al.*, 2001a], in the lines of the scenario put forward by *Latif and Barnett* [1994; 1996] for the North Pacific. Similarly, the buoyancy forcing of the ocean driven by the NAO through its impact on Labrador Sea convection, and its possible subsequent feedback on the NAO through changes in the path of the separated Gulf Stream has been studied in observations and a simple model by *Joyce et al.* [2000]. These studies can generally be understood in a forced delayed oscillator framework, as widely used for ENSO, with the ocean circulation providing the delay (typically set by the propagation time of long Rossby waves across the North Atlantic - from 5 to 10 years, depending on latitude - or set by advection, again typically 5 to 10 years for subpolar-subtropical gyre exchange), and intrinsic NAO variability the source of stochastic forcing (Figure 12). Although some realistic atmosphere-ocean GCMs support this type of active coupling between anomalous ocean gyres and the NAO [e.g., *Grötzner et al.*, 1998], it must be emphasized that not all coupled GCMs show this behavior over the North Atlantic. Rather, some exhibit a purely passive oceanic response to unaltered NAO variability [e.g., *Zorita and Frankignoul*, 1997; *Frankignoul et al.*, 2001a], even when idealized in their settings [e.g., *Kravtsov and Robertson*, 2002]. As discussed by *Latif* [1998], this might reflect the different sensitivity to SST of the atmospheric component of the coupled models. It may also reflect differences in the mean oceanic and atmospheric states simulated, since the latter may impact the models' ability to represent anomalous ocean heat transport and its influence on the NAO.

It is interesting to observe a reduced energy level in SST or SLP at low frequency (Figure 10) whereas an enhanced energy level was predicted in the absence of ocean dynamics (section 3.1). This might be understood as an additional damping effect introduced by the ocean circulation on low-frequency oceanic and atmospheric temperature anomalies. In the illustrative model of anomalous ocean current devel-

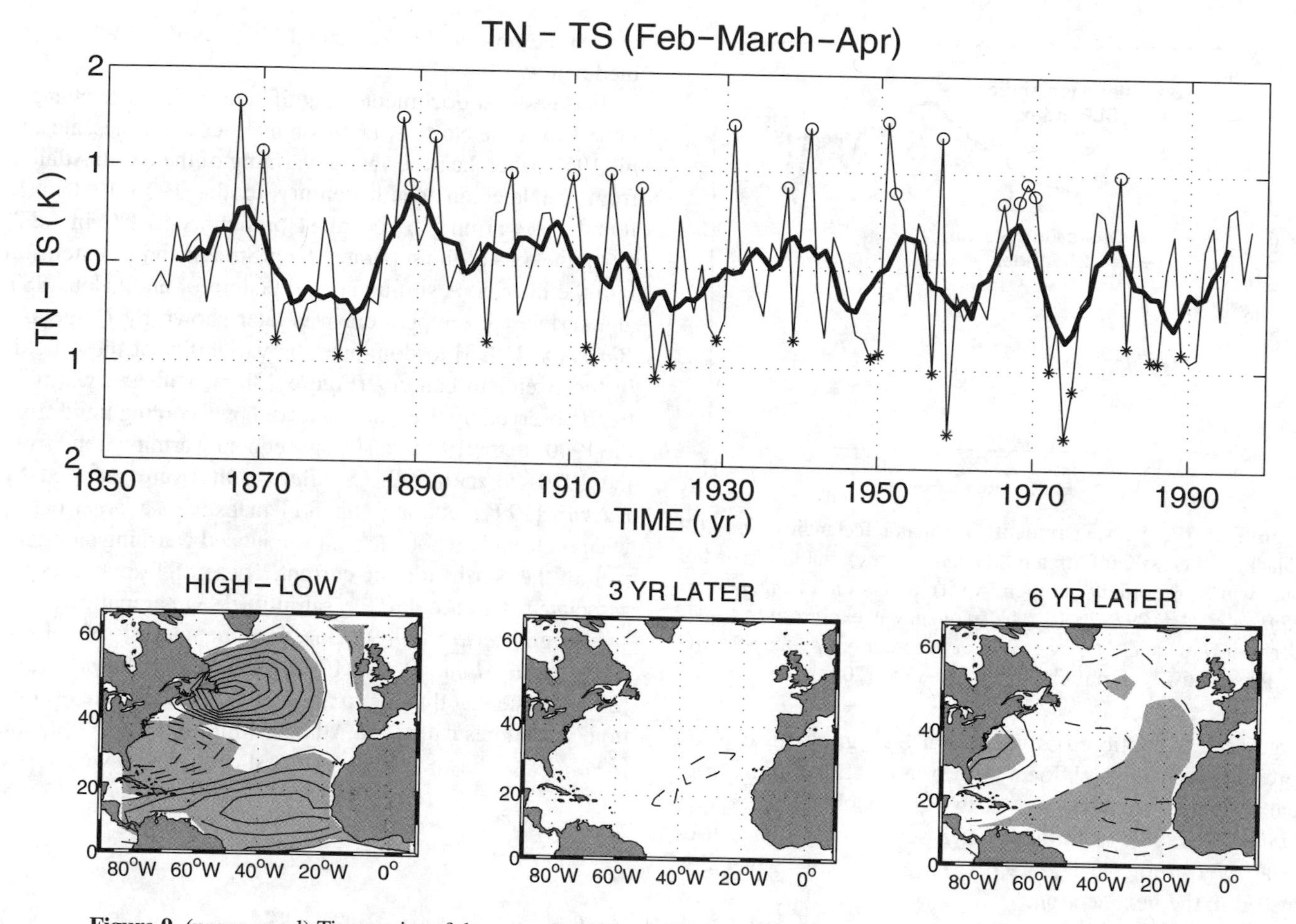

Figure 9. (upper panel) Time series of the cross Gulf Stream SST index ΔT (in *K*, February through April averaged, raw time series thin, 6-yr running mean thick). (bottom panel) Composite maps for SST anomalies, based on years where the ΔT index is high and low, as indicated by the circles and stars in the upper panel. The left panel indicates the large-scale SST anomalies associated simultaneously with $\Delta T > 0$, while the middle and right panels indicate the SST anomalies 3 years and 6 years after strong $\Delta T > 0$ events. The SST anomalies are contoured every 0.2K (dashed when negative), and the shading indicates where the composites are significant within the 95% confidence level, as deduced from a Student *t*-test. All SST data are from *Kaplan et al.* [1997]. From *Czaja and Marshall* [2001].

oped by *Marshall et al.* [2001a], anomalous ocean heat transport is directed from the warm to the cold anomaly at timescales longer than the oceanic adjustment time to NAO wind forcing. It thus acts on SST as an anomalous down gradient heat flux at these timescales [*Czaja and Marshall*, 2001; Figure 12]. A similar feature is found when considering the role of mean advection in the model of *Saravanan and McWilliams* [1998], with atmospheric and oceanic temperature anomalies in spatial quadrature for sufficiently strong advection and long timescales, thus 'pulled away' from thermal equilibration by ocean currents (not shown).

It is certainly an oversimplification to attribute the bulk of the decadal variability displayed by the tripole to air-sea interactions in the vicinity of the intergyre boundary alone. Indeed, an alternative explanation has been invoked, where the decadal timescale is driven from the tropical Atlantic through local unstable air-sea interactions [*Chang et al.*, 1997]. In this scenario, tropical Atlantic variability would then remotely impact the middle latitude centers of action of the SST tripole [*Okumura et al.*, 2001] and the NAO (section 3.3). Some observational support for this hypothesis exists. *Rajagopalan et al.* [1998; see also *Tourre et al.*, 1999] found evidence of a significant coherence on decadal timescales between tropical Atlantic SSTs and the NAO signature in SLP (Icelandic low and Azores high). Both the northern (5°N–20°N) and the southern (5°S–15°S) SST anomalies are involved, as is their difference (the Atlantic cross-equatorial SST index), with signs consistent with the remote forcing discussed earlier (section 3.3). Nevertheless, the coherence may also reflect a passive response of the

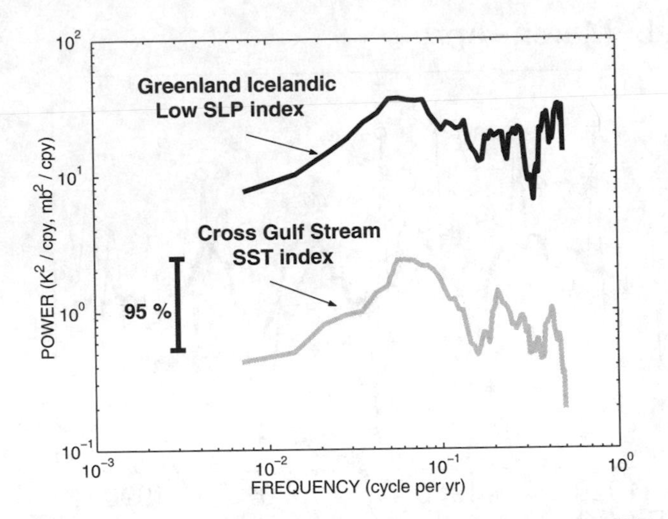

Figure 10. Power spectrum of Greenland Icelandic Low SLP (black) and cross Gulf Stream SST index (grey), both averaged in late winter (February through April) based on the data from *Kaplan et al.* [1997; 2000]. The frequency is expressed in cycles per year (*cpy*) and the power in K^2/cpy and mb^2/cpy for SST and SLP respectively. From *Czaja and Marshall* [2001].

large-scale Atlantic cross equatorial SST gradient to NAO forcing, either through local air-sea interactions in the tropical Atlantic connecting the two hemispheres [*Xie and Tanimoto*, 1998], or through basin-scale changes in meridional overturning [*Yang*, 1999]. The latter are further discussed in the next section.

4.2. NAO/Ocean Circulation Interaction on a Basin-Scale

Various coupled GCMs or ocean only integrations have shown substantial variability of the MOC at interdecadal timescales, with associated current, temperature and salinity anomalies extending all the way to the equator [e.g., *Delworth et al.*, 1993; *Timmerman et al.*, 1998; *Häkkinen*, 1999; *Eden and Willebrandt*, 2001; *Eden and Jung*, 2001; *Dong and Sutton*, 2001]. The dynamics of such basin-scale changes were discussed in detail in *Visbeck et al.* [this volume]. They involve tropical extra-tropical oceanic teleconnections set rapidly (months) by boundary Kelvin waves [e.g., *Kawase, 1987*; *Johnson and Marshall*, 2002] as a response to changes in the rate of deep-water formation at high latitudes, and the subsequent generation of long planetary Rossby waves along the eastern boundary of the North and South Atlantic [*Wajowicz*, 1986]. Further delay between changes in buoyancy forcing at high latitudes and oceanic changes elsewhere might also be introduced by oceanic advection and storage [e.g., *Marotzke and Klinger*, 2000]. The possibility that changes in meridional oceanic heat transport might be actively involved in (not simply responding to) the observed interdecadal NAO variability is examined below.

Bjerknes first documented significant basin-scale changes in SST over the North Atlantic on interdecadal timescales. In his 1958 paper, he describes a warming of the North Atlantic from the late nineteenth century to the 1920–1930s. He found the warming to be most pronounced (2°C in SST) along the axis of the separated Gulf Stream and its extension into the interior. A similar spatial pattern of anomalous SST and surface air temperature was later shown by *Deser and Blackmon* [1993] to dominate the variability of these fields in the twentieth century (Figure 13b,c), with the warming trend observed by Bjerknes followed by a cooling trend from the 1940s to the 1970s, and a subsequent warming trend from the 1970s to the 1990s. Similar results were reported by *Kushnir* [1994], although his SST maps have a larger basin-wide scale and do not show a pronounced warming along the path of the North Atlantic current. These SST anomalies are associated with sea surface salinity (SSS) anomalies of the same sign [*Levitus*, 1989]. Based on the analysis of various proxy data, *Mann et al.* [1998] and *Delworth and Mann* [2000] argue that this approximately 70-year climate oscillation, sometimes called the Atlantic Multidecadal Oscillation [*Kerr*, 2000], has occurred over past centuries as well.

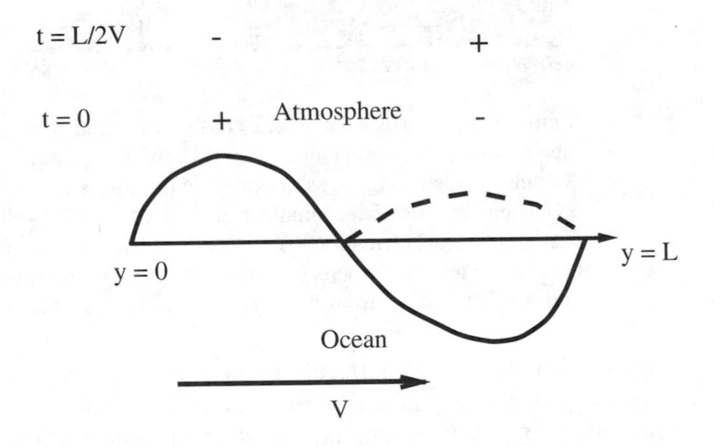

Figure 11. Schematic of ocean-atmosphere interactions occurring in presence of a mean oceanic velocity V and a dipolar surface NAO forcing of meridional scale L. The mean current advects the initial dipole (thick continuous line) northward once it has been generated through NAO surface forcing at $t = 0$ (the sign of the forcing is indicated by the plus-minus signs). At $t = L/2V$, the SST anomaly has thus become that shown in thick dashed line, if damping if not too strong. Of all timescales associated with the climate noise NAO forcing, that corresponding to L/V will reinforce this SST anomaly at $t = L/2V$, as indicated by the change of signs in the forcing. This leads to enhanced variance of the SST dipole and NAO near the 'advective resonant timescale' L/V. From *Saravanan and McWilliams* [1998].

A striking feature of Figure 13 is that the SST anomalies do not seem to be a local response to changes in windstress as commonly found on shorter timescales. Rather, they are found well upstream of the dominant surface wind changes. In addition, for a warming along the separated Gulf Stream, an anomalous cyclonic circulation centered at 40°N is observed, with reduced westerlies and trade winds to the north and south respectively. Such weakening of the winds favors the development of a negative SST anomaly along the Gulf Stream extension, as a result of the weakening of the subtropical gyre, rather than a positive SST anomaly as observed. This may be an indication that the MOC, rather than the atmosphere, is driving the basin-scale warming and cooling at these long timescales. The SST pattern (Figure 13b) might reflect the local convergence of ocean heat transport between 30–50°N, with the associated warming of the atmosphere inducing a downstream low pressure anomaly at the surface (but see below for a cautionary note on this type of interpretation), as found in simple linear baroclinic models of the tropospheric response to heating [*Hoskins and Karoly*, 1981].

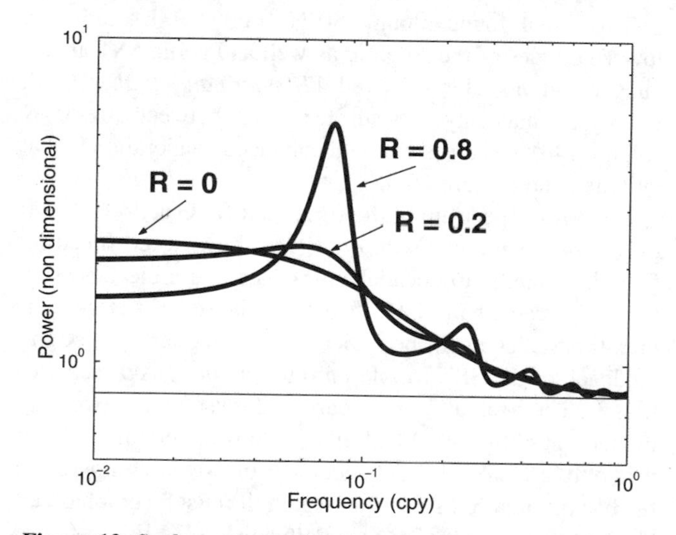

Figure 12. Surface wind τ (or atmospheric temperature) spectra from the delayed oscillator model of *Marshall et al.* [2001a]: $\tau = N - fT_o$ and $dT_o/dt = -\lambda T_o - R\lambda T_o(t - t_d/2) - \alpha N$ where N is a stochastic forcing, α a scaling constant, R measures the strength of anomalous ocean heat transport, T_o is SST, f is the strength of the SST feedback on the surface winds, and λ measures local damping of SST anomalies through air-sea interactions. The spectrum for τ is given for various choices of R (weak and strong impact of ocean heat transport, as indicated on the plot). The thin continuous line indicates the climate noise spectrum ($\tau = N$), while the limit of no ocean dynamics ($R = 0$) is another version of the 'reduced thermal damping' spectrum of Figure 5 (thick black line). Note the reduced energy level at long timescales, as the strength of the ocean circulation increases. The frequency is expressed in cycles per year. The delay timescale t_d was set to $t_d = 10$ years. Other parameters are taken from *Czaja and Marshall* [2000].

Kushnir [1994] also found this SST/sea level pressure association for the subsequent cooling period (1970–1984 minus 1950–1964), with anomalous high pressure along 45°N. Interestingly, Bjerknes' [1958] map for the earlier warming trend (1925–1932 minus 1890–1897) shows anomalously warm SST associated with positive pressure anomalies in middle latitudes, i.e. opposite to that of *Deser and Blackmon* [1993] and *Kushnir* [1994], but similar to the pressure response found in the coupled simulations of *Timmermann et al.* [1998; Figure 14] and *Delworth et al.* [1993]. These disparities reflect the difficulties of inferring ocean and atmospheric dynamics soley from analysis of observations. Recent numerical experiments by *Sutton and Mathieu* [submitted] and *D'Andréa et al.* [in preparation] further emphasize this limitation. They show that the atmosphere/slab ocean mixed layer model response to a prescribed oceanic heat flux convergence along the separated Gulf Stream displays only small equilibrium SST anomalies over the region of oceanic forcing. We will come back to this important issue later. Note also that the composite analysis in Figure 13, although using decadal averages separated by more than a decade, does not ensure that the ocean and atmosphere are in dynamical equilibrium, making even more difficult a dynamical interpretation.

As mentioned earlier, *Delworth et al.* [1993] and *Delworth and Mann* [2000] describe large variations of the overturning circulation in the GFDL R15 coupled model at timescales of about 50 yr, which have anomalous SST and SSS patterns similar to those seen in observations [*Kushnir*, 1994; *Levitus*, 1989], i.e. of similar sign over the North Atlantic basin. In the GFDL model, the MOC variability was interpreted as a damped oscillation sustained by stochastic atmospheric forcing [*Griffies and Tziperman*, 1995; *Delworth and Greatbatch*, 2000], as was also proposed by *Selten et al.* [1999] and *Saravanan et al.* [2000] in simpler coupled ocean-atmosphere models. This type of oscillation was found recently to emerge through a linear unstable mode of the ocean circulation [*Huck and Vallis*, 2001; *te Raa and Dijkstra*, 2002], a form of longwave baroclinic instability setting the interdecadal timescale [*Colin de Verdière and Huck*, 1999; *Huck et al.*, 2001]. Thus, although stochastic atmospheric forcing is important in sustaining the interdecadal variability of the MOC in the GFDL simulation, active coupling with the atmosphere is not crucial. Nevertheless, as the MOC strengthens and drives basin-scale warming of the North Atlantic, the GFDL model atmosphere shows a weak tendency to go preferentially towards a positive NAO index phase.

A more fundamentally coupled mechanism was proposed by *Timmermann et al.* [1998] to describe the interdecadal variability in the ECHAM3/LSG simulation, involving changes in the MOC on periods around 35 years. Although its signature in SST, SSS and SLP resembles that of

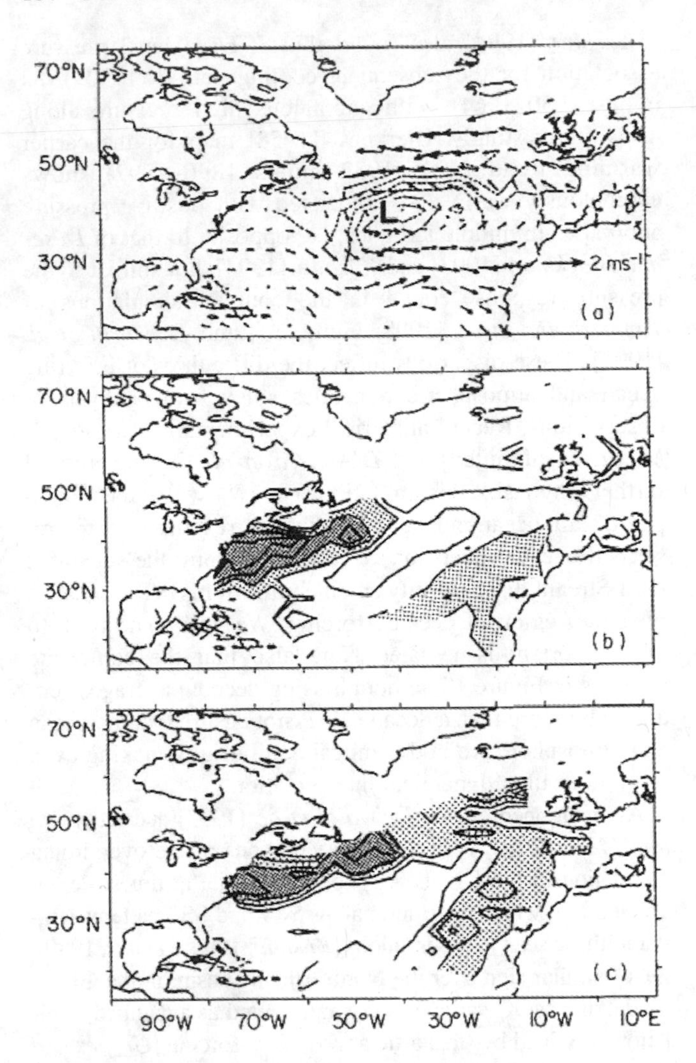

Figure 13. Difference between the periods 1939–1968 and 1900–1929 of winter (a) sea level pressure and wind, (b) SST, and (c) surface air temperature. In (a) the contour interval is 0.5 mb, with negative contours dashed. The lowest pressure anomaly is -3 mb. Wind scale is indicated in lower right. In (b) light shading indicates values between 0.8°C and 1°C; heavy shading indicates values greater than 1°C. Contour interval is 0.2°C. In (c) light shading indicates values between 0.6°C and 0.8°C; heavy shading indicates values greater than 0.8°C. Contour interval is 0.2°C. From *Deser and Blackmon* [1993; their Figure 7].

Delworth et al. [1993], the NAO-like response of the atmosphere seems most important to the generation of SSS anomalies (through surface freshwater water flux and Ekman transport off Newfoundland and Greenland) and their subsequent impact on density gradients and the large-scale overturning circulation (see their Figure 20). The origin of the 35-yr timescale is not simple, but factors such as the accumulation time for salinity anomalies and the response of the MOC to them are key factors.

Figure 14 illustrates the spatial structure and phase relationships between these variables, based on simultaneous regression maps onto a band-passed MOC index. The positive NAO index signature is seen in the SLP regression map (Figure 14a), associated with basin wide positive anomalous SST (Figure 14b). As mentioned earlier for its observational counterpart (Figure 13), this association is different from that found at decadal and shorter timescales [see *Visbeck et al.*, this volume] and suggests that the SST anomalies reflect enhanced oceanic heat transport rather than local atmospheric forcing. Indeed the regression map for surface currents (Figure 14e) shows a stronger Gulf Stream and North Atlantic current, with anomalous southward Ekman currents present along ~55°N. These changes in surface currents reflect not only the persistence of prior increases in deep convection south of Greenland and prior decreases in the Greenland-Iceland-Norwegian Sea (Figure 14f), but also partly the spin-up of the subtropical gyre induced by the strengthening of surface winds. Note that the positive NAO index phase (Figure 14a) is consistent with the enhanced SST gradient found along ~50°N (Figure 14b), although lower latitudes of the Atlantic as well as Pacific SST anomalies could also be involved [*Timmerman et al.*, 1998]. Finally, anomalously salty surface waters between 30°–55°N (Figure 14d) are consistent with enhanced evaporation in that latitude band (Figure 14c).

The power spectrum of the simulated NAO at 500 mb displayed by *Timmermann et al.* [1998] shows a red spectrum from interannual to decadal timescales and a clear peak in the interdecadal band (30–35 years). The redness at interannual timescales might be indicative of a moderate to strong feedback of the SST tripole onto the model NAO (see section 3). The peak at 30–35 years might reflect the two-way interaction of the NAO with the overturning circulation, but it is only a factor of 2 enhancement of power compared to the background red spectrum. Again, it must be emphasized that not all coupled GCMs show this behavior. For instance, *Robertson* [2001] found no such departures from climate noise when analyzing both the ECHAM4/OPYC and the CCM3/NCOM coupled models.

4.3. Summary

The Atlantic Ocean responds strongly to NAO changes on a broad spectrum of timescales. The climate noise scenario sees these fluctuations in ocean circulation as a passive response to (unaltered) intrinsic atmospheric variability. The presence of strong, intrinsic background NAO variability with a weak dynamical SST feedback leads to the conclusion that NAO variability is at best modulated by the Atlantic Ocean, rather than reflecting coupled dynamics like ENSO.

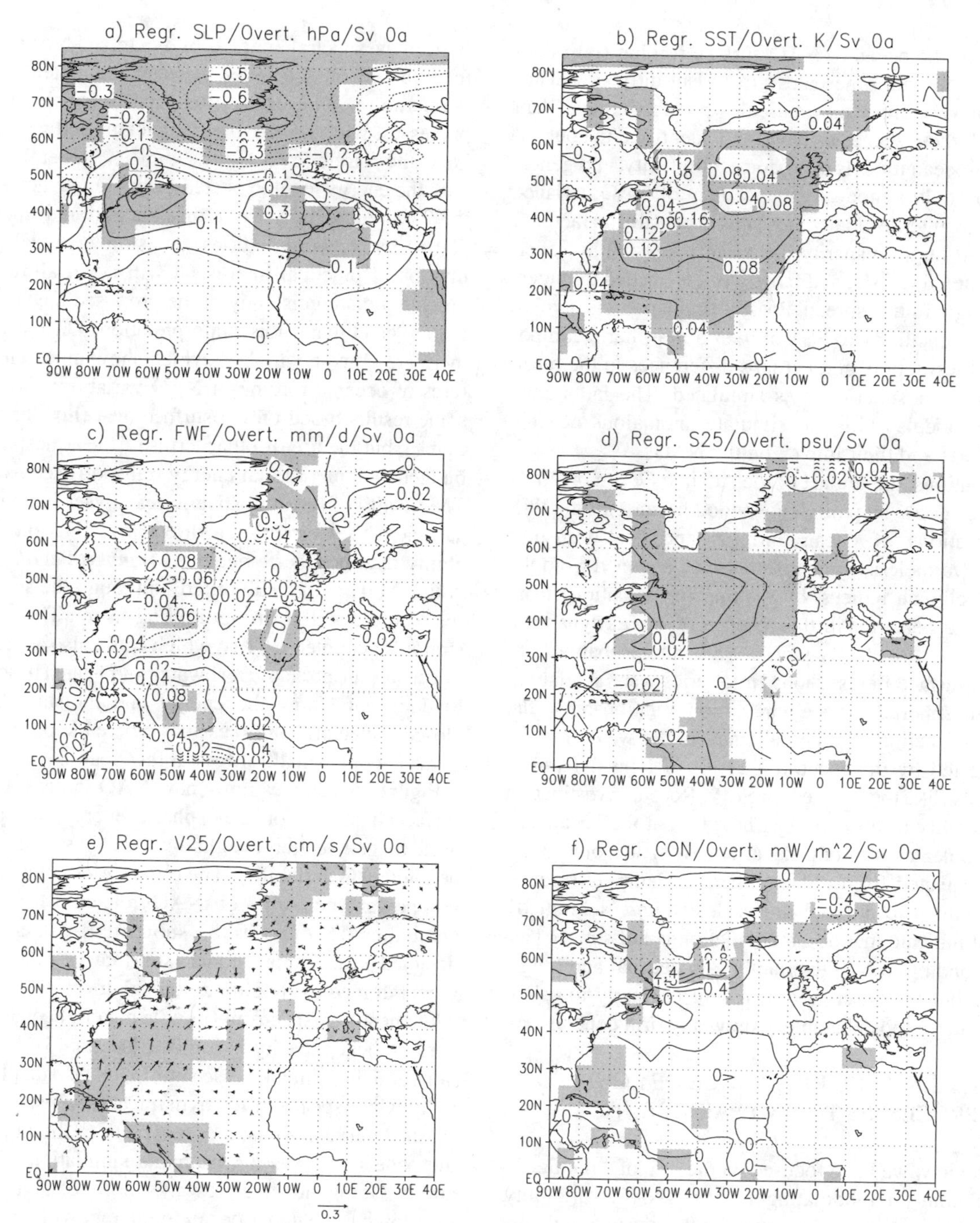

Figure 14. Simultaneous regression map of various quantities onto a band-passed (25–45 years) filtered meridional overturning index. See text for a description. From *Timmermann et al.* [1998; their Figure 14].

Analysis of simple models suggests that ocean heat transport variability can introduce preferred (decadal and interdecadal) timescales to which deviations from climate noise (section 2) and/or reduced thermal damping (section 3) are expected.

Ocean heat transport changes can arise through the advection of temperature anomalies by the mean ocean circulation. The corresponding advection speed combined to a length scale determined by intrinsic atmospheric processes then sets an 'advective resonance' timescale (at most a decade). Alternatively, ocean heat transport changes can arise through anomalous advection of mean temperature gradients. The associated anomalous currents can reflect the adjustment of the ocean to NAO surface forcing (introducing typically decadal timescales), or reflect intrinsic oceanic modes of variability (at timescales possibly longer than decadal). All these scenarios have been identified in coupled atmosphere-ocean models of various complexity, and generally result in only a modest enhancement of NAO variance at decadal to interdecadal timescales (typically a factor 2 increase). A large disparity of results exists, however, between coupled GCMs, and more work is needed to understand the origin of this spread. Possible explanations involve the different sensitivity to SST of their atmospheric component, model resolution, and perhaps differences in the mean oceanic and atmospheric states simulated. The latter may impact the models ability to simulate anomalous oceanic heat transports and their impact on the NAO.

In observations, the impact of the Atlantic Ocean on NAO variability can only be partly understood because of the limited observations of ocean currents and heat transport changes. Nevertheless, some insights can be gained from the analysis of climate indices, by looking for deviations from their expected behavior in the absence of ocean dynamics. Such departures are clearly seen in SST (and probably even more in subsurface fields if long time series were available), in SST-atmospheric surface circulation relationships that change as a function of timescale, and in the weak decadal peak suggested by the spectral analysis of long dynamical NAO indices (like those based on SLP). Recent investigation of the variability of observed net heat flux at the ocean surface also indicates a strong departure from the blue noise spectrum expected from reduced thermal damping (section 3) at interannual to decadal timescales [*Czaja*, submitted]. This might indicate that ocean circulation is not simply passively responding to NAO variability on decadal and longer timescales, but the observational record is clearly too short and the evidence too debatable to draw any firm conclusions.

5. ASSESSING THE RELEVANCE OF COUPLED PROCESSES TO NAO VARIABILITY

In this review, we have focused on the role of Atlantic air-sea interactions in modulating intrinsic NAO variability. Despite various mechanisms by which the ocean could conceivably impact the atmosphere over the Atlantic sector, the modeling and observational evidence of their relevance to NAO variability appears much weaker than the zero order description based on climate noise. We now make some suggestions about what could be done in the future to better identify signatures of Atlantic air-sea interactions in observational and modeling studies of NAO variability.

It is clear from this chapter that the impact of coupled processes on NAO variability should not be evaluated solely from the spectral analysis of dynamical NAO indices, like those based on sea level pressure, geopotential height or temperature. The reason is that intrinsic atmospheric processes, with a related white background spectrum, dominate the variability of these indices, generally consistent with the energy level expected from climate noise. Although the weak departure from this background spectrum displayed by the observations and simple model predictions may be of practical interest, it will never allow us to convincingly distinguish any of the proposed mechanisms from a simpler climate noise interpretation. On the other hand, there are climate variables which show much clearer signatures of oceanic forcing of NAO variability: we mentioned some results based on net surface heat flux variations (section 4), but the use of other terms of the atmospheric energy budget (like meridional energy transport) seems promising. The quality of the atmospheric reanalyses is now becoming sufficient to warrant such direct investigation of the variability in atmospheric budgets [*Trenberth et al.*, 2002; *Boer et al.*, 2001; *Czaja*, submitted], although the short length of the record is still a strong limiting factor. This is not a constraint for coupled GCMs, which could also address changes in ocean transports of freshwater and heat. These are crucial to understand how the energy and moisture budgets are closed, especially on timescales where the oceanic tendency terms become small (at least a few years).

Figure 15 sketches how 'new NAO indices' based on net surface heat flux or atmospheric energy transport can be used to gain better insight into the physical mechanisms behind NAO variability. The observations suggest a slightly red power spectrum for NAO related surface heat flux anomalies [*Czaja*, 2002, submitted], as schematically depicted in the upper panel (solid curve). An atmospheric model forced with climatological mean SST produces a white heat flux spectrum (middle panel), because an infinite oceanic heat capacity is implied (section 3). Both observations and this climate noise simulation also display an essentially white spectrum for dynamical NAO spectra like SLP (dashed curves). Thus, a comparison of the upper and middle panels suggests a reasonable simulation of NAO variability. Going, however, towards more realism (lower panel) by allowing the upper ocean to interact with the atmosphere (through coupling the AGCM to a slab ocean mixed layer) leads to a very different (blue) spectrum for surface heat flux, far from that observed (upper panel). This illustrates a caveat in the climate noise interpretation (middle panel), namely that the good simulation of the heat flux spectrum is obtained for an unphysical reason (the infinite oceanic heat capacity). The 'new NAO indices' thus appear to contain more striking signatures of the mechanisms behind NAO variability, as Figure 15 suggests the similarity of observed

SLP (or other dynamical NAO indices) and net surface heat flux spectra is not trivial. It was suggested in section 4 that this similarity could reflect the impact of oceanic advection on NAO variability.

It also emerges from section 3.1 that one should not consider the troposphere and the upper-ocean separately. Large-scale anomalies in atmospheric circulation inevitably interact with the upper ocean through exchange of heat and moisture at the sea surface and through Ekman advection. It thus seems natural to consider the troposphere and the upper ocean as a single system, possibly interacting with the geostrophic ocean circulation on interannual and longer timescales. Recent experiments by *Sutton and Mathieu* [submitted] emphasize this point [see also *D'Andréa et al.*, in preparation]. By forcing an AGCM coupled to a slab ocean mixed layer with strengthened ocean heat transport convergence along the separated Gulf Stream (so-called 'Q-flux' forcing), they obtained an equilibrium response of a surface low pressure and warm SST anomaly *downstream* of the forcing region. Within the forcing region itself, SST anomalies were unremarkable. The downstream SST anomaly thus developed both as a response to the anomalous ocean heat transport and the surface heat flux changes associated with the developing atmospheric low pressure anomalies. Solely based on the knowledge of the equilibrium SST and SLP anomaly, one would fail to interpret the response as driven by ocean changes along the separated Gulf Stream.

This tight association between the atmosphere and the upper-ocean also has implications for interpreting recent simulations of decadal NAO variability from the time-history of observed SST and sea ice anomalies [AMIP-type integrations, *Venzke et al.*, 1999; *Rodwell et al.*, 1999; *Mehta et al.*, 2000; *Latif et al.*, 2000; *Hoerling et al.*, 2001; *Robertson et al.*, 2001; *Cassou and Terray*, 2001a,b; *Terray and Cassou*, submitted]. If, on the one hand, we take the extreme view that changes in ocean circulation are not responsible for the generation of the observed SST tripole anomaly, the good simulation of the NAO index could merely reflect the positive feedback of the tripole on the NAO [*Bretherton and Battisti*, 2000]. If, on the other hand, ocean circulation changes are involved in driving the tripole SST anomalies, then the whole approach of the AMIP-type integrations is misleading [*Sutton and Mathieu*, submitted].

Another important need is to gain a better understanding of the differences between the two phases of NAO circulation anomalies. Unpublished observational results by *Molteni and Pavan* [personal communication] suggest that a negative NAO index phase is associated with reduced net diabatic heating in the atmosphere, while a positive NAO index phase corresponds to a thermally forced regime, with strong diabatic heating. The analyses of *Cheng and Wallace* [1993] and *Peng et al.* [2001] show east-west wavetrain-like structures in geopotential height associated with a positive NAO index phase, but essentially a localized meridional dipole in the negative index phase (perhaps consistent with *Molteni and Pavan*). Whether the interaction of the upper ocean and the atmosphere is needed or not to reproduce this asymmetry is an open issue.

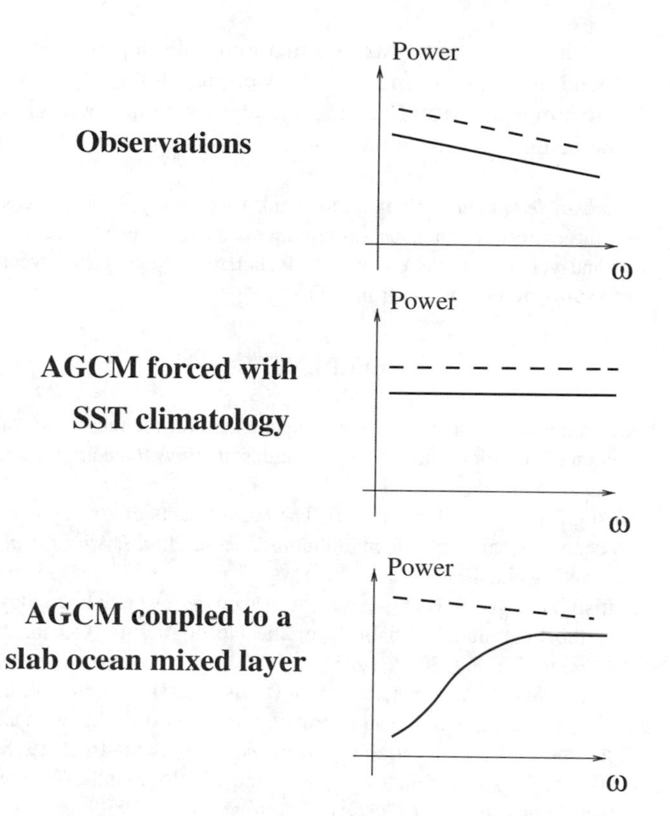

Figure 15. Schematic of observed (top panel) and simulated (middle and bottom panels) net surface heat flux (continuous line) and SLP (dashed line) power spectra as a function of frequency ω.

Finally, some questions addressed in the context of NAO variability and air-sea interactions in this review appear as more fundamental:

- is the degree of coupling between oceanic and atmospheric fluctuations solely a consequence of the sensitivity of the atmosphere to SST changes?
- or does the degree of coupling also reflect the partition of the total (atmosphere + ocean) meridional energy transport between the ocean and the atmosphere, with strong coupling when each component plays an equal role (as in the tropics), and weak coupling when one component dominates (as in the extra tropics)?
- is the impact of cryospheric and land-vegetation anomalies on NAO variability stronger than that of the Atlantic ocean?

It is hoped that analysis of a hierarchy of coupled ocean-ice-land-atmosphere models, as well as the analysis of future atmospheric and oceanic reanalysis products will give some insight into these problems.

Acknowledgments. We wish to thank four anonymous referees and the editors for their useful comments and suggestions on the original version of this chapter. A. Robertson and A. Czaja were both supported by grants from NOAA.

REFERENCES

Alexander, M. A., and C. Deser, A mechanism for the recurrence of wintertime midlatitude SST anomalies, *J. Phys. Oceanogr.*, *25*, 122–137, 1995.

Barsugli, J. J., and D. S. Battisti, The basic effects of atmosphere-ocean thermal coupling on midlatitude variability, *J. Atmos. Sci.*, *55*, 477–493, 1998.

Battisti, D. S., U. S. Bhatt, and M. A. Alexander, A modeling study of the interannual variability of the North Atlantic Ocean, *J. Climate*, *8*, 3067–3083, 1995.

Bhatt, U. S., M. A. Alexander, D. S. Battisti, D. Houghton, and L. Keller, Atmosphere-ocean interaction in the North Atlantic: near sea surface climate variability, *J. Climate*, *11*, 1615–1632, 1998.

Bjerknes, J., The recent warming of the North Atlantic, *Rossby memorial volume, in The Sea in Motion*, 65–73, 1958.

Bjerknes, J., Synoptic survey of the interaction of sea and atmosphere in the North Atlantic, *Geophysica Norvegia*, in memory of Vilhem Bjerknes, XXIV, 3, 115–145, 1962.

Bjerknes, J., 1964. Atlantic air-sea interaction, *Advances in Geophysics*, *10*, Academic Press, 1–82, 1964.

Bladé, I., The influence of midlatitude Ocean-Atmosphere coupling on the low - frequency variability in a GCM: Part 1. No tropical SST forcing, *J. Climate*, *10*, 2087–2106, 1997.

Boer, G. J., S. Fourest, and B. Yu, The signature of the annular modes in the moisture budget, *J. Climate*, *14*, 3655–3665, 2001.

Box, G. E. P., G. M. Jenkins, and G. C. Reinsel, *Time Series Analysis: Forecasting and Control*. 3rd ed., Prentice Hall, 1993.

Bresch, D., Coupled flow and SST patterns of the North Atlantic, Ph. D. Thesis, Swiss Federal Institute of Technology, 119 pp, 1998.

Bretherton, C. S., C. S. Smith, and J. M. Wallace, 1992. An intercomparison of methods for finding coupled patterns in climate data, *J. Climate*, *5*, 541–560, 1992.

Bretherton, C. S., and D. S. Battisti, An interpretation of the results from atmospheric general circulation models forced by the time history of the observed sea surface temperature distribution, *Geophys. Res. Lett.*, *27*, 767–770, 2000.

Cassou, C., and L. Terray, Oceanic forcing of the wintertime low frequency atmospheric variability in the North Atlantic European sector: a study with the ARPEGE model, *J. Climate*, *14*, 4266–4291, 2001a.

Cassou, C., and L. Terray, Dual influence of Atlantic and Pacific SST anomalies on the North Atlantic/Europe winter climate, *Geophys. Res. Lett.*, *28*, 3195–3198, 2001b.

Cayan, D., Latent and sensible heat flux anomalies over the Northern oceans: driving the sea surface temperature, *J. Phys. Oceanogr.*, *22*, 859–881, 1992.

Cessi, P., Thermal feedback on wind stress as a contributing cause of climate variability, *J. Climate*, *13*, 232–244, 2000.

Cessi, P., and F. Primeau, Dissipative selection of low frequency modes in a reduced-gravity basin, *J. Phys. Oceanogr.*, *31*, 127–137, 2001.

Cessi, P., and F. Paparella, Excitation of basin modes by ocean-atmosphere coupling, *Geophys. Res. Lett.*, *28*, 2437–2440, 2001.

Chang, E. K. M., The influence of Hadley circulation intensity changes on extratropical climate in an idealized model, *J. Atmos. Sci.*, *52*, 2006–2024, 1995.

Chang, P., L. Ji, and H. Li, A decadal climate variation in the tropical Atlantic Ocean from thermodynamic air-sea interactions, *Nature*, *385*, 516–518, 1997.

Cheng, X., and J. M. Wallace, Cluster analysis of the Northern Hemisphere wintertime 500 hPa height field: spatial patterns, *J. Atmos. Sci.*, *50*, 2674–2696, 1993.

Colin de Verdière, A., and T. Huck, Baroclinic instability: an oceanic wavemaker for interdecadal variability, *J. Phys. Oceanogr.*, *29*, 893–910, 1999.

Costa E. D., and A. Colin de Verdière, 2002. Extended canonical correlation analysis of North Atlantic SST and SLP, *Quart. J. Roy. Meteor. Soc.*, in press, 2002.

Curry, R. G., and M. S. McCartney, Ocean gyre circulation changes associated with the North Atlantic Oscillation, *J. Phys. Oceanogr.*, *31*, 3374–3400, 2001.

Czaja A., and C. Frankignoul, Influence of the North Atlantic SST anomalies on the atmospheric circulation, *Geophys. Res. Lett.*, *26*, 2969–2972, 1999.

Czaja A., and J. Marshall, On the interpretation of AGCMs response to prescribed time-varying SST anomalies, *Geophys. Res. Lett.*, *27*, 1927–1930, 2000.

Czaja A., and J. Marshall, Observations of atmosphere-ocean coupling in the North Atlantic, *Quart. J. Roy. Meteor. Soc.*, *127*, 1893–1916, 2001.

Czaja A., and C. Frankignoul, Observed impact of North Atlantic SST anomalies on the North Atlantic Oscillation, *J. Climate*, *15*, 606–623, 2002.

Davis, R., Predictability of sea surface temperature and sea level pressure anomalies over the North Pacific Ocean, *J. Phys. Oceanogr.*, *8*, 249–266, 1976.

Delworth, T., North Atlantic interannual variability in a coupled ocean - atmosphere model, *J. Climate*, *9*, 2356–2375, 1996.

Delworth, T., S. Manabe, and R. J. Stouffer, Interdecadal variations in the thermohaline circulation in a coupled ocean - atmosphere model, *J. Climate*, *6*, 1993–2010, 1993.

Delworth, T., and M. E. Mann, Observed and simulated multidecadal variability in the North Atlantic, *Clim. Dyn.*, *16*, 661–676, 2000.

Delworth, T. L., and R. J. Greatbatch, Multidecadal thermohaline circulation variability driven by atmospheric surface flux forcing, *J. Climate*, *13*, 1481–1495, 2000.

Deser, C, and M. L. Blackmon, Surface climate variations over the North Atlantic during winter: 1900–1989, *J. Climate*, *10*, 393–408, 1993.

Deser, C, and M. S. Timlin, Atmosphere-ocean interaction on weekly timescales in the North Atlantic and Pacific, *J. Climate*, *10*, 393–408, 1997.

Dewar, W. K., On ocean dynamics in mid-latitude climate, *J. Climate*, *14*, 4380–4397, 2001.

Dong, B. W., and R. T. Sutton, The dominant mechanisms of variability in Atlantic ocean heat transport in a coupled ocean-atmosphere GCM, *Geophys. Res. Lett.*, *28*, 2445–2448, 2001.

Döös, B. R., The influence of exchange of sensible heat with the Earth's surface on the planetary flow, *Tellus*, *2*, 133–147, 1962.

Drévillon, M., L. Terray, P. Rogel, and C. Cassou, Mid-latitude Atlantic SST influence on european winter climate variability in the NCEP reanalysis, *Clim. Dyn.*, *18*, 331–344, 2001.

Drijfhout, S. S., A. Kattenberg, R. J. Haarsmaa, and F. M. Selten, The role of the ocean in midlatitude, interannual to decadal timescale climate variability in a coupled model, *J. Climate*, *14*, 3617–3630, 2001.

Eden C., and T. Jung, North Atlantic interdecadal variability: oceanic response to the North Atlantic Oscillation (1865–1997), *J. Climate*, *14*, 676–691, 2001.

Eden C., and J. Willebrandt, Mechanisms of interannual to decadal variability in the North Atlantic circulation, *J. Climate*, *14*, 2266–2280, 2001.

Ferreira, D., C. Frankignoul, and J. Marshall, Coupled ocean-atmosphere dynamics in a simple midlatitude climate model, *J. Climate*, *14*, 3704–3723, 2001.

Feldstein, S. B., The timescale, power spectra, and climate noise properties of teleconnection patterns, *J. Climate*, *13*, 4430–4440, 2000.

Frankignoul, C., Sea surface temperature anomalies, planetary waves and air-sea feedbacks in the middle latitude, *Rev. of Geophys.*, *23*, 357–390, 1985.

Frankignoul, C., and K. Hasselmann, Stochastic climate models, part II: application to sea-surface temperature variability and thermocline variability, *Tellus*, *29*, 289–305, 1977.

Frankignoul, C., A. Czaja, and B. L'Hévéder, Air-sea feedback in the North Atlantic and surface boundary conditions for ocean models, *J. Climate*, *11*, 2310–2324, 1998.

Frankignoul, C., E. Kestenare, N. Sennéchael, G. de Coëtlogon, and F. D'Andréa, On decadal-scale ocean - atmosphere interactions in the extended ECHAM1/LSG climate simulation, *Clim. Dyn.*, *16*, 333–354, 2001a.

Frankignoul, C., G. de Coëtlogon, T. M. Joyce, and S. Dong, Gulf stream variability and ocean-atmosphere interactions, *J. Phys. Oceanogr.*, *31*, 3516–3529, 2001b.

Gill, A. E., *Atmosphere-Ocean Dynamics*, Academic Press, 662 pp., 1982.

Goodman J., and J. Marshall, A model of decadal middle-latitude atmosphere-ocean coupled modes, *J. Climate*, *12*, 621–641, 1999.

Goodman J., Interannual middle-latitude atmosphere-ocean interactions, Ph.D., Massachusetts Institute of Technology, 151 pp., 2001.

Griffies, S. M., and E. Tziperman, A linear thermohaline oscillator driven by stochastic atmospheric forcing, *J. Climate*, *8*, 2440–2453, 1995.

Grötzner, A., M. Latif, and T. P. Barnett, A decadal climate cycle in the North Atlantic Ocean as simulated by the ECHO coupled GCM, *J. Climate*, *11*, 831–847, 1998.

Häkkinen S., Variability of the simulated meridional heat transport in the North Atlantic for the period 1951–1993, *J. Geophys. Res.*, *104*, 10991–11007, 1999.

Halliwell, G., Simulation of North Atlantic decadal/multidecadal winter SST anomalies driven by basin-scale atmospheric circulation anomalies, *J. Phys. Oceanogr.*, *28*, 5–21, 1998.

Halliwell, G. R., and D. A. Mayer, Frequency response properties of forced climatic SSS anomaly variability in the North Atlantic, *J. Climate*, *9*, 3575–3585, 1996.

Held, I. M., S. W. Lyons, and S. Nigam, Transients and the extratropical response to El Niño, *J. Atmos. Sci.*, *46*, 163–174, 1989.

Hendon, H. H., and D. L. Hartmann, Stationary waves on a sphere: sensitivity to thermal forcing, *J. Atmos. Sci.*, *39*, 1906–1920, 1982.

Hoerling, M. P., and M. Ting, Organization of extratropical transients during El Niño, *J. Climate*, *7*, 745–766, 1994.

Hoerling, M. P., J. W. Hurrell, and T. Xu, Tropical origins for recent North Atlantic climate change, *Science*, *292*, 90–92, 2001.

Hoskins, B. J., Modelling of the transient eddies and their feedback on the mean flow, in *Large Scale Dynamical Processes in the Atmosphere*, R. P. Pearce and B. J. Hoskins, Eds., 160–199, 1983.

Hoskins, B. J., and D. J. Karoly, The steady linear response of a spherical atmosphere to thermal and orographic forcing, *J. Atmos. Sci.*, *36*, 1179–1196, 1981.

Hoskins, B. J., and P. D. Sardeshmukh, A diagnostic study of the dynamics of the northern hemisphere winter of 1985–1986, *Quart. J. Roy. Meteor. Soc.*, *113*, 759–778, 1987.

Hoskins, B. J., and P. J. Valdes, On the existence of storm-tracks, *J. Atmos. Sci.*, *47*, 1854–1864, 1990.

Hou, A. Y., The influence of tropical heating displacement on the extratropical climate, *J. Atmos. Sci.*, *50*, 3553–3570, 1993.

Hou, A. Y., Hadley circulation as a modulator of the extratropical climate, *J. Atmos. Sci.*, *55*, 2437–245, 1998.

Hou, A. Y., and A. Molod, Modulation of dynamic heating in the winter extratropics associated with the cross-equatorial Hadley circulation, *J. Atmos. Sci.*, *52*, 2609–2626, 1995.

Huck, T., G. K. Vallis, and A. Colin de Verdière, On the robustness of the interdecadal modes of the thermohaline circulation, *J. Climate*, *14*, 940–963, 2001.

Huck, T., and G. K. Vallis, Linear stability analysis of the three-dimensional thermally-driven ocean circulation: application to interdecadal oscillations, *Tellus*, *53A*, 526–545, 2001.

Hurrell, J. W., Decadal trends in the North Atlantic Oscillation: regional temperatures and precipitation, *Science*, *269*, 676–679, 1995.

Hurrell, J. W., G. Ottersen, Y. Kushnir, and M. Visbeck, An overview of the North Atlantic Oscillation, this volume.

Jiang, S., F.-F., Jin, and M. Ghil, Multiple equilibria, periodic and aperiodic solutions in wind-driven, double gyre, shallow-water model, *J. Phys. Oceanogr.*, *25*, 764–786, 1995.

Jin F., A theory of interdecadal climate variability of the North Pacific Ocean-atmosphere system, *J. Climate*, *10*, 1821–1835, 1997.

Johnson, H. L., and D. P. Marshall, A theory for the surface Atlantic response to thermohaline variability, *J. Phys. Oceanogr.*, *32*, 1121–1132, 2002.

Joyce, T. M., C. Deser, and M. A. Spall, The relation between decadal variability of subtropical mode water and the North Atlantic Oscillation, *J. Climate*, *13*, 2550–2569, 2000.

Kalnay E. et al., The NCEP/NCAR 40-year reanalysis project, *Bull. Am. Meteorol. Soc.*, *103*, 18567–18589, 1996.

Kaplan A., Y. Kushnir, M. Cane, and B. Blumenthal, Reduced space optimal analysis for historical datasets: 136 years of Atlantic sea surface temperatures, *J. Geophys. Res.*, *102*, 27835–27860, 1997.

Kaplan A., Y. Kushnir, and M. Cane, Reduced space optimal interpolation of historical marine sea level pressure: 1854–1992, *J. Climate*, *13*, 2987–3002, 2000.

Kawase, M., Establishment of deep ocean circulation driven by deep-water production, *J. Phys. Oceanogr.*, *17*, 2294–2317, 1987.

Kerr, R. A., A North Atlantic climate pacemaker for the centuries, *Science*, 288, 1984–1986, 2000.

Krahman, G., Visbeck, M., and G. Reverdin, Formation and propagation of temperature anomalies along the North Atlantic current, *J. Phys. Oceanogr.*, *31*, 1287–1303, 2001.

Kravtsov, S., and A. W. Robertson, Midlatitude ocean-atmosphere interaction in an idealized coupled model, *Clim. Dyn.*, in press, 2002.

Kushnir Y., Interdecadal variations in North Atlantic sea surface temperature and associated atmospheric conditions, *J. Climate*, *7*, 141–157, 1994.

Kushnir, Y., and I. Held, Equilibrium atmospheric responses to North Atlantic SST anomalies, *J. Climate*, *9*, 1208–1220, 1996.

Kushnir, Y., W. A. Robinson, I. Bladé, N. M. J. Hall, S. Peng, and R. T. Sutton, Atmospheric GCM response to extratropical SST anomalies: Synthesis and evaluation, *J. Climate*, *15*, 2233–2256, 2002.

Latif, M., and T. P. Barnett, Causes of decadal climate variability in the North Pacific/North Atlantic sector, *Science*, *266*, 634–637, 1994.

Latif, M., and T. P. Barnett, Decadal climate variability over the North Pacific and North America: dynamics and predictability, *J. Climate*, *9*, 2407–2423, 1996.

Latif, M., Dynamics of interdecadal variability in coupled ocean-atmosphere models, *J. Climate*, *11*, 602–624, 1998.

Latif, M., K. Arpe, and E. Roeckner, Oceanic control of decadal North Atlantic sea level pressure variability in winter, *Geophys. Res. Lett.*, *27*, 727–730, 2000.

Leith, C. E., The standard error of time - average estimates of climatic means, *J. Appl. Meteor.*, *12*, 1066–1069, 1973.

Levitus, S., Interpentadal variability of salinity in the upper 150 m of the North Atlantic Ocean, 1970–1974 versus 1955–1959, *J. Geophys. Res.*, *94*, 9679–9685, 1989.

Madden, R. A., A quantitative approach to long-range prediction, *J. Geophys. Res.*, *86*, 9817–9825, 1981.

Manabe, S., and R. Stouffer, Low frequency variability of surface air temperature in a 1000-year integration of a coupled ocean-atmosphere-land model, *J. Climate*, *9*, 376–393, 1996.

Mann, M. E., R. S. Bradley, and M. K. Hughes, Global-scale temperature patterns and climate forcing over the past six centuries, *Nature*, *392*, 779–787, 1998.

Marotzke, J., and B. Klinger, The dynamics of equatorially asymmetric thermohaline circulations *J. Phys. Oceanogr.*, *30*, 955–970, 2000.

Marshall, J., and D. K. So, Thermal equilibration of planetary waves, *J. Atmos. Sci.*, *47*, 963–978, 1990.

Marshall, J., H. Johnson, and J. Goodman, A study of the interaction of the North Atlantic Oscillation with the ocean circulation, *J. Climate*, *14*, 1399–1421, 2001a.

Marshall, J., Y. Kushnir, D. Battisti, P. Chang, A. Czaja, J. Hurrell, M. Mc Cartney, Saravanan, and M. Visbeck, Atlantic climate variability, *Int. J. Climatol.*, *21*, 1863–1898, 2001b.

Meacham, S. P., Low-frequency variability in the wind-driven circulation, *J. Phys. Oceanogr.*, *30*, 269–293, 2000.

Mehta, V., M. Suarez, J. V. Manganello, and T. D Delworth, Oceanic influence on the North Atlantic Oscillation and associated Northern Hemisphere climate variations: 1959–1993, *Geophys. Res. Lett.*, *27*, 121–124, 2000.

Molteni, F., Atmospheric low-frequency variability and the role of diabatic processes, Proceedings of the international school of Physics Enrico Fermi, *The use of EOS for studies of Atmospheric physics*, Gille and Visconti, Eds., 1992.

Moron, V., and R. Vautard, and M. Ghil, Trends, interdecadal and interannual oscillations in global sea surface temperatures, *Clim. Dyn.*, *14*, 545–569, 1998.

Münnich, M., M. Latif, S. Venzke, and E. Maier-Reimer, Decadal oscillations in a simple coupled model, *J. Climate*, *11*, 3309–3319, 1998.

Neelin, J. D., and W. Weng, Analytical prototypes for ocean-atmosphere interaction in midlatitudes. Part I: coupled feedbacks as a sea surface temperature dependent stochastic process, *J. Climate*, *12*, 697–721, 1999.

Nilsson, J., Spatial reorganization of SST anomalies by stationary atmospheric waves, *Dyn. of Atmos. Oce.*, *34*, 1–21, 2001.

Nobre, P., and J. Shukla, Variations of sea surface temperature, wind stress, and rainfall over the tropical Atlantic and South America, *J. Climate*, *9*, 2464–2479, 1996.

Nogues-Paegle, J., A. W. Robertson, C. R. Mechoso, Relationship between the North Atlantic Oscillation and river flow regimes of South America. *Proceedings of the 25th Annual Climate Diagnostics and Prediction Workshop*, Palisades, New York, 323–326, 2000.

Okumura, Y., S-P. Xie, A. Numaguti, and Y. Tanimoto, Tropical Atlantic air-sea interaction and its influence on the NAO, *Geophys. Res. Lett.*, *28*, 1507–1510, 2001.

Palmer, T. N., and Z. Sun, A modeling and observational study of the relationship between sea surface temperature in the north west Atlantic and the atmospheric general circulation, *Quart. J. Roy. Meteor. Soc.*, *111*, 947–975, 1985.

Peña, M., E. Kalnay, and M. Cai, The life span of intraseasonal atmospheric anomalies: dependence on the phase relationship with the ocean, WMO/CAS working group on numerical experiments: Progress in atmospheric and oceanic modelling. WMO, Geneva, 2001.

Peng, S., and J. S. Whitaker, Mechanisms determining the atmospheric response to midlatitude SST anomalies, *J. Climate*, *12*, 1393–1408, 1999.

Peng, S., W. A. Robinson, and S. Li, 2002. North Atlantic SST forcing of the NAO and relationships with intrinsic hemispheric variability, *Geophys. Res. Lett.*, in press, 2002.

Percival, D. B., and A. T. Walden, Spectral analysis for physical applications: multitaper and conventional univariate techniques, 583 pp., Cambridge University Press, 1993.

Primeau, F., and P. Cessi, Coupling between wind driven currents and midlatitude storm tracks, *J. Climate*, *14*, 1243–1261, 2001.

Rajagolapan, B., Y. Kushnir, and Y. Tourre, Observed decadal midlatitude and tropical Atlantic climate variability, *Geophys. Res. Lett.*, *25*, 3967–3970, 1998.

Robertson, A. W., and C. R. Mechoso, Interannual and decadal cycles in river flows of southeastern South America, *J. Climate*, *11*, 2570–2581, 1998.

Robertson, A. W., C. R. Mechoso, and Y. J. Kim, The influence of Atlantic sea surface temperature anomalies on the North Atlantic Oscillation, *J. Climate*, *13*, 122–138, 2000.

Robertson, A. W., On the influence of ocean-atmosphere interaction on the Arctic Oscillation in two general circulation models, *J. Climate*, *14*, 3240–3254, 2001.

Robinson, W. A, Review of WETS-The workshop on extra-tropical SST anomalies. *Bull. Am. Meteorol. Soc.*, *81*, 567–577, 2000.

Rodwell, M. J., D. P. Rowell, and C. K. Folland, Oceanic forcing of the wintertime North Atlantic Oscillation and European climate, *Nature*, *398*, 320–323, 1999.

Saravanan, R., Atmospheric low-frequency variability and its relationship to midlatitude SST variability: studies using the NCAR climate system model, *J. Climate*, *11*, 1386–1404, 1998.

Saravanan, R., and J. C. McWilliams, Advective ocean-atmosphere interaction: an analytical stochastic model with implications for decadal variability, *J. Climate*, *11*, 165–188, 1998.

Saravanan, R., G. Danabasoglu, S. C. Doney, and J. C. Mc Williams, Decadal variability and predictability in the midlatitude ocean-atmosphere system, *J. Climate*, *13*, 1073-1097, 2000.

Sardeshmukh, P. D., and B. J. Hoskins, The generation of global rotational flow by steady idealized tropical divergence, *J. Atmos. Sci.*, *45*, 1228–1251, 1988.

Selten, S. M., R. J. Haarsma, and J. D. Opsteegh, On the mechanism of North Atlantic decadal variability, *J. Climate*, *12*, 1956–1973, 1999.

Shutts, G. J., Some comments on the concept of thermal forcing, *Quart. J. Roy. Meteor. Soc.*, *113*, 1387–1394, 1987.

Stephenson, N. C., V. Pavan, and R. Bojariu, Is the North Atlantic Oscillation a random walk? *Intl. J. Climatol.*, *20*, 1–18, 2000.

Sutton R. T., and M. R. Allen, Decadal predictability of North Atlantic sea surface temperature and climate, *Nature*, *388*, 563–567, 1997.

Sutton, R. T., W. A. Norton, and S. P. Jewson, The North Atlantic Oscillation - What role for the Ocean? *Atm. Sci. Lett.*, *(doi:10.1006/asle.2000.0018)*, 2001.

Sweet, W., R. Fett, J. Kerling, and P. La Violette, Air-sea interaction effects in the lower troposphere across the North wall of the Gulf Stream, *Mon. Weather Rev.*, *109*, 1042–1052. 1981.

Taylor, A. H., and J. A. Stephens, The North Atlantic Oscillation and the latitude of the Gulf Stream, *Tellus*, *50A*, 134–142, 1998.

te Raa, L. A., and H. A. Dijkstra, Instability of the thermohaline ocean circulation on interdecadal time scales, *J. Phys. Oceanogr.*, *32*, 138–160, 2002.

Thompson, D. W. J., and J. M. Wallace, Annular modes in the extratropical circulation. Part I: month-to-month variability, *J. Climate*, *13*, 1000–1016, 2000.

Thompson, D., S. Lee, and M. P. Baldwin, Atmospheric processes governing the North Atlantic Oscillation/Northern hemisphere annular mode, this volume.

Timmerman, A., M. Latif, R. Voss, and A. Grötzner, Northern hemispheric interdecadal variability: a coupled air-sea mode, *J. Climate*, *11*, 1906–1931, 1998.

Ting, M., The stationary wave response to a midlatitude SST anomaly in an idealized GCM, *J. Atmos. Sci.*, *48*, 1249–1275, 1991.

Ting, M., and S. Peng, Dynamics of early and middle winter atmospheric responses to the Northwest Atlantic SST anomalies, *J. Climate*, *8*, 2239–2254, 1995.

Tourre Y. M., Rajagolapan B. and Y. Kushnir, Dominant patterns of climate variability in the Atlantic Ocean during the last 136 years, *J. Climate*, *12*, 2285–2299, 1999.

Trenberth K. E., D. Stepaniak, and J. M. Caron, Interannual variations in the atmospheric heat budget, *J. Geophys. Res.*, *10.1029/2000JD000297*, 2002.

Venzke, S., M. R. Allen, R. T. Sutton, and D. P. Rowell, The atmospheric response over the North Atlantic to decadal changes in sea surface temperatures, *J. Climate*, *12*, 2562–2584, 1999.

Visbeck, M., H. Cullen, G. Krahman, and N. Naik, An ocean model's response to North Atlantic Oscillation – like wind forcing, *Geophys. Res. Lett.*, *25*, 4521–4524, 1998.

Visbeck, M., R. Curry, B. Dickson, E. Chassignet, T. Delworth, and G. Krahman, The ocean's response to North Atlantic Oscillation variability, this volume, 2002.

von Storch, H., and F. W. Zwiers, *Statistical Analysis in Climate Research*, Cambridge University Press, 499 pp. 1999.

Wajsowicz, R.C., Adjustment of the ocean under buoyancy forces, II: the role of planetary waves, *J. Phys. Oceanogr.*, *16*, 2115–2136, 1986.

Wallace, J. M., C. Smith, and Q. Jiang, 1990. Spatial patterns of Atmosphere-Ocean interaction in the Northern hemisphere, *J. Climate*, *3*, 990–998, 1990.

Watanabe, M., and M. Kimoto, Tropical-extratropical connection in the Atlantic atmosphere-ocean variability, *Geophys. Res. Lett.*, *26*, 2247–2250, 1999.

Watanabe, M., and M. Kimoto, On the persistence of decadal SST anomalies in the North Atlantic, *J. Climate*, *13*, 3017–3028, 2000a.

Watanabe, M., and M. Kimoto, Ocean Atmosphere thermal coupling in the North Atlantic: a positive feedback, *Quart. J. Roy. Meteor. Soc.*, *126*, 3343–3369, 2000b.

Woodruff, S. D., R. J. Slutz, R. L. Jenne, and P. M. Steurer, A comprehensive ocean-atmosphere data set, *Bull. Am. Meteorol. Soc.*, *68*, 1239–1250, 1987.

Wunsch C., The interpretation of short climate records, with comments on the North Atlantic Oscillation and Southern Oscillations, *Bull. Am, Meteorol. Soc.*, *80*, 245–255, 1999.

Xie, S-P., and Y. Tanimoto, A pan-Atlantic decadal oscillation, *Geophys. Res. Lett.*, *25*, 2185–2188, 1998.

Yang, J., A linkage between decadal climate variations in the Labrador sea and the tropical Atlantic ocean, *Geophys. Res. Lett.*, *26*, 1023–1026, 1999.

Zorita, E., and C. Frankignoul, Modes of North Atlantic decadal variability in the ECHAM1/LSG coupled atmosphere-ocean general circulation model, *J. Climate*, *10*, 183–200, 1997.

A. Czaja, Department of Earth, Atmospheric and Planetary Sciences, Massachusetts Institute of Technology, 77 Massachusetts Ave., Cambridge, MA, 02139-4397, USA. czaja@ocean.mit.edu

A. W. Robertson, International Research Institute for Climate Prediction, P.O. Box 1000, Palisades, NY, 10964-8000, USA. awr@iri.columbia.edu

T. Huck, Laboratoire de Physique des Océans, Université de Bretagne Occidentale, 29285 Brest, France. thuck@univ-brest.fr

On the Predictability of North Atlantic Climate

Mark J. Rodwell

Hadley Centre, Met Office, Bracknell, U.K.

Our inability to predict, more than a few days ahead, the winter storms of the North Atlantic sector suggests that climate prediction for this region is likely to be difficult. Early studies attempted to find predictability in the observational record based on knowledge of the sea surface temperature (SST). This is not a trivial task as the unpredictable storms can themselves strongly force the SSTs and Atlantic circulation, thus leaving a confusing signal in the data. Later, computational "models" of the atmosphere, forced with prescribed SSTs, were introduced to investigate the roles of tropical and extratropical SSTs and to determine "potential predictability". More recently, coupled ocean-atmosphere models have allowed an investigation into the predictive role of the ocean circulation and of "coupled modes" of ocean-atmosphere variability. Here an analysis of the observations and the average of 10 atmospheric model simulations show a local atmospheric response to subtropical SSTs, a winter North Atlantic Oscillation (NAO) response to a "tripole" pattern in North Atlantic SSTs and a summer anticyclonicity signal over the UK. The predictive correlation skill for the winter NAO may be as high as 0.45 but this appears to depend on the strength of decadal variability. Some predictability may also come from tropical Pacific SSTs and volcanic forcing although a predictive role for solar forcing is less clear. An example is given which shows how the levels of predictability found could be of benefit to the end user. Individual simulations of the atmospheric and coupled models display weaker ocean-forced links than observed. This provides some hope that model developments could lead to improved seasonal and longer timescale forecasts.

1. INTRODUCTION

1.1. What do we Mean by "Predictability"?

If we could predict with total accuracy that it will be raining in Paris five years from now, we would have a *perfectly deterministic forecast* in its most extreme sense. Clearly, we cannot make such a forecast because "chaos" [the growth of errors arising from uncertainty in the present state of the climate system, *Lorenz*, 1963] and forecast model errors (due to, for example, the finite resolution of the computational model of the atmosphere) will not permit it. At present, such deterministic forecasts can only be made for a few days ahead. Recently, *Orrell et al.* [2002] have cast some doubt on the commonly held perception that chaos is the current limiting factor for deterministic numerical weather prediction (NWP). If, instead, model errors dominate at present, there may be scope for making longer deterministic forecasts that even provide some skill for seasonal-mean forecasts from knowledge of the initial atmospheric conditions. Nevertheless, for useful seasonal and longer-range forecasts, it is clear that other techniques will need to be employed.

The *ensemble technique* involves making multiple model forecasts all starting from slightly different initial conditions. If the starting conditions for each forecast differ from each other by less than our uncertainty in the current state of the climate system, then each is an equally valid prediction. Ensembles can be used to estimate the likelihood of a particular forecast and the uncertainty associated with it.

The North Atlantic Oscillation:
Climatic Significance and Environmental Impact
Geophysical Monograph 134
Published in 2003 by the American Geophysical Union
10.1029/134GM08

If we define *deterministic predictability* to be when the forecast uncertainty is within the margin of acceptable error, then a prediction (based on an ensemble forecast for example) with 100% certainty that next summer's European rainfall average will be between 30 and 50% above normal can also be classed as deterministic. Such a forecast is a more plausible, if somewhat doubtful, possibility. It would be likely to rely on the role of quasi-external "boundary forcings" such as sea-surface temperatures, soil moisture or the radiative effects of a recently erupted volcano [*Robock*, 2000].

A more likely situation may be that an ensemble of forecasts would span the entire range of possible outcomes but with the *probability density function* (pdf) being a maximum around a particular outcome. In this case, we may be able to say, for example, with 70% certainty that the North Atlantic Oscillation (NAO) will be negative on average next winter or over the next 5 winters. Such a forecast is known as a *probabilistic forecast.*

When assessing predictability, investigating climate variability or when validating climate models, it is useful to focus on specific aspects of the highly complex ocean-atmosphere system. One traditional way of doing this is to specify the observed sea-surface temperatures (SSTs) and use these to "drive" an atmosphere-only model. Such simulations are sometimes referred to as "AMIP-style" integrations after the international Atmospheric Model Intercomparison Project. By comparing the atmospheric model simulation with the observed atmospheric circulation, we can estimate the *potential predictability* of the atmosphere. Such simulations generally give *hindcast* information since the SSTs are required to be known before the simulation is made. They can give forecast information if, for example, one assumes that the present SST anomalies (*i.e.,* differences from the long-term climatological mean) will persist for the next few months. Since SSTs will never be forecast perfectly, potential predictability is likely to exceed *real predictability*.

Low frequency coupled ocean-atmosphere modes that rely on feedbacks between the ocean circulation and the atmosphere [e.g., *Latif and Barnett*, 1994; *Grötzner et al.*, 1998] may also afford some long-range predictability. This would perhaps most likely be probabilistic in nature.

1.2. Who Could Benefit From Predictability?

Assuming that a forecast for some aspects of North Atlantic regional climate is possible at some timescales (and there is observational and model evidence that this may be true – see below), the question remains as to how *useful* such a forecast is likely to be. Seasonal timescale forecasts are of potential use to the agricultural [*Mysterud et al.*, this volume], fishery [*Drinkwater et al.*, this volume] and utility [see section 7; *Hurrell et al.,* this volume] sectors. Decadal forecasts are of potential use to major infrastructure projects such as building reservoirs or sea-defenses. At decadal and longer timescales, forecasts that have skill over-and-above that offered by anthropogenic forcings could become important to the "climate change community" and to policy makers. The European Community-funded PREDICATE project was set-up to investigate the feasibility and utility of decadal forecasts.

To gain the full benefit of a long-range forecast, which has only modest skill in itself, one needs to tailor the forecast to each particular user by incorporating their specific requirements and cost:loss ratios into the prediction itself. We will see that such tailoring can greatly enhance the value of a forecast.

For the North Atlantic region, there are specific topics, which are of regional and global importance. The oceanic thermohaline circulation (THC) in the North Atlantic involves the surface transport of saline water northward, the cooling and sinking of this water at high latitudes and a deep-water southward return flow [*Visbeck et al.*, this volume]. This circulation plays a major role in the hemispheric transport of heat from the equator to the high latitudes and warms the adjacent continental land masses by 2-3°C above what they would be in the absence of this ocean circulation [*Vellinga and Wood*, 2002]. A collapse of the THC would have major global consequences. Although generally thought unlikely, its high impact makes it an important feature to attempt to predict. Projects such as the recently funded RAPID proposal are aiming to better understand and predict the THC. Due to the quite large uncertainties involved in representing the THC in present climate models, an assessment of its predictability is not made here.

2. OBSERVATIONAL EVIDENCE OF REAL PREDICTABILITY

Bjerknes [1964] wrote that "changes in intensity of the oceanic circulation are mainly dictated by changes in the atmospheric circulation, and the resulting changes in the temperature field of the ocean surface must in turn influence the thermodynamics of the atmosphere". He argued that, for the planet to remain in thermal equilibrium, changes in the extratropical oceanic poleward heat transport (predominantly in the North Atlantic) would have to be matched by opposite changes in the atmospheric poleward heat transport; implying systematic changes in the extratropical atmospheric circulation in response to extratropical changes in the ocean and, hence, atmospheric predictability.

Namias [1964] suggested that negative SST anomalies in the vicinity of Atlantic Ocean Weather Ship "Charlie" (35.5°W, 52.5°N) were probably the cause of persistent blocking in the northeast Atlantic during the period mid 1958 to early 1960.

Since this time, many researchers have looked for signals in the observations that would imply an active forcing role for the North Atlantic. Such signals are difficult to isolate owing to the very strong short-timescale atmospheric forcing of the ocean by heat and momentum fluxes [*Czaja et al.*, this volume].

Ratcliffe and Murray [1970] presented evidence, based on 1886 – 1968 data (excluding 1940-1946), of a relationship between Newfoundland SST anomalies and pressure anomalies a month later over the North Atlantic and northern Europe. For example, Ratcliffe and Murray associated mean sea-level pressure (MSLP) anomalies of ±5 hPa over Scandinavia in October with SST anomalies of the order of ∓1°C off Newfoundland in September. During summer, similar SST anomalies correlated with a ± 3 hPa MSLP anomaly over Norway, Scotland and Iceland. Ratcliffe and Murray were also able to find relationships between Newfoundland SST anomalies and temperatures and rainfall over Europe. From inspection of their diagrams, temperature anomalies tend to be associated with anomalous meridional advection (of mean temperatures) whereas rainfall anomalies correlate with anomalous zonal (westerly) flow, presumably associated with anomalous moisture fluxes from the Atlantic. In addition to being somewhat sensitive to season, their atmospheric response was sensitive to the longitude of the SST anomaly center. Such sensitivity may suggest a larger number of degrees of freedom in the analysis and therefore reduced statistical confidence. The results of *Barnett* [1984] tended to contradict those of *Ratcliffe and Murray* [1970], showing only limited skill in predicting Eurasian temperatures from tropical and east Atlantic SSTs, with no skill from north west Atlantic SSTs.

Kushnir [1994], following *Bjerknes* [1964], investigated the timescale dependence of the relationships between the atmosphere and Atlantic Ocean. At interannual timescales, the relationship was explained by the atmosphere forcing the SST field through anomalous surface heat fluxes (particularly latent heat fluxes associated with evaporation) in the extratropics and a combination of anomalous heat fluxes and upwelling in the subtropics. *Hasselmann* [1976] and *Frankignoul and Hasselmann* [1977] had previously suggested that this was the dominant physics with the ocean simply responding to, and temporally integrating, the "stochastic" forcing by the atmosphere. However, Kushnir demonstrated that the relationship seen at interannual timescales broke-down at interdecadal timescales. Interdecadal SST changes were thought to be associated with the THC [*McCartney and Talley*, 1984] and the Gulf-stream/gyre circulation [*Greatbatch et al.*, 1991]. South of 50°N, the wintertime atmosphere actually appeared to attenuate the interdecadal SST anomaly rather than maintain it. By analyzing tendencies in SST and MSLP, Kushnir concluded that the strong rise around 1970 in 15-year mean MSLP over the area 20°-40°W, 40°-50°N was *preceded* by a cooling in extratropical North Atlantic SSTs. *Deser and Blackmon* [1993] suggested that there could also be an atmospheric response to high-latitude SST anomalies.

Czaja and Frankignoul [1999] used "lagged singular value decomposition" (LSVD) to look for oceanic forcing of North Atlantic climate in observational data. Singular value decomposition, applied to two fields (such as SST and 500 hPa atmospheric geopotential height, written here as Z500), identifies pairs of patterns, each pair consisting of an SST pattern and a Z500 pattern, that maximally co-vary in time. By taking the SST field leading the atmosphere field by a few months, it is possible to demonstrate "cause and effect". This is possible because monthly-mean SSTs show longer persistence than monthly-mean Z500. SSTs a month or so ahead are not affected by the later Z500 data but can be used as a predictor for the SSTs during the period of the Z500 data. They found statistically significant patterns for spring and autumn suggesting that the North Atlantic does force the local climate in these seasons and there may be some real predictability to be derived from this. *Rodwell* [2001] and *Rodwell and Folland* [2002a,b], extended the technique of Czaja and Frankignoul to investigate this possible predictability and to validate air-sea interaction in climate models. They also looked briefly at the role of other SST regions in forcing North Atlantic Climate. See section 6 for more details.

3. ATMOSPHERIC MODEL EVIDENCE OF POTENTIAL PREDICTABILITY

Here, we discuss the potential predictability revealed by atmospheric modeling studies. We also discuss briefly the mechanisms and physics that may be important for such predictability. *Czaja et al.* [this volume] give a more detailed account of aspects of the physics.

Using "super anomalies" of ±12°C, early studies by *Chervin et al.* [1976], *Chervin et al.* [1980] and *Kutzbach et al.* [1977] were able to detect local and hemispheric effects of placing SST anomalies in the North Pacific over and above the noise of internal variability. Whether the true response is linear enough to allow these results to be scaled down to more realistic levels is still a subject of research.

Based on simplified modeling studies, *Frankignoul* [1985] suggested that the magnitude of Z500 anomalies associated with realistic midlatitude SST anomalies might be at most 10-30 m. Since the standard deviation of Z500 in midlatitudes ranges between 40-110 m on the monthly timescale and 20-80 m on the seasonal timescale, it may be that midlatitude SST anomaly effects are negligible on monthly timescales, but may have some influence at seasonal (and longer) timescales. *Palmer and Sun* [1985] modeled the effect of large but more realistic idealized SST anomalies in Ratcliffe and Murray's region. They further scaled down the role of midlatitude SST anomaly forcing to perhaps 10-20 m.

Many of the proposed mechanisms by which SSTs can force a response in the (North Atlantic) climate have a common feature. This is the modulation by the SSTs of surface heat fluxes and, therefore, heating or cooling (diabatic forcing) of the atmosphere. To demonstrate the link from diabatic forcing to circulation, *Hendon and Hartmann* [1982] showed that the linear wave response to strong shallow midlatitude thermal forcing (350 Wm^{-2} in the column mean) involved upper tropospheric height anomalies of up to 200m.

More recently *Peterson et al.* [2002] forced 30-member ensembles of a simple atmospheric model [based on the primitive equation model of *Hoskins and Simmons*, 1975; *Hall*, 2000] with "observed" diabatic forcing. When this forcing was applied globally, they were able to reproduce very accurately the observed variability of the NAO (the ensemble mean correlation was 0.79 but each individual simulation was, on average, nearly as good). When the observed diabatic forcing was applied in just the extratropics with climatological (average) forcing elsewhere, the correlation was 0.55. When only the tropical forcing (30°S-30°N) was applied, the correlation was 0.39 but, in agreement with *Hoerling et al.* [2001], these integrations appeared to capture the long-term trend in the NAO. These results suggest that both tropical and extratropical diabatic forcing play a role in the climate of the North Atlantic region.

The Clausius-Clapeyron equation suggests that surface heat fluxes may be more sensitive to changes in tropical SSTs than they are to changes in the cooler SSTs of the midlatitudes. Indeed, *Davis et al.* [1997] demonstrated the increased sensitivity of the tropical circulation to SST forcing. Tropical forcing of the North Atlantic climate would also rely on the existence of "teleconnections" that link the tropics to the extratropics [*Horel and Wallace*, 1981]. The pioneering work of *Rowntree* [1976], who used several variants of an atmospheric model run for 80 days, suggested that warm tropical and sub-tropical east Atlantic SSTs (up to 2.5K anomalies) could have played a role in the 1962/63 blocked winter in Europe. His two experiments with real initial atmospheric data had a consistent "response" as far north as 60°N over the North Atlantic with a low pressure anomaly of –7 hPa at 45°N, high pressure anomalies further north and a weaker, southward shifted, north Atlantic stormtrack. This was somewhat similar to the observed negative NAO conditions that prevailed during the winter of 1962/63. However, a third experiment, started from an isothermal atmosphere, gave an opposite change for the 41-80 day mean response. Based on subsequent work [e.g., *Visbeck et al.*, 1998], Rowntree's observational support from MSLP composites based on SSTs near Mindelo (Cape Verde Islands) may now be viewed as primarily reflecting atmospheric forcing of the ocean.

It could be argued that the extratropical diabatic forcing, that plays a key role in the results of *Peterson et al.* [2002], is simply consistent with the local atmospheric circulation, rather than forcing it. It could also be argued that the extratropical diabatic forcing is partly a response to the tropical forcing. However, surface heat fluxes are sensitive not only to SSTs but also to the strength of the low-level winds. Where these winds are stronger (generally in the midlatitudes), the surface heat flux sensitivity to SST could be enhanced. Indeed, the *global* maxima in January surface latent heat fluxes occur in the Gulf Stream and Kuroshio regions where fluxes peak at over 250 Wm^{-2}. Surface sensible heat fluxes also peak in the high latitudes with up to 75 and 100 Wm^{-2} in these two regions, respectively [*Josey et al.*, 1998]. Equally, observed diabatic forcing calculated using a residual method applied to the thermodynamic energy equation shows values in the (rather broader) stormtrack regions of the North Atlantic and Pacific of around 225 Wm^{-2} during northern winter [*Hoskins et al.*, 1989]. These values are comparable with the maximum values seen in the tropics at this time, at the 5° latitude x 5° longitude resolution. In idealized AGCM experiments, *Rodwell et al.* [1999] found that realistic SST anomalies in the form of a North Atlantic "tripole" could lead to changes in the surface evaporation rate in the extratropical as well as the tropical North Atlantic region of the order of 30 Wm^{-2}. These changes were seen to affect, fairly locally, atmospheric diabatic forcing by a similar magnitude (at least as far north as 45°N). Hence, it is possible that extratropical SSTs also play a forcing role in North Atlantic climate [see also *Czaja et al.,* this volume].

If the extratropical SSTs do play a forcing role then, since horizontal advection (i.e. transport by the flow) is important in the extratropical thermodynamic energy balance, the response to midlatitude diabatic forcing is likely to be highly sensitive to the vertical profile of heating [*Hoskins and Karoly*, 1981] – the bulk of which tends to occur below 700 hPa – and the profile of mean zonal wind [*Lindzen et al.*, 1982]. The intimate relationship between diabatic forcing, the mean circulation over the North Atlantic and transients (such as storms) is well documented [e.g., *Rogers*, 1990; *Ting and Lau*, 1993; *Watanabe and Kimoto*, 1999]. *Peng and Whitaker* [1999] showed, with linear baroclinic and stormtrack models, that the effect of transients was sensitive to the background state. The same extratropical (Pacific) heating anomaly produced only slightly different anomaly circulation (streamfunction) responses for January and February background states but the anomalous transient eddy forcing consistent with the heating-induced response was very different. With a particular configuration of heating and stormtrack, the transients could enhance and make "equivalent barotropic" the initial baroclinic response to low-level heating (a strengthened ridge downstream of the heating anomaly). With a different configuration, the anomalous transient forcing acted elsewhere, suggesting little positive feedback. Hence it is clear that a better understanding of the long timescale predictability of North Atlantic climate is likely to involve further study of short timescale variability. Coincidentally, the predictability of the statistics of (extreme) transient events is itself becoming an important goal of long-range forecasting.

"Analysis of variance" of an ensemble of atmospheric model simulations forced with observed SSTs can estimate the percentage of atmospheric variance that is forced by the SSTs. It is a method of quantifying potential predictability. For the North Atlantic region, the method [e.g., *Davies et al.*, 1997; *Rowell*, 1998] suggests that perhaps 15% of the variance in seasonal-mean MSLP is "due to" forcing from SSTs. As an example, Plate 1a shows the results of the analysis of variance applied to a 10-member ensemble of simulations with the Hadley Centre's HadAM3 AGCM. The analysis is applied to December – February (DJF) means for the period 1948/9 – 1998/9. In the subtropics (~10°N – 30°N) the percentage of variance explained by SST forcing ranges from 30 to 60%. This can be thought of as reflecting the variations in the MSLP that are common to all the simulations. The remaining percentage (40 – 70%) is associated with internal atmospheric variability and likely to be unpredictable. In the midlatitudes, the percentage of variability explained by SST forcing drops to 10 – 30%. Some areas show insignificant values (blocked in white) based on this 10-member, 51-year ensemble.

It is possible to calculate the potential predictability on longer timescales [*Rowell and Zwiers*, 1999]. Plate 1b shows that at decadal timescales (i.e. timescales greater than ~6 years), the percentage of DJF MSLP variance explained by SST forcing is higher, with values of over 80% in the subtropics and, for this ensemble, values of around 40% for both the Azores and Iceland regions (the main MSLP centers of action for the classical NAO). Some ensembles of HadAM3 show values as high as 50% for these locations.

Does this imply higher real predictability at this longer timescale? It possibly does not. Analysis of variance applied to AGCM ensembles is a very useful way of investigating the role of forcing by SSTs (and sea-ice). However, to better understand the implications of the above results, a number of points need to be stressed in relation to the method of analysis of variance and the use of AGCM simulations:

- The above analysis of variance results (unfiltered and decadal) assume that the model is perfect: an actual (imperfect) model could over-estimate or under-estimate the true role of the ocean. An important question to address is "how accurately does the atmospheric model respond to SST?" This is important because, with such a large proportion of the atmospheric variance not due to SST forcing, a model climatology could look reasonable even if it did not represent well the link from SSTs. Note however, that at present, model extratropical biases can be as large as one standard deviation of interannual variability [*Palmer et al.*, 2000].
- Because such atmospheric model simulations are forced with observed SSTs rather than predicted SSTs, they can only identify potential predictability [*Rodwell et al.*, 1999; *Mehta et al.*, 2000] which is likely, for a perfect model, to exceed real predictability. Results from *Bretherton and Battisti* [2000] imply that, if there is a link from SST to the NAO (however weak) and if this link involves similar SST patterns to those of the much stronger link from NAO to SST, then one should be able to reproduce well (possibly too well) the evolution (but not the magnitude) of the NAO by taking a sufficiently large AGCM ensemble. The relative importance of the direct forcing of SSTs by the atmosphere at short timescales [*Visbeck et al.*, 1998] and the modulation of SSTs by ocean circulation changes (e.g., gyre or thermohaline circulation changes and perhaps El Niño-associated variability) at longer timescales [*Delworth et al.*, 1993; *Sutton and Allen*, 1997; *Marshall et al.*, 2001] is of key importance for assessing real predictability.
- On the other hand, in the fully coupled ocean-atmosphere system, atmospheric thermal anomalies can modify surface heat and moisture fluxes and SSTs so as to reduce the rate at which the atmospheric thermal anomalies are damped (the SSTs get closer to the atmospheric temperatures and as the difference gets smaller, so do the heat fluxes and the damping). In AGCM experiments, no such affect is felt by the SSTs since they are prescribed and hence atmospheric thermal anomalies may be unduly damped, reducing their variance, persistence and possibly predictability. This effect has been demonstrated to be quite small [J. Hurrell, personal communication] and is thought not to significantly alter estimates of seasonal to interannual predictability [*Kushnir et al.*, 2002].
- If one considers longer than seasonal timescales, like the decadal timescales in Plate 1b, the percentage of variance

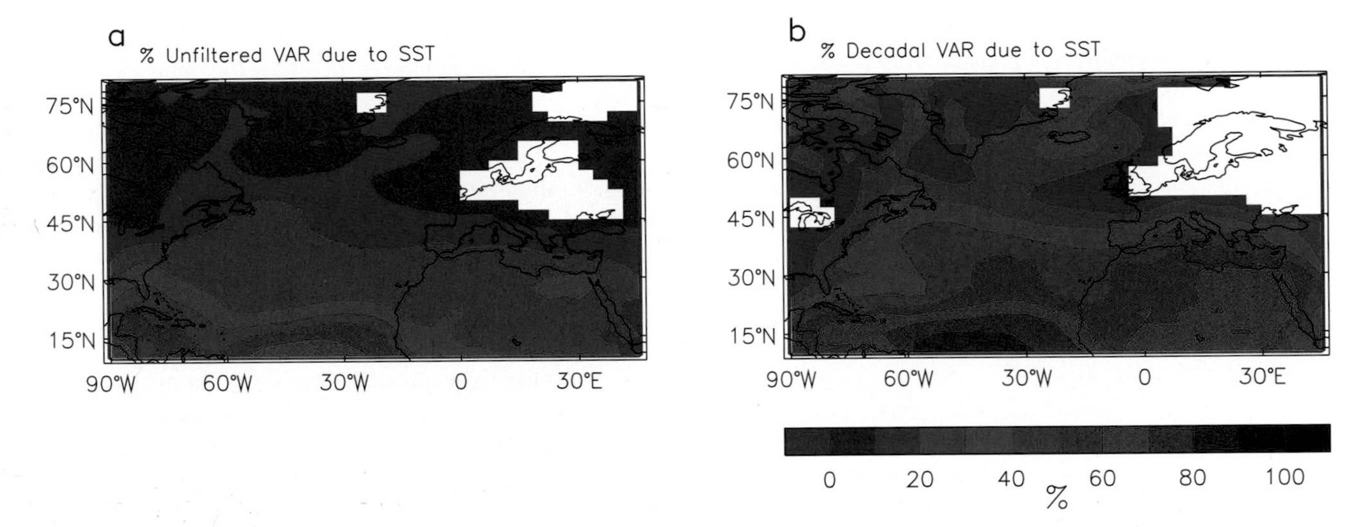

Plate 1. (a) Percentage of variance of December – February mean sea-level pressure (MSLP) that is due to forcing by sea-surface temperature (and sea-ice), estimated by an analysis of variance method using data for the winters 1948/9 to 1998/9 from a 10-member ensemble of HadAM3 AGCM simulations forced with HadISST1.0 SST and sea-ice. White areas show where values do not significantly exceed zero at the 10% significance level. (b) As (a) but for the percentage of decadal variance due to SST (and sea-ice) forcing.

explained is likely to increase. One reason for this expectation is that, at longer timescales, more of the atmospheric internal variability is averaged away but the strength of the forcing from the ocean remains approximately unchanged so that the ratio of "signal to noise" increases. However, the gap between potential predictability and real predictability is also likely to increase since one is now assuming that SSTs are perfectly predictable over decades and not just seasons. On the other hand it is also possible that, at longer timescales, ocean circulation changes could affect SSTs in a manner that is more able to excite changes in the atmosphere. Such processes could lead to an enhancement of real, as well as potential, predictability at longer timescales. In addition, other boundary forcings such as those due to anthropogenic emissions and variations in insolation, may have a relatively stronger role to play at longer timescales.

Whilst the percentage of atmospheric variance explained by SST forcing does not directly relate to predictability, it is nevertheless one important indicator of air-sea interaction. Whether the "true" figure for, say seasonal-mean MSLP variance, is nearer 15% or 30% will have a strong bearing on North Atlantic climate prediction and its usefulness to society. Yet we do not know even to this accuracy what the true figure is. For the Azores region, a range of models (including the ECHAM4 AGCM from the Max-Planck Institut, ARPEGE3 from Metéo France, and HadAM3) shows that the percentage of decadal DJF MSLP variance explained by SST forcing ranges between insignificant and about 35% [*L. Terray*, personal communication]. Interestingly, the local maximum over Iceland shown in Plate 1b is a common trait of all of these models. Research is required to better quantify the direct forcing role of the oceans. In addition to the use of models, the availability of longer observational datasets together with estimates of their associated errors would help in this regard.

4. COUPLED MODEL EVIDENCE OF REAL PREDICTABILITY

Over-and-above the potential predictability of North Atlantic climate that may derive from the simple inertia of the ocean (for example the slow decay of oceanic "mixed layer" thermal anomalies) there is the possibility that preferred modes of variability, perhaps involving positive feedbacks between the ocean and the atmosphere, could also lead to some predictability. Here, we briefly discuss two such modes [see also *Czaja et al.*, this volume].

The decadal coupled mode hypothesis of *Latif and Barnett* [1994], extended to the North Atlantic region by *Grötzner et al.* [1998], involved a positive feedback between the NAO and North Atlantic SSTs at short (seasonal to interannual) timescales and a negative feedback involving NAO-forced changes in the intensity of the subtropical gyre at longer (decadal) timescales. Such a mode could explain the peak at decadal timescales of the observed NAO power spectrum and, if real, could give rise to some predictability for the North Atlantic region. However, *Visbeck et al.* [1998], who investigated NAO-like windstress forcing of the North Atlantic Ocean (using a full Atlantic ocean model), found little lagged SST response south of about 45°N. They concluded that this result diminished the possibility for a delayed oscillator type coupled mode of variability arising from the subtropical ocean. In addition, since the proposed atmospheric feedback to the ocean may involve surface heat fluxes that are in opposition to those involved in the original forcing of the atmosphere, it may be difficult to achieve a positive feedback at short timescales. Nevertheless, it is possible that aspects of this proposed coupled mode, such as the role of the Gulf Stream, may help explain the "red" character of the observed NAO power spectrum.

Delworth et al. [1993] analyzed time-lagged regressions of salinities and temperatures in the sinking region of the thermohaline circulation in a 600-year simulation of a (fairly coarse resolution) coupled model. They suggested a mechanism for THC variability whereby a maximum in the subpolar gyre circulation (associated via geostrophic arguments to a minimum in column-mean temperatures in the sinking region but increasing *surface* temperatures and salinities due to gyre advection) leads, around 10 years later, to a maximum in the THC (associated with high salinities and sinking, increasing column mean temperatures and thus a weakening of the subpolar gyre). SSTs and surface heat fluxes out of the ocean are maximal at this time and the speculation was that this cooling could also be important for generating ocean convection. By the same arguments, the minimum in the gyre is achieved about 10 years after the THC maximum, followed by a minimum in the THC and then back to the maximum in the subtropical gyre. The total period for one such THC oscillation is then around 40-50 years. In reality, Delworth et al. noted that THC variability is quite irregular. They suggested that the irregularity was primarily due to oceanic processes. The ability of the NAO to drive decadal subpolar gyre anomalies [*e.g., Visbeck et al.*, 1998] suggests that atmospheric processes may also be important. Interestingly, *Delworth et al.* [1993] found a relationship between the THC and surface temperatures that matched closely the relationship *Kushnir* [1994] found in his observational analysis. However, the possibility for atmospheric predictability remains to be quantified.

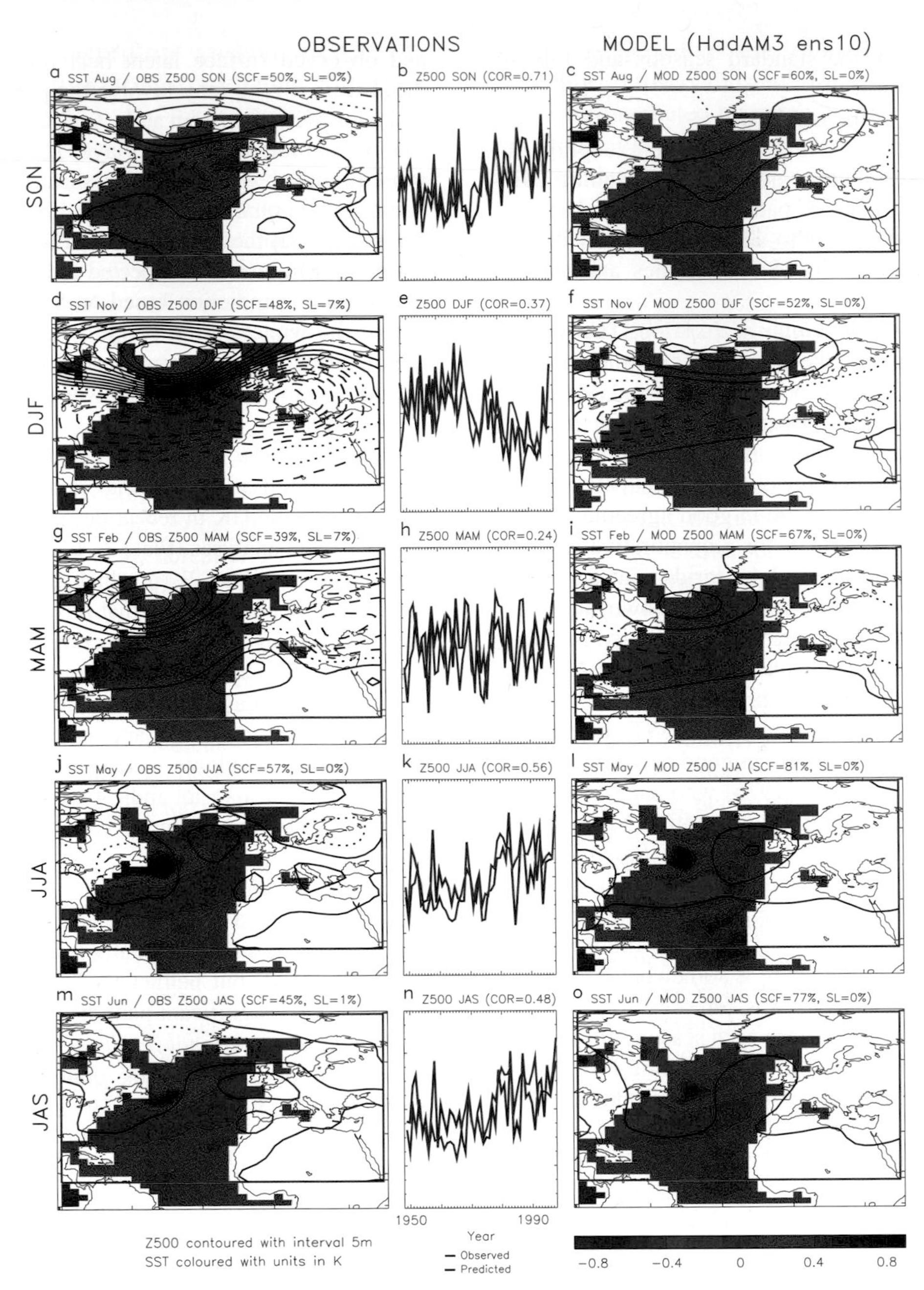

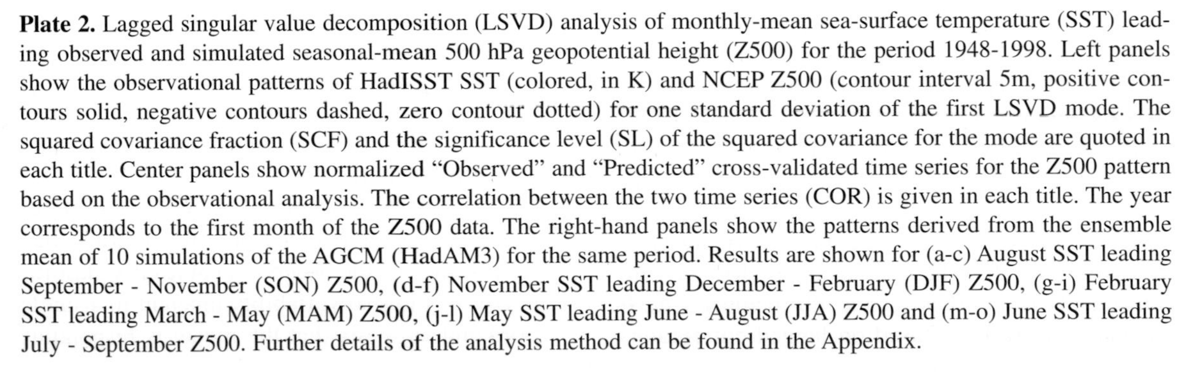

Plate 2. Lagged singular value decomposition (LSVD) analysis of monthly-mean sea-surface temperature (SST) leading observed and simulated seasonal-mean 500 hPa geopotential height (Z500) for the period 1948-1998. Left panels show the observational patterns of HadISST SST (colored, in K) and NCEP Z500 (contour interval 5m, positive contours solid, negative contours dashed, zero contour dotted) for one standard deviation of the first LSVD mode. The squared covariance fraction (SCF) and the significance level (SL) of the squared covariance for the mode are quoted in each title. Center panels show normalized "Observed" and "Predicted" cross-validated time series for the Z500 pattern based on the observational analysis. The correlation between the two time series (COR) is given in each title. The year corresponds to the first month of the Z500 data. The right-hand panels show the patterns derived from the ensemble mean of 10 simulations of the AGCM (HadAM3) for the same period. Results are shown for (a-c) August SST leading September - November (SON) Z500, (d-f) November SST leading December - February (DJF) Z500, (g-i) February SST leading March - May (MAM) Z500, (j-l) May SST leading June - August (JJA) Z500 and (m-o) June SST leading July - September Z500. Further details of the analysis method can be found in the Appendix.

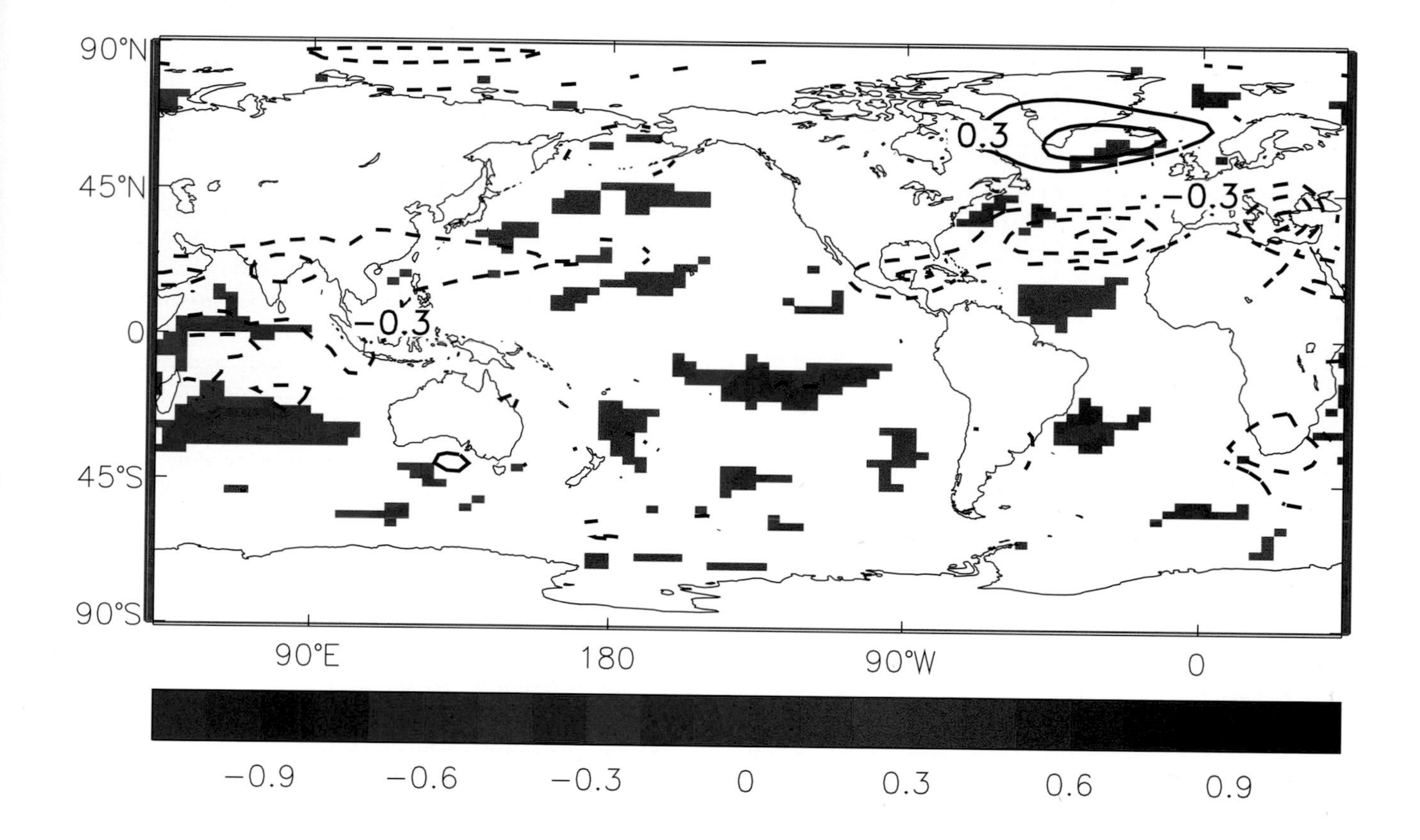

Plate 3. Grid-point May SST and DJF Z500 correlations with the cross-validated May SST time series from the observational May SST / DJF Z500 lagged SVD analysis. Homogeneous SST correlations are colored, heterogeneous Z500 correlations are contoured with contour interval 0.1. Only correlations significant at the 10% level using a 1-sided t-test are plotted.

forcing with approximately the correct patterns. The magnitudes of the Z500 patterns are similar for the observations and the model ensemble mean in the subtropics. This is in agreement with the conclusion that at least 60% of AGCM seasonally averaged MSLP variability in the tropics and subtropics is forced by SSTs (Plate 1a). Away from the tropics and subtropics, magnitudes of the model patterns become progressively weaker relative to those of the observations so that at about 50°N, they have around half the magnitude of the observational patterns. Possible reasons for this weakening include the "contamination" of the observational patterns by atmospheric internal variability although the lack of ocean-atmosphere feedbacks with an AGCM experiment and a weak model extratropical response to SST may also be important.

- *Both atmospheric models give similar results:* The patterns from the present model (HadAM3) are remarkably similar to those obtained from the previous version of the model (HadAM2b, Figure 8 in RF02a). This robustness gives further confidence that the patterns do represent a true SST-forced signal. The agreement between models for the SON season is interesting because of the lack of agreement in the extratropics with the observational results. The suggestion is that either there is a systematic error that is robust across these two AGCMs or that internal variability in the observations is leading to a different (possibly still valid) extratropical response being highlighted.
- *Response patterns throughout the annual cycle*: For all seasons, the model ensemble-mean patterns are highly significant. The year appears to be split between two pairs of patterns: the extratropical "north-south" (NAO) pattern seen in the cold half of the year NDJ to MAM (e.g., Plate 2f, i), and the extratropical "east-west" pattern seen in the warm half of the year MJJ to JAS (e.g., Plate 2l, o). A transition from one set of patterns to the other is seen in the intervening seasons. In the observational results, the east-west pattern is robust from JJA to ASO. The north-south pattern appears in NDJ, DJF and MAM but statistical confidence is weaker for JFM and FMA. The suggestion is that one realization of the observations has so much atmospheric internal variability that the ocean-forced signal can sometimes be obscured.
- *Predictability*: The results do indicate a degree of seasonal predictability throughout the annual cycle although skill levels are quite low. For example, a correlation skill of 0.45 for a prediction of DJF Z500 based on the previous May SSTs. There are also (possibly large) errors associated with the correlation skill estimates. For example, the SON correlation skill estimate is 0.71 (Plate 2a, b) but the corresponding estimate in RF02a (using somewhat different datasets and SST regions) was 0.44. Such seasonal skill levels may be useful and, indeed, there has been strong interest in the Met Office's experimental winter NAO forecast based on the May SST analysis. If low-frequency variability in the ocean (such as variability of the thermohaline circulation) has an appreciable influence on low-frequency SSTs, then the linkages identified here could also be important for decadal timescale predictability.
- *Origin of predictability:* One way to assess how much of the predictability, seen in the centre column panels of Plate 2, is associated with interannual rather than longer timescale variations is to correlate the time series of interannual changes. RF02a found that, for SON, the correlation of interannual changes was as high as the correlation for the full time series. Interannual changes also accounted for a large part of the MAM COR value. For winter, they found that decadal variations play a dominant role in the predictability. This does *not* necessarily mean that there is no seasonal predictability for winter, just that any seasonal predictability is associated with lower-frequency changes in SST – possibly associated with gyre or THC variability.

Note that extensive further tests were done on these results, including removing more than one year in the cross-validation process, checking for auto-correlation of the individual time series and investigating all SST lead-times up to 8 months in RF02b and 13 months in RF02a. The results of these tests, documented in RF02a, do not change the conclusions drawn above.

- *Model validation*: To aid with model validation and to look for signs of multi-decadal variability in air-sea interaction, the same analysis was applied to the 28 50-year periods of the control simulation of the OAGCM. The results were summarized in Figure 2 of RF02b. There was a possible indication of multidecadal variability in coupling but the overwhelming conclusion was that the OAGCM does not capture an ocean-to-atmosphere link as strongly as seen in the observations over the period 1948-1998. The periods of the OAGCM simulation that displayed the same level of low-frequency DJF NAO variability as that seen in the recent observations displayed no better predictability than the others.

Figure 1 shows all the COR values from the observations (filled circles), the OAGCM (plus signs) and AGCM (diagonal crosses) for SST leading Z500 by 2 months (e.g., Aug SST/SON Z500). Negative values imply no link is detected and, therefore, no predictability. It is clear that the observational COR values are generally higher than those for both models from SON to JFM (cold season). From AMJ to ASO (warm season),

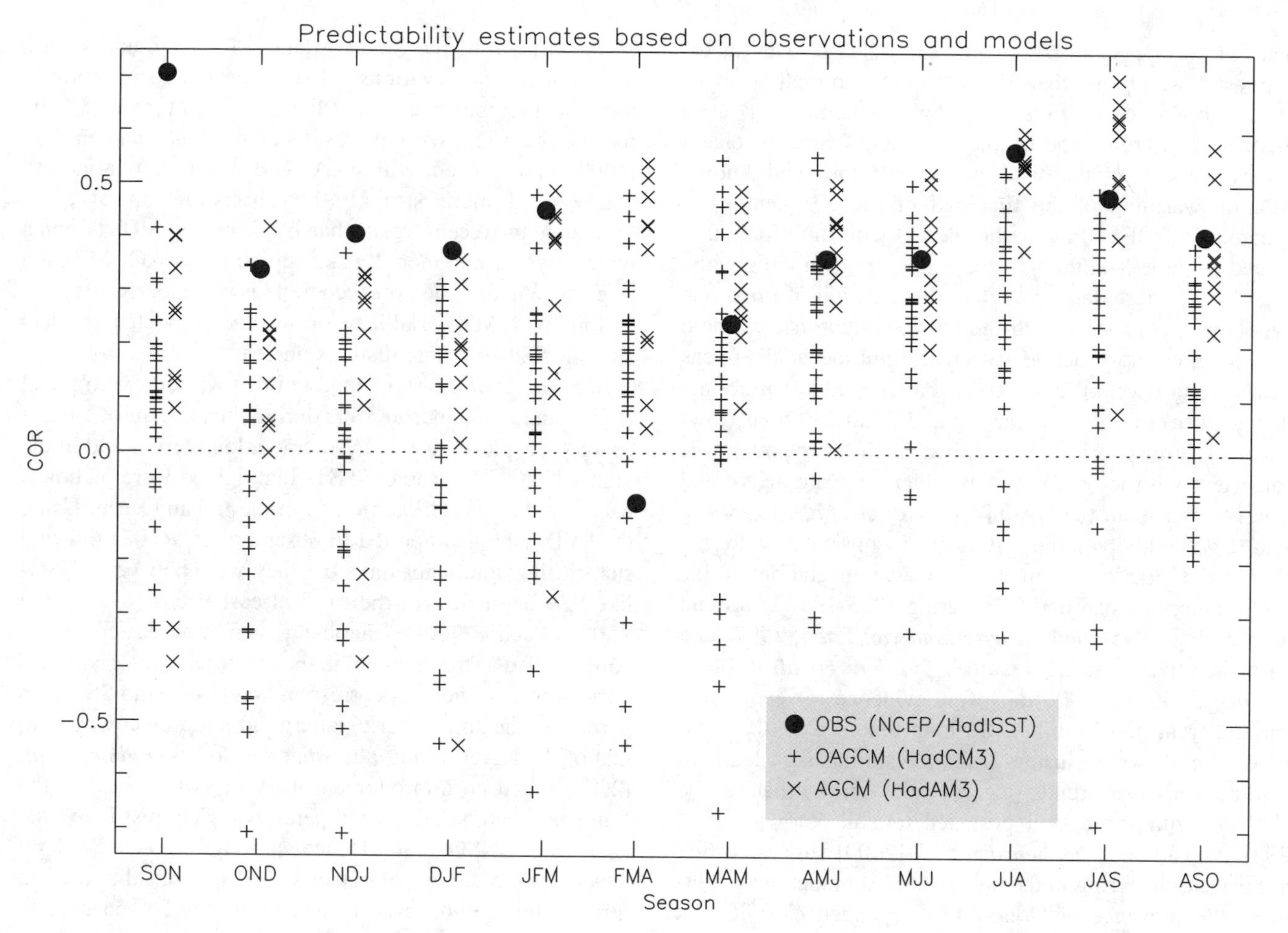

Figure 1. Cross-validated correlations (CORs, from the LSVD analysis – described in detail in the appendix) obtained from the observations (circles), OAGCM (plus signs) and AGCM (diagonal crosses) when SSTs lead Z500 by two months.

the observations are generally within both model COR ranges but still above their average. The observed COR value for FMA is negative and exceeded on average by both models. The reason for this negative observational COR value appears to be because the first two LSVD modes for this season are not well distinguished, both explaining a similar percentage of covariability. This means that during the cross-validation process, it is possible for the order of the first two modes to be changed (and possibly even the patterns altered). If the second mode is not physical, this will degrade the cross-validated correlation.

If the OAGCM were a perfect model, the autumn and winter results would indicate that the observations are particularly unusual. For example, they would pass, at the ~96% confidence level, a test which claimed they were outside the true distribution (since they fall outside a 28-member sample). If one discounts the possibility that the observations are particularly unusual, then the conclusion must be that the OAGCM is not perfect and that improvements could be made for the autumn and winter seasons. On the other hand, it is interesting to note that the AGCM COR values tend to occupy the upper ranges of the OAGCM COR values. Since the AGCM simulations are forced with the recent observed SSTs, this may suggest that the recent observed period *has* involved unusually strong air-sea interactions. However, it could also simply reflect the differences between AGCM and OAGCM experimental design. Although one cannot determine conclusively the reason for the high observational results in autumn and winter, the likelihood that there is scope for model improvement is sufficiently strong to warrant further investigation.

5.3. North Atlantic SST: The Longer Record

RF02a repeated the May SST/DJF MSLP analysis for the period 1902/03-1947/48 but found an apparent loss of signal that did not appear to be unequivocally due to data uncertainties. It was found that the first two "empirical orthogonal eigenfunctions" (EOFs) of North Atlantic SST were markedly different. The tripole and a Gulf Stream-related pattern appear in the later period but the well known Atlantic warming of the first half of the 20th century is apparent in EOF1 of the earlier period with the tripole displaced to EOF2. Although this may simply highlight the need for accurate data spanning a long period of time, one hypothesis, consistent with the above results, is that there can be alternately active (inactive) multidecadal epochs with strong (weak) gyre and NAO decadal variability, strong (weak) air-sea interaction and relatively high (low) seasonal atmospheric predictability. Such an hypothesis is consistent with ice-core proxies which indicate active and inactive phases of the NAO [*Appenzeller et al.*, 1998; see also *Cook*, this volume]. It is also consistent with the observed reduction of interannual autocorrelation of the NAO index from around 0.38 during 1935-1994 to around 0.10 during 1895-1944 [*Hurrell and van Loon*, 1997] and with idealized model results of *Eden et al.* [2002]. *Stephenson et al.* [2000] also find evidence for epoch-like variability in predictability. On long time scales, the presence of multi-year clusters (runs) of similar sign leads to marked long-term trends (e.g., decadal trends) that can be hindcast with some skill provided reliable knowledge of SSTs is available. *Stephenson et al.* [2000] suggested that multi-year clustering in the historical NAO index was related to the presence of "long-range dependence", with the NAO having more variance on longer time scales than could be expected from either a white noise or a Hasselmann red noise process. For example, decadal means of the NAO SLP index from 1864-1998 were found to have almost twice the variance expected for a short-range process such as white noise or first order red noise. They indicate that such long-term effects could be exploited to make long-range predictions of the NAO.

5.4. The Role of SSTs Outside the North Atlantic Region

So far, the analysis has attempted to focus on the role of local SSTs. However, as pointed out in section 5.2, it has not always been possible to do this and the role of other SSTs could not be completely discounted. Through tropical-extratropical interactions, other SST regions could very readily have an influence on North Atlantic regional climate [*Qin and Robinson*, 1995].

Robertson et al. [2000] suggested, using multivariate and modeling results for the last half of the 20th century, that South Atlantic SSTs may play a role in forcing the winter NAO. Using a similar analysis to that used above, but with SSTs confined to the South Atlantic, RF02a found such a signal in the observations but it appeared to be entirely dependent on the existence of linear trends in the data. This may be consistent with the results of *Watanabe and Kimoto* [1999] who showed, with an AGCM, that NAO variability was excited more strongly by observed tropical SSTs (including the recent trend) than by simulated SSTs (when a mixed layer ocean model was coupled to the AGCM in the tropics). Whether the observed trend is particularly effective at forcing NAO variability or whether modeled tropical Atlantic SSTs are unrealistic is unclear.

Hoerling et al. [2001] suggested that warming of tropical Pacific and Indian ocean SSTs during the last half of the last century has played a role in forcing a long-term trend in the winter NAO. The tropical SSTs highlighted were predominantly in the central Pacific and in the Indian Ocean. Using the SVD approach applied to observations, RF02a did find statistically significant links but they were between ENSO-like SST anomalies in the tropical east Pacific and, in the Z500, a Pacific-North-American pattern over eastern North America and a strong signal in the tropics and over northern Africa (positive heights corresponding to El Niño SSTs). In particular, the atmospheric pattern did not project well onto that of the NAO. Using atmospheric models, *Palmer et al.* [2000] found no improvement in Brier skill scores for the European region 850hPa temperatures when restricting the analysis to ENSO years. Hence, although tropical SSTs do appear to force an extratropical response, the precise form of this response is unclear (and may be sensitive to the real or modeled mean atmospheric circulation in the extratropics).

6. OTHER PREDICTORS, OTHER PREDICTANDS

6.1. A Solar Effect on the NAO?

After anthropogenic effects, solar effects generally attract the most attention in the climate change community, but have not done so in the context of the NAO [although see *Gillett et al.*, this volume]. This may seem surprising looking at the rightmost part of Figure 2 [W. Ingram, personal communication], which plots the winter NAO index against solar irradiance as estimated by *Hoyt and Schatten* [1993]. The curves look clearly related and their correlation is 0.71, much higher than that of the NAO with greenhouse gas forcing. While this is not the sort of response to changing solar heating one might naively expect from the climate system, modeling studies [*Haigh*, 1999] find that changing solar irradiance, with the resulting ozone changes, consistently tend to produce quite specific and local changes in the zonal-mean zonal wind. If this were all the data we had, it would be a strong case for solar effects on the NAO and, therefore, NAO predictability given the partial predictability of the "11-year" (actually 8-13 year) solar cycle.

There is, however, considerable earlier data, even if coverage is not as good. During the preceding 30 years (between the 2 vertical bars in Figure 2) the correlation is –0.4 with both the overall trend and the shorter-term variability in opposition. Over 121 years the correlation is only 0.14. The relationship is also sensitive to which estimate of solar irradiance is used: with that of *Lean et al.* [1995] the correlation for the recent period drops from 0.71 to 0.37. (Direct measurements of solar irradiance can only be made by satellites: they cover only the last couple of decades and have problems of homogeneity. Variation in solar irradiance on these timescales is associated with variations of solar magnetic activity. *Lean et al.* [1995] assume this relationship and derive an estimate whose longer timescales are linked to the *amplitude* of the sunspot cycle. However, *Hoyt and Schatten* [1993] find this physically implausible and derive an estimate whose longer timescales are linked to the length of the sunspot cycle).

The conclusion must be that a solar effect on the NAO on decadal timescales is much less clear than the initial data may have suggested.

These results are consistent with those of *Shindell et al.* [2001a] who found an extremely weak Northern Hemisphere annular mode response to the 11-year solar cycle. At longer centennial timescales, *Shindell et al.* [2001b] suggest that there can be an annular mode and NAO response to insolation changes involving interactions with stratospheric ozone and tropical SSTs. However, it is not clear that such forcing mechanisms can provide predictability for North Atlantic climate. Further details of troposphere/stratosphere links and the response to ozone changes can be found in *Thompson et al.* [this volume] and *Gillett et al.* [this volume], respectively.

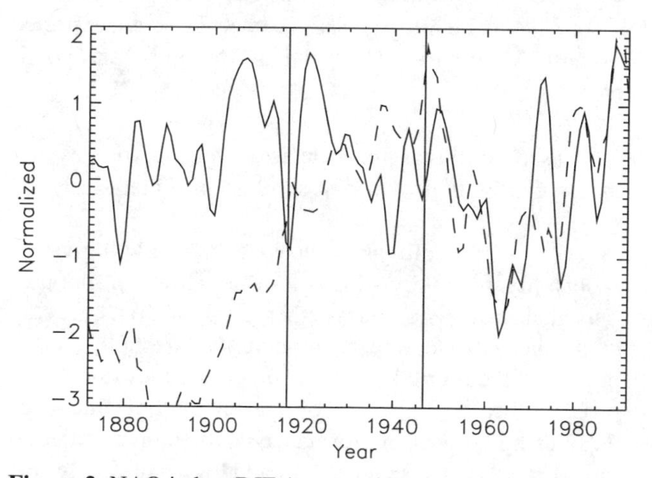

Figure 2. NAO index, DJF Azores minus Iceland mean sea-level pressure (solid), and solar irradiance as estimated by *Hoyt and Schatten* [1993] (dashed) 1872-1992. Both time series are normalized over the period 1947-92.

6.2. A Volcanic Effect on the NAO?

It is possible that volcanic eruptions can produce a predictable signal in the climate over the following two years. Volcanic aerosols associated with a tropical eruption tend to enhance the stratospheric equator-to-pole temperature gradient. In his review on the climate response to volcanoes, *Robock* [2000] argues that this leads to a stronger midlatitude jet stream and a warming of the Northern Hemisphere continents in winter. Warming over North America, Europe and Siberia was observed and modeled following the Eruption of Mount Pinatubo in 1991. This (indirect) warming is stronger than the direct radiative cooling effect that dominates in summer. *Gillett et al.* [this volume] discuss in more detail the role of stratospheric volcanic aerosol.

6.3. Predictability from Climate Trends?

A reviewer noted that seasonal forecast skill for the United States has increased during the 1990s, largely due to the use of the "trend" and that this "trend" effect is yielding seasonal skill equal to or larger than the ENSO effect. Presumably, such trends may be associated with global or local man-made climate change, or with low-frequency natural variability. Further discussion of the role of anthropogenic emissions on the NAO can be found in *Osborn et al.* [1999], *Collins et al.* [2001], and *Gillett et al.* [this volume].

6.4. Predicting Other Quantities

Here, the focus has been on the predictability of mean sea-level pressure and 500 hPa atmospheric geopotential heights. These are quite large-scale quantities. Potential predictability, based on an analysis of variance, can also be estimated for other meteorological variables. For precipitation and surface temperature, there is some indication (based on simulations of HadAM3) of decadal potential predictability for western Europe with perhaps 20% of decadal variance in these quantities associated with SST forcing. Further model validation is required to more accurately quantify this potential predictability.

7. USEFUL PREDICTABILITY

So far, we have identified some potential for North Atlantic climate predictability. An important question remains: would these (limited) levels of predictive skill be useful to anyone? "Usefulness" is, in fact, quite user-specific. Depending on the vulnerability to a particular aspect of climate, one user may be able to improve their profits from the knowledge of even a small change in the odds of a particular climate "event" occurring. Another user would gain little benefit. One procedure for determining the usefulness of a climate forecast, based on *Palmer et al.* [2000], is outlined below:

- First, the user identifies the climate factors which affect their profitability. For example, Figure 3, derived from *van den Berg* [1994], shows the demand for natural gas in the Netherlands as a function of temperature. Typically, a 1°C temperature fall results in a 6% increase in demand.
- The user defines a climate event 'E' that they are vulnerable to. E could be, for example, that there are 10 frost days during the spring, that there is no rainfall for 7 consecutive days in summer or that the mean winter temperature is 1°C or more colder than normal.
- The user calculates the loss, L, that they would suffer if an event E occurred and they hadn't taken action to avert the loss or insure against it. The user also calculates the cost, C, of taking action to avert or insure against the loss. For example, during periods of unexpected demand, the price of gas on the open market can rise dramatically. The supplier who fails to predict or insure, at a cost C, against this can face a considerable "double whammy" loss, L, due to high customer demand and high source costs. These cost/loss values can be inserted into the action matrix:

A=

Action taken	Event occurs: No	Event occurs: Yes
No	0	L
Yes	C	C

- The user calculates the upper and lower bounds of the savings they could make by using a climate forecast:

For a (rather optimistic) lower bound on their expected *expense*, suppose that we have a perfect deterministic climate forecast. Here, the user only takes action (at cost C) when we know an event will occur and they never incur a loss, L. Over many realizations, the fraction of times that the event, E, does occur will approach its climatological probability, p_{clim}, and so the expected expense is simply:

$$M_{per} = p_{clim} \cdot C$$

For an upper bound on the expected expense, suppose that only the climatological (average) probability, (p_{clim}),

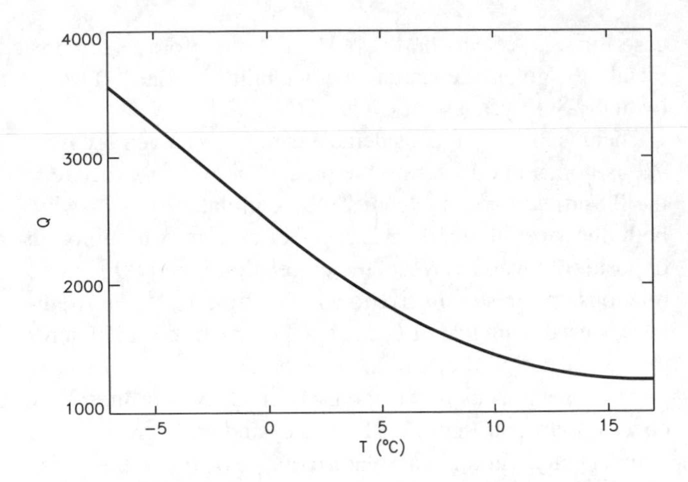

Figure 3. Dutch natural gas demand as a function of temperature. Derived from results of *van den Berg* [1994].

is available as a forecast. Here, action is either always taken, at an expense C, or never taken, at an expected expense $p_{clim} \cdot L$, whichever of these is the smallest. Hence:

$$M_{clim} = \min(p_{clim} \cdot L, C)$$

One of the events E that *Palmer et al.* [2000] investigated was that the Northern Hemisphere 850mb winter-mean temperature would be 1°C or more below normal. For this, they calculated from observations that $p_{clim} \approx 0.2$. Here we construct a concrete example of a particular user and apply this to *Palmer et al.*'s results. Suppose that for such an event a user would incur a loss L=$5000 unless they took mitigating action at a cost C=$1000. For this example, $M_{per} = p_{clim} \cdot C$ = $200 and $M_{clim} = \min(p_{clim} \cdot L, C)$= $1000. For a forecast to be of any use, it must have an expected expense, M, less than M_{clim} and as close to M_{per} as possible.

- The user optimizes their decision making process by incorporating their own cost and loss information:

The ensemble climate forecast technique will give the probability, p, that event E will occur. However, the user must make a decision on whether to take action. One way to do this is to define a "probability threshold", p_t. If $p>p_t$, then the user takes action, otherwise they don't. The problem then becomes one of optimizing the value of p_t.

By using the past performance of the climate forecast model and the user-specific cost and loss values, it is possible to calculate the expected expense for a range of threshold values. The value of p_t that is finally chosen is the one that minimizes the expected expense. For the event that Northern Hemisphere 850mb winter-mean temperature would be 1°C or more below normal, and C=$1000, L=$5000, the optimal threshold is calculated to be p_t=0.3. With this threshold, the long term statistics of the user-specific model are given by the forecast probability matrix:

P=

Action taken	Event occurs: No	Event occurs: Yes
No	0.57	0.07
Yes	0.23	0.13

The user's expected expense is simply M=P·A=$710. If the event was for European temperatures alone, then M=$880 with M_{per} and M_{clim} approximately unchanged. Since $M<M_{clim}$, there is clear potential for savings to be made for this user.

Note with this optimal solution, which combines climate model information and the user's cost and loss values, the matrix P shows that nearly 2 out of 3 times that the user decides to take mitigating action, the event does not happen. Although intuitively such a forecast would seem quite poor, it is of potential use to this specific user.

This example represents a best-case scenario. For example, the models used were atmosphere-only models with a perfect knowledge of the underlying SSTs, the users were interested in large area forecasts (North Hemisphere-average or European-average temperatures), and the cost:loss ratio was optimized. Nevertheless, it does demonstrate the process that one would go through in determining a user's exposure to climate and shows how even modest climate forecast skill can benefit some users.

For more complicated decisions, with multiple possible outcomes, the ensemble forecast pdf could be used in the optimization process. See, e.g., *Barnston et al.* [2000] for a de-scription of how one can make use of forecast distributions.

8. CONCLUSIONS AND DISCUSSION

A lagged singular value decomposition technique has been applied to observations, 10 simulations of an atmospheric model (HadAM3) and 1400 years of an ocean-atmosphere model (HadCM3) simulation in order to assess the predictability of North Atlantic climate and to validate the low-frequency variability in models. Model and observational results suggest that:

- The winter NAO responds to a "tripole" pattern in North Atlantic SSTs. There may be some predictability (predictive skill correlation of around 0.45) of the winter NAO from a knowledge of the previous May SSTs.
- Summer anticyclonicity over the UK is seen downstream of warm SST anomalies off Newfoundland and is possibly also related to warm subtropical North Atlantic SSTs or SSTs elsewhere.
- Analysis of the 10 individual simulations of the atmospheric model and the 28 50-year periods of the ocean-atmosphere model tend to confirm that the modeled atmosphere responds too weakly to North Atlantic SST forcing, particularly in autumn and winter.
- Other SST regions may have an impact on the winter North Atlantic Climate and, therefore, on predictability. El Niño is strongly implicated, although perhaps the response does not involve the NAO. A possible South Atlantic role is dependent on the long-term trend.
- It is possible that there could be multi-decadal variability in North Atlantic climate predictability, which is related to the strength of coupled decadal oscillations.
- The high (decadal) correlation between the NAO and insolation over the last half of the 20th century tends to break-down when the longer record is examined. Although it has been reported that insolation may be able to force the NAO on longer timescales, this may not yield predictability for the North Atlantic climate.

The suggestion from this work is that there could be some seasonal predictability of North Atlantic climate based on a knowledge of SSTs in the North Atlantic and tropical Pacific. Although the skill of such forecasts is likely to be quite small, it is possible that they could be of benefit if tailored to the end user's exposure to climate. If low-frequency variations in the deep ocean affect low-frequency variations in North Atlantic SST, as has been suggested, then this seasonal predictability could project and even amplify onto longer timescales too.

For the future, the use of ensembles of coupled model forecasts (already underway) will help ascertain the levels of real predictive skill to be gained. Such studies either assume that the model is perfect, and look at the spread of the ensemble with time, or they include an "assimilation" of observed ocean data and look for predictability of the real atmosphere. Further validation of climate models is also urgently required. To this end, better and longer historical datasets would be beneficial, as would a better knowledge of the present deep ocean (for example obtained from ARGO float measurements).

Finally, for posterity and fun, an experimental forecast based on the May SST results reported above would predict a positive NAO for the winter of 2002/03! For details, see http://www.metoffice.gov.uk/research/seasonal/regional/nao.

9. APPENDIX

A maximal covariance analysis, using the technique of singular value decomposition, SVD, [see e.g., *Bretherton et al.*, 1992] was applied to monthly-mean sea-surface temperatures (SSTs) and seasonal-mean 500 hPa atmospheric geopotential height (Z500). A lag was introduced so that, for example, August SSTs and September – November (SON) Z500 were jointly analyzed. In an analogous way to empirical orthogonal eigenfunction (EOF) analysis, two fields ($\mathbf{S}(t)$, here SST, and $\mathbf{Z}(t)$, here Z500) are decomposed into a sum of spatially normalized patterns ($\mathbf{p}_k$ and $\mathbf{q}_k$) multiplied by corresponding time series ($a_k(t)$ and $b_k(t)$):

$$\mathbf{S} = \sum_k a_k(t)\mathbf{p}_k \qquad \mathbf{Z} = \sum_k b_k(t)\mathbf{q}_k$$

with p_1 and q_1 (the only patterns of interest here) chosen to maximize the covariance $\langle a_1, b_1 \rangle \equiv \sigma_1$. An important quantity is the first squared covariance fraction:

$$\mathrm{SCF} = \sigma_1^2 / \sum_k \sigma_k^2$$

which reflects how much of the covariability the first "mode" explains. Using temporally shuffled data, a 100-member Monte Carlo test, based on the (squared) covariance of the first mode, $\sigma_1{}^2$, was used to calculate a "significance level" (SL) for the existence of a true physical connection. This test follows that of *Czaja and Frankignoul* [1999] although other test statistics can be chosen such as SCF itself [*Iwasaka and Wallace*, 1995].

The usefulness of any forecast system based on lagged SVD (LSVD) applied to a single observed or modeled realization was assessed using a "cross-validation" (*Livezey*, 1995) technique. The year to be hindcast, t, is removed from the original data to give $\{S_{i\neq t}\}$, $\{Z_{i\neq t}\}$ and the LSVD analysis is applied to these reduced datasets to produce patterns $\mathbf{p}_1{}^{(t)}, \mathbf{q}_1{}^{(t)}$. "Cross-validated time series" are then defined for each value of t by the scalar product:

$$a_1'(t) = \mathbf{S}_2 \cdot \mathbf{p}_1{}^{(t)} \qquad b_1'(t) = \mathbf{Z}_t \cdot \mathbf{p}_1{}^{(t)}$$

The cross-validated time series based on the leading field (SST) is referred to as the 'predicted time series'. The cross-validated time series based on the lagging (Z500) field is referred to as the 'observed time series'. The correlation between the two cross-validated time series;

$$\mathrm{COR} \equiv r_{a_1' b_1'}$$

is used to assess the skill of the forecast. COR is also a test of the physical significance of the patterns. Note that tests have been made by removing multiple years in the cross-validation process (RF02a). The results do not alter the conclusions.

Acknowledgments. The author would like to thank Chris Folland, Rowan Sutton and the anonymous referees for useful comments about this paper. This work has been funded jointly by the UK Government Meteorological Research Programme, and the EC SINTEX and PREDICATE projects.

REFERENCES

Alexander, M. A. and C. Deser, A mechanism for the recurrence of wintertime midlatitude SST anomalies, *J. Phys. Oceanogr., 25*, 122-137, 1995.

Appenzeller, C., T. Stocker and M. Anklin, North Atlantic Oscillation dynamics in Greenland ice cores, *Science, 282*, 446-449, 1998.

Barnett, T. P., Statistical prediction of seasonal air temperature over Eurasia, *Tellus, 36A*, 132-146, 1984.

Barnston, A. G., Yuxiang He and D. A. Unger, A forecast product that maximizes utility for state-of-the-art seasonal climate prediction, *Bull. Amer. Meteorol. Soc., 81*, 1271-1279, 2000.

Bjerknes, J., Atlantic air-sea interaction, *Adv. in Geophys, 10*, 1-82, 1964.

Bretherton, C. S. and D. S. Battisti, An interpretation of the results from atmospheric general circulation models forced by the time history of the observed sea surface temperature distribution, *Geophys. Res. Lett., 27(6)*, 767-770, 2000.

Bretherton, C. S., C. Smith and J. M. Wallace, An intercomparison of methods for finding coupled patterns in climate data, *J. Climate, 5*, 541-560, 1992.

Chervin, R. M., W. M. Washington and S. Scheider, Testing the statistical significance of the response of the NCAR general circulation model to North Pacific Ocean surface temperatures, *J. Atmos. Sci., 33*, 413-423. 1976.

Chervin, R. M., J. E. Kutzbach, D. D. Houghton and R. G. Gallimore, Response of the NCAR general circulation model to prescribed changes in the ocean surface temperature, II, Midlatitude and subtropical changes, *J. Atmos. Sci., 37*, 308-332, 1980.

Collins, M, S. F. B. Tett and C. Cooper, The internal climate variability of HadCM3, a version of the Hadley Centre coupled model without flux adjustments, *Clim. Dyn., 17*, 61-81, 2001.

Cook, E. R., Multi-proxy reconstructions of the North Atlantic Oscillation (NAO) index: A critical review and a new well-verified winter NAO index reconstruction back to AD 1400, this volume.

Czaja, A. and C. Frankignoul, Influence of North Atlantic SST on the atmospheric circulation, *Geophys. Res. Lett., 26(19)*, 2969-2972, 1999.

Czaja, A., A. W. Robertson, and T. Huck, The role of Atlantic ocean-atmosphere coupling in affecting North Atlantic Oscillation variability, this volume.

Davies, J. R., D. P. Rowell and C. K. Folland, North Atlantic and European seasonal predictability using an ensemble of multi-decadal AGCM simulations, *Int. J. Climatol., 17*, 1263-1284, 1997.

Delworth, T., S. Manabe and R. J. Stouffer, Interdecadal variations of the thermohaline circulation in a coupled ocean-atmosphere model, *J. Climate, 6*, 1993-2011, 1993.

Deser, C. and M. L. Blackmon, Surface climate variations over the North Atlantic Ocean during winter: 1900-1989, *J. Climate, 6*, 1743-1753, 1993.

Drinkwater, K. F., A. Belgrano, A. Borja, A. Conversi, M. Edwards, C. H. Greene, G. Ottersen, A. J. Pershing, and H. Walker, The response of marine ecosystems to climate variability associated with the North Atlantic Oscillation, this volume.

Eden, C., R. J. Greatbatch and Jian Lu, Prospects for decadal prediction of the North Atlantic Oscillation (NAO), *Geophys. Res. Lett., 29*, 2002.

Frankignoul, C. and K. Hasselmann, Stochastic climate models, Part II: Application to sea-surface temperature anomalies and thermocline variability, *Tellus, 29*, 289-305, 1977.

Frankignoul, C., Sea-surface temperature anomalies, planetary waves and air-sea feed-back in the middle latitudes, *Rev. Geophys., 23*, 357-390, 1985.

Gillett, N. P., H. F. Graf, and T. J. Osborn, Climate Change and the North Atlantic Oscillation, this volume.

Gordon, C., C. Cooper, C. A. Senior, H. Banks, J. M. Gregory, T. C. Johns, J. F. B. Mitchell and R. A. Wood, The simulation of SST, sea ice extents and ocean heat transports in a version of the Hadley Centre coupled model without flux adjustments, *Clim. Dyn., 16*, 147-168, 2000.

Greatbatch, R. J., A. F. Fanning, A. D. Goulding and S. Levitus, A diagnosis of interpentadal circulation changes in the North Atlantic, *J. Geophys. Res., 96*, 22009-22023, 1991.

Grötzner, A. M., M. Latif and T. P. Barnett, A decadal climate cycle in the North Atlantic Ocean as simulated by the ECHO coupled GCM, *J. Climate, 11*, 831-847, 1998.

Haigh, J. D., A GCM study of climate change in response to the 11-year solar cycle, *Quart. J. Roy. Meteor. Soc., 125*, 871-892, 1999.

Hall, N. M. J., A simple GCM based on dry dynamics and constant forcing, *J. Atmos. Sci., 57*, 1557-1572, 2000.

Hasselmann, K., Stochastic climate models. Part I: Theory, *Tellus, 28*, 473-485, 1976.

Hendon, H. H. and D. L. Hartmann, Stationary waves on a sphere: Sensitivity to thermal feedback, *J. Atmos. Sci.*, 1906-1920, *39*, 1982.

Hoerling, M., J. Hurrell and T. Xu, Tropical origins for recent North Atlantic climate change, *Science, 292*, 90-92, 2001.

Horel, J. D. and J. M. Wallace, Planetary-scale atmospheric phenomena associated with interannual variability of sea-surface temperature in the equatorial Pacific, *Mon. Wea. Rev., 109*, 813-829, 1981.

Hoskins, B. J., H. H. Hsu, I. N. James, M. Matsutani, P. D. Sardeshmukh and G. H. White, *Diagnostics of the global atmospheric circulation based on ECMWF analyses 1979-1989*, 223 pp., World Climate Research Programme, WCRP-27, 1989.

Hoskins, B. J. and D. Karoly, The steady linear response of a spherical atmosphere to thermal and orographic forcing, *J. Atmos. Sci., 38*, 1179-1196, 1981.

Hoskins, B. J. and A. J. Simmons, A multi-layer spectral model and the semi-implicit method, *Quart. J. Roy. Meteorol. Soc., 101*, 637-655, 1975.

Hoyt, D. V. and K. H. J. Schatten, A discussion of plausible solar irradiance variations, 1700-1992, *J. Geophys. Res., 98*, 18895-18906, 1993.

Hurrell, J. W. and H. van Loon, Decadal variations in climate associated with the North Atlantic Oscillation, *Climate Change., 36*, 301-326, 1997.

Iwasaka, N. and J. M. Wallace, Large scale air sea interaction in the Northern Hemisphere from a view point of variations of surface heat flux by SVD analysis, *J. Meteorol. Soc. Japan, 73(4)*, 781-794, 1995.

Josey, S. A., E. C. Kent and P. K. Taylor, *The Southampton Oceanography Centre (SOC) ocean–atmosphere heat, momentum and freshwater flux atlas,* 55pp, *SOC report No. 6,* 1998.

Kalnay, E., et al., The NCEP/NCAR 40 Year Reanalysis Project. *Bull. Amer. Meteorol. Soc., 77(3)*, 437-471, 1996.

Kushnir, Y. Interdecadal variations in North Atlantic sea surface temperatures and associated atmospheric conditions, *J. Climate, 7*, 141-157, 1994.

Kushnir, Y., W. A. Robinson, I. Bladé, N. M. J. Hall, S. Peng and R. T. Sutton, Atmospheric GCM response to extratropical SST anomalies: Synthesis and evaluation, *J. Climate, 15,* 2233-2256, 2002.

Kutzbach J. E., R. M. Chervin and D. D. Houghton, Response of the NCAR general circulation model to prescribed changes in ocean surface temperature, *J. Atmos. Sci., 34*, 1200-1213, 1977.

Latif, M. and T. P. Barnett, Causes of decadal variability over the North Pacific and North America, *Science, 266*, 634-637, 1994.

Lamb, P. J., On the mixed-layer climatology of the north and tropical Atlantic, *Tellus A, 36*, 292-305, 1984.

Lean J., J. Beer and R. Bradley. Reconstruction of solar irradiance since 1610: Implications for climate change. *Geophys. Res. Lett., 22*, 3195-3198, 1995.

Lindzen, R. S., T Aso and D. Jacqmin, Linearized calculations of stationary waves in the atmosphere, *J. Meteorol. Soc. Jap., 60*, 66-77, 1982.

Livezey, R. E., The evaluation of forecasts, in *Analysis of climate variability*, edited by H. von Storch and A. Navarra, pp. 334, Springer, Berlin, 1995.

Lorenz, E. N., Deterministic nonperiodic flow, *J. Atmos. Sci., 20*, 130-141, 1963.

McCartney, M. S. and L. D. Talley, Warm-to-cold water conversion in the northern North Atlantic Ocean, *J. Phys. Oceanogr., 14*, 922-935, 1984.

Marshall, J., H. Johnson and J. Goodman, A study of the interaction of the North Atlantic Oscillation with ocean circulation, *J. Clim., 14*, 1399-1421, 2001.

Mehta, V., M. Suarez, J. Manganello and T. Delworth, Oceanic influence on the North Atlantic Oscillation and associated Northern Hemisphere climate variations: 1959 1993, *Geophys. Res. Lett., 27(1)*, 121-124, 2000.

Mysterud, A., N. C. Stenseth, N. G. Yoccoz, G. Ottersen, and R. Langvatn, The response of terrestrial ecosystems to climate variability associated with the North Atlantic oscillation, this volume.

Namias, J., Seasonal persistence and recurrence of European blocking during 1958-1960, *Tellus, 16*, 394-407, 1964.

Orrell, D., L. Smith, J. Barkmeijer and T. Palmer, Model error in weather forecasting, *Nonlinear Processes in Geophys., in press*, 2002.

Osborn, T. J., K. R. Briffa, S. F. B. Tett, P. D. Jones and R. Trigo, Evaluation of the North Atlantic Oscillation as simulated by a coupled climate model, *Climate Dynamics, 15*, 685-702, 1999.

Palmer, T. N., Č. Branković and D. S. Richardson, A probability and decision-model analysis of PROVOST seasonal multi-model ensemble integrations, *Quart. J. Roy. Meteor. Soc., 126*, 2013-2033, 2000.

Palmer, T. N. and Z. Sun, A modelling and observational study of the relationship between sea surface temperature in the north-west Atlantic and the atmospheric general circulation, *Quart. J. Roy. Meteor. Soc., 111*, 947-975, 1985.

Peng, S. and J. S. Whitaker, Mechanisms determining the atmospheric response to midlatitude SST anomalies, *J. Climate, 12*, 1393-1408, 1999.

Peterson, K. A., R. J. Greatbatch, Jian Lu, H. Lin and J. Derome, Hindcasting the NAO using diabatic forcing of a simple AGCM, *Geophys. Res. Lett., 29,* 10.1029/2001GL014502, 2002.

Pope, V. D., M. L. Gallani, P R Rowntree and R. A. Stratton, The impact of new physical parametrizations in the Hadley Centre climate model - HadAM3, *Climate Dyn., 16*, 123-146, 2000.

Qin J. C. and W. A. Robinson, The impact of tropical forcing on extratropical predictability in a simple global model. *J. Atmos. Sci., 52*, 3895-3910, 1995.

Ratcliffe, R. A. S. and R. Murray, New lag associations between North Atlantic sea temperatures and European pressure, applied to long-range weather forecasting, *Quart. J. R. Meteor. Soc., 96*, 226-246, 1970.

Robertson, A., C. Mechoso and Y.-J. Kim, The influence of Atlantic sea surface temperature anomalies on the North Atlantic Oscillation, *J. Climate, 13*, 122-138, 2000.

Robock, A., Volcanic eruptions and climate, *Rev. Geophys., 32*, 191-219, 2000.

Rodwell, M. J., D. P. Rowell and C. K. Folland, Oceanic forcing of the winter North Atlantic Oscillation and European climate, *Nature, 398*, 320-323, 1999.

Rodwell, M. J., Atlantic air-sea interaction revisited, in *Meteorology at the Millennium*, edited by R. P. Pearce, 333 pp., Academic Press. 2001.

Rodwell, M. J. and C. K. Folland, Atlantic air-sea interaction and seasonal predictability, *Quart. J. Roy. Meteor. Soc., 128*, 1413-1443 2002a.

Rodwell, M. J. and C. K. Folland, Atlantic air-sea interaction and model validation, *Annals of Geophysics, in press*, 2002b.

Rogers, J. C., Patterns of low-frequency monthly sea level pressure variability (1899-1986) and associated wave cyclone frequencies, *J. Climate, 3*, 1364-1379, 1990.

Rowell, D. P., Assessing potential seasonal predictability with an ensemble of multidecadal GCM simulations, *J. Climate, 11, 109-120*, 1998.

Rowell, D. P. and F. W. Zwiers, The global distribution of sources of atmospheric decadal variability and mechanisms over the tropical Pacific and southern North America, *Climate Dynamics, 15*, 751-772, 1999.

Rowntree, P., Response of the atmosphere to a tropical Atlantic ocean temperature anomaly, *Q. J. R. Meteorol. Soc., 102*, 607-625, 1976.

Shindell D. T., G. A. Schmidt, R. L. Miller and D. Rind, Northern Hemisphere winter climate response to greenhouse gas, ozone, solar, and volcanic forcing, *J. Geophys. Res., 106*, 7193-7210, 2001a.

Shindell D. T., G. A. Schmidt, M. E. Mann, D. Rind and A. Waple, Solar forcing of regional climate change during the Maunder Minimum, *Science, 294*, 2149-2152, 2001b.

Sutton, R. T. and M. R. Allen, Decadal predictability of north Atlantic sea surface temperature and climate, *Nature, 338*, 563-567, 1997.

Stephenson, D.B., V. Pavan and R. Bojariu, Is the North Atlantic Oscillation a random walk?, *Int. J. Climatol., 20*, 1-18, 2000.

Thompson, D. W. J., S, Lee, and M. P. Baldwin, Atmospheric Processes governing the Northern Hemisphere Annular Mode/North Atlantic Oscillation, this volume.

Ting, M and N-C Lau, A diagnostic and modeling study of the monthly mean wintertime anomalies appearing in a 100-year GCM experiment, *J. Atmos. Sci., 50*, 2845-2867, 1993.

van den Berg, W. D., The role of various weather parameters and the use of worst-case forecasts in prediction of gas sales, *Met. App., 1*, 33-37, 1994.

Vellinga, M. and R. A. Wood, Global climatic impacts of a collapse of the Atlantic thermohaline circulation. *Clim. Change, 54(3)*, 251-267, 2002.

Visbeck, M., H. Cullen, G. Krahmann and N. Naik, An ocean model's response to North Atlantic Oscillation-like wind forcing, *Geophys. Res. Lett., 25(24)*, 4521-4524, 1998.

Visbeck, M., E. P. Chassignet, R. G. Curry, T. L. Delworth, R. R. Dickson, and G. Krahmann, The ocean's response to North Atlantic Oscillation variability, this volume.

Watanabe, M. and M. Kimoto, Tropical-extratropical connection in the Atlantic atmosphere-ocean variability, *Geophys. Res. Lett., 26*, 2247-2250, 1999.

Watanabe, M. and M. Kimoto, On the persistence of decadal SST anomalies in the North Atlantic, *J. Climate, 13*, 3017 3028, 2000.

Mark J. Rodwell, Met. Office, Hadley Centre, London Road, Bracknell RG12 2SY, U.K.

mark.rodwell@metoffice.com

Climate Change and the North Atlantic Oscillation

Nathan P. Gillett

School of Earth and Ocean Sciences, University of Victoria, Victoria, BC, Canada

Hans F. Graf

Max Planck Institute for Meteorology, Hamburg, Germany

Tim J. Osborn

Climate Research Unit, University of East Anglia, Norwich, UK

Over recent decades the boreal winter index of the North Atlantic Oscillation (NAO) has exhibited an upward trend, corresponding to lowered surface pressure over the Arctic and increased surface pressure over the subtropical North Atlantic. This trend has been associated with over half the winter surface warming in Eurasia over the past thirty years, as well as strong regional trends in precipitation over Western Europe. Several studies have shown this trend to be inconsistent with simulated natural variability. Most climate models simulate some increase in the winter NAO index in response to increasing concentrations of greenhouse gases, though the modeled changes are generally smaller than those seen in the real atmosphere. The two other principal anthropogenic forcings, sulphate aerosol and stratospheric ozone depletion, are generally found to have little significant effect on the NAO. Natural forcings may have also had an impact on the atmospheric circulation: volcanic aerosols induce the westerly (positive index) phase of the NAO in the 1–2 years following major eruptions, and multi-decadal changes in the NAO have also in part been attributed to changes in solar irradiance. These natural forcings, however, are unlikely to account for a substantial component of the recently observed positive NAO index trend: it is most likely to be the result of increasing greenhouse gas concentrations. Experiments using climate models forced only with changes in tropical sea surface temperatures suggest that at least part of this trend may be due to remote forcing from the tropics. Some authors have argued that greenhouse gas-induced changes in the meridional temperature gradient in the lower stratosphere may be responsible for the upward NAO index trend, but overall the mechanism of response to greenhouse gases remains open to debate.

1. INTRODUCTION

In recent decades the boreal winter North Atlantic Oscillation (NAO) index has increased markedly, with surface pressure in the winter season falling over Iceland by around 7 hPa over the past thirty years. This change has been associated with over half the Eurasian winter surface temperature increase over the same period [*Hurrell,* 1996], and much of the observed trend in precipitation over Western Europe. Thus if we are to make accurate climate predictions, we need to understand the relationship between climate change and the NAO. In this chapter we consider both how external forcing may induce changes in the NAO, and how such NAO changes manifest themselves in other climate variables, such as surface temperature. We start by asking whether the observed winter upward trend in the NAO index is consistent with natural variability. Estimating

The North Atlantic Oscillation:
Climatic Significance and Environmental Impact
Geophysical Monograph 134

10.1029/134GM09

natural variability from the recent observed record is difficult because it is so short and is contaminated by anthropogenic forcing. Instead, estimates of internal variability may be obtained from historical observations [*Jones et al.*, this volume], paleoclimate data [*Cook*, this volume], or from long control simulations of climate models. Aside from an upward trend in the NAO index, we also ask whether the mode of variability is itself changing, for example by a shift in the northern lobe of the NAO pattern [*Ulbrich and Christoph,* 1999].

Having assessed observed changes in the NAO, we also consider how this mode might vary in response to specific external climate forcings by reviewing a wide range of climate model simulations, and we assess the extent to which the simulated changes are consistent with those observed. Over the past century the largest change in tropospheric radiative forcing has been due to changes in anthropogenic greenhouse gases; thus, we first examine evidence of an NAO response to this forcing change. Stratospheric ozone depletion and tropospheric sulphate aerosol both also have significant climate impacts, but the literature on the atmospheric circulation response to these anthropogenic forcings is more limited. Of course changes in the NAO may not only be forced by anthropogenic climate forcings: natural forcings could be important too. Changes in solar irradiance and stratospheric aerosol from explosive volcanism have both been linked to changes in tropospheric circulation, so their impacts on the NAO are also reviewed.

Whilst climate model simulations may allow us to attribute observed changes in the NAO to external forcings, they do not necessarily lead to an improved physical understanding of these changes. Thus, we also ask by which mechanisms these external climate forcings influence the NAO. Though there have been several studies that attempt to answer this question, their conclusions differ.

Various definitions of the NAO index are used in the studies we review. The simplest use differences in sea level pressure (SLP) at two stations [e.g., between Gibraltar and Iceland; see *Jones et al.*, this volume]. Alternatively, hemispheric definitions can be used, derived by projecting SLP onto an Empirical Orthogonal Function (EOF) pattern, for instance of November-April monthly SLP northward of 20°N [*Thompson and Wallace,* 1998]. The latter is often referred to as the Arctic Oscillation index, or Northern Hemisphere Annular Mode index [*Thompson et al.*, this volume]. However, since these all represent essentially the same phenomenon, and to avoid confusing readers, we refer to them as simply the NAO index. When we wish to draw a distinction, we refer to the former as a "station-based" NAO index, and the latter as an "EOF-based" NAO index.

2. OBSERVED CHANGES IN THE NAO

The focus of this chapter is on linkages between the NAO and climate change, here defined as the climate response to external forcings, natural as well as anthropogenic. We are interested in the observed fluctuations of the NAO for three reasons: (i) has the NAO exhibited variations that imply the influence of external forcings?; (ii) are the observed NAO changes consistent with the simulated response to external forcings?; and (iii) are forced or unforced NAO variations masking or enhancing the signal of anthropogenic climate change?

The first question is effectively a "detection" question: can we detect an "unusual" change in the NAO index, where "unusual" might be in reference to the past range of observed variation, or to the range of natural variation generated internally within the climate system (or, more strictly still, generated internally within the atmosphere). Considering the former context, and focusing on the NAO index behavior over the past 30 years, *Thompson et al.* [2000] concluded that the observed positive trend between January and March is significant compared to its own internal variability, assuming that variability is uncorrelated. The upward trend in the monthly NAO index between November and January is also statistically significant compared to a red noise model [*Gillett et al.*, 2001; *Feldstein*, 2002]. Other statistical models may yield different results [*Wunsch*, 1999], though the model used must fit the data if it is to be considered valid [*Trenberth and Hurrell*, 1999].

The current instrumental record is insufficient in length to validate the realism of statistical models of lower frequency NAO behavior [although see *Jones et al.,* this volume]. Extended estimates based on climate proxies of the past variations in the NAO index were compared with recent observed NAO variations by *Jones et al*. [2001], who found that neither the recent, highly positive NAO index values of the 1990s nor the change from the low index values of the 1960s appear to be unique in the context of longer records [see also *Cook*, this volume]. Extended records based on climate proxies are, however, liable to substantial uncertainties [e.g, *Cook*, this volume] and also measure a climate that has been subject to various external forcings. It is worthwhile considering the more restricted context of comparing recent changes with the range of unforced variability, for which we must use unforced climate model simulations to obtain an estimate. Figure 1 shows a comparison of observed NAO winter trends over the past century with those simulated in a control integration of a general circulation model (GCM), using a hemispheric EOF-based NAO index. For almost all trend lengths between 20 and 60 years the observed trend is outside the 5–95% range of simulated internal variability.

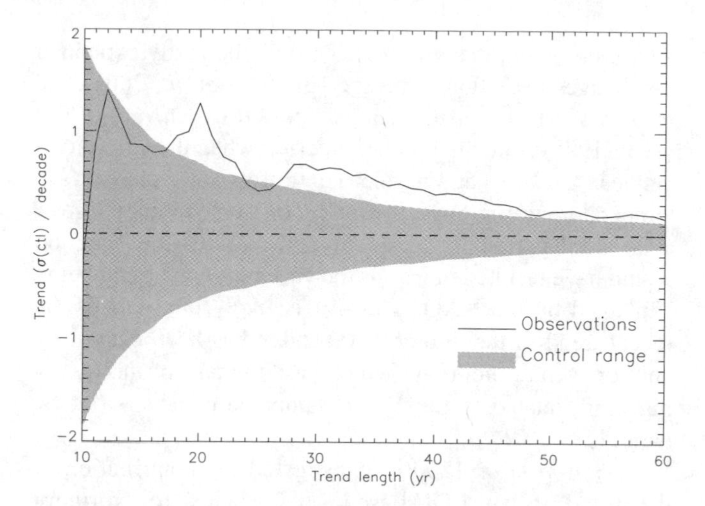

Figure 1. A comparison of the trend in the observed December-February EOF-based NAO index with the corresponding trend in HadCM2 control. The solid line shows the observed trend as a function of the length of time over which it is measured, always ending in 1997. The grey band shows the 5%–95% range of trends in 1091 years of HadCM2 control. Adapted from *Gillett et al.* [2000].

Osborn et al. [1999] demonstrated that observed (positive) 30-yr winter NAO index trends beginning between 1960 and 1967 were outside the 95% range of variability simulated during a 1400-year control run of the HadCM2 coupled climate model. In fact, over these periods observed trends exceeded all simulated 30-yr trends in the HadCM2 control, even those beginning in a period of anomalously low NAO index values. *Osborn* [2002] has extended this study to consider the variability simulated during (shorter) control runs of six other climate models in addition to HadCM2. Figure 2 shows the 95% ranges (from the 2.5 to 97.5 percentiles) obtained when results from these seven simulations are averaged, together with the minimum and maximum of the simulated ranges. These are based on distributions of 30-year winter trends from control simulations, and are compared in Figure 2 with the 30-year trends in the observed NAO index computed in a sliding window. Recent 30-year trends are clearly outside the mean of the simulated natural variability ranges, and those centered around 1981 (e.g., 1966-1995) are also outside the most extreme of the natural variability ranges. The widest ranges are from the two models (CCSR/NIES and NCAR PCM) that simulate interannual variance of the NAO that is more than twice as strong as that observed. The extreme ranges are likely, therefore, to be overestimates.

Our analysis of internally generated climate variability leads us to conclude that increases in the winter NAO index over recent decades are either due in part to some external forcing, or all seven climate simulations are deficient in their levels of interdecadal variability. While the latter possibility cannot be excluded, we note that the models are not in general deficient in their simulated levels of interannual NAO variability, and *Osborn et al.* [1999] point out that the simulated interdecadal NAO variability was similar to that observed prior to the 1950s. Thus it would seem likely that some response to external forcing is present in the observed NAO index record. The evidence from paleoclimate data suggests that this forcing may not be solely of anthropogenic origin [e.g., *Cook*, this volume].

Our understanding of the relationship between the NAO and climate change is complicated by the facts that (i) most external forcings influence the surface temperature field as well as (possibly) the atmospheric circulation and the NAO, and (ii) variations in the NAO influence surface temperature.

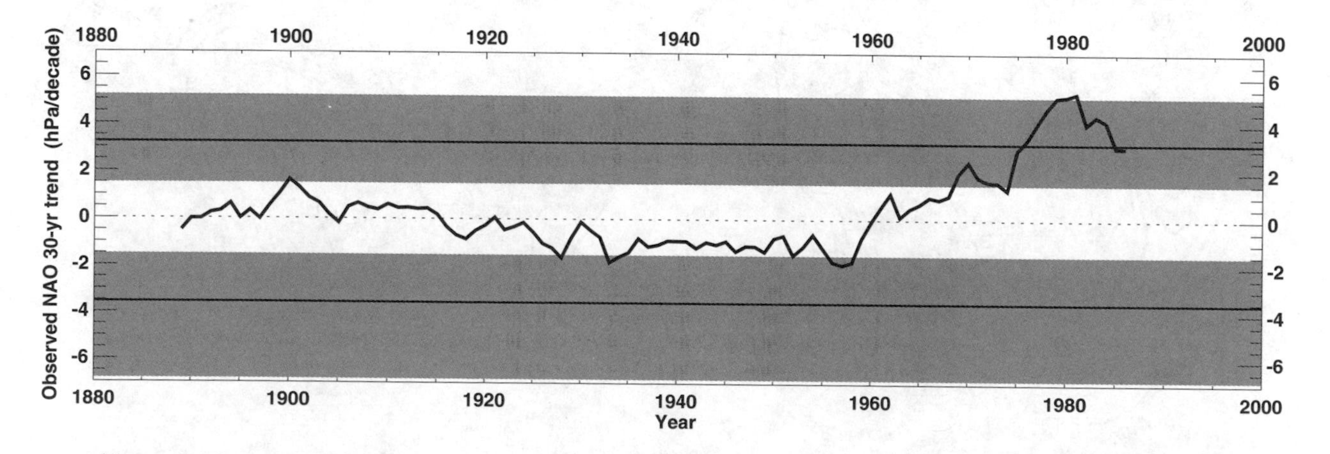

Figure 2. 30-year trend (black line, expressed as hPa/decade) in the observed EOF-based NAO index, computed in a sliding window and plotted against the central year of the window. Grey shading shows the ranges of 2.5 and 97.5 percentiles of 30-year trends computed during seven multi-century control simulations, with the solid horizontal lines indicating the mean 2.5 and 97.5 percentiles computed across the seven models.

The correlation pattern between the winter NAO index and the winter surface temperature field contains regions of both positive and negative influence (Figure 3), but the positive (generally land) regions dominate when averaged over the Northern Hemisphere [*Hurrell,* 1996]. The result is a strong positive correlation between the winter NAO index and the average Northern Hemisphere extratropical winter temperature [see also *Jones et al.*, this volume]. Based on the period 1935-1994, Hurrell [1996] found that the winter NAO variations explain 31% of the interannual variability of the mean extratropical Northern Hemisphere winter surface temperature, and 48% of the recent (1977-1994 average) warming (inclusion of Pacific circulation changes raised this to 71%).

Consequently, it is possible that observed NAO changes (which might be due to natural externally-forced or internally-generated variability) have caused us to overestimate the magnitude of anthropogenically-induced warming. There are, however, three important points to note regarding this statement. First, Northern Hemisphere winter temperatures are but one contributor to the trend in the global, annual mean surface temperature, which has increased 0.6±0.2°C over the past century [*Folland et al.*, 2001]. Second, *Osborn et al.* [1999] note that the strength of the relationship between the NAO and the northern extratropical mean surface temperature has varied through time [see also *Jones et al.*, this volume]. This can be partly explained by changes in station coverage [*Broccoli et al.*, 2001] and also by the fact that the analysis period of *Hurrell* [1996] (1935-1994) coincided with the period when the correlation between the NAO and northern extratropical winter surface temperature was strongest [*Jones et al.*, this volume]. Use of a different period for developing the regression equation would have resulted in less of the variability and trend being explained by the NAO. Third, it is likely that part of the recent trend in the winter NAO index has been caused by anthropogenic forcing, in which case the part of the warming trend related to the NAO cannot be considered to be natural.

Thompson et al. [2000] investigated the contribution of the trend in an EOF-based NAO index to Northern Hemisphere warming. This signal is similar to that shown in Figure 3b, with strong warming over much of Eurasia, and explains ~30% of January-March warming over the entire Northern Hemisphere in the past forty years. *Gillett et al.* [2000] note that HadCM2 does not simulate an upward trend in the NAO index in response to a prescribed increase in greenhouse gases. Some authors [e.g., *Shindell et al.*, 1999a] have suggested that such a failing may invalidate the results of climate change detection studies using this model [e.g., *Tett et al.*, 1999]. However, *Gillett et al.* [2000] show that, using 50 years of surface temperature records, green-

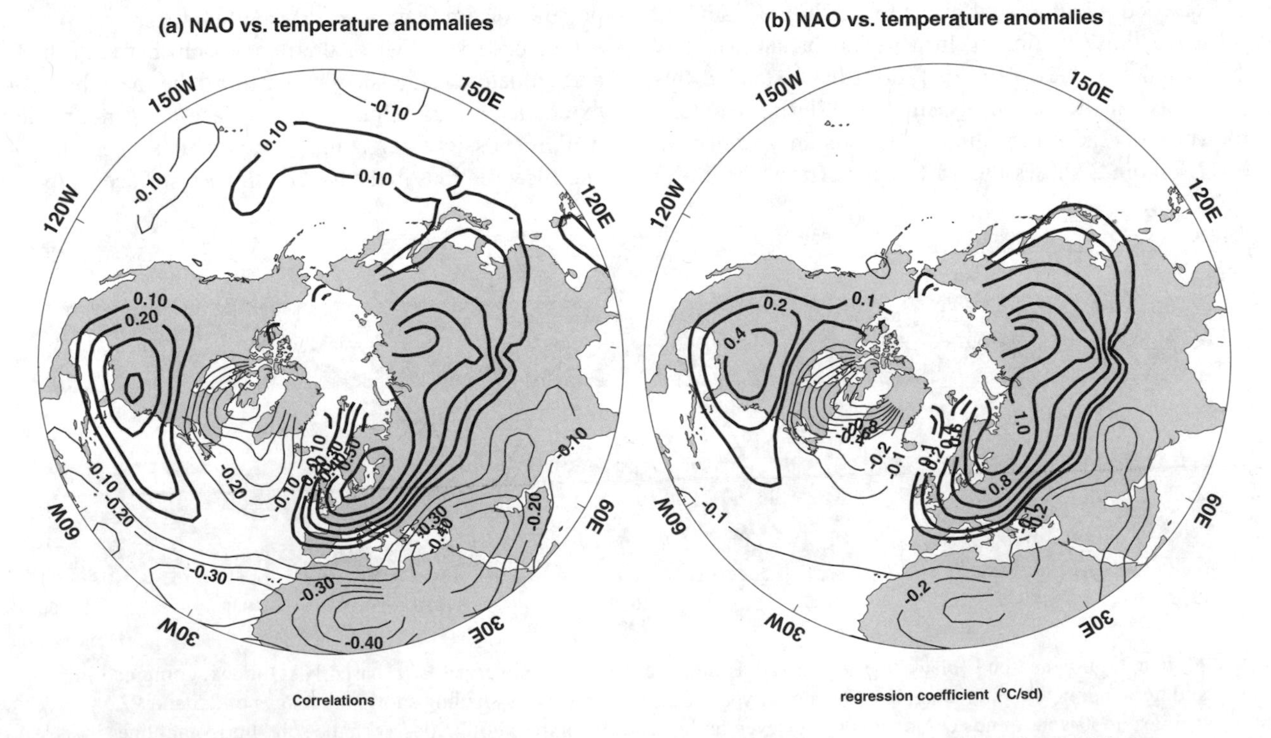

Figure 3. Correlation (a) and regression coefficient (b) of near surface air temperature on the NAO index.

house gas and sulphate aerosol influences may still be separately detected, even when the component of temperature change associated with the trend in the NAO is disregarded. This is likely to be due largely to the temperature changes observed in the tropics and Southern Hemisphere, which are not associated with a change in the NAO.

3. THE RESPONSE OF THE NAO TO FORCING CHANGES

3.1. Greenhouse Gases

Increases in anthropogenic greenhouse gas concentrations represent the largest external forcing on the climate system over the instrumental period of record [*Ramaswamy et al.*, 2001]. Thus, if we are to attempt to identify the cause of the recent increase in the winter NAO index, it is natural that we investigate this influence first.

When the recent upward trend in the NAO index first became apparent in the mid-1990s, several authors speculated that increasing greenhouse gas concentrations were a contributing factor [e.g., *Palmer*, 1993; *Hurrell*, 1995; *Graf et al.*, 1995]. The dominant effect of greenhouse gas increases on the atmosphere is a warming of the troposphere and a cooling of the stratosphere [e.g., *Cubasch et al.*, 2001], but the influence on the NAO is still uncertain. In this section we start by reviewing the response of the NAO to greenhouse gas increases in some coupled general circulation models, and go on to examine the proposed mechanisms by which this response might come about. Note that existing climate models have limited horizontal and vertical resolution that may limit their ability to properly simulate features such as baroclinic disturbances or stratospheric processes, both of which could be important in the NAO response to external forcing [*Thompson et al.*, this volume]. Despite this shortcoming, we devote considerable attention to model-based studies since GCMs are the best tools available to investigate these issues.

3.1.1. Coupled model results.

The first modeling studies to discuss the influence of greenhouse gases on the tropospheric circulation appeared about a decade ago. *Graf et al.* [1995] speculate that greenhouse gases may have contributed to the observed upward trend in the boreal winter NAO index, and they test this hypothesis by forcing a coupled ocean-atmosphere model (ECHAM1) with increased carbon dioxide concentrations and other anthropogenic forcings. However, they found no significant circulation response in their model, though they speculate that this may be due to model failings, such as its lack of modeled chemistry. In contrast, using a later version of the model (ECHAM4), *Ulbrich and Christoph* [1999] found a significant change towards the positive index phase of the NAO (using an EOF-based index), which they noted was linearly proportional to the applied radiative forcing. They also noted a northeastward shift of the northern center of action of the NAO in response to increasing greenhouse gas concentrations. *Osborn et al.* [1999] examined changes in a station-based NAO index in HadCM2 with prescribed increases in greenhouse gases, and found a decrease in the NAO index. However, other studies using this model have since found either no significant change [*Gillett et al.*, 2000] or a weak increase [*Osborn*, 2002] depending on the definition of the NAO index used.

At about the same time, two other studies used an EOF-based index to examine the circulation response to forcing in two different GCMs. *Fyfe et al.* [1999] found a significant increase in the NAO index in response to a combination of greenhouse gas and sulphate aerosol forcing in the Canadian Centre for Climate Modelling and Analysis (CCCma) model. However, although the pattern of SLP trends projected positively onto the hemispheric NAO pattern in their analysis, a station-based NAO index showed no significant increase. *Shindell et al.* [1999a] found that a 9-level model with its upper boundary at 10 hPa showed no NAO response to an increase in greenhouse gases, whilst a second model with 23 levels and an upper boundary at 0.002 hPa showed an increase in the NAO index of an amplitude similar to that observed. This prompted them to argue that greenhouse gases induce a change in the NAO toward its positive index phase through a stratospheric mechanism, though other models without a well-resolved stratosphere have since also been found to produce a similar response.

Thanks in part to the high levels of interest in both climate change and the NAO, circulation changes in response to enhanced greenhouse gases have now been examined in all of the major coupled general circulation models (Table 1). With the exception of HadCM2, the GISS 9-level model, and the NCAR CSM, all these models show at least some increase in the winter NAO index in response to enhanced greenhouse gas concentrations. Note that unlike the real world, these simulations generally include greenhouse gas changes only, prescribed with differing rates of change. However, since anthropogenic greenhouse gases have been the dominant climate forcing over the past century, and since NAO changes are generally found to be linearly dependent on the instantaneous radiative forcing [*Ulbrich and Christoph*, 1999; *Gillett et al.*, 2002a], we would expect the sign, if not the magnitude, of the response to be comparable with observed changes. The recently-published IPCC Third Assessment Report concluded that there was not yet a

Table 1. North Atlantic Oscillation index response to greenhouse gas increases in several general circulation models. Upward arrows indicate an increase in the NAO index, corresponding to a decrease in Arctic SLP, in response to increasing greenhouse gas concentrations (and changes in sulphate aerosol in the case of the CCCma model). A '.' indicates either no significant change, or that the sign of the change has been shown to be dependent on the definition of the NAO index. Note that authors use various different EOF and station-based measures of the NAO, and vary in the rigor of their estimates of the significance of the trend. Adapted from *Gillett et al.* [2002a].

Model	NAO response	Source	Model description
HadCM2	.	1, 2, 3, 4, 12	14
HadCM3	↑	4,12	15
ECHAM3	↑	4, 5	16
ECHAM4	↑	2, 4, 6, 7, 12	6
CCCma	↑	8,12	17
GISS-S	↑	4, 9, 10	18
GISS-T	.	9, 10	9
CSIRO	↑	11,12	19
CCSR	↑	11,12	20
NCAR-CSM	.	7, 12	21
NCAR-PCM	↑	12	22
GFDL	↑	13	23

Sources: 1, *Gillett et al.* [2000]; 2, *Zorita and González-Rouco* [2000]; 3, *Osborn et al.* [1999]; 4, *Gillett et al.* [2002a]; 5, *Paeth et al.* [1999]; 6, *Ulbrich and Christoph* [1999]; 7, *Robertson* [2001]; 8, *Fyfe et al.* [1999]; 9, *Shindell et al.* [1999a]; 10, *Shindell et al.* [2001b]; 11, E. Zorita (pers. comm.); 12, *Osborn* [2002]; 13, *Stone et al.* [2001]; 14, *Johns et al.* [1997]; 15, *Gordon et al.* [2000]; 16, *Voss et al.* [1998]; 17, *Flato et al.* [2000]; 18, *Shindell et al.* [1998]; 19, *Gordon and O'Farrell* [1997]; 20, *Emori et al.* [1999]; 21, *Boville and Gent* [1998]; 22, *Washington et al.* [2000]; 23, *Stouffer and Manabe* [1999]. GISS-S denotes the 23 level GISS model with a model top at 0.002 hPa, and GISS-T the 9 level GISS model with a model top at 10 hPa.

consistent picture of the ability of coupled models to reproduce the observed upward trend in the NAO index [*Cubasch et al.*, 2001], but there is now a growing consensus amongst modeling groups that greenhouse gases induce some increase in the NAO index.

We now turn our attention to the spatial patterns of the simulated SLP trends. Figure 4a [*Osborn*, 2002] shows the average of the first EOF of winter SLP computed separately from the Atlantic sectors of seven coupled GCM control simulations. Most models show broadly realistic patterns of interannual, unforced variability, and this is reflected in the 7-model mean. One consistent difference between the simulated (Figure 4a) and observed (Figure 4b) EOF is the stronger simulated covariance between the North Pacific and the North Atlantic SLP. Such a pattern is only obtained in the observed data by performing an EOF analysis on the hemispheric SLP field [e.g., *Deser,* 2000; *Ambaum et al.*, 2001].

Figure 5 [*Osborn*, 2002] shows the 7-model mean of the trends in boreal winter SLP simulated under increasing greenhouse gas concentrations (a similar scenario was applied to each model), together with an indication of inter-model agreement. Two points are apparent: First, there are large inter-model variations in the patterns simulated. Such trend patterns have been shown to be very sensitive to the details of the model physics. For example *Williams et al.* [2001] found that the relatively zonally symmetric trend pattern simulated by HadCM3 in response to enhanced greenhouse gases was transformed into a response pattern much more like that of HadCM2, with a decrease in pressure over the North Pacific, simply by a small change in the critical relative humidity for cloud formation, and a change in the model's boundary layer scheme. Second, despite these large variations, the models generally show a decrease in pressure over the Arctic region, and some increases at lower latitudes. Such patterns project onto the positive (westerly) phase of the NAO. Thus, if the model EOF patterns (Atlantic-sector only) are used to measure the evolution of the NAO index under enhanced greenhouse forcing, then the winter NAO index shows a positive trend in all seven climate models (Figure 6). Though the magnitude of this trend is highly model-dependent, the consistency in the sign of the change provides some confidence that the circulation response to greenhouse gas forcing is likely to be an enhancement of the westerly circulation in the North Atlantic sector. Note that, if the NAO index is measured by a simple pressure difference between two points (e.g., Gibraltar and Iceland), the spread of modeled responses is greater than if an EOF-based index is used, and some models even show a decrease in such an index [e.g., *Osborn et al.*, 1999].

Often climate model simulations are carried out with different scenarios of increasing greenhouse gas concentrations,

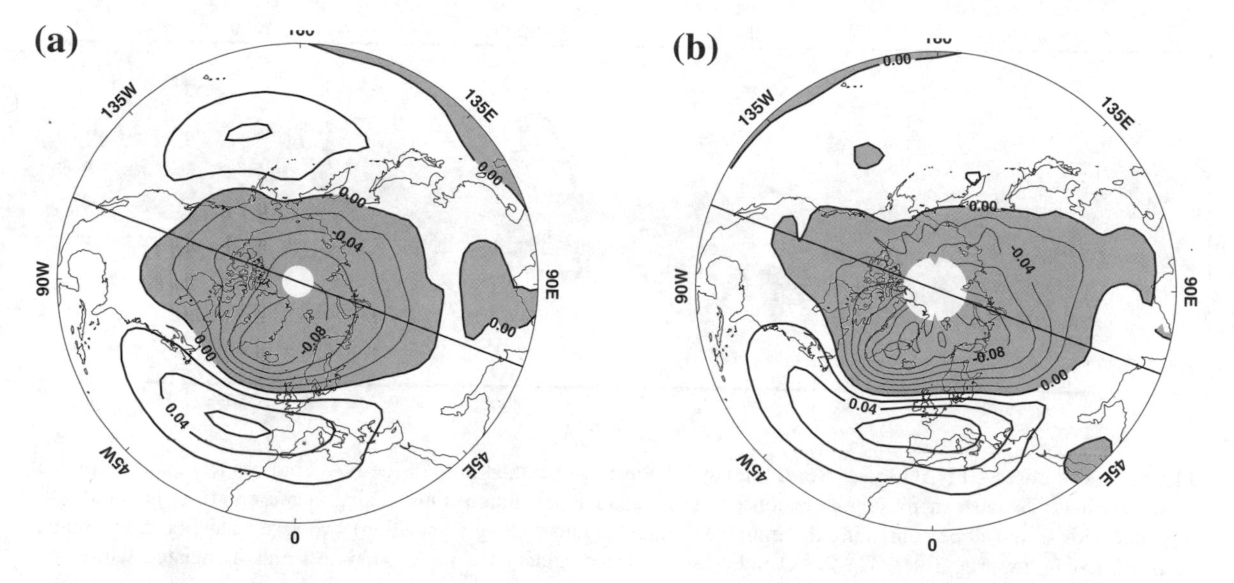

Figure 4. The leading EOF of the winter SLP field computed over the Atlantic sector (below the black line) and then extended to identify covariances across the whole hemisphere: (a) the mean of seven climate model control run EOFs (HadCM2, HadCM3, ECHAM4/OPYC, CCCma, CCSR/NIES, CSIRO and NCAR PCM); (b) the observed EOF.

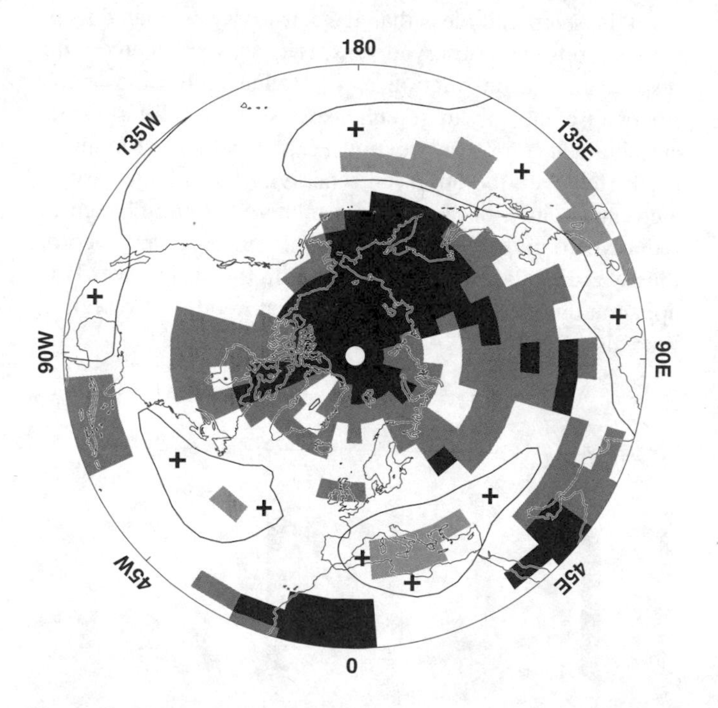

Figure 5. The trend in winter SLP computed from seven climate model simulations under increasing greenhouse gas forcing (1% per annum compounded increase in [CO_2] after 1990) averaged to produce a single mean trend pattern. Only the zero isoline is shown, with '+' to indicate the regions with positive SLP trends (and, therefore, SLP trends are negative elsewhere). Black boxes are drawn where all 7 models exhibit negative trends, dark grey boxes where 6 out of 7 models exhibit negative trends, and pale grey boxes where 6 out of 7 models exhibit positive trends.

and the forcing rarely increases linearly with time. Further, changes in the NAO index are, in some cases, difficult to distinguish from climate noise. How then, are we to quantitatively compare the NAO response in different integrations and different models? *Ulbrich and Christoph* [1999] noted that in a greenhouse gas forced integration of ECHAM4, the NAO index varies linearly with the applied forcing, suggesting that the ratio between the NAO change and the radiative forcing at the tropopause remains constant. *Gillett et al.* [2002] showed that this is also the case in HadCM3 and ECHAM3. This result allows us to make objective comparisons of the sensitivity of the NAO in a range of models by regressing the NAO index against a reconstruction of the radiative forcing. Averaging over different integrations with different time histories of greenhouse gas variations allows us to further reduce uncertainties.

Figure 7 compares the NAO sensitivity in a range of GCMs forced with greenhouse gas changes only, using an EOF-based index [*Gillett et al.*, 2002a]. These sensitivities are essentially the mean regression coefficients of the index with respect to a reconstruction of the radiative forcing at the tropopause. In simple terms they represent how much the NAO index changes per Wm^{-2} of radiative forcing. The associated uncertainty intervals were derived using simulated control variability. Thus if the uncertainty range for a particular model does not include zero, that model has a significant NAO response to greenhouse gas increases. Using a definition of the NAO pattern based on the first EOF of HadCM3 control variability, all the models bar HadCM2 and CCCma show a significant increase in the

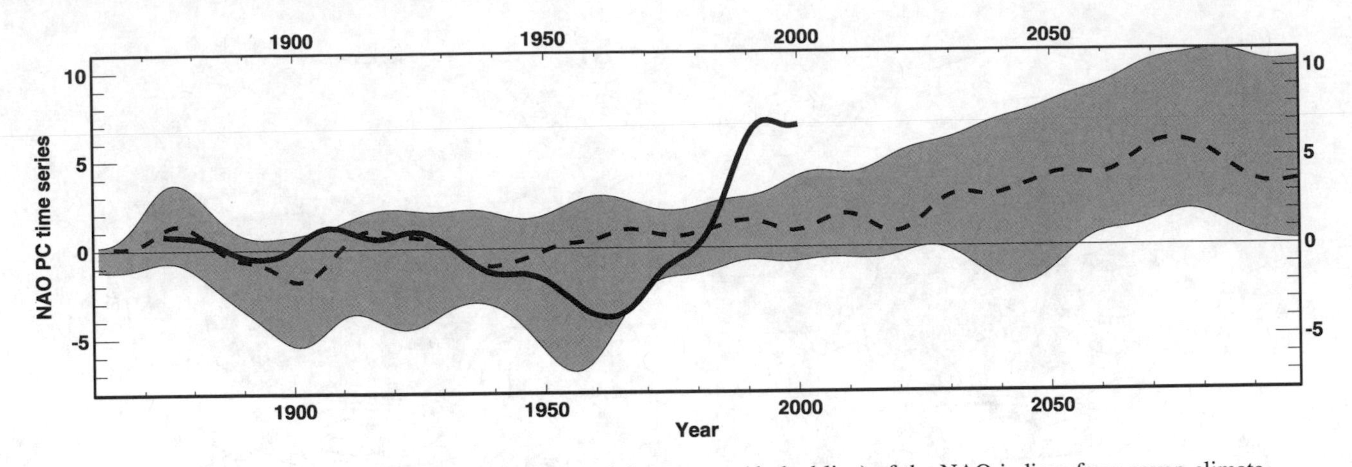

Figure 6. The observed NAO index (solid line) and the average (dashed line) of the NAO indices from seven climate model simulations under increasing greenhouse gas forcing (1% per annum compounded increase in [CO_2] after 1990), together with an envelope containing the individual model simulations (grey shading). All series have been smoothed with a 30-year low-pass filter. The NAO index is the scaled principal component time series associated with each model's leading EOF of the Atlantic-sector SLP field (defined during the control run of each model).

NAO index in response to greenhouse gas increases. Figure 8 shows equivalent sensitivities, but for a station-based NAO index. Again most of the models show a positive NAO index response to the forcing, though in this case more of the uncertainty ranges include zero. Note that the CCCma integration shown here had greenhouse gas forcing only, unlike the ensemble used by *Fyfe et al.* [1999], which also had sulphate aerosol forcing. Applying our analysis to the latter, we too find an NAO index increase, though this result is not significantly different from that for the greenhouse gas only integration.

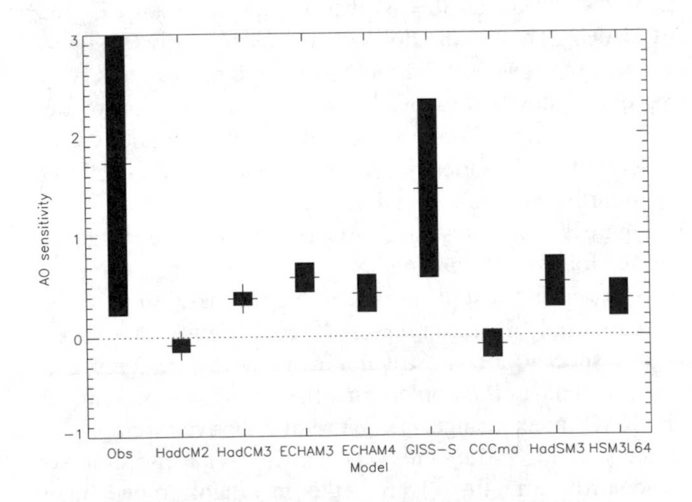

Figure 7. The mean sensitivity of a hemispheric EOF-based NAO index to net radiative forcing at the tropopause due to greenhouse gases in eight GCMs and observations. Black bars show 5-95% uncertainty ranges estimated using control variability. Adapted from *Gillett et al.* [2002a].

Although most of the GCMs included in Figures 7 and 8 simulate a positive NAO sensitivity to greenhouse gas forcing, this is generally less than the sensitivity estimated from observations. The observed sensitivity was calculated with respect to a reconstruction of the radiative forcing at the tropopause due only to greenhouse gases, but similar results are obtained if a reconstruction of total radiative forcing is used. However, the observed sensitivity has a large associated uncertainty, because of the relatively small change in radiative forcing over the length of the observed record, thus we cannot attribute significance to this difference. This apparent underestimate of the observed trend is also seen in time histories of the NAO index shown in Figure 6. Here the

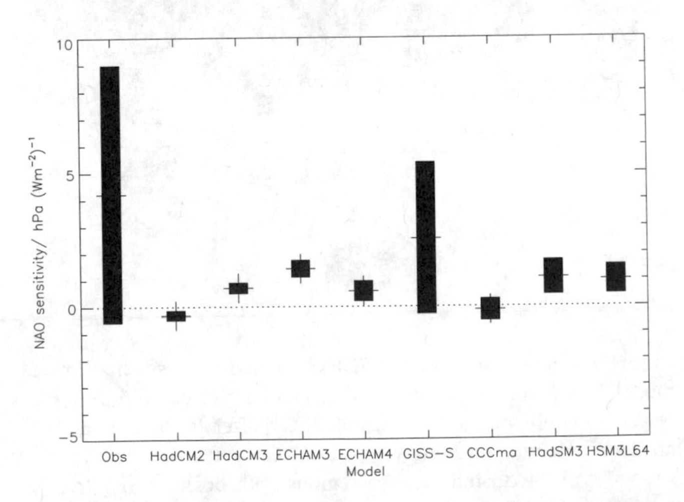

Figure 8. The mean sensitivity of a station-based NAO index to net radiative forcing at the tropopause due to greenhouse gases in eight GCMs and observations. Adapted from *Gillett et al.* [2002a].

observed index is seen to increase outside the range of indices simulated in the forced simulations.

One model that appears to be in closer agreement with observations is the GISS-S stratosphere-resolving model. This model has a much higher upper boundary than most GCMs, with 23 levels extending up to 85km. However, this estimate is based only on the first 60 years of a single greenhouse gas forced integration of the model, prior to the time when *Shindell et al.* [1999a] report that the response saturates. If later years of the integration or additional ensemble members incorporating greenhouse gas and ozone changes are included in the analysis, then the estimate of the NAO sensitivity is considerably reduced [*Gillett et al.*, 2002a]. Furthermore, an enhancement of sensitivity with improved stratospheric resolution is not a feature common to all models. HadSM3 shows no such increase in its NAO sensitivity when its stratospheric resolution is increased and its upper boundary is raised to over 80 km (compare the sensitivities of HadSM3 and HadSM3-L64 in Figure 7) [*Gillett et al.*, 2002b]. Such a test of the sensitivity of the modeled NAO response to stratospheric resolution has not been carried out with any other models to date.

3.1.2. Mechanisms

3.1.2.1. Regime paradigms

Palmer [1993] discusses the role of the North Atlantic Oscillation in climate change, and speculates that increased CO_2 may force it towards its positive phase. However, he contends that this is not because the circulation response to CO_2 looks like the NAO in any direct way. He argues that the system responds in this way primarily because the NAO is a dominant mode of natural variability. Using a variant of the Lorenz model that includes a "forcing" term, he demonstrates that the response of this system is not in the direction of the forcing, but rather close to the dominant mode of natural variability. *Palmer* [1999] and *Corti et al.* [1999] take this argument further by claiming to find observational evidence that Northern Hemisphere circulation changes are manifested through changes in the occupation probabilities of naturally occurring "regimes." These regimes are generally defined as preferentially visited atmospheric states, which are manifested as multiple maxima in histograms of the NAO index or other circulation indices. This viewpoint is, however, contested by *Hsu and Zwiers* [2001] who argue that the regimes identified are not statistically significant, with the possible exception of the Cold Ocean Warm Land (COWL) [*Wallace et al.,* 1995] pattern. *Stone et al.* [2001] also argue against a "regime" view of SLP changes, based on integrations of the GFDL model forced with changing greenhouse gas concentrations. They identify no "regimes" in this model, and they also find a purely linear shift in the distribution of an NAO-like index under increasing greenhouse gas concentrations. This result is also reflected in that of *Gillett et al.* [2002a], who find that the NAO index increases linearly with radiative forcing in HadCM3 and does not exhibit multiple regimes.

However, even in the absence of quasi-stationary regimes, if physical processes act to amplify the natural variability of a particular pattern, then they may also amplify any component of the greenhouse gas response which projects onto that pattern. *Graf et al.* [1995] argue that climate change forces changes in naturally occurring modes, which they identify through physical arguments. They argue that the low latitude greenhouse effect and other anthropogenic forcing produce changes in a coupled troposphere-stratosphere baroclinic mode.

An alternate paradigm would suggest that the physical mechanisms underlying the natural mode of variability and the response to anthropogenic forcing are distinct, and it is perhaps only a consequence of the Earth's approximate zonal symmetry that the pattern of SLP trends is similar to the pattern of natural variability. Until we obtain a better understanding of the physical mechanisms underlying both the variability and the trends in SLP, it may be impossible to determine which view more closely reflects the real world.

3.1.2.2. Sea surface temperature modulated changes

Several authors have suggested that sea surface temperatures (SSTs) may be critical in modulating the NAO, either in the North Atlantic [*Rodwell et al.*, 1999] or in the tropical Indian and Pacific oceans [*Hoerling et al.*, 2001; see also *zaja et al.*, this volume]. *Rodwell et al.* [1999] prescribe the observed, time-varying history of global SSTs in an ensemble of simulations with an atmospheric GCM, and they are able to explain around half of the observed boreal winter trend in the NAO index. They further argue that SST variations in the North Atlantic are of greater relevance. *Mehta et al.* [2000] and *Latif et al.* [2000] obtained similar results with different models. However, other authors have argued that this result could also come about from an ocean that essentially responds passively to stochastic atmospheric forcing [*Bretherton and Battisti*, 2000; *Czaja et al.*, this volume]. *Bretherton and Battisti* [2000] used a model that consists of two time-dependent equations describing a stochastically forced atmosphere and a mixed-layer ocean. It is first integrated in a coupled configuration with a single realization of atmospheric stochastic noise: this is taken to represent the "observed" record. Ocean temperatures from this integration of the model are then prescribed in an ensemble of integrations with an atmosphere-only configuration, and the resulting ensemble mean atmospheric temperatures were found to be highly correlated with the original "observed" temperatures. They thus demonstrated that a model of this type could have significant hindcast skill in an ensemble mean, even though the ocean was only responding passively to atmospheric forcing.

More recently *Hoerling et al.* [2001] argued that the upward trend in the boreal winter NAO index is linked to a warming of tropical SSTs, particularly over the Indian and Pacific oceans. By prescribing observed SSTs since 1950 over the tropics, and fixed seasonally varying climatological SSTs elsewhere, they were able to explain roughly half the magnitude of observed mid-tropospheric height trends in the Northern Hemisphere. This was similar to the trend explained when global observed SSTs were specified. Although NAO anomalies could certainly force changes in tropical Atlantic SSTs, they would be less likely to influence tropical SSTs in the Indian and Pacific oceans. Thus simulated changes that project onto the NAO pattern may really be a response to SST forcing in this experiment, and the objections raised by *Bretherton and Battisti* [2000] to the *Rodwell et al.* [1999] experiment may not apply. *Hoerling et al.* [2001] ascribe this response to changes in tropical rainfall, latent heating, and hence changes in the driving of the extratropical circulation. Note that these studies do not necessarily indicate that the ocean itself plays a strong physical role in determining NAO variability. Prescribed SSTs in a model closely constrain the temperature of the whole tropospheric column, thus it may be that warming tropical SSTs are just a reflection of greenhouse gas forcing in the atmosphere, and it is this atmospheric change that is more directly responsible for the upward trend in the NAO index. These results do, however, suggest that greenhouse gas forcing could modulate the NAO merely through changes in the tropical circulation.

3.1.2.3. Meridional temperature gradient changes in the upper troposphere - lower stratosphere

The first order effect of greenhouse gas forcing is a warming of the troposphere, and a cooling of the stratosphere. In particular, models predict an enhanced warming of the tropical upper troposphere: this is the signature we expect of a moist adiabat. On the basis of radiative calculations and some models, we expect the stratospheric polar vortex to cool in response to enhanced greenhouse gas concentrations. Since the tropopause tilts downwards from the tropics to the pole, these effects would together combine to increase the meridional temperature gradient in the upper troposphere – lower stratosphere region. Many models show such a temperature response to enhanced greenhouse gases [*Cubasch et al.*, 2001]. This in turn might be expected to enhance the vertical shear of zonal wind in the mid-latitude tropopause region, and strengthen the stratospheric vortex.

Shindell et al. [1999a] and others have argued that this strengthening of the vortex may be augmented by a planetary wave feedback, whereby the changed zonal wind profile in the mid-latitude tropopause region acts to deflect planetary-scale Rossby waves equatorwards. These waves are responsible for driving the meridional circulation of the stratosphere, hence also the descent in the region of the pole, which induces a dynamical warming. An equatorward deflection of the waves would thus weaken the stratospheric meridional circulation and reduce dynamical heating of the pole, leading to a further cooling of the vortex: a positive feedback.

On average, circulation anomalies on short timescales have been observed to appear first at around the 10-hPa level, and at the surface around three weeks later [*Baldwin and Dunkerton*, 1999]. This apparent downward propagation has been shown to be significant compared to a noise model [*Gillett et al.*, 2001], though the physical mechanism underlying downward propagation through the troposphere is not well understood [*Thompson et al.,* this volume]. Some authors thus argue that a change toward a stronger zonal circulation in the stratosphere could cause an enhancement of the zonal circulation at the surface, and hence a change toward the positive index phase of the NAO [*Shindell et al.*, 1999a; *Hartmann et al.*, 2002; *Thompson et al.,* this volume].

There are, however, some aspects of this mechanism that are not in agreement with other modeling studies. First, some stratosphere-resolving models predict that planetary wave changes in response to enhanced greenhouse gas concentrations will act against radiative changes and lead to a weakening of the Arctic winter vortex [e.g., *Rind et al.*, 1990; *Schnadt et al.*, 2002; *Gillett et al.*, 2002b]. Second, at least one model shows an increase in the NAO index even though its stratospheric vortex weakens [*Gillett et al.*, 2002b], suggesting that a tropospheric mechanism might be responsible for the NAO trend. More work is clearly needed to resolve these issues.

3.2. Tropospheric Aerosols

The direct and indirect effects of tropospheric sulphate aerosol together make up the second largest radiative forcing after greenhouse gases on the climate over the last century [*Ramaswamy et al.*, 2001]. Sulphate aerosol has a direct radiative effect through the scattering and absorption of solar and infrared radiation, and an indirect effect through induced changes in cloud properties. Sulphate aerosol has had a large cooling effect on surface temperatures, likely counteracting much of the greenhouse warming over the past century [*Tett et al.*, 1999]. Thus one might expect that this forcing agent has also had an influence on the NAO. Several authors have examined trends in the NAO index in integrations of climate models incorporating the effects of changes in anthropogenic sulphate aerosol [*Fyfe et al.*, 1999; *Osborn et al.*, 1999; *Gillett et al.*, 2000]. However, *Fyfe et al.* [1999] and *Gillett et al.* [2000] look only at runs incorporating both greenhouse gas increases and sulphate aerosol increases, thus the influence of the sulphate aerosol cannot be separately identified. *Osborn et al.* [1999] compare changes in the NAO index in ensembles of HadCM2 integrations with both greenhouse gas and sulphate aerosol increases, and with greenhouse gas increases only. They note no significant differences between the NAO responses.

HadCM2 was recently integrated with sulphate aerosol changes only (M. Wehner, personal communications), allowing the response to this forcing to be better separated from the response to greenhouse gases. Based on a four-member ensemble over the past 140 years, we found that winter SLP changes were generally not significant in the northern extratropics, but showed a small decrease over parts of the tropics and southern middle latitudes, and an increase over Southeast Asia, a strong aerosol source region. The simulated SLP response to sulphate aerosol forcing was derived for two more GCMs (CCCma and HadCM3) by differencing SLP changes in ensembles forced with both greenhouse gas and sulphate aerosol changes, and ensembles forced with greenhouse gas changes only. Both models showed few regions of significant SLP response. In recent work, *Chung et al.* [2002] prescribed changes in radiative heating due to aerosol changes over the Indian Ocean region, as measured by the INDOEX experiment, in the NCAR Community Climate Model (CCM3). They found a large dynamical response to this forcing over the tropics, with a northward shift of the intertropical convergence zone, and changes to the subtropical jetstream. Preliminary results also suggest that increases in the South Asian haze may have contributed to the upward trend in the boreal winter NAO index, though the response appears to be very sensitive to the geographical extent of the forcing in the model (Chung, personal communication). Thus, overall, most studies show that anthropogenic sulphate aerosol is unlikely to have had a large influence on the NAO, although the work on the topic must be viewed as preliminary.

3.3. Stratospheric Ozone Depletion

The substantial reduction of lower stratospheric ozone content over the last two decades has received considerable attention in view of its possible contribution to climate change [e.g., *Bengtsson et al.*, 1999; *Forster*, 1999; *Randel and Wu*, 1999; *Langematz*, 2000]. Model simulations and observational evidence show that ozone depletion clearly leads to reduced stratospheric temperatures over both poles mainly in late winter and spring, when sunlight comes back to the polar night area [e.g., *Ramaswamy et al.*, 1996; *Graf et al.*, 1998]. Since a colder polar stratosphere leads to a stronger and more stable polar vortex, which may be linked to a positive index phase of the NAO [*Perlwitz and Graf*, 1995; *Hartmann et al.*, 2002], some model studies have examined the contribution of Arctic ozone depletion to the increase of the winter NAO index [*Graf et al.*, 1998; *Volodin and Galin*, 1999; *Shindell et al.*, 2001b]. While *Volodin and Galin* [1999] claimed that tropospheric circulation and temperature anomalies in the years 1989-94 relative to 1977-88 can be explained by the reduction in Arctic ozone, *Graf et al.* [1998] and *Shindell et al.* [2001b] suggest that ozone depletion plays only a minor role for the months with the strongest NAO index trends, when compared with the effect of the increasing greenhouse gas concentrations. In accordance with *Hartmann et al.* [2000], both authors find that ozone reduction acts in the same direction as greenhouse gases, i.e. strengthening the positive index phase of the NAO, but only at an amplitude which is not sufficient to explain observations. *Gillett et al.* [submitted] found that, in the Southern Hemisphere, stratospheric ozone depletion induced a significant change in the Southern Hemisphere Annular Mode [see *Thompson et al.*, this volume] in HadSM3 in austral summer, a result also obtained by *Sexton* [2001]. However, this change was not reflected in the Northern Hemisphere, where circulation changes due to the smaller changes in ozone concentrations were not found to be statistically significant.

3.4. Volcanic Activity

There are only a few studies directly linking the NAO with volcanic eruptions. The interest in this relationship mainly arose from the need to explain the winter warming observed over Northern Hemisphere continents after strong eruptions. The SO_2 injected into the stratosphere during such volcanic events is oxidized within a few weeks to sulphuric acid, which, due to its very low saturation pressure, is rapidly condensed into submicron droplets that interact with solar and terrestrial radiation. The direct radiative effect of these particles leads to surface cooling and warming in the stratosphere lasting up to 3 years. This was clearly demonstrated with radiative-convective models [*Hansen et al.*, 1978], with Energy Balance models [e.g., *Schneider and Mass*, 1975] and also with early GCM studies [*Hunt*, 1977; *Hansen et al.*, 1988; *Rind et al.*, 1992]. In a comprehensive study into the radiative effects of the Mt. Pinatubo eruption, *Stenchikov et al.* [1998] showed that the main forcing of the atmosphere is due to effects in the visible and near-infrared part of the solar spectrum. Here the volcanic aerosol absorbs incoming solar radiation that leads to a deficit at the surface of the order of several Wm^{-2}. Infrared effects at the surface are at least one order of magnitude smaller. However, in the stratosphere the absorption of longwave radiation from below adds to stratospheric warming by as much as absorption of shorter wavelengths.

It was first suggested by *Groisman* [1985] and in earlier Russian literature [see references in *Groisman*, 1985] that over central Russia and North America warm winters occurred after strong volcanic eruptions. Such positive temperature anomalies were also found in a GCM study by *Graf et al.* [1992], who simulated a high latitude eruption by forcing the ECHAM1 model with a mid to high latitude reduction in solar radiation only. In a later "perpetual January" simulation, *Graf et al.* [1993] also included the effect of absorption of longwave radiation and showed, for an idealized stratospheric aerosol anomaly of the Pinatubo eruption (1991), that mid-tropospheric circulation anomalies and surface temperature patterns generated by the model are in very good agreement with observations. *Robock and Mao* [1992] presented a systematic analysis of the global temperature anomalies after all violent eruptions since Krakatoa (1883). They found, after removal of El Niño effects, similar patterns after all 12 eruptions included in their study.

Kodera [1994], *Graf et al.* [1994] and *Kelly et al.* [1996] demonstrated that observed mid-troposphere geopotential height anomalies after eruptions since the 1950s resemble the mid-tropospheric signature of the positive index phase of the NAO in the boreal winter. *Perlwitz and Graf* [1995] and *Kodera et al.* [1996] suggested that this pattern is the positive phase of the leading mode of the coupled tropospheric-stratospheric circulation, which occurs when the stratospheric polar vortex in winter is very strong. The associated temperature pattern at the surface corresponds to that observed after volcanic eruptions by *Robock and Mao* [1992]. *Graf et al.* [1994] suggested that volcanic eruptions exaggerate the positive phase of a naturally occurring variability mode similar to the NAO.

Graf et al. [1993] first proposed a mechanism for the exaggeration of the positive index phase of the NAO by volcanic stratospheric aerosol. They suggested that the absorption of near-infrared solar and longwave terrestrial radiation by the volcanic aerosol leads to strong heating in the stratosphere at low latitudes. Since, over the winter pole, solar radiation is absent and long-wave irradiance from the surface is smaller than at low latitudes, the meridional temperature gradient in the stratosphere is strengthened, which makes the polar vortex of the winter hemisphere stronger and less susceptible to vertically propagating planetary waves. Hence a stronger zonal wind is induced, which further deflects planetary waves equatorwards: a positive feedback. The effect is much stronger in the Northern Hemisphere, since in the Southern Hemisphere the mean state of the polar vortex is already close to radiative equilibrium.

This interpretation was shared by *Kodera et al.* [1996], *Shindell et al.* [1999b] and others. However, the model simulations by *Graf et al.* [1992], which neglected the stratospheric warming effects and only accounted for the reduction in shortwave radiation, showed similar effects in tropospheric circulation and near surface temperature. Hence, the differential heating in the stratosphere is probably not the only possible factor that may induce the positive phase of NAO and a strong polar vortex. *Stenchikov et al.* [submitted] performed a specific experiment series to study this. They studied the response of the NAO, using an EOF-based index, to aerosols and observed ozone changes in the stratosphere after the June 15, 1991 Mount Pinatubo eruption using the SKYHI GCM. An enhanced positive phase of the NAO is reproduced in the model when forced with aerosols. Experiments with albedo changes, but without aerosol absorption and its associated stratospheric heating, show as strong an NAO response as with the total aerosol forcing. *Stenchikov et al.* [submitted] suggested that aerosol stratospheric warming in the tropical lower stratosphere is not the dominant NAO-forcing mechanism. Stratospheric aerosols can also induce tropospheric cooling, which is strongest in low latitudes especially in winter. This could then influence the NAO directly through a tropospheric mechanism, or via the stratosphere through a reduction in the tropospheric meridional temperature gradient, which may lead to a decrease of the mean zonal energy and amplitudes of planetary waves in the troposphere. The corresponding decrease in wave driving in the lower stratosphere may then cause a strengthening of the polar vortex.

3.5. Solar Forcing

Many studies have examined links between 11-yr solar cycle variability and the tropospheric climate, though observed links are generally stronger in the stratosphere. Changes in irradiance are generally largest in the ultra-violet, thus they may be expected to have particularly strong impacts on ozone and stratospheric temperatures. *Kodera* [1995] identified a correlation between the strength of the polar vortex in the Northern Hemisphere winter stratosphere and the 11-year solar cycle, a stronger vortex being associated with higher solar irradiance. *Labitzke and van Loon* [2000] also identified a strengthening of the winter vortex in periods of high solar irradiance.

Several authors have also attempted to examine the impact of solar cycle variability on the atmospheric circulation using GCMs. In a model with prescribed changes in solar irradiance and corresponding stratospheric ozone changes, but limited stratospheric resolution and prescribed sea surface temperatures, *Haigh* [1996] found a slight increase in the strength of the westerlies through most of the Northern Hemisphere extratropical troposphere when solar irradiance was increased. She also found a corresponding decrease in SLP over the Arctic, and an increase further south. In a second study incorporating spectrally resolved changes in solar irradiance and solar-induced changes in the distribution of stratospheric ozone, *Haigh* [1999] found a similar pattern of SLP response, with a decrease over the Arctic, and an increase over the northern mid-latitudes.

Recently, *Shindell et al.* [2001a] argued that model sea surface temperatures must be allowed to adjust in order to properly simulate the tropospheric response to changed solar forcing. By comparing output from two equilibrium integrations of the GISS GCM coupled to a mixed layer ocean model with reconstructed spectrally-resolved solar forcing representative of 1680 and 1780 (Figure 9), they concluded that the main regional climate changes associated with the Maunder Minimum (a period of decreased solar irradiance in the late seventeenth century) were due to a solar-induced shift towards the negative phase of the NAO. They demonstrated that the pattern of temperature response they simulated is similar to that derived by regressing 20-yr lagged paleo-temperature measurements onto a reconstruction of solar forcing. They associated this change with solar-induced changes in tropical sea surface temperatures, and argue that this is why their earlier studies using fixed sea surface temperatures showed no such response [*Shindell et al.*, 1999b; *Shindell et al.*, 2001b].

Thus, overall we conclude that there is considerable evidence that enhanced solar forcing strengthens the winter stratospheric vortex, perhaps because of enhanced heating in the tropical stratosphere due to the associated ozone increases. There is also weaker evidence linking changes in solar irradiance to changes in the NAO, particularly on multidecadal timescales. However, reconstructions of solar irradiance [e.g., *Lean et al.*, 1995] do not show a large change over the past forty years. Thus, even if solar forcing has caused multi-decadal NAO variations in the past, it is unlikely to account for the recent upward trend in the NAO index.

4. SUMMARY

Boreal winter indices of the NAO have shown a positive trend over the last forty years [e.g., *Hurrell et al.,* this volume]. While paleoclimate reconstructions of the NAO index indicate that the recently observed upward trend is rare, but possibly not unique over the past 600 years [*Cook*, this volume], significance testing against a red noise model [*Thompson et al.*, 2000; *Gillett et al.*, 2001], or comparisons with model control variability [*Osborn et al.*, 1999; *Gillett et al.*, 2000] indicate that the trend is outside the range of internal variability. If the trend in the NAO index is not due to internal variability, what then is it due to?

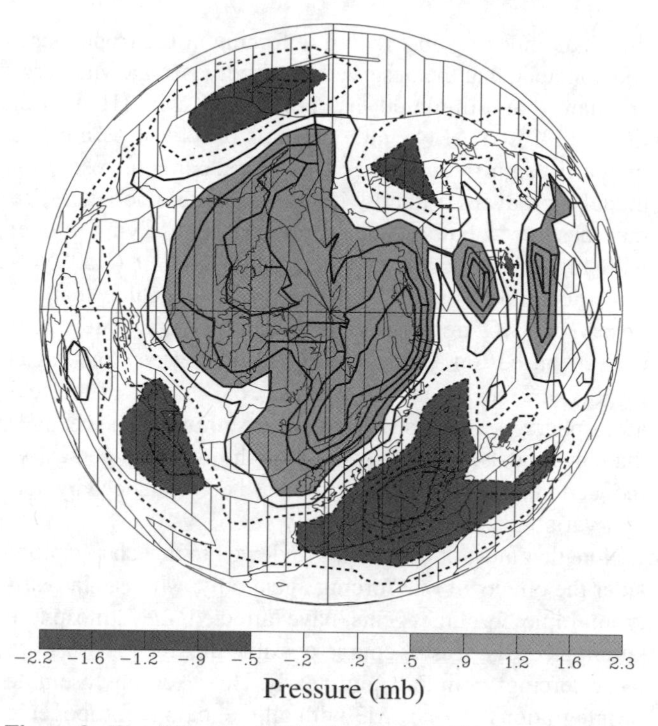

Figure 9. The difference in SLP between equilibrium simulations with spectrally-resolved solar irradiance reconstructed for 1680 and 1780 (higher SLP over the Arctic in 1680). The simulations were performed with the GISS model. Reprinted with permission from *Shindell et al.* [2001a]. Copyright 2001, American Association for the Advancement of Science.

The impacts of various anthropogenic and natural forcing agents on the NAO have been examined in the literature, in particular greenhouse gases, stratospheric ozone, tropospheric sulphate aerosol, volcanic aerosol, and solar irradiance changes. There is considerable disagreement about the relative roles of all these forcings, but most authors agree that greenhouse gases are likely to be at least partly responsible for the long-term trend in the boreal winter NAO index. The influence of prescribed increases in greenhouse gases on the NAO has been examined in all the major climate models. With some exceptions, an increase in the winter NAO index is generally found as greenhouse gas forcing is increased. Some authors argue that the change in the NAO is not limited to a simple increase in its index, but that the locations of the centers of action of the pattern move [*Ulbrich and Christoph*, 1999]. *Gillett et al.* [2002a] regress the NAO index against radiative forcing at the tropopause, and conclude that the observed sensitivity is somewhat larger than that simulated by several GCMs (HadCM3, ECHAM3, ECHAM4), although this difference is not statistically significant. The GISS stratosphere-resolving model reproduces a trend with a magnitude much closer to that observed, but another stratosphere-resolving model with higher horizontal and vertical resolution (HadSM3-L64) does not. Thus, overall, we might conclude that the observed NAO trend is somewhat larger than one might expect based on models forced with greenhouse gas increases. Thus it may be that the NAO index in the real atmosphere is more sensitive to greenhouse gas changes than in models, or it may be that the anthropogenically-induced trend in the observations has been enhanced by natural variability over the past forty years.

Note that most of the studies reviewed in this chapter consider the effects of one forcing agent only, whereas in reality multiple forcing agents have affected the atmosphere simultaneously. It is of course possible that the responses to these forcings combine nonlinearly. However, an ensemble of integrations of HadCM3 with all the main anthropogenic and natural forcings did not reproduce an NAO index trend comparable to that observed [*Stott et al.*, 2000]. Thus, at least in HadCM3, a nonlinear combination of responses does not explain the discrepancy between the simulated and observed NAO trend.

Several mechanisms have been proposed to explain the observed increase in the winter NAO index. Some authors [e.g., *Palmer*, 1999] argue that a system's response to external forcing will always resemble its leading mode of variability. This idea relies on the fact that the processes that enhance a mode of natural variability are likely also to enhance any component of a forced response which projects onto that mode. Thus, the argument suggests that the response of the Northern Hemisphere circulation to anthropogenic forcing is likely to resemble its leading mode of natural variability, the NAO. If this is the case, then it is important first to understand the mechanisms underlying the natural variability of the mode [*Czaja et al.,* this volume; *Thompson et al.*, this volume]. An alternative paradigm would suggest that the mechanisms underlying the natural variability and the forcing response are independent, and that it is a coincidence that the associated SLP patterns are similar.

Shindell et al. [1999a] argue that the upward trend in the NAO index has been forced by changes in the stratospheric circulation. Greenhouse gases warm the tropical upper troposphere and cool the polar stratosphere radiatively at the same height, leading to an enhancement of the equator-to-pole temperature gradient in the tropopause region, strengthening zonal winds in the lower stratosphere and the polar vortex. They also suggest that this effect is augmented by a planetary wave feedback. Although the exact mechanism is not clear, this enhancement of the stratospheric circulation is then presumed to lead to an increase in the strength of the tropospheric westerlies. While a strengthening of both surface and stratospheric westerlies has been observed, some model simulations of the response to increasing concentrations of greenhouse gases show an enhancement of the tropospheric westerlies, but a weakening of the Arctic stratospheric vortex [*Gillett et al.*, 2002b]. This suggests that at least in some models the stratospheric and tropospheric changes may be independent, and the processes responsible for the NAO change may have their origins outside the stratosphere.

One remaining influence to which NAO changes have been attributed is the warming of tropical SSTs. *Hoerling et al.* [2001] were able to reproduce an upward trend in the winter NAO index in a GCM by prescribing observed changes in tropical SSTs. Observed interdecadal variations were also reproduced, though the amplitude of the 12-member ensemble mean changes was only about half of that observed. They ascribe most of the NAO change to warming in the Pacific and Indian Oceans. Since SSTs are closely coupled to temperatures throughout the tropospheric column, it may be that the ocean has simply "recorded" the influence of increasing greenhouse gases, and the temperature of the tropical troposphere could have more directly forced the observed NAO changes.

Overall we conclude that there is considerable evidence that remote forcing from the tropics has played a role in inducing the upward trend in the NAO index. While the state of the stratosphere may influence the surface circulation on sub-annual timescales [*Thompson et al.*, this volume], it remains to be determined whether it has played a significant role in inducing long-term trends in surface climate. We feel that changes in the NAO have been convincingly shown not to be associated with changes in the occupation probabilities of distinct "regimes", but the importance of the physical mechanisms underlying the natural variability of the NAO in determining its response to external forcing remains to be determined. Thus considerable work remains to be done before we fully understand the physical origins of the observed upward trend in the boreal winter NAO index.

Thompson et al. [2000] ascribe 50% of January-March warming over Eurasia over the past thirty years to the trend in the NAO index, and even larger fractions of the trends in precipitation in some regions, particularly the increase in precipitation over Northwest Europe, and the decrease over Southern Europe. *Thompson and Wallace* [2001] also show that the NAO is linked to the occurrence of extremes in a range of surface climate variables around the Northern

Hemisphere, and that its upward trend has contributed to a decrease in the severity of winters over most of this region. Thus, if the NAO index continues to increase in response to enhanced greenhouse gases, it is likely to have significant implications for Northern Hemisphere winter climate. If current climate models adequately resolve the response to anthropogenic forcing of this mode of variability, then these effects will already be incorporated into projections of future climate change. However, if as some authors suggest [e.g., *Shindell et al.*, 1999a], the response of this mode to external forcing is not well-simulated by most GCMs, then predictions of regional Northern Hemisphere climate change are likely to be unreliable. Thus, if we are to make accurate predictions of regional climate change, it is important to ensure that the NAO and its response to external forcings are adequately represented in the models we use.

Acknowledgments. We thank Peter Stott (U.K. Met Office) for the provision of HadCM2 and HadCM3 data, and Dáithí Stone for his comments on the manuscript. We are also grateful for the useful comments and suggestions of four anonymous reviewers. NPG was funded in Oxford by a CASE studentship from the U.K. NERC and Met Office, and in Victoria by NSERC and CFCAS under the Canadian CLIVAR program. TJO was funded by the U.K. NERC (GR3/12107).

REFERENCES

Ambaum, M.H.P., B. J. Hoskins, and D. B. Stephenson, Arctic Oscillation or North Atlantic Oscillation?, *J. Climate, 14*, 3495-3507, 2001.

Baldwin, M. P., and T. J. Dunkerton, Propagation of the Arctic Oscillation from the stratosphere to the troposphere, *J. Geophys. Res.*, *104*, 30 937–30 946 1999.

Bengtsson, L., E. Roeckner, and M. Stendel, Why is the global warming proceeding much slower than expected?, *J. Geophys. Res.*, *104*, 3865–3876, 1999.

Boville, B. A., and P. R. Gent, The NCAR Climate System Model, version one, *J. Climate, 6,* 1115-1130, 1998.

Bretherton, C. S., and D. S. Battisti, An interpretation of the results from atmospheric general circulation models forced by the time history of the observed sea surface temperature distribution, *Geophys. Res. Lett.*, *27*, 767–770, 2000.

Broccoli, A. J., T. L. Delworth, and N. C. Lau, The effect of changes in observational coverage on the association between surface temperature and the Arctic Oscillation, *J. Climate*, *14*, 2481–2485, 2001.

Chung, C. E., V. Ramanathan, and J. T. Kiehl, Effects of the South Asian absorbing haze on the Northeast monsoon and surface-air heat exchange, *J. Climate*, *17*, 2462-2476, 2002.

Cook, E. R., Multi-proxy reconstructions of the North Atlantic Oscillation (NAO) index: A critical review and a new well-verified winter NAO index reconstruction back to AD 1400, this volume.

Corti, S., F. Molteni, and T. N. Palmer, Signature of recent climate change in frequencies of natural atmospheric circulation regimes, *Nature*, *398*, 799–802, 1999.

Cubasch, U., et al., Projections of future climate change, in *Climate Change 2001, The Scientific Basis*, edited by J. T. Houghton, et al., pp. 525–582, Cambridge Univ. Press, 2001.

Czaja, A., A. W. Robertson, and T. Huck, The role of North Atlantic ocean-atmosphere coupling in affecting North Atlantic Oscillation variability, this volume.

Deser, C., On the teleconnectivity of the Arctic Oscillation, *Geophys. Res. Lett.*, *27*, 779-782, 2000.

Emori, S., T. Nosawa, A. Abe-Ouchi, A. Numaguti, M. Kimoto, and T. Nakajima, Coupled ocean-atmosphere model experiments of future climate change with an explicit representation of sulphate aerosol scattering, *J. Meteorol. Soc. Jpn., 77*, 1299-1307, 1999.

Feldstein, S. B., The recent trend and variance increase of the Annular Mode, *J. Climate, 15*, 88-94, 2002.

Flato, G. M., G. J. Boer, W. G. Lee, N. A. McFarlane, G. Ramsden, M. C. Reader, and A. J. Weaver, The Canadian Centre for Climate Modelling and Analysis global coupled model and its climate, *Climate Dyn., 16*, 451-467, 2000.

Folland, C. K., et al., Observed climate variability and change, in *Climate Change 2001, The Scientific Basis*, edited by J. T. Houghton, et al., pp. 99-181, Cambridge Univ. Press, 2001.

Forster, P. T. D., Radiative forcing due to stratospheric ozone changes, 1979-1997, using updated trend estimates, *J. Geophys. Res.*, *104*, 24395–24399, 1999.

Fyfe, J. C., G. J. Boer, and G. M. Flato, The Arctic and Antarctic Oscillations and their projected changes under global warming, *Geophys. Res. Lett.*, *26*, 1601–1604, 1999.

Gillett, N. P., M. R. Allen, R. E. McDonald, C. A. Senior, D. T. Shindell, and G. A. Schmidt, How linear is the Arctic Oscillation response to greenhouse gases?, *J. Geophys. Res.*, *107,* 10.1029/2001JD000589, 2002a.

Gillett, N. P., M. R. Allen, and K. D. Williams, The role of stratospheric resolution in simulating the Arctic Oscillation response to greenhouse gases, *Geophys. Res. Lett.*, *29*(10), 2001GL01444, 2002b.

Gillett, N. P., M. P. Baldwin, and M. R. Allen, Evidence for nonlinearity in observed stratospheric circulation changes, *J. Geophys. Res.*, *106*, 7891–7901, 2001.

Gillett, N. P., G. C. Hegerl, M. R. Allen, and P. A. Stott, Implications of changes in the Northern Hemisphere circulation for the detection of anthropogenic climate change, *Geophys. Res. Lett.*, *27*, 993–996, 2000.

Gordon, C., C. Cooper, C. A. Senior, H. Banks, J. M. Gregory, T. C. Johns, J. F. B. Mitchell, and R. A. Wood, The simulation of SST, sea ice extents and ocean heat transports in a version of the Hadley Centre coupled model without flux adjustments, *Climate Dyn., 16*, 147-168, 2000.

Gordon, H. B., and S. P. O'Farrell, Transient climate change in the CSIRO coupled model with dynamic sea ice, *Mon. Wea. Rev., 125*, 875-907, 1997.

Graf, H. F., Arctic radiation deficit and climate variability, *Climate Dyn.*, *7*, 19–28, 1992.

Graf, H. F., I. Kirchner, and J. Perlwitz, Changing lower stratospheric circulation: The role of ozone and greenhouse gases, *J. Geophys. Res.*, *103*, 11 251–11 261, 1998.

Graf, H. F., I. Kirchner, A. Robock, and I. Schult, Pinatubo eruption winter climate effects - Model versus observations, *Climate Dyn.*, *9*, 81–93, 1993.

Graf, H. F., J. Perlwitz, and I. Kirchner, Northern Hemisphere tropospheric midlatitude circulation after violent volcanic eruptions, *Contrib. Atmos. Phys.*, *67*, 3–13, 1994.

Graf, H. F., J. Perlwitz, I. Kirchner, and I. Schult, Recent northern winter climate trends, ozone changes and increased greenhouse gas forcing, *Contrib. Phys. Atmos.*, *68*, 233–248, 1995.

Groisman, P. Y., Regional climate consequences of volcanic eruptions, *Meteorol. Hydrol.*, *4*, 39–45, 1985.

Haigh, J. D., The impact of solar variability on climate, *Science*, *272*, 981–984, 1996.

Haigh, J. D., A GCM study of climate change in response to the 11-year solar cycle, *Q. J. R. Meteorol. Soc.*, *125*, 871–892, 1999.

Hansen, J., I. Fung, A. Lacis, D. Rind, S. Lebedeff, R. Ruedy, G. Russell, and P. Stone, Global climate changes as forecast by Goddard Institute for Space Studies Three-Dimensional Model, *J. Geophys. Res.*, *93*, 9341–9364, 1988.

Hansen, J. E., W. C. Wang, and A. A. Lacis, Mount Agung eruption provides test of a global climatic perturbation, *Science*, *199*, 1065–1068, 1978.

Hartmann, D. L., J. M. Wallace, V. Limpasuvan, D. W. J. Thompson, and J. R. Holton, Can ozone depletion and global warming interact to produce rapid climate change?, *P. Natl. Acad. Sci.*, *97*, 1412–1417, 2000.

Hoerling, M. P., J. W. Hurrell, and T. Y. Xu, Tropical origins for recent North Atlantic climate change, *Science*, *292*, 90–92, 2001.

Hsu, C. J., and F. Zwiers, Climate change in recurrent regimes and modes of Northern Hemisphere atmospheric variability, *J. Geophys. Res.*, *106*, 20145–20159, 2001.

Hunt, B. G., A simulation of the possible consequences of a volcanic eruption on the general circulation of the atmosphere, *Mon. Wea. Rev.*, *105*, 247–260, 1977.

Hurrell, J. W., Decadal trends in the North Atlantic Oscillation: Regional temperatures and precipitation, *Science*, *269*, 676–679, 1995.

Hurrell, J. W., Influence of variations in extratropical wintertime teleconnections on Northern Hemisphere temperature, *Geophys. Res. Lett.*, *23*, 655–668, 1996.

Hurrell, J. W., Y. Kushnir, G. Ottersen, and M. Visbeck, An overview of the North Atlantic Oscillation, this volume.

Johns, T. C., R. E. Carnell, J. F. Crossley, J. M. Gregory, J. F. B. Mitchell, C. A. Senior, S. F. B. Tett, and R. A. Wood, The second Hadley Centre coupled ocean-atmosphere GCM: model description, spinup and validation, *Climate Dyn.*, *13*, 103-134, 1997.

Jones, P. D., T. J. Osborn, and K. R. Briffa, The evolution of climate over the last millennium, *Science*, *292*, 662–667, 2001.

Jones, P. D., T. J. Osborn, and K. R. Briffa, Pressure-based measures of the North Atlantic Oscillation (NAO): A comparison and an assessment of changes in the strength of the NAO and in its influence on surface climate parameters, this volume.

Kelly, P. M., P. Jia, and P. D. Jones, The spatial response of the climate system to explosive volcanic eruptions, *Intl. J. Climatol.*, *16*, 537–550, 1996.

Kodera, K., Influence of volcanic eruptions on the troposphere through stratospheric dynamical processes in the Northern Hemisphere winter, *J. Geophys. Res.*, *99*, 1273–1282, 1994.

Kodera, K., On the origin and nature of the interannual variability of the winter stratosphere circulation in the Northern Hemisphere, *J. Geophys. Res.*, *100*, 14077–14087, 1995.

Kodera, K., M. Chiba, H. Koide, A. Kitoh, and Y. Nikaidou Y, Interannual variability of the winter stratosphere and troposphere in the Northern Hemisphere, *J. Meteorol. Soc. Jpn.*, *74*, 365–382, 1996.

Labitzke, K., and H. van Loon, The QBO effect on the solar signal in the global stratosphere in the winter of the Northern Hemisphere, *J. Atmos. Sol.-Terr. Phys.*, *62*, 621–628, 2000.

Latif, M., K. Arpe, and E. Roeckner, Oceanic control of decadal North Atlantic sea level pressure variability in winter, *Geophys. Res. Lett.*, *27*, 727–730, 2000.

Lean, J., J. Beer, and R. Bradley, Reconstruction of solar irradiance since 1610: Implications for climate change, *Geophys. Res. Lett.*, *22*, 3195–3198, 1995.

Mehta, V. M., M. J. Suarez, J. V. Manganello, and T. L. Delworth, Oceanic influence on the North Atlantic Oscillation and associated Northern Hemisphere climate variations: 1959–1993, *Geophys. Res. Lett.*, *27*, 121–124, 2000.

Osborn, T. J., The winter North Atlantic Oscillation: roles of internal variability and greenhouse gas forcing. *CLIVAR Exchanges*, *25*, 54-58, 2002.

Osborn, T. J., K. R. Briffa, S. F. B. Tett, P. D. Jones, and R. M. Trigo, Evaluation of the North Atlantic Oscillation as simulated by a coupled climate model, *Climate Dyn.*, *15*, 685–702, 1999.

Paeth, H., A. Hense, R. Glowienka-Hense, S. Voss, and U. Cubasch, The North Atlantic Oscillation as an indicator for greenhouse-gas induced regional climate change, *Climate Dyn.*, *15*, 953–960, 1999.

Palmer, T. N., A nonlinear dynamical perspective on climate change, *Weather*, *48*, 314–326, 1993.

Palmer, T. N., A nonlinear dynamical perspective on climate prediction, *J. Climate*, *12*, 575–591, 1999.

Perlwitz, J., and H. F. Graf, The statistical connection between tropospheric and stratospheric circulation of the Northern Hemisphere in winter, *J. Climate*, *8*, 2281–2295, 1995.

Ramaswamy, V., et al., Radiative forcing of climate change, in *Climate Change 2001, The Scientific Basis*, edited by J. T. Houghton, et al., pp. 349–416, Cambridge Univ. Press, 2001.

Ramaswamy, V., M. D. Schwarzkopf, and W. J. Randel, Fingerprint of ozone depletion in the spatial and temporal pattern of recent lower-stratospheric cooling, *Nature*, *382*, 616–618, 1996.

Randel, W. J., and F. Wu, Cooling of the Arctic and Antarctic polar stratospheres due to ozone depletion, *J. Climate*, *12*, 1467–1479, 1999.

Rind, D., N. K. Balachandran, and R. Suozzo, Climate change and the middle atmosphere. Part II: The impact of volcanic aerosols, *J. Climate*, *5*, 189-220, 1992.

Rind, D., R. Suozzo, N. K. Balachandran, and M. J. Prather, Climate change and the middle atmosphere. Part I: The doubled CO_2 climate, *J. Atmos. Sci.*, *47*, 475–494, 1990.

Robertson, A. W., Influence of ocean-atmosphere interaction on the Arctic Oscillation in two general circulation models, *J. Climate*, *14*, 3240–3254, 2001.

Robock, A., and J. P. Mao, Winter warming from large volcanic eruptions, *Geophys. Res. Lett.*, *19*, 2405–2408, 1992.

Rodwell, M. J., D. P. Powell, and C. K. Folland, Oceanic forcing of the wintertime North Atlantic Oscillation and European climate, *Nature*, *398*, 320–323, 1999.

Schnadt, C., M. Dameris, M. Ponater, R. Hein, V. Grewe, and B. Steil, Interaction of atmospheric chemistry and climate and its impact on stratospheric ozone, *Climate Dyn.*,*18*, 501-517, 2002.

Schneider, S. H., and C. Mass, Volcanic dust, sunspots and temperature trends, *Science*, *190*, 741–746, 1975.

Sexton, D. M. H., The effect of stratospheric ozone depletion on the phase of the Antarctic Oscillation, *Geophys. Res. Lett.*, *28*, 3697–3700, 2001.

Shindell, D. T., R. L. Miller, G. A. Schmidt, and L. Pandolfo, Simulation of recent northern winter climate trends by greenhouse-gas forcing, *Nature*, *399*, 452–455, 1999a.

Shindell, D. T., D. Rind, N. K. Balachandran, J. Lean, and P. Lonergan, Solar cycle variability, ozone, and climate, *Science*, *284*, 305–308, 1999b.

Shindell, D. T., D. Rind, and P. Lonergan, Increased polar stratospheric ozone losses and delayed eventual recovery owing to increasing greenhouse-gas concentrations, *Nature, 392*, 589-592, 1998.

Shindell, D. T., G. A. Schmidt, M. E. Mann, D. Rind, and A. Waple, Solar forcing of regional climate change during the Maunder Minimum, *Science*, *294*, 2149–2152, 2001a.

Shindell, D. T., G. A. Schmidt, R. L. Miller, and D. Rind, Northern Hemisphere winter climate response to greenhouse gas, ozone, solar, and volcanic forcing, *J. Geophys. Res.*, *106*, 7193–7210, 2001b.

Stenchikov, G. L., I. Kirchner, A. Robock, H. F. Graf, J. C. Antuna, R. G. Grainger, A. Lambert, and L. Thomason, Radiative forcing from the 1991 Mount Pinatubo volcanic eruption, *J. Geophys. Res.*, *103*, 13837–13857, 1998.

Stone, D. A., A. J. Weaver, and R. J. Stouffer, Projection of climate change onto modes of atmospheric variability, *J. Climate*, *14*, 3551–3565, 2001.

Stott, P. A., S. F. B. Tett, G. S. Jones, M. R. Allen, J. F. B. Mitchell, and G. J. Jenkins, External control of 20th century temperature by natural and anthropogenic forcings, *Science*, *290*, 2133–2137, 2000.

Stouffer, R. J., and S. Manabe, Response of a coupled ocean-atmosphere model to increasing atmospheric carbon dioxide: sensitivity to the rate of increase, *J. Climate, 12*, 2224-2237, 1999.

Tett, S. F. B., P. A. Stott, M. R. Allen, W. J. Ingram, and J. F. B. Mitchell, Causes of twentieth century temperature change, *Nature*, *399*, 569–572, 1999.

Thompson, D. W. J., S. Lee, and M. P. Baldwin, Atmospheric processes governing the Northern Hemisphere Annular Mode/North Atlantic Oscillation, this volume.

Thompson, D. W. J., and J. M. Wallace, The Arctic Oscillation signature in the wintertime geopotential height and temperature fields, *Geophys. Res. Lett.*, *25*, 1297–1300, 1998.

Thompson, D. W. J., and J. M. Wallace, Regional climate impacts of the Northern Hemisphere Annular Mode, *Science*, *293*, 85–89, 2001.

Thompson, D. W. J., J. M. Wallace, and G. C. Hegerl, Annular modes in the extratropical circulation. Part II: Trends, *J. Climate*, *13*, 1018–1036, 2000.

Trenberth, K. E., and J. W. Hurrell, Comments on "The interpretation of short climate records with comments on the North Atlantic and Southern Oscillations," *Bull. Am. Meteorol. Soc.*, *80*, 2721–2722, 1999.

Ulbrich, U., and M. Christoph, A shift of the NAO and increasing storm track activity over Europe due to anthropogenic greenhouse gas forcing, *Climate Dyn.*, *15*, 551–559, 1999.

Volodin, E. M., and V. Y. Galin, Interpretation of winter warming on Northern Hemisphere continents in 1977-94, *J. Climate*, *12*, 2947–2955, 1999.

Voss, R., R. Sausen, and U. Cubasch, Periodically synchronously coupled integrations with the atmosphere-ocean general circulation model ECHAM3/LSG, *Climate Dyn., 14*, 249-266, 1998.

Washington, W. M., et al., Parallel climate model (PCM) control and transient simulations, *Climate Dyn., 16*, 755-774, 2000.

Williams, K. D., C. A. Senior, and J. F. B. Mitchell, Transient climate change in the Hadley Centre models: The role of physical processes, *J. Climate*, *14*, 2659–2674, 2001.

Wunsch, C., The interpretation of short climate records, with comments on the North Atlantic and Southern Oscillations, *Bull. Am. Meteorol. Soc.*, *80*, 245–255, 1999.

Zorita, E., and F. González-Rouco, Disagreement between predictions of the future behavior of the Arctic Oscillation as simulated in two different climate models: Implications for global warming, *Geophys. Res. Lett.*, *27*, 1755–1758, 2000.

N. P. Gillett, School of Earth and Ocean Sciences, University of Victoria, PO Box 3055, Victoria, BC, V8W 3P6, Canada.
gillett@uvic.ca

H. F. Graf, Max-Planck-Institut für Meteorologie, Bundesstrasse 55, 20146 Hamburg, Germany.
graf@dkrz.de

T. J. Osborn, Climatic Research Unit, University of East Anglia, Norwich NR4 7TJ, UK.
t.osborn@uea.ac.uk

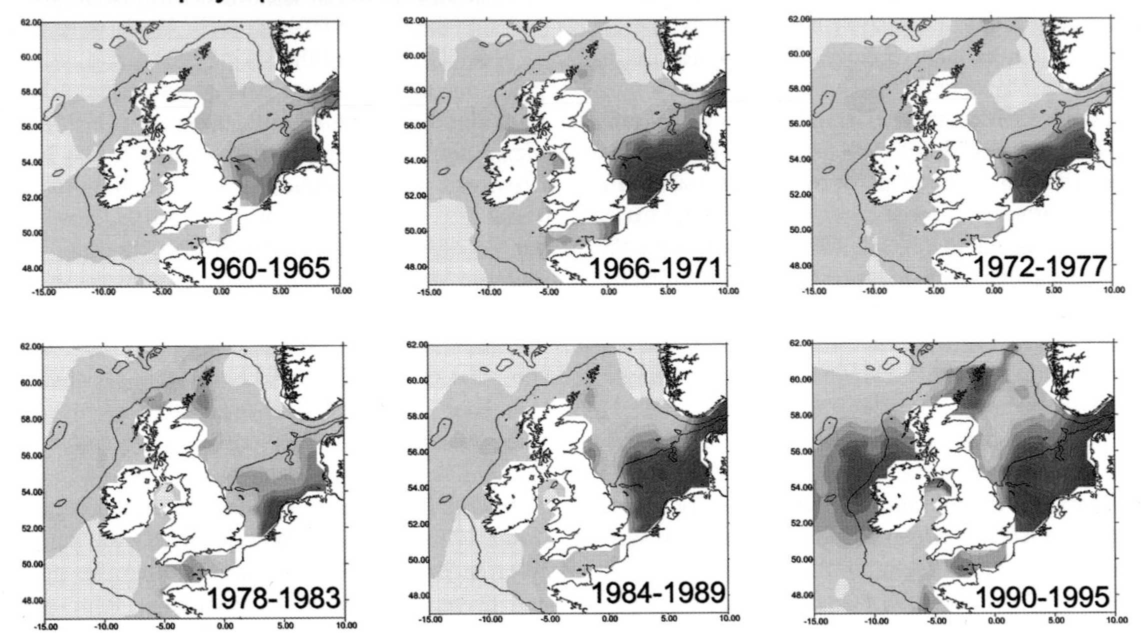

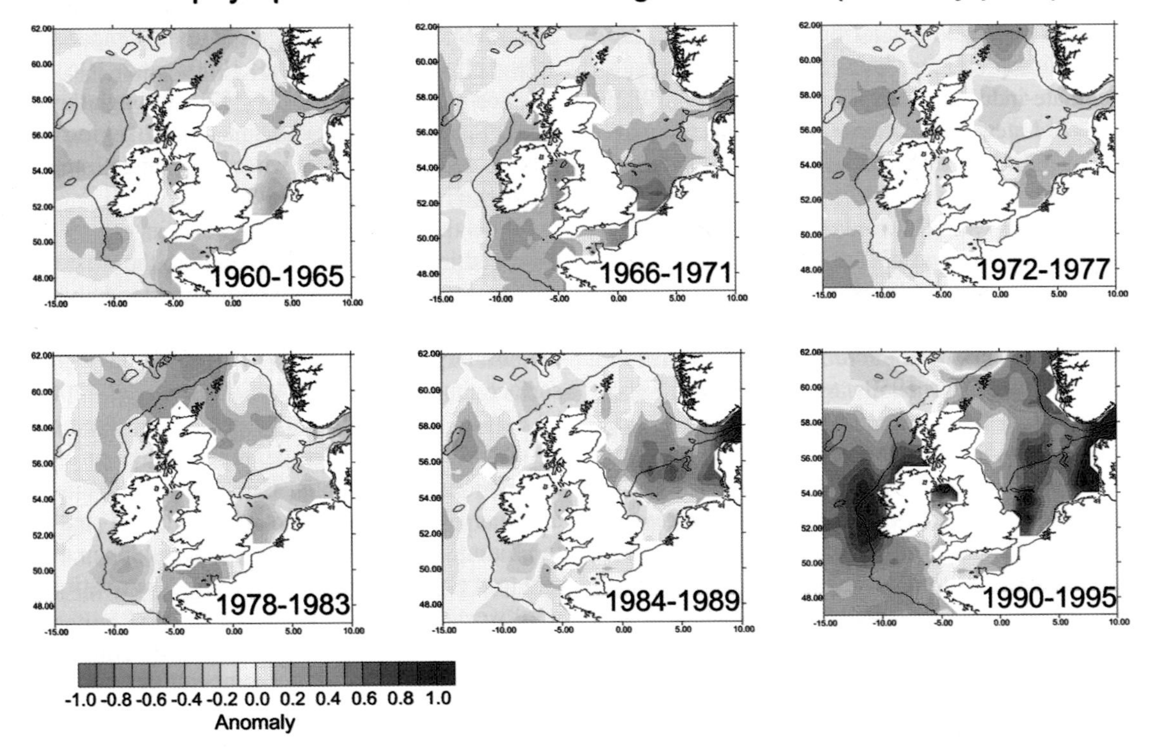

Plate 1. Geostatistical estimates of the mean spatial distribution of phytoplankton color in six-year periods from 1960-1995 (top six panels). Anomaly maps show the mean spatial distribution of phytoplankton color minus the long-term mean (bottom six panels). From *Edwards* [2000].

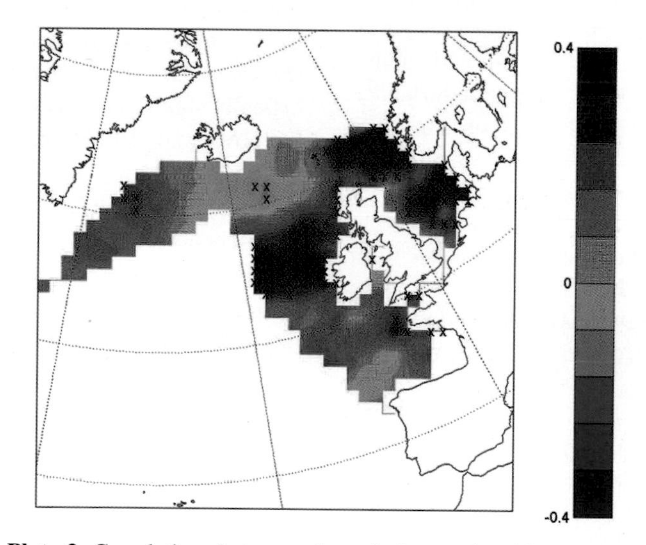

Plate 2. Correlations between phytoplankton color (biomass) and the NAO index. The crosses indicate significant correlations (p=≤0.01). From *Edwards* [2000].

lead to an earlier spring bloom. While an earlier bloom could not be detected from an analysis of the CPR data in the Northeast Atlantic, the monthly temporal resolution was considered insufficient to detect the expected small change [*Edwards et al.*, 2001]. An earlier stratification or increased intensity in stratification also would be expected to lead to a change in the phytoplankton community structure. Indeed, there is evidence from the CPR data that this occurred. Dinoflagellates appeared earlier and were more abundant in the 1990s than in the 1960s [*Edwards et al.*, 2001]. They also disappeared earlier in the year, with the result that their duration remained relatively unchanged. *Edwards et al.* [2001] also observed that long-term trends in dinoflagellate abundance in the Northeast Atlantic and the North Sea were positively correlated with the NAO index, whereas diatom abundance was negatively correlated. This provides nominal support for the contention that the temperature influence on the plankton may be through its influence on stratification. This result contrasts with that of *Irigoien et al.* [2000], however, who found a positive relationship of the NAO index with May-April concentrations of diatoms in a 50 m deep station off Plymouth in the English Channel from 1993–1999. They argued that the deeper mixed layer caused by high wind stresses during a high NAO index and reduced light from increased cloud cover would tend to favor diatoms. The cause of the differences between the two studies remains unclear.

Barton et al. [2002] have examined the color index from the CPR data throughout the North Atlantic. They noted an apparent relationship between the color index and the NAO index but found that the detrended time series were not significantly correlated. They suggested that the long-term trends in the phytoplankton color index are likely linked to the NAO index, but the correlation of the two time series at higher frequencies are not strong.

NAO-induced variability also extends into the coastal embayments. One of the best studied is Gullmar Fjord on the west coast of Sweden. *Lindahl et al.* [1998] found that primary production at a monitoring site at the mouth of the fjord was positively related to the NAO. They suggested that stronger winds during a positive NAO index led to an increase in the northward transport of low salinity waters from the Kattegat area. These waters enhance the nutrient concentrations at the mouth of the fjord, leading to higher primary production. *Belgrano et al.* [1999] showed that the NAO index could account for approximately 45% of the variance in the May primary production and 63% in the population growth between April and May in Gullmar Fjord. In a statistical modelling study of primary production in the Fjord, *Belgrano et al.* [2001] found that the NAO was one of the significant explanatory climate variables at a lag of one month. The other variables included nutrient concentrations and stratification related indices such as density, precipitation and winds.

During the late 1980s, large phytoplankton blooms were observed along the Swedish coast with the main species being toxic flagellates. These blooms coincided with a positive NAO index and, for the period 1986–1996, the NAO index accounted for over 90% of the variability in the abundance of the three toxic *Dinophysis* species found in the fjord [*Belgrano et al.*, 1999]. The NAO also affects the plankton biodiversity in Gullmar Fjord. Biodiversity of a community is usually measured as the number of species or species richness, S, and their relative abundances known as species evenness, E [*Magurran*, 1988]. Adding the ln(E) to the ln(S) produces the biodiversity information function, H [*Buzas and Hayek*, 1996]. As part of the present study, a statistically significant positive correlation was found between the NAO and H for Gullmar Fjord. Between the early and late 1990s there was a noticeable shift from larger to smaller phytoplankton in Gullmar Fjord, especially an increase in dinoflagellates. This corresponded in time to a change from very high NAO to low NAO index values and is a similar response to that reported by *Edwards et al.* [2000] for the North Sea and the Northeast Atlantic but opposite to that reported by *Irigoien et al.* [2000] in the English Channel.

On the other side of the Atlantic, changes in the phytoplankton species composition at Narragansett Bay have also been linked to climate variations [*Karentz and Smayda*, 1998; *Smayda,* 1998]. Major changes were observed among the boreal-Arctic diatom bloom species, *Detonula confervacea* and *Thalassiosira nordenskioeldii*, and the diatoms, *Skeletonema costatum* and *Asterionellopsis glacialis*. The shifts in species abundance were in part related to a warming trend in annual SST beginning in the early 1960s associated with the increasing positive trend in the NAO index, as well as to increasing copepod grazing pressure [*Durbin and Durbin*, 1992].

Primary production determined from C^{14} measurements at a site southeast of Bermuda was negatively related to the NAO index [*Bates*, 2001]. In this location, during periods of low NAO index, storm tracks shift southward, more frequent outbreaks of cold air come off North America, surface waters cool, and the winter mixed layer deepens [*Dickson et al.*, 1996]. *Bates* [2001] suggested that the deeper mixed layer during low NAO index years led to elevated nutrient levels in the near-surface layer, which ultimately led to higher primary production. This argument linking winds and primary production is opposite that often given for the more northern areas where increased mixing has been suggested as leading to reduced production. This difference was explained by *Dutkiewicz et al.* [2001] from modelling studies. They showed that the relationship between mixing and production depends upon the ratio of the Sverdrup's critical depth in spring (h_c) to the mixed layer depth (h_m) at the end of the winter. Where h_c/h_m is approximately 1, such as in the subtropics, strong mixing does lead to enhanced production through nutrient enhancement, but in subpolar regions where the ratio is <<1, increased mixing should lead to lower production because of reduced light levels.

The above examples have focused upon linkages between the phytoplankton and the NAO through air-sea fluxes. However, the phytoplankton community is also affected by circulation changes related to variability in the NAO index. For example, during the late 1980s when the NAO index was strongly positive and the strength of the westerly winds increased in the Northeast Atlantic, there was an increase of oceanic inflow into the North Sea. Subsequently, the phytoplankton biomass in the North Sea was the highest ever recorded [*Edwards et al.*, 2001]. Phytoplankton biomass in 1989 was nearly three standard deviations above its long-term mean. At this time there was also an influx of unprecedented numbers of oceanic species into the North Sea, including the diatom *Thalassiothrix longissima* and a succession of short-lived exceptional phytoplankton blooms occurred. Many of the phytoplankton species recorded their peak abundance one to three months in advance of their normal seasonal peak. The inflowing Atlantic water is thought to have possibly brought a pulse of oceanic-derived nutrients, which when coupled with very mild atmospheric conditions, produced an exceptionally favorable period for phytoplankton growth [*Edwards et al.*, 2001].

3. ZOOPLANKTON

3.1. Background

Zooplankton are small, passively floating animals that generally feed on phytoplankton, or in some cases, other smaller zooplankton. They are important food for most fish larvae and some small adult fish. There are numerous species of zooplankton, each adapted to particular oceanographic characteristics. In the North Atlantic, one of the most important zooplankton species is *Calanus finmarchicus* because of its pan-Atlantic distribution and large biomass. It has been estimated to reach up to 92% of the total zooplankton biomass in Icelandic waters [*Gislason and Astthorsson*, 1995]. In the Northwest Atlantic, on continental shelves off the United States, this species, together with *Pseudocalanus* spp. and *Centropages typicus*, account for 75% of the total zooplankton abundance [*Sherman et al.*, 1987; 1998]. In addition, *C. finmarchicus* has been identified as the prominent prey for fish larvae such as cod and haddock in the Barents Sea [*Helle*, 1994], off Iceland [*Astthorsson and Gislason*, 1995] and in U.S. waters [*Marak*, 1960; *Sherman et al.*, 1981]. The abundance of herbivorous (phytoplankton eating) zooplankton, such as *C. finmarchicus*, usually follows a similar cycle to that of the local phytoplankton production cycle but lagged slightly. Carnivorous zooplankton, which eat other smaller zooplankton, tend to be tuned to the abundance cycles of their particular prey. Zooplankton generally produce one to several generations per year and many species diapause through the winter. Some, such as *C. finmarchicus*, spend this diapause period in deep waters in the open ocean. Awaking in early spring, they ascend to the surface where they feed on the spring bloom, mate and reproduce.

During the last 20 years there has been increasing interest among marine scientists in understanding the relationship between zooplankton and climate because of the fact that most marine fish and invertebrates feed-on zooplankton during some stage of their life. Early studies documented large fluctuations in zooplankton in the Northeast Atlantic and the North Sea from the CPR records [*Colebrook*, 1972; *Gieskes and Kraay*, 1977]. Long-term patterns of variability in the zooplankton were found to be similar over large geographic areas of the North Atlantic, and often across trophic states [*Colebrook et al.*, 1984; *Aebischer et al.*, 1990; *Cushing*, 1990]. The magnitude of the spatial scale was the first suggestion that hydrographic or climatic factors most likely drive interannual variability in these marine species.

3.2. Relationship with NAO

Fromentin and Planque [1996] were the first to document linkages between NAO variability and interannual changes in zooplankton in the North Atlantic. They found significant correlation between the NAO index and two major copepod species in the eastern Northeast Atlantic and the North Sea. A negative relationship (R^2 = 0.58) with *C. finmarchicus* was attributed to the association of the NAO to the wind and SST. It was argued that higher wind stresses under a high NAO index lead to reduced stratification, lower phytoplankton, and ultimately lower abundance of zooplankton because of less food. The warmer temperatures were also considered to inhibit production of this cold-water species. A weaker but positive relationship (R^2 = 0.28) of the NAO was found with the abundance of *C. helgolandicus* with a delay of one year. Differences between the responses of the two species were attributed to differences in seasonal cycles, temperature affinities and geographical locations coupled with the spatial heterogeneity in the wind and temperature responses to NAO forcing and competition between the two species.

Planque and Reid [1998], based upon the results of *Fromentin and Planque* [1996], generated a regression model of the abundance of *C. finmarchicus* from the NAO using data from 1958–1992 (Figure 1). They went on to predict the abundances for the years 1993 to 1997. There was reasonable agreement between predicted and observed abundances for 1993–1995, but when the NAO index dropped precipitously in 1996, they predicted a corresponding increase in the *C. finmarchicus* abundances. The observed abundance, on the other hand, fell to its lowest value in the entire time series. Acknowledging a lack of understanding as to why such a large discrepancy existed, they speculated that it might have been due to very low overwintering stocks. They suggested this could have been a consequence of a prolonged period of poor years or a lag response in the establishment of circulation to colonize favorable feeding grounds following the new phase of the NAO. Another possibility is related to the unusual eastward shift in the position of both the Icelandic Low and Azores High in 1996, which might have caused a different response than under normal low NAO index conditions when the pressure systems were located more towards the central North Atlantic.

Several studies have shown a strong positive relationship between the interannual variability in the latitudinal position of the northern edge or "wall" of the Gulf of Stream and zooplankton indices around the United Kingdom, including the central North Sea [*Taylor and Stephens*, 1980; *Taylor et al.*, 1992; *Hays et al.*, 1993; *Taylor*, 1995]. The lack of any lag and similar relationships between the Gulf Stream and zooplankton from a freshwater lake (Windermere) in the northwest of England lead to the conclusion that the connection must be via the atmosphere. That link was later shown to be through the NAO. *Taylor and Stephens* [1998] noted a strong positive relationship between the NAO index and the Gulf Stream position such that the "north wall" was located farther north following high NAO index years and farther south when the NAO index is low. *Curry and McCartney* [2001] found that the Gulf Stream transport covaries with its position such that a more northward location occurs during increased transport and a southward location when the Stream weakens. They too noted the positive relationship between the NAO index and the Gulf Stream transport, and therefore its position. *Planque and Taylor* [1998] suggested that the relationship between the NAO (and hence the Gulf Stream) and zooplankton around the U.K. was due to several mechanisms. In the case of *C. finmarchicus*, these included direct advection of zooplankton into the North Sea through increased Atlantic inflow as well as changes in stratification and temperature that affected the timing and intensity of the spring phytoplankton bloom and subsequently the zooplankton.

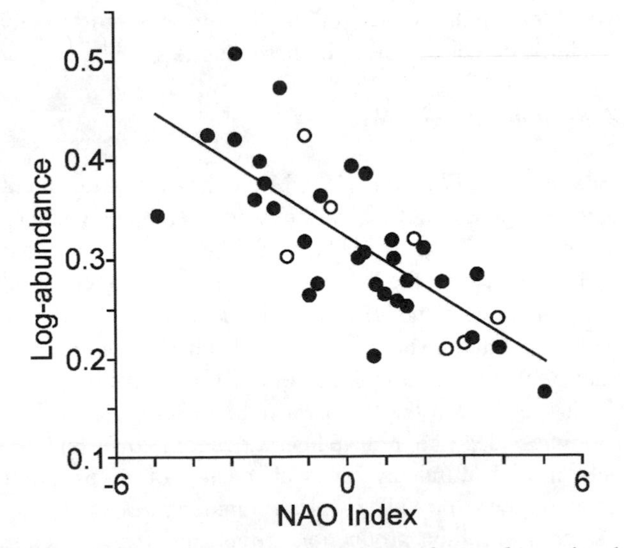

Figure 1. Log-abundance of *Calanus finmarchicus* in the Northeast Atlantic against the NAO winter index for the period 1958-1995 used to generate the linear regression model to predict the *Calanus* abundances. Open circles indicate data for 1958-1961 and 1993-1995 that were not included in the analysis by *Fromentin and Planque* [1996]. Taken from *Planque and Reid* [1998].

Studies of zooplankton along the CPR transect that crosses the Bay of Biscay, the Celtic Sea and the English Channel during 1979–1995 were carried out by *Beaugrand et al.* [2000]. They found a negative relationship between the NAO index and the zooplankton (primarily copepod) abundance in the English Channel, similar to that for *C. finmarchicus* in the North Sea and Northeast Atlantic found by *Fromentin and Planque* [1996]. *Beaugrand et al.* [2000] argued that the stronger winds associated with a high NAO index produce greater wind-induced turbulence leading to more intense vertical mixing. While higher levels of turbulence increase contact rates between the zooplankton and its prey [*Rothschild and Osborn*, 1988] and hence feeding rates [*Mackenzie and Leggett*, 1991; *Sundby et al.*, 1994], they also increase metabolic rates [*Alcaraz et al.*, 1994]. High metabolic rates also result from the higher temperatures associated with a high NAO index. *Beaugrand et al.* [2000] proposed that the increased feeding does not compensate for these higher metabolic costs. The resulting lower net energy consumption translates into higher mortality of adult copepods and lower fecundity, implying fewer eggs and a delay in their spring production. No relationship was found between the NAO and zooplankton in the Celtic Sea or the Bay of Biscay. Although not commented on by *Beaugrand et al.* [2000], this may be due to the observed southward reduction in the effect of the NAO on the atmospheric and physical oceanic variables in this region as discussed by *Planque and Taylor* [1998].

NAO-associated changes in current transport have also been suggested as a control on zooplankton abundances. Persistent anomalies in the wind field associated with the NAO causes variability in the direction and strength of surface currents. Off northern Norway, a high NAO index is associated with stronger northeastward flow of warmer Atlantic water into the Barents Sea [*Ådlandsvik and Loeng*, 1991; *Dickson et al.*, 2000]. *Helle and Pennington* [1999] have shown that in this region, the abundance of zooplankton (primarily *C. finmarchicus*) is positively related to the volume of the Atlantic inflow and that this effect is propagated through the food web up to at least juvenile cod. A similar relationship between zooplankton and Atlantic inflow is suggested for the North Sea [*Planque and Taylor*, 1998; *Stephens et al.*, 1998; *Gallego et al.*, 1999; *Heath et al.*, 1999; *Edwards et al.*, 1999].

Variability in both the extent of the large-scale convection and the circulation of the deep waters in the North Atlantic are associated with the NAO [*Dickson*, 1997]. Such changes appear to affect zooplankton populations over long time scales. The North Sea population of *C. finmarchicus* is seeded from the Faroe-Shetland Channel, where it overwinters in Norwegian Sea Deep Water. Changes in convective intensity over the latter decades of the 1900s have been associated with a decrease of the volume of Norwegian Sea Deep Water [*Schlosser et al.*, 1991; *Hansen et al.*, 2001]. *Heath et al.* [1999] proposed that this reduction in the volume of *C. finmarchicus* overwintering habitat is partly responsible for the decline in abundance of the species observed in the North Sea since the late 1950s. They further suggested that such changes occur on decadal time scales but that interannual fluctuations in the NAO, which are accompanied by immediate changes in the northwesterly winds, do not necessarily lead to corresponding changes in the abundance of *C. finmarchicus* in the North Sea.

Beare and McKenzie [1999] noted significant changes in the seasonality of both the NAO and the abundance of stage 5 and 6 of *C. finmarchicus* in the North Sea off northeastern Scotland from the CPR surveys beginning in 1967 (Figure 2). The *C. finmarchicus* population densities were relatively high between 1958 and 1965 but collapsed in 1967 and did not recover through to the mid-1990s. Associated with this collapse was a change in the bimodal pattern of abundance. The earlier high abundances were associated with a dominant spring peak whereas after the collapse, the autumn peak dominated until the mid-1980s. Using the NAO index of *Jones et al.* [1997], *Beare and McKenzie* [1999] noted increasing abundance in winter and decreasing in spring and they contended that this was a major contributor to the collapse of *C. finmarchicus* in 1967 although they were unable to provide the mechanism. They speculated that it might be related to changes in physical conditions or currents in April when these copepods are ascending to the surface waters after their winter diapause.

Conversi et al. [2001] investigated the relationship between the NAO index and *C. finmarchicus* (copepodites 5 and adults) abundance from the CPR records in the Gulf of Maine for the period 1961-1991. They found a positive relationship at all scales studied. The seasonal cycle of *C. finmarchicus* showed a clearer cycle (less variability) and there was higher overall abundance during high NAO versus low NAO index years. There was a 30-year increasing linear trend (accounting for 39% of the total variability of *C. finmarchicus* abundance), confirming the pattern described by *Jossi and Goulet* [1993]. Similar increasing trends were observed in the Gulf of Maine SST (39% of total variability) and the NAO index (20% of total variability). Cross correlation of detrended series showed the variations in the NAO index

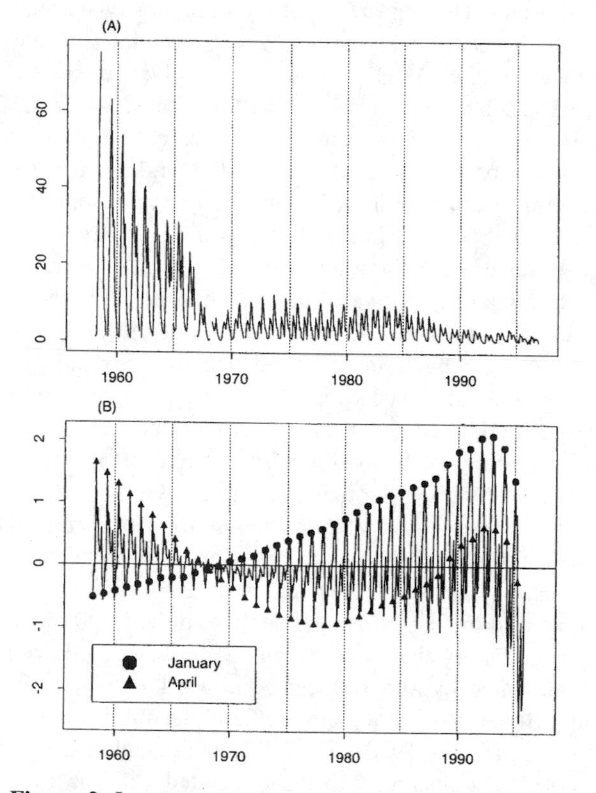

Figure 2. Long-term and seasonal changes in (a) *Calanus finmarchicus* abudance to the north-west of Scotland (57° to 61.4°N, 0° to 4°W) and (b) the North Atlantic Oscillation index. Both time series were analyzed using decomposition techniques that separate the long-term trend and seasonal structure. Taken from *Beare and McKenzie* [1999].

were positively correlated with and preceded the winter SST fluctuations by 2 years (R^2 = 0.22) and the *C. finmarchicus* summer abundance by 4 years (R^2 = 0.25). Winter sea surface temperatures preceded summer *C. finmarchicus* by 2 years (R^2 = 0.38). A stepwise regression indicated that the winter SST alone accounted for a third of the observed variability in *C. finmarchicus* summer abundance.

Greene and Pershing [2000], using GLOBEC (GLOBal ocean ECosystems dynamics) field data, found that *C. finmarchicus* abundance in the Gulf of Maine was related to variations in bottom temperature. They suggested this was because the species is tightly coupled to the conditions of the Slope Water [see also Marine Ecosystem Response to Climate in the North Atlantic *MERCINA*, 2001]. Slope Water occupies the area between the continental shelf and the Gulf Stream and penetrates into the Gulf at depth through the deep gullies and channels. The Slope Water characteristics adjacent to the Gulf vary on time scales of years to decades between cold (4°–8°C), fresh conditions indicative of a northern (Labrador Current) origin and warm (8°–12°C), salty waters of a southern (North Atlantic Current) origin. The appearance of the Labrador-type Slope Water off the Gulf of Maine appears to be related to the NAO [*Worthington*, 1964; *Marsh et al.*, 1999; *Drinkwater et al.*, 1999]. During the decade of the 1960s, when the NAO index was predominantly negative, cold Labrador Slope Water extended as far south as the Middle Atlantic Bight. As a result, both bottom water temperatures and *C. finmarchicus* abundance in the Gulf of Maine were relatively low. In contrast, during the 1980s when the NAO index was predominantly positive, warm Slope Water lay off the shelf, the temperature of the lower layers of the Gulf was warm, and *C. finmarchicus* abundances were relatively large. The Labrador-type Slope Water during this period was largely confined to an area north of the Laurentian Channel. During each southward intrusion of Labrador Slope Water along the Scotian Shelf to the Gulf of Maine after 1980, *C. finmarchicus* abundance in the Gulf declined in subsequent years. For example, the very low abundances of *C. finmarchicus* during 1998 and early 1999 are believed to be linked to the largest annual decline in the NAO index in the past century that occurred in 1996. It was followed two years later by the furthest southward extension of Labrador Slope Water along the shelf and the coldest waters in the Gulf since the 1960s. The CPR data revealed several arctic boreal zooplankton species associated with the 1998 southward excursion of the Labrador Slope Water, including the southern most record of *C. hyperboreus* [*Johns et al.*, 2001].

The differences and similarities in the relationship between the NAO index and *C. finmarchicus*, as observed in the western Gulf of Maine and the eastern Atlantic including and North Sea are striking. *C. finmarchicus* abundance in the Northeast Atlantic tends to be lower during high NAO index years and the overall 30-year trend of *Calanus* is downward while that of the NAO index is upward. In the Gulf of Maine, the opposite occurs as the relationship between the NAO index and *Calanus* abundance is positive. To verify whether the relationship between the NAO index and *C. finmarchicus* abundance found in the eastern and western North Atlantic corresponded to opposite patterns in the long-term variations of the copepods' populations, Conversi, Licandro and Ibanez (unpublished data) compared *C. finmarchicus* time series sampled on the opposite sides of the ocean. CPR abundance data of *C. finmarchicus,* stages 5–6, from the Gulf of Maine and from the Northern North Sea (area A2), over the 38 year period 1961–1998 were converted to the same units (ln(counts)/10 n.mile). The first modes of temporal variations of the two series represent, respectively, 25.8% of the temporal variability of *Calanus* populations in the Gulf of Maine and 26.4% off western Ireland and represent the general trends of the series. An out-of-phase relationship becomes evident when comparing the first mode of *C. finmarchicus* in the Gulf of Maine with that in the North Sea. Generally, years when *C. finmarchicus* increased in the western Atlantic corresponded to years in which it decreased in the North Sea, and vice versa. The correlation (r = -0.28) between detrended series for the period Feb. 1978 – Dec. 1998 (a period with reduced number of missing data) confirms the inverse relationship but indicates that it is relatively weak. The year-to-year changes in *C. finmarchicus* abundance that are coupled across the North Atlantic are considered to be a result of atmospheric forcing linked to the NAO.

4. BENTHOS

Benthos refers to those animals living on or beneath the sea floor. Although there has been less research conducted on the benthos compared to phytoplankton and zooplankton, several studies provide evidence of relationships between the NAO and benthic populations. *Cohen and McCartney* [2000], examining chemical and structural variations in the skeletons of small brain corals (*Diploria labyrinthiformis*) collected at 50-ft depth on the south-east edge of the Bermuda platform, constructed seasonally-resolved oxygen isotope records and examined changes in skeletal density (calcification rate). These parameters were strongly correlated with the instrumental record of the NAO over a 40-year period covering the late 1900s.

On the opposite side of the Atlantic, *Nordberg et al.* [2000], working with benthic foraminiferal records between

1930 and 1996, showed a significant change in faunal composition during the mid-1970s. In those years, *Stainforthia fusiformis*, an opportunistic indicator of low oxygen environments in the Scandinavian fjords, became the most common Skagerrak species in the foraminiferal fauna assemblage. The timing of this faunal change coincided with a severe low oxygen event in the Gullmar Fjord, which *Nordberg et al.* [2000] related to changes in the NAO. Positive NAO index values in the early 1970s were marked by strong westerly winds in the Skagerrak region, thereby preventing the exchange of bottom-water in the fjords and decreasing the oxygenation of the sea floor and its benthic community. Working in the same area, *Tunberg and Nelson* [1998] monitored soft sediment macrofauna in depths from 10 to 300 m for periods of 12 to 20 years. The variability in macrobenthic abundance was in phase over the upper 100 m but out of phase with that at 300 m. Abundances and biomasses varied with a 7 to 8 year periodicity, which approximated that of the NAO index (7.9 years as determined from spectral analysis). The NAO index over the period 1970 to 1990 was positively correlated with Skagerrak deep-water (600 m) temperatures and negatively correlated with stream flow from western Sweden. Stream flow, in turn, was positively correlated with benthic abundance and biomass at stations down to 100 m, but negatively correlated with bottom water oxygen content. *Tunberg and Nelson* [1998] proposed that the NAO association with the benthos occurs by bottom-up control of the population through influences on primary production. They further suggested that climatic variability in the region is likely the most important factor in controlling the variability in the benthos.

Hagberg and Tunberg [2000] continued the work in Gullmar Fjord and the Swedish Skagerrak. They compared a 7 to 13 year data series (covering 1983-1995) of mean macrobenthic abundance data from eight stations (25 to 118 m) to the freshwater runoff to the fjord, temperature at 600 m in the Skagerrak and the NAO index. The macrobenthic abundances at 3 of the innermost fjord stations were positively correlated with the NAO index ($R^2 = 0.6\text{-}0.7$, $p<0.05$) with a delay of up to 1 year. The authors proposed that this was due to stronger stratification as a result of higher freshwater runoff under a low NAO index. This in turn reduced primary production and hence there was less food input to the benthos (Figure 3). At the three stations outside the fjord and the remaining two stations inside the fjord, the highest correlations of the macrobenthic abundances were with Skagerrak temperatures at 600 m. The authors suggested the negative correlations with temperature might be related to the NAO. This could occur, they stated, through periodic upwelling of deep colder water, rich in dissolved inorganic nutrients, resulting in an increase in primary production and subsequently more food for the benthos.

Kroencke et al. [1998] examined long-term (1938–1995) macrofauna data in the subtidal zone (10 to 20 m) off Norderney, one of German's East Frisian barrier islands. They found strong positive correlations between abundance, species number and to a lesser extent the biomass of macrofauna in spring and the NAO index. They suggested that the mediator is probably the SST in late winter and early spring. Mild meteorological conditions, probably acting in conjunction with eutrophication, were believed to result in the observed increase in total biomass from 1989–1995. The authors concluded that climate variability explains most of the interannual variability in macrozoobenthos off Norderney.

5. FISH

5.1. Atlantic Cod (Gadus morhua L)

Atlantic cod is an ideal candidate for examination of the relationship between the NAO and fish because of its pan-Atlantic distribution and the fact that climate variations have been shown to affect recruitment [*Planque and Frédou*, 1999], growth [*Brander*, 1994; 1995] and distribution [*Jensen and Hansen*, 1931; *Rose et al.*, 2000]. The disadvantage is that they are heavily exploited. This has led to a general decline in numbers, especially during the last half of the 1900s, which sometimes has made it difficult to separate climate from fisheries effects. In spite of this, several studies have provided convincing evidence of NAO associated variability of cod.

Cod eggs, larvae and early juveniles are generally distributed in the upper water column where they are free-floating before settling on the bottom as half-year olds. It is principally during these early life stages that the year-class strength of cod (the number of fish that reach commercial size) is determined [*Sundby et al.*, 1989; *Myers and Cadigan*, 1993]. This is also a stage at which climate is considered to have its most profound effect [*Cushing*, 1966; *Ellertsen et al.*, 1989; *deYoung and Rose*, 1993; *Dickson and Brander*, 1993; *Ottersen et al.*, 1994; *Ottersen and Sundby*, 1995].

The south and central Barents Sea is a highly productive region, being home to one of the largest cod stocks in the North Atlantic, the Arcto-Norwegian or Northeast Arctic cod. Recruitment varies extensively with the ratio between strong and weak year classes evaluated at age 3 being about 15 [*Ottersen*, 1996], and as high as 70 based on early juveniles [*Ottersen and Sundby*, 1995]. Interannual variability in the thermal conditions in the areas of the Barents Sea inhabited by cod are principally determined by winter conditions [*Ottersen and Stenseth*, 2001] through changes in the temperature and quantity of the Atlantic inflow from the southwest [*Loeng*, 1991] and regional air-sea heat exchanges [*Ådlandsvik and Loeng*, 1991; *Loeng et al.*, 1992]. The impact of interannual and decadal shifts in sea temperatures in the Barents Sea on cod recruitment has been well documented with high recruitment associated with warm years [*Sætersdal and Loeng*, 1987; *Ottersen et al.*, 1998; *Ottersen and Loeng*, 2000]. This is thought to be because of (i) higher primary production due a larger ice-free area, (ii) a larger influx of zooplankton carried by the increased Atlantic inflow and (iii) higher temperatures that promote higher biological activity at all trophic levels [*Sakshaug*, 1997].

Ottersen and Stenseth [2001] demonstrated a positive association between the NAO and Barents Sea cod recruitment (1970–1998). They evaluated several statistical models that predicted year-class strength from climate variables during the winter the year-class was spawned. The single most important variable was the NAO index that alone accounted for 53% of the recruitment variability. The mechanistic link (Figure 4) was considered to be through effects on regional sea temperatures and food availability. Higher NAO index values are associated with warmer temperatures through both increased atmospheric heat transfer to the ocean and an increased Atlantic inflow. The latter transports more *Calanus finmarchicus* into the Barents Sea and hence more food for cod. On the other hand, the year-class strength of cod in the much warmer North and Irish Seas is negatively related to both the NAO index and temperature. This is believed to result from a limitation in energy resources neces-

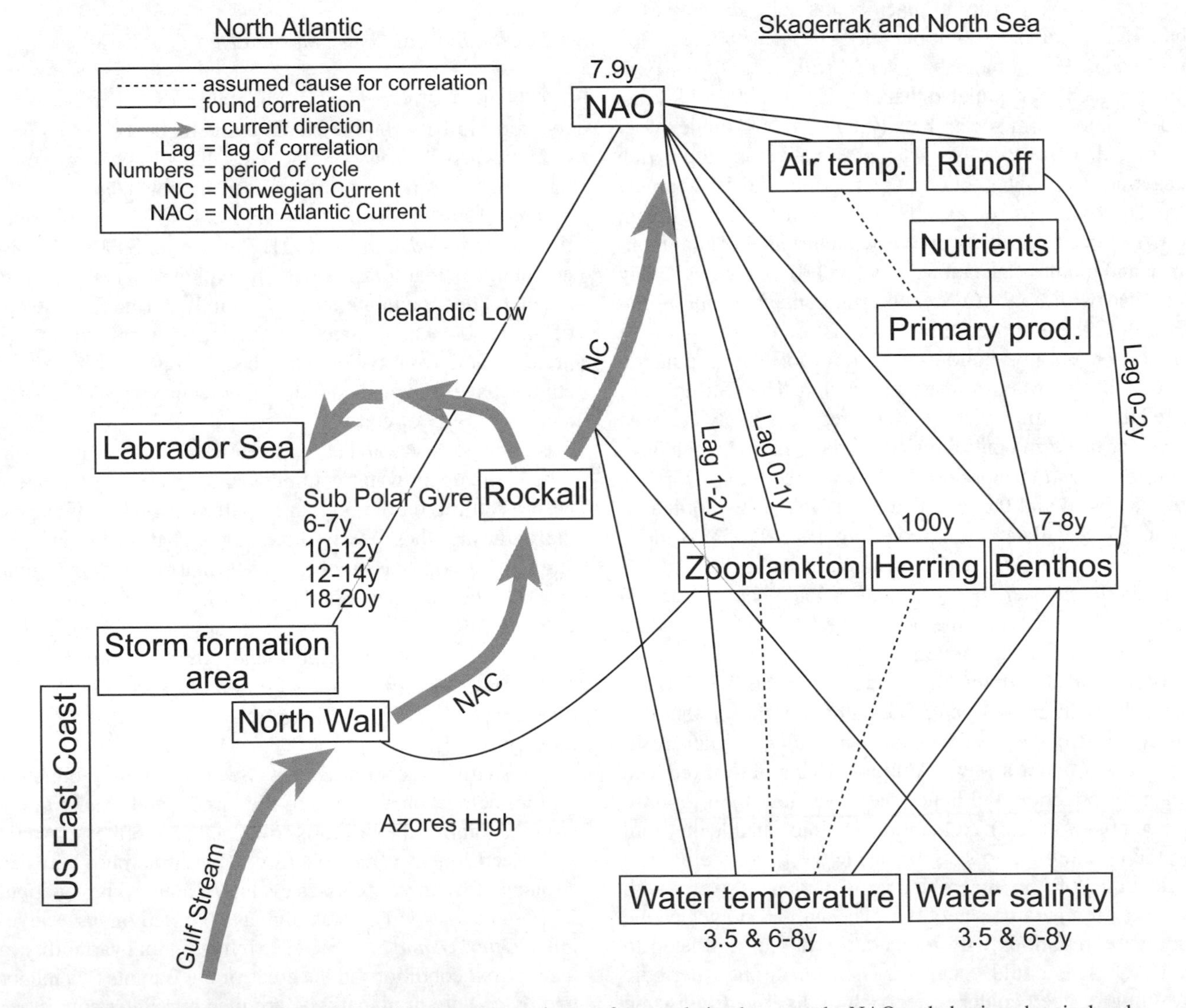

Figure 3. Schematic diagram describing the assumed cause for correlation between the NAO and physico-chemical and biological interactions courtesy of J. Hagberg, Department of Marine Ecology, Goteborg University, Sweden and based upon *Hagberg and Tunberg* [2000]. Note the number of years indicate the dominant periods of variability.

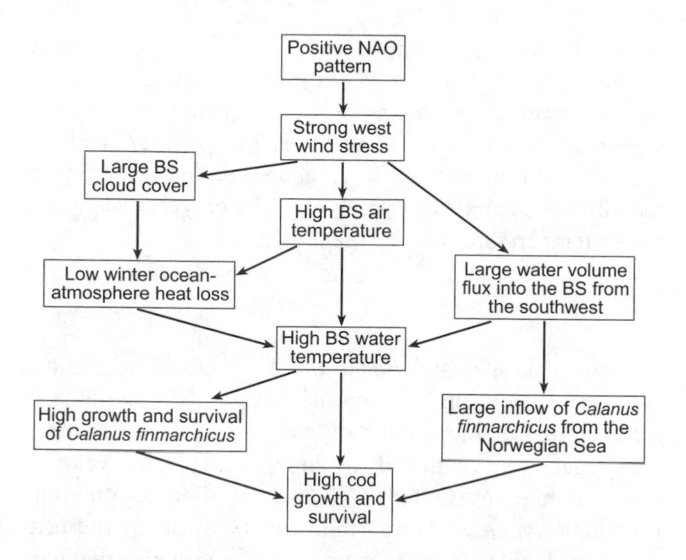

Figure 4. Mechanisms linking the NAO to variability in Barents Sea (BS) oceanography and ecology. A positive NAO index phase results in increased westerly winds over the North Atlantic. This increases BS water temperatures through enhanced volume flux of relative warm Atlantic water from the southwest, higher air temperatures, and increased cloud cover. Higher BS water temperature improves growth and survival of cod larvae both directly through faster development rates and indirectly through regulating the production of their main prey, nauplii of the copepod Calanus finmarchicus. Increased inflow from the zooplankton rich Norwegian Sea further increases availability of food for the cod larvae. High food availability for larval and juvenile fish results in higher growth rates and greater survival through the vulnerable stages when year-class strength is determined. Modified from *Ottersen and Stenseth* [2001].

sary to achieve higher metabolic rates during warm years [*Planque and Fox*, 1998]. These results are consistent with *Planque and Fredou* [1999] who found that recruitment was positively related to temperature for stocks occupying relatively cold waters and negative if in relatively warm waters.

Dippner and Ottersen [2001] also related cod recruitment to large-scale climate variability and Barents Sea temperatures. They showed that the temperature anomalies at the Kola section across the southern Barents Sea are significantly correlated to the anomalies of the NAO index. Furthermore, a statistically significant CCA (canonical correlation analysis) correlation was found between the Kola section temperature and both the number of 0-group cod (R^2 = 0.44, unlagged) and recruitment measured at age 3 (R^2 = 0.37, lag of 2 years).

Opposing year-class strengths of cod between the eastern and western regions of the North Atlantic was hypothesised [*Izhevskii*, 1964; *Templeman*, 1972], and was suggested to be related to the NAO [*Rodionov*, 1995]. Links between strongly negative NAO index events and good recruitment and growth for the Northern cod stock off southern Labrador and northern Newfoundland in Canada were discussed by *Mann and Drinkwater* [1994] while positive NAO index anomalies have been linked to favorable conditions for Arcto-Norwegian cod [*Ottersen and Stenseth*, 2001; *Ottersen et al.*, 2001]. Both regions are characterised by sea temperatures towards the lower end of the overall range inhabited by cod. The NAO index accounts for approximately 50% of the interannual variability in atmospheric, oceanic and sea-ice indices both in the Labrador Sea Region [*Drinkwater and Mountain*, 1997] and the Barents Sea [*Ottersen and Stenseth*, 2001], but the signs of the correlation are opposite [e.g., *Hurrell et al.*, this volume]. Years of high NAO index values produce cold temperatures in the Newfoundland-Labrador area and warm temperatures in the Barents Sea and visa versa. Such inverse fluctuations in Barents and Labrador Sea temperatures were pointed out by *Izhevskii* [1964] and the association of the NAO with this "seesaw" pattern in temperatures was demonstrated by *van Loon and Rogers* [1978]. Recruitment in both areas tends to be higher in warmer years than in colder years, thus during a high NAO index recruitment is good in the Barents Sea and poor in the Labrador while during low NAO index years, the reverse is true [*Ellertsen et al.* 1989; *DeYoung and Rose*, 1993].

Cod growth also is linked to NAO variability. *Brander* [1994; 1995] showed that temperature accounts for both the mean differences in size-at-age for cod throughout the North Atlantic and the interannual variability in mean size within individual stocks. Higher size-at-age occurs under warmer conditions for most stocks. Consistent with this, *Drinkwater* [2002] showed that the NAO accounted for over 50% of the variability in growth increment between 3 and 5 year olds from the Northern Cod stock off Newfoundland (Figure 5).

Changes in climate patterns associated with the NAO also affect predator-prey interactions. In the Barents Sea, an increase in the basic metabolic rate of cod, associated with higher temperature during years of high NAO index values, results in an increase in the consumption of capelin (*Mallotus villosus*) by 100 thousand tonnes per degree centigrade [*Bogstad and Gjøsæter*, 1994].

The effects of NAO-associated events on cod can be sustained for several years [*Ottersen et al.*, 2001]. For example, the increase in survival of Arcto-Norwegian cod through the vulnerable early stages during warm, high NAO index years historically results in stronger year classes in later years. As such year-classes mature, the number of spawners tends to remain higher-than-normal, enhancing the potential for good recruitment to the next generation. Furthermore, if

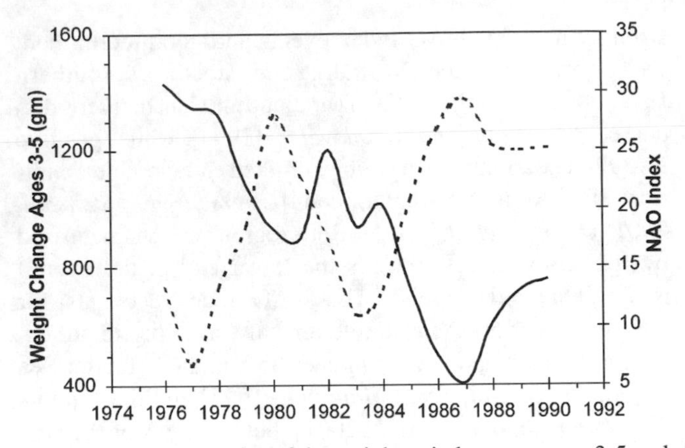

Figure 5. The time series of the weight gain between ages 3-5 and the three year-average of the NAO index for the equivalent years. Redrawn from *Drinkwater* [2002].

individuals in a cohort of Arcto-Norwegian cod are larger-than-average as half-year olds, they tend to remain large as they grow older and the cohort abundance tends to be high [*Ottersen and Loeng*, 2000].

5.2. Herring (Clupea harengus) and Sardines (Sardina pilchardus)

Small pelagic fishes such as sardine and herring are widespread and represent about 20–25% of the total annual catch of the world fisheries. Most are highly mobile and have short, plankton-based food chains, with a few species feeding directly upon phytoplankton. They are also short-lived (3–7 years, except herring) and highly fecund. Catch records of several hundreds of years for herring and sardines in northern Europe show that the fisheries oscillate between times of very high yields and other times when the fish were totally absent. These periods of high and low catches varied spatially. Such fluctuations occur on decadal time scales and in part can be explained as a response to different regimes of prevailing wind directions corresponding to related phases of the NAO [*Alheit and Hagen*, 1997; 2001]. Herring stocks off Bohuslän on the Swedish West coast, off southwestern England, in the eastern English Channel and the Bay of Biscay were favored during periods of low NAO index when the westerly winds were shifted to the south and the sea temperatures in these regions were low. In contrast, the Norwegian spring spawning herring, the sardines of southwestern England and the sardines caught by the French fleet in the English Channel exhibit high catches during the high index phase of the NAO, i.e. when the westerly winds intensified and high local temperatures prevailed.

Guisande et al. [2001] showed that higher sardine recruitment off northern Spain occurs during low NAO index years. During such conditions, the winds tend to be more southerly, bring warmer-than-normal temperatures to this region as well as promoting onshore Ekman transport. The higher temperatures result in faster growth rates and the onshore drift in greater larval retention inshore. Lower recruitment occurs during high NAO index years due to the colder temperatures and offshore drift.

5.3. Tuna and Other Large Pelagics

Santiago [1997] and *Borja and Santiago* [2001] examined the relationship between the NAO index with tuna in the eastern Atlantic for the period 1969–1995. The mean recruitment was estimated for three stocks during years of low and high NAO index (Figure 6). For bluefin tuna (*Thunnus thynnus*) in the eastern Atlantic, mean recruitment during high NAO index situations was near double that during low NAO index conditions. The opposite occurred in the case of northern albacore (*Thunnus alalunga*), with recruitment during high NAO index years being approximately half that of low NAO index years. These differences were estimated to be statistically significant. In contrast, there was no difference in recruitment of bluefin tuna from the western Atlantic between the two phases of the NAO. Standard correlation analysis confirmed a negative relationship between the NAO index and recruitment of northern albacore ($R^2 = 0.52$, $p<0.05$) but no statistically significant relationship ($p>0.05$) was found for either eastern or western bluefin tuna.

Borja and Santiago [2001] suggested that the mechanism linking the NAO to albacore recruitment is through the temperature of the spawning-overwintering area. Also, during positive NAO index winters, the storm activity increases, especially in a narrow band following the main eastern U.S.

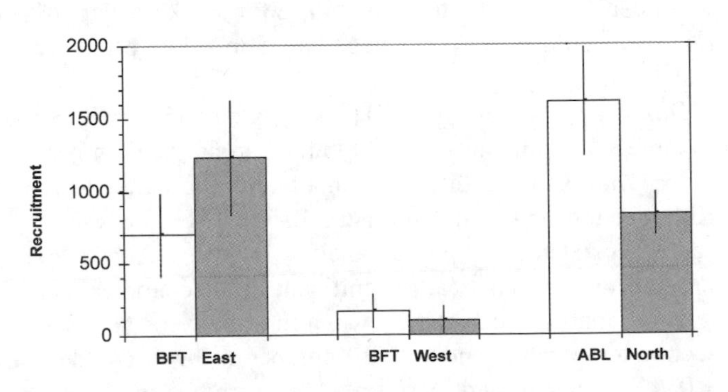

Figure 6. Mean recruitment of eastern and western bluefin (times 1000) and northern albacore (times 10000) estimated for low NAO index years (1969, 1970, 1977, 1979) and high NAO index years (1973, 1981, 1983, 1989, 1990, 1992, 1993, 1994, 1995). Low NAO index situations are expressed as blank boxes; high NAO index situations as filled boxes. 95% confidence intervals are indicated.

coastal baroclinic zone, which increases mixed-layer depth [*Dickson et al.*, 1996]. The stronger winds are thought to be responsible for increased upwelling and mixing, eventually leading to a higher concentration of food organisms.

Opposite to the albacore tuna, the relationship between the NAO index and eastern bluefin tuna is positive but the NAO only accounts for 13% of the recruitment variability. However, this rises to 49% when the recruitment lags the NAO index by 1 year. The different sign of the responses of the two species can be explained by differences in their overwintering areas. Whereas the albacore overwinters along the western side of the Atlantic, the eastern bluefin overwinters on the eastern side (along the Moroccan coast and in the Canary Islands). The NAO effect on SST differs in sign between the two overwintering areas *Krovnin* [1995].

Another possible factor linking the NAO to the eastern bluefin recruitment is the observation of *Fromentin and Planque* [1996] that the abundance of *C. helgolandicus* was significantly higher during high NAO index years than during low NAO index years and the degree of significance increased when the NAO index lagged the zooplankton by one year. If the abundance of this dominant zooplankton group is compared with bluefin year-class strength estimated for the same year, the relationship is statistically significant, suggesting a possible direct connection between zooplankton abundance and recruitment success of bluefin.

Climate variability seems to play a minor role in the recruitment success of western bluefin tuna, at least during the analyzed period. It should be noted, however, that the number of western bluefin has been extremely low since the 1970s, which could account for a low signal to noise ratio.

Other links between the NAO and changes in abundance and distribution of large pelagic fish have been observed. The Norwegian bluefin fishery developed after the Second World War, peaked in 1952 (11,400 t) and collapsed in the mid-1960s [*Tiews*, 1978] coinciding with a period of low NAO index. The same collapse was observed in the Danish and German fisheries in the North Sea that dropped from 2,400 t in 1952 to less than 100 t by the mid-1960s. According to *Tiews* [1978], it was the lack of recruit year classes to the Northeast Atlantic tuna fishery that led to the absence of bluefin tuna in the central North Sea after 1962.

Mejuto and de la Serna [1997] found a statistically significant relationship ($p<0.05$) between North Atlantic swordfish (*Xiphias gladius*) year-class strength and the NAO index. As in the case of northern albacore, high NAO index years were associated with low recruitment and low NAO index years with high recruitment levels. The model explains 33% of the variability of swordfish recruitment. Standardized catch per unit fishing effort indices of age 1 corresponding to the Spanish longline fleet from 1983–1995 were used as estimates of swordfish year-class strength.

Cushing [1982] refers to an increased abundance of bluefin tuna in the Northeast Atlantic during a period of warming, between the 1920s and 1950, and bluefin tuna, together with swordfish, appeared off the Faroe Islands and Iceland. This period was typified by high NAO winter index values. After disappearing in the 1960s during low NAO index years, the bluefin again appeared in high latitude waters during the 1990s. In the western Mediterranean, the abundance of age-0 bluefin in 1996, when the NAO index fell dramatically to the lowest value since 1969, was extremely low [*de la Serna*, 1997].

5.4. Atlantic Salmon (Salmo salar)

Atlantic salmon is another pan-Atlantic species. It is affected by climate variability in all of its many stages, from the parr through the smolt, marine post-smolt and mature stages. Parr is the name for juvenile salmon in freshwater, smolt is the juvenile stage adapted to life in the ocean and post-smolt is the name for juveniles during their first year in the ocean. The similarity in the return rates of salmon to different and wide spread rivers throughout the North Atlantic indicates that the highest mortality affecting Atlantic salmon populations occurs in the marine phase. The largest component of the natural mortality tends to occur during the first year at sea, with the result that the post-smolt period emerges as the critical stage for salmonids [*Pearcy*, 1992]. The temperature and productivity of the coastal waters that the salmon enter as they leave their rivers varies from year to year and may be critical in determining their ultimate survival rate. However, post-smolts rapidly migrate through coastal waters, and commence large-scale oceanic migrations. During their marine phase, salmon achieve upwards of 90% of their somatic growth and will reach sexual maturity before making their return migration to their native river [*Dickson and Turrell*, 2000]. Long-term patterns of stock abundance for regional and continental stock complexes, which are defined by post-smolt survival, are often associated with climate forcing [*Friedland*, 1998].

Reddin and Shearer [1987] demonstrated that the abundance of salmon off West Greenland was related to the area enclosed by 4°C to 10°C in the Northwest Atlantic. Quantitative thermal habitat concepts were further developed and applied to European salmon by *Friedland* [1998] who found that survival rates of one-seawinter and two-seawinter salmon in the North Sea were correlated with the area of 8°-10°C water in May. The first study to link variability in the thermal habitat indices of Atlantic salmon to the NAO, was that by *Friedland et al.* [1993]. They found that during years of high NAO index, the 4°C isotherm was positioned to the south and east of its mean location in the area of the Labrador Sea, Davis Strait and Norwegian Sea and conversely in low index years it was located to the north and west. The thermal habitat index of Atlantic salmon was shown to covary with the NAO, with the thermal habitat index decreasing during the years of positive NAO index and expanding during negative index phases [*Friedland et al.*, 1993; *Dickson and Turrell,* 2000].

Dickson and Turrell [2000] suggest that salmon in European waters are linked to the NAO variability in many different ways, including at first entry to the sea, while leaving the European shelf and in distant oceanic waters. During the last few decades of the 1900s, the NAO index correlates with SST and winds in the coastal waters around the U.K., both key variables likely to determine the environment of fish leaving the European Shelf for middle and distant waters [*Dickson and Turrell*, 2000; *Ottersen et al.*, 2001]. *Dickson and Turrell* [2000] admit that although the factual basis for ascribing importance to any particular set of environmental properties as a control on salmon is scanty, there is strong circumstantial evidence that salmon variability is linked to the NAO.

5.5. Winter Flounder (*Pseudopleuronectes americanus*)

The positive phase of the winter NAO index leads to warmer and wetter winters along the U.S. east coast [*Hurrell*, 1995; *Shindell et al.*, 1999]. From 1960 to 1990, as the winter NAO index moved from persistent negative to persistent positive values, the winter temperatures increased in Narragansett Bay, Rhode Island by 3°C. This in turn altered marine food webs [*Keller et al.*, 1999] and may have impacted negatively upon the winter flounder, the formerly dominant commercial fish [*Keller and Klien-MacPhee*, 2000]. Warmer winters result in smaller winter-spring phytoplankton blooms, an observation experimentally reproduced in marine mesocosms [*Keller et al,* 1999]. During the past 25 years winter flounder abundances in southern New England have been in decline. One hypothesis is that warmer sea water temperatures result in more of the winter marine phytoplankton bloom being consumed in the water column by pelagic food chains, with reduction in the amount of fixed carbon available to benthic (bottom dwelling) food chain members, such as flounder. In contrast to the winter flounder, Atlantic herring stocks, which feed in the upper water column, have been on the increase. Another hypothesis is that temperature increases affect predation and survival of winter flounder during critical early life stages [*Keller and Klein-MacPhee*, 2000].

There is still debate, however, about how much of the observed decline is due to warmer winters and how much may be attributed to heavy fishing pressure. The physiology and ecology of winter flounder provides some interesting clues. The winter flounder is a former dominant member of the bottom dwelling fish community in southern New England [*Cooper and Chapleau*, 1998]. Most adult fish migrate into inshore waters in the late fall and early winter, and spawn in late winter and early spring when seawater temperatures are quite cold [*Klein-MacPhee*, 1978]. To accomplish this feat, winter flounder make use of unique antifreeze proteins found in a number of polar fish which allow them to survive cold temperatures, even as low as -1.9°C [*Wen and Laursen*, 1992]. Winter flounder spawning occurs at night in the upper portions of estuaries. Eggs are attached to the bottom. Hatching rate, larval development rate, and mortality rates due to predation are all temperature dependent. Variations in egg and larval survival during the first year determine the age-1 year-class strength. Observations suggest that a significant component of the decline in winter flounder abundance in southern New England is associated with a shift from a period with cold winters and sea water temperatures in Southern New England during the 1960s, into a period of relatively warmer winters during the following three decades. February sea temperatures from the three years prior to recruitment of age 1 winter flounder is associated with about 70% of the interannual variation in the abundance (year-class strength) of age-1 winter flounder in Niantic Bay. The series of warm winters such as experienced in southern New England during the 1990s is clearly unfavorable for winter flounder.

6. MARINE DISEASES

Climate variability and change is also associated with variations in the geographic range of marine diseases [*Harvell et al.*, 1999]. Off the eastern seaboard of the United States, a positive NAO index and the associated winter warming of coastal marine waters favor a northward extension of marine species typically found to the south. For example, in the past several decades, *Dermo* disease in oysters has progressively moved northward from off the mid-Atlantic states into New England [*Ford*, 1996; *Cook et al.,* 1998]. Also, at lower latitudes in the North Atlantic, there are disturbances in coral ecosystems, involving coral bleaching and a variety of coral diseases that are associated with NAO variations [*Barber et al.,* 2001; *Sherman*, 2001; *Sherman and Epstein*, 2001; *Hayes et al.,* 2002]. *Hayes et al.* [2002] and *Barber et al.* [2001] advanced the hypothesis that observed increases in aeolian dust transport into the western North Atlantic, associated with increased persistence of a positive phase of the winter NAO index, may "release" coral disease organisms from an iron–limited state.

7. WHALES

Present day right whales (*Balaena glacialis*) in the western North Atlantic rely almost exclusively on feeding grounds in the Gulf of Maine/western Scotian Shelf region [*Winn et al.*, 1986]. These are a small remnant of a much larger population that existed on both sides of the North Atlantic but that were drastically reduced through harvesting [*Reeves and Mitchell*, 1986; *Aguilar*, 1986]. The population growth rate of the right whales declined from the 1980s to the 1990s [*Caswell et al.*, 1999]. As reported by *O'Connell* [2001], Wood's Hole scientists Caswell and Fujiwara examined the effects of shipping, gillneting, the Southern Oscillation and the NAO on the survival rates of the right whales and found that the NAO index was the most important factor.

The Gulf of Maine/Western Scotian Shelf region presents right whales with a highly variable feeding environment. Physically, the region lies within a transient oceanographic transition zone, located between cold subpolar waters influenced by the Labrador Current to the northeast and warm temperate waters influenced by the Gulf Stream to the south (Loder et al., 2001; MERCINA, 2001). The transitions that occur within this zone are not only physical, as reflected in hydrographic changes, but also biological, as reflected in the changes in composition and relative abundance of plankton (see zooplankton section). The shifting nature of this transition zone makes the Gulf of Maine and Western Scotian Shelf regions especially vulnerable to climate-driven changes in North Atlantic circulation patterns.

Since *C. finmarchicus* is the principal source of nutrition for right whales in the region [*Kenney et al.*, 1986; *Wishner et al.*, 1995], it has been proposed that the response of right whale populations to climate variability may be mediated by trophic interactions with this prey species [*Kenney et al.*, 1986]. Although the mechanisms underlying the climate-driven changes in *C. finmarchicus* abundance are not fully resolved, they appear to be linked to the NAO through advective processes from the Slope Water [*Greene and Pershing*, 2000; *MERCINA*, 2001; see section 3].

Since consistent data were first collected in 1982, declines in right whale calving rates have generally tracked declines in *C. finmarchicus* abundance [*Greene et al.*, unpublished data]. From 1982 to 1992, calving rates were relatively stable with a mean total rate of 12.4 $\pm$ 0.9 (standard error) calves per year, consistent with the relatively high abundance of *C. finmarchicus*. From 1993 to 2001, calving rates exhibited two major declines, and the mean rate dropped to 11.2 calves per year and became more variable ($\pm$ 2.7 calves per year). They followed large declines in *C. finmarchicus* abundance although their timing varied. During the first event in the early 1990s, the lower calving rates occurred two years after *C. finmarchicus* abundances fell. During the second event in the late 1990s, calving rates exhibited a steep decline in the same year that abundances began to fall. Although these differences cannot presently be explained, some hypotheses based on right whale reproductive physiology and behavior may help reconcile the different responses.

Right whale reproductive physiology typically requires at least three years between births – one year for lactation, one year to amass fat stores to support the next pregnancy, and one year during the pregnancy [*Knowlton et al.*, 1994]. Hence, feeding conditions over several years are likely integrated when determining if a given female will reproduce or not. Since the first multi-year decline in calving rates occurred two years after a period of relatively stable reproduction and good feeding conditions, the time-lagged response may have required two years of poor feeding conditions before taking effect. When *C. finmarchicus* abundance increased in the mid-1990s, many females in the right whale population had not given birth recently and were available for reproduction. Hence, when good feeding conditions returned, calving rates nearly doubled during 1996 and 1997. This rapid increase in reproduction limited the number of females available for reproduction in the immediate years following and, in combination with the poor feeding conditions during the late 1990's, calving rates plummeted from 1998 to 2000. When *C. finmarchicus* abundance increased again in 2000, many females in the right whale population had not given birth recently and were available for reproduction. With the combination of many females available for reproduction and good feeding conditions, the annual calving rate reached an historical high in 2001.

8. SEABIRDS

Few published studies of the relationship between the NAO and seabirds exist. An exception is that by *Thompson and Ollason* [2001]. Using 50 years of data on the northern fulmar (*Fulmarus glacialis*) collected on Eynhallow in Scotland, they showed a statistically significant negative relationship between the proportion of breeding adult fulmars present at the colony each summer and the NAO index. In addition, the hatching success and the fledging success in the summer were both negatively related to the previous winter's NAO index. The variances accounted for by the NAO were between 10–20% ($p \leq 0.05$). The authors suggested that all three fulmar indices might be linked to the NAO through *C. finmarchicus* as the latter increase in the North Sea during a negative NAO index phase. More zooplankton was assumed to lead to an increase in the crustaceans and small fish eaten by fulmars.

The Seabird Ecology Working Group within ICES (Anonymous, 1998) had earlier examined the relationship between the NAO index and the seabird breeding numbers and success in the Northeastern Atlantic (United Kingdom and German Wadden Sea). From 1983 through 1996, there were no significant correlations between the NAO index and breeding population sizes or breeding success of seven species (guillemot (*Uria aalge*), razorbill (*Alca torda*), puffin (*Fratercula arctica*), fulmar, gannet (*Sula bassana*), shag (*Phalacrocorax aristotelis*) and kittiwake (*Rissa tridactyla*)) at several U.K. seabird colonies. In the Wadden Sea, no significant relationship was found between the NAO index and breeding numbers of cormorant (*Phalacrocorax* spec.), Arctic tern (*Sterna paradisaea*) or common tern (*Sterna hirundo*). However, there were significant correlations ($p<0.02$) between the NAO and the breeding number of several other species , including fulmars, herring gulls (*Larus argentatus*), lesser black-backed gulls (*Larus fuscus*), common gulls (*Larus canus*), black-headed gulls (*Larus ridibundus*), kittiwakes, sandwich terns (*Sterna sandvicensis*), guillemots and razorbills. These time series consisted of 44 years of data from the early 1950s to the early to mid-1990s for all species except the sandwich tern, where data were available from 1907 to 1996. The variance accounted for by the NAO ranged from 7 to 32%. These relationships may occur through the influence of the NAO on food resources; however, the feeding requirements and general feeding ecology of these species are so diverse that the authors thought this highly unlikely.

9. SUMMARY

Over the past decade, numerous investigations have established links between the NAO index and the biology of the North Atlantic. Our review has included studies of the changes in the biomass and species composition of phytoplankton and zooplankton, the biomass, distribution and growth of several commercial species of fish, the abundance of benthos, the spread of marine diseases, the survival rates of right whales and the hatching success of certain seabirds. These studies span the North Atlantic and include the deep ocean basins, the continental shelves and coastal embayments. This reflects the extent of the physical oceanographic responses to the NAO-associated forcing [*Visbeck et al.*, this volume] through which the links to the biology occur. The relationship to the NAO results in large-scale coherency between certain biological phenomena. However, different physical responses in different regions lead to differences in the biological responses. For example, during a high NAO index phase, the Barents Sea warms through increased heat exchange with the atmosphere and increased Atlantic inflow. These result in higher abundances of *C. finmarchicus* and increased recruitment of cod [*Ottersen et al.*, 2001]. On the other side of the Atlantic, off Newfoundland, a high NAO index is associated with stronger northwesterly winds, cold conditions and generally poor recruitment of cod [*Mann and Drinkwater*, 1994]. Such relationships are not limited to fish but have also been shown to hold for zooplankton, with opposite trends in the abundance of *C. finmarchicus* on both sides of the Atlantic (see section 3).

Some of the studies presented in this review are of short duration or have significant long-term trends, which were not removed in the statistical analyses. This sometimes brings into question the statistical reliability of these studies. However, the sheer number of studies linking the NAO to biological variability in the North Atlantic, plus that they occur at different trophic levels, for different species, and different aspects (abundance, distribution, growth, speciation, etc.), is convincing evidence that the biology is definitely linked to the climate variability associated with the NAO. The majority of the studies have been carried out on fish, or the lower levels of the food chain, i.e. phytoplankton and zooplankton. This is due to the availability of long-term time series that are needed to establish such relationships with any statistical reliability. One of the reasons for the large quantity of the work on phytoplankton and zooplankton has been the CPR data. They confirm the importance of maintaining consistent, long-term monitoring programs.

While it is clear that biotic changes are linked to NAO variability, much work still remains in determining the precise mechanisms through which the linkages occur. Many hypotheses have been proposed. For zooplankton, the association with the NAO is often assumed to be mediated through phytoplankton and for higher trophic levels, through the zooplankton. The responsible mechanism ultimately must be linked to changes in the physical environment, however. These usually occur either from changes in advection by ocean currents or through regional atmospheric forcing.

Changes in the geographical distribution and abundance of plankton through advective processes have been well demonstrated [*Edwards et al.*, 1999; *Johns et al.*, 2001; *MERCINA*, 2001; *Ottersen and Stenseth*, 2001]. Advection can also lead to changes in the water properties, such as temperature, or vertical structure (i.e., depth of the mixed layer). For phytoplankton abundance, the NAO is typically assumed to be mediated through changes in meteorological forcing, such as temperature or wind mixing. The phytoplankton response to variability in these variables can be complex, however. For example, higher temperatures can raise phytoplankton production directly by increasing turnover rates but can suppress production through increased stratification, if nutrients are limiting. On the other hand, stratification early in the production season can be beneficial by limiting the depth to which the phytoplankton cells are mixed. Wind affects are generally assumed to be through their affect on vertical mixing. The response to wind mixing varies, however, depending upon the ratio of the mixed layer depth to the critical depth in spring [*Dutkiewicz et al.*, 2001]. This results in increased primary production from deeper mixing in the subtropics but decreased production in the subpolar gyre. No one process is able to explain phytoplankton production linkages to the NAO throughout the North Atlantic. While studies in the Northeast Atlantic and North Sea presently suggest temperature to be the overriding factor [*Edwards et al.*, 2001], wind is suggested as the principal mechanism off Bermuda [*Bates*, 2001]. Even where temperature is thought to be the dominant factor, it is often unclear whether it is through increased turnover rates, temperature effects on stratification or some other process. Sorting out the precise mechanisms and relationships between the NAO and phytoplankton is extremely important, given that it is often assumed that the links between zooplankton and the NAO are through the phytoplankton.

Other hypotheses linking zooplankton to the NAO, besides through phytoplankton and advection, include direct temperature effects on development times of the zooplankton and the effects of wind on the contact rates between the zooplankton and their food through turbulent mixing. For fish and other higher trophic levels, the links of abundance levels to the NAO are often considered to occur through the association between the NAO and zooplankton production. Growth effects are generally felt to be through temperature, while distribution effects for phytoplankton, zooplankton and fish are usually hypothesized to occur through either advection or from hydrographic changes that favor certain species over others.

Given that the NAO is not directly responsible for the changes observed in the biological components of the ecosystem, but rather are usually driven by local physical changes associated with NAO variability, one may ask, why is the NAO index so useful in accounting for changes in the marine ecosystem? Why not use the local characteristics instead? First, in some cases long-term data of the necessary local physical variables are unavailable and thus the NAO index provides an excellent proxy. Second, the NAO has been found to account for as much, and in some cases more, of the variance of biological phenomena than local physical climate indices. This may be because the NAO is linked to changes in several physical characteristics of a particular ecosystem, including its hydrographic characteristics, mixed-layer depth, or circulation patterns. In such cases, the NAO captures more of the overall physical variability than any individual local climate index. In such cases it can be considered as an integrator of the local climate changes. Third, the NAO provides a conceptual framework and a broader understanding of the observed changes in the local physical environment.

While the NAO research has helped to advance the field of climate-induced impacts on the marine ecosystems of the North Atlantic Ocean, future strides will require not only expanding the list of NAO-associated linkages but also going beyond the correlations and establishing the precise mechanisms through which the NAO acts. Critical to achieving these objectives are the continuance of the present long-term data sets such as the CPR surveys and the establishment of new time series, such as for benthos. The challenge is to continue improving our understanding of the links of climate with the marine ecosystem to allow us to predict what might happen under future climate change scenarios.

Acknowledgements. The authors would like to thank those of our colleagues with whom we have collaborated and those providing comments on an earlier draft of the chapter including M. Huber, D. Nacci, C. Wigand and two anonymous reviewers. Finally we would like to thank J. Hurrell for his comments, encouragement and patience.

REFERENCES

Ådlandsvik B., and H. Loeng, A study of the climatic system in the Barents Sea, *Pol. Res., 10*, 45–49, 1991.

Aebischer, N. J., J. C. Coulson, and J. M. Colebrook, Parallel long-term trends across four marine trophic levels and weather, *Nature, 347*, 753–755, 1990.

Aguilar, A., A review of old Basque whaling and its effect on the right whales (Eubalaena glacialis) of the North Atlantic, *Rep. Int. Whal. Commn. (Spec. Issue), 10*, 191–199, 1986.

Alcaraz, M., E. Saiz, and A. Calbet, Small-scale turbulence and zooplankton metabolism: effects of turbulence on heartbeat rates of planktonic crustaceans, *Limnol. Oceanogr., 39*, 1465–1470, 1994.

Alheit, J., and E. Hagen, Long time climate forcing of European herring and sardine populations, *Fish. Ocean., 6*, 130–139, 1997.

Alheit, J., and E. Hagen, The effect of climatic variation on pelagic fish and fisheries, in *History and Climate. Memories of the Future*, edited by P. D. Jones, A. E. J. Ogilvie, T. D. Davies and K. R. Briffa, pp. 247–265, Kluwer Academic/Plenum Publishers, New York, 2001.

Anonymous, Evidence for decadal scale variations in seabird population ecology and links with the North Atlantic Oscillation, p. 29-32, in *Oceanography Committee. Report of the Working Group on Seabird Ecology*, pp. 29–32, ICES CM 1998/C:5, 1998.

Astthorsson, O. S., and A. Gislason, Long-term changes in zooplankton biomass in Icelandic waters in spring, *ICES J. Mar. Sci., 52*, 657–668, 1995.

Bates, N. R., Interannual variability of oceanic CO_2 and biogeochemical properties in the Western North Atlantic subtropical gyre, *Deep-Sea Res. II, 48* , 1507–1528, 2001.

Barber R. T., A. K. Hilting, and M. L. Hayes, The changing health of coral reefs, *Human and Ecological Risk Assessment, 7*, 1255–1270, 2001.

Barton, A. D., C. H. Greene, B. C. Monger, and A. J. Pershing, The continuous plankton recorder survey and North Atlantic Oscillation: interannual to multi-decadal scale patterns of phytoplankton variability in Northwest Shelf, Northeast Shelf and Central North Atlantic ecosystems, *Prog. Oceanogr.*, in press, 2002.

Baumgartner, T. R., A. Soutar, and V. Ferreira-Bartrina, Reconstruction of the history of Pacific sardine and northern anchovy populations over the past two millennia from sediments of the Santa Barbara Basin, California, *CalCOFI Rep., 33*, 24–40, 1992.

Beare, D. J., and E. McKenzie, Connecting ecological and physical time-series: the potential role of changing seasonality, *Mar. Ecol. Prog. Ser., 178*, 307–309, 1999.

Beaugrand, G., F. Ibañez, and P. C. Reid, Spatial, seasonal and long-term fluctuations of plankton in relation to hydroclimatic features in the English Channel, Celtic Sea and Bay of Biscay, *Mar. Ecol. Prog. Ser., 200*, 93–102, 2000.

Belgrano, A, O. Lindahl, and B. Hernroth, North Atlantic Oscillation (NAO), primary productivity and toxic phytoplankton in the Gullmar Fjord, Sweden (1985–1996), *Proc. R. Soc. Lond. B., 266*, 425–430, 1999.

Belgrano, A., B. A. Malmgren, and O. Lindahl, Application of artificial neural networks (ANN) to primary production time-series data, *J. Plank. Res., 23*, 651–658, 2001.

Bogstad, B., and H. Gjøsæter, A method for estimating the consumption of capelin by cod in the Barents Sea, *ICES J. Mar. Sci., 51*, 273–280, 1994.

Borja, A., and J. Santiago, *Does the North Atlantic Oscillation control some processes influencing recruitment of temperate tunas?* ICCAT SCRS/01/33, 19 pp., 2001.

Brander, K. M., Patterns of distribution, spawning, and growth in North Atlantic cod: the utility of inter-regional comparisons, *ICES Mar. Sci. Symp., 198*, 406–413, 1994.

Brander, K. M., The effect of temperature on growth of Atlantic cod (*Gadus morhua* L.), *ICES J. Mar. Sci., 52*, 1–10, 1995.

Brander, K., Effects of climate change on cod (*Gadus morhua*) stocks, in *Global Warming: Implications for Freshwater and Marine Fish,* edited by C. M. Wood and D. G. McDonald, pp. 259–278, Soc. Exp. Biol. Sem. Ser., 61, 1996.

Buzas, M. A., and L. C. Hayek, Biodiversity resolution: an integrated approach, *Biodiv. Lett., 3*, 40–43, 1996.

Caswell, H., M. Fujiwara, and S. Brault, Declining survival probability threatens the North Atlantic right whale, *Proc. Natl. Acad. Sci. USA, 96*, 3308–3313, 1999.

Colebrook, J. M., Variability in the distribution and abundance of the plankton, *ICNAF Spec. Pub., 8*, 167–184, 1972.

Colebrook, J. M., The continuous plankton recorder survey: Automatic data processing methods, *Bull. Mar. Ecol., 8*, 123–142, 1975.

Colebrook, J. M., Continuous plankton records: phytoplankton, zooplankton and environment, North-East Atlantic and North Sea, 1958-1980, *Oceanol. Acta, 5*, 473–480, 1982.

Colebrook, J. M., G. A. Robinson, H. G. Hunt, J. Roskell, A. W. G. John, H. H. Bottrell, J. A. Lindley, N. R. Collins, and N. C. Halliday, Continuous plankton records: A possible reversal in the downtrend in the abundance of the plankton of the North Sea and the Northeast Atlantic, *J. Cons. Int. Explor. Mer., 41*, 304–306, 1984.

Conversi, A., S. Piontkovski, and S. Hameed, Seasonal and interannual dynamics of *Calanus finmarchicus* in the Gulf of Maine (Northeastern US shelf) with reference to the North Atlantic Oscillation, *Deep-Sea Res. II, 48*, 519–539, 2001.

Cook, T., M. Folli, J. Klinck, S. Ford, and J. Miller, Increasing sea surface temperature and northward spread of *Perkinsus marinus* (Dermo) disease epizootics in oysters, *Estuar. Coast. Shelf Sci., 46*, 587–597, 1998.

Cooper, J. A., and F. Chapleau, Monophyly and intrarelationships of the family *Pleuronectidae* (*Pleuronectiformes*), with a revised classification, *Fish. Bull., U.S., 96*, 686–726, 1998.

Curry, R. G., and M. S. McCartney, Ocean gyre circulation changes associated with the North Atlantic Oscillation, *J. Phys. Oceanogr., 31*, 3374–3400, 2001.

Cushing, D. H., Biological and hydrographic changes in British Seas during the last thirty years, *Biol. Rev., 41*, 221–258, 1966.

Cushing, D. H., *Climate and fisheries*, 373 pp., Academic Press, London, 1982.

Cushing, D. H., Recent studies on long-term changes in the sea, *Freshwater Biology, 23*, 71–84, 1990.

de la Serna, J. M., E. Alot, and P. Rioja, *Nota sobre el reclutamiento de atún rojo (Thunnus thynnus L. 1758) en el Mediterráneo Occidental durante el año 1996*, ICCAT SCRS/97/81, 1997.

de Young, B., and G. A. Rose, On recruitment and distribution of Atlantic cod (*Gadus morhua*) off Newfoundland, *Can. J. Fish. Aquat. Sci., 50*, 2729–2740, 1993.

Dickson, R. R., From the Labrador Sea to global change, *Nature, 386*, 649–650, 1997.

Dickson, R. R., and K. M. Brander, Effects of a changing windfield on cod stocks of the North Atlantic, *Fish. Oceanogr., 2*, 124–153, 1993.

Dickson, R. R., and W. R. Turrell, The NAO: the dominant atmospheric process affecting oceanic variability in home, middle and distant waters of European salmon, in *The Ocean Life of Atlantic Salmon-Environmental and Biological Factors Influencing Survival*, edited by D. Mills, pp. 92–115, Fishing News Books, Oxford, U.K., 2000.

Dickson, R. R., P. M. Kelly, J. M. Colbrook, W. S. Wooster, and D. H. Cushing, North winds and production in the eastern North Atlantic, *J. Plank. Res., 10,* 151–169, 1988.

Dickson, R., J. Lazier, J. Meincke, P. Rhines, and J. Swift, Long-term coordinated changes in the convective activity of the North Atlantic, *Prog. Oceanogr., 38,* 241–295, 1996.

Dickson R. R., T. J. Osborn, J. W. Hurrell, J. Meincke, J. Blindheim, B. Ådlandsvik, T. Vinje, G. Alekseev, and W.

Maslowski, The Arctic Ocean response to the North Atlantic Oscillation, *J. Clim., 13*, 2671–2696, 2000.

Dippner, J., and G. Ottersen, Cod and climate variability in the Barents Sea, *Clim. Res., 17*, 73–82, 2001.

Drinkwater, K. F., A review of the role of climate variability in the decline of northern cod, *Amer. Fish. Soc. Symp., 32*, 113-130, 2002.

Drinkwater, K. F., and D. B. Mountain, Climate and Oceanography in *Northwest Atlantic Groundfish: Perspectives on a Fishery Collapse*, edited by J. G. Boreman, B. S. Nakashima, J. A. Wilson and R. L. Kendell, pp. 3–25, American Fisheries Society, Bethesda, Maryland, 1997.

Drinkwater, K .F., D. B. Mountain, and A. Herman, Variability in the slope water properties off eastern North America and their effects on the adjacent shelves, *ICES C.M. 1999/O:08*, pp. 26, 1999.

Durbin, A. G., and E .G. Durbin, Seasonal changes in size frequency distribution and estimated age in the marine copepod *Acartia hudsonica* during a winter-spring diatom bloom in Narragansett Bay, *Limnol. Oceanogr, 37*, 379–392, 1992.

Dutkiewicz, S., M. Follows, J. Marshall, and W. W. Gregg, Interannual variability of phytoplankton abundances in the North Atlantic, *Deep-Sea Res. II, 48*, 2323–2344, 2001.

Edwards, M., Large-scale temporal and spatial patterns of marine phytoplankton and climate variability in the North Atlantic, Ph.D. thesis, 243 pp., University of Plymouth, 2000.

Edwards, M., A. W. G. John, H. G. Hunt, and J. A. Lindley, Exceptional influx of oceanic species into the North Sea late 1997, *J. Mar. Biol. Assoc. U.K., 79*, 737–739, 1999.

Edwards, M., P. C. Reid, and B. Planque, Long-term and regional variability of phytoplankton biomass in the Northeast Atlantic (1960–1995), *ICES J. Mar. Sci., 58*, 39–49, 2001.

Ellertsen, B., P. Fossum, P. Solemdal, and S. Sundby, Relation between temperature and survival of eggs and first-feeding larvae of northeast Arctic cod (*Gadus morhua* L.), *Rapp. P.-v. Cons. Int. Explor. Mer, 191*, 209–219, 1989.

Ford, S. E., Range extension by the oyster parasite *Perkinsuys marinus* into the northeastern United States: response to climate change?, *J. Shellfish Res., 15*, 45–56, 1996.

Friedland, K., Marine temperatures experienced by postsmolts and the survival of Atlantic salmon, *Salmo salar* L., in the North Sea area, *Fish. Oceanogr., 7*, 22–34, 1998.

Friedland, K. D., D. G. Reddin, and J. F. Kocik, Marine survival of North American and European Atlantic salmon: effects of growth and environment, *ICES J. Mar. Sci., 50*, 481–492, 1993.

Fromentin, J. M., and B. Planque, *Calanus* and environment in the eastern North Atlantic. II. Influence of the North Atlantic Oscillation on *C. finmarchicus* and *C. hegolandicus, Mar. Ecol. Prog. Ser.,134*, 111–118, 1996.

Gallego, A., J. Mardaljevic, M. R. Heath, D. Hainbucher, and D. Slagstad, A model of the spring migration into the North Sea by *Calanus finmarchicus* overwintering off the Scottish continental shelf, *Fish. Ocean., 8 (Suppl. 1)*, 107–125, 1999.

Gieskes, W. W. C., and G. W. Kraay, Continuous plankton records: changes in the plankton of the North Sea and its eutrophic southern Bight from 1948 to 1975, *Neth. J. Sea Res., 11*, 334–364, 1977.

Gislason, A., and O. S. Astthorsson, Seasonal cycles of zooplankton southwest of Iceland, *J. Plank. Res., 17*, 1959–1976, 1995.

GLOBEC, Global Ocean Ecosystem Dynamics Implementation Plan, *IGBP Report 47*, GLOBEC Report 13, pp. 1–207, 1999.

Greene, C. H., and A. J. Pershing, The response of *Calanus finmarchicus* populations to climate variability in the Northwest Atlantic: Basin-scale forcing associated with the North Atlantic Oscillation (NAO), *ICES J. Mar. Sci., 57*, 1536–1544, 2000.

Guisande, C., J. M. Cabanas, A. R. Vergara and I. Riveiro, Effect of climate on recruitment success of Atlantic Iberian sardine *Sardina pilchardus*, *Mar. Ecol. Prog. Ser., 223*, 243–250, 2001.

Hagberg, J., and B. G. Tunberg, Studies on the covariation between physical factors and the long-term variation of the marine soft bottom macrofauna in Western Sweden, *Est. Coast. Shelf Sci., 50*, 373–385, 2000.

Hansen, B., W. R. Turrell, and S. Østerhus, Decreasing overflow from the Nordic Seas into the Atlantic Ocean through the Faroe Bank Channel since 1950, *Nature, 411*, 927–930, 2001.

Hansen, P. M., Studies on the biology of the cod in Greenland waters, *Rapp. P. v. Réun. Cons. Int. Explor. Mer, 123*, 1–83, 1949.

Hardy, A. C., Ecological investigations with the continuous plankton recorder: object, plan and methods, *Hull Bull. Mar. Ecol., 1*, 1–57, 1939.

Harvell, C. D., et al., Emerging marine diseases - climate links and anthropogenic factors, *Science, 285*, 1505–1510, 1999.

Hayes, M. L., J. Bonaventura, T. P. Mitchell, J. M. Prospero, E. A. Shinn, F. Van Dolah, and R. T. Barber, How are climate and emerging diseases functionally linked?, *Hydrobiologia,* in press, 2002.

Hays, G. C., and A. J. Warner, Consistency of towing speed and sampling depth for the continuous plankton recorder, *J. Mar. Biol. Assoc. U.K., 73*, 967–970, 1993.

Hays, G. C., M. C. Carr, and A. H. Taylor, The relationship between Gulf Stream and copepod abundance derived from the continuous plankton recorder survey: separating biological signal from sampling noise, *J. Plank. Res., 15*, 1359–1373, 1993.

Heath, M. R., et al, Climate fluctuations and the spring invasion of the North Sea by *Calanus finmarchicus*, *Fish. Ocean., 8 (Suppl. 1)*, 163–176, 1999.

Helle, K., Distribution of early juvenile Arcto-Norwegian cod (*Gadus morhua* L.) in relation to food abundance and watermass properties, IC*ES Mar. Sci. Symp., 198*, 440–448, 1994.

Helle, K., and M. Pennington, The relation of the spatial distribution of early juvenile cod (*Gadus morhua* L.) in the Barents Sea to zooplankton density and water flux during the period 1978–1984, *ICES J. Mar. Sci., 56*, 12–27, 1999.

Hurrell, J. W., Decadal trends in the North Atlantic Oscillation: regional temperatures and precipitation, *Science, 169*, 676–679, 1995.

Hurrell, J. W., Y. Kushnir, G, Ottersen, and M. Visbeck, An overview of the North Atlantic Oscillation, this volume.

Irigoien, X., R. P. Harris, R. N. Head, and D. Harbour, North Atlantic Oscillation and spring bloom phytoplankton composition in the English Channel, *J. Plank. Res., 22*, 2367–2371, 2000.

Izhevskii, G. K., Forecasting of oceanological conditions and the reproduction of commercial fish, Moskva. Moscow, Pishcepromizdat, 1964.

Jensen, Ad. S., and P. M. Hansen, Investigations on the Greenland cod (*Gadus callarias L.*), *Rapp. P.-v. Reun. Cons. int. Explor. Mer., 72*, 1–41, 1931.

Johns, D. G., M. Edwards, and S. D. Batten, Arctic boreal plankton species in the Northwest Atlantic, *Can. J. Fish. Aquat. Sci., 58*, 2121–2124, 2001.

Jones, P. D., T. Jonsson, and D. Wheeler, Extension to the North Atlantic Oscillation using early instrumental pressure observations from Gibraltar and south-west Iceland, *Int. J. Climatol., 17*, 1433–1450, 1997.

Jossi, J. W., and J. R. Goulet, Zooplankton trends: US north-east shelf ecosystem and adjacent regions differ from north-east Atlantic and North Sea, *ICES J. Mar. Sci., 50*, 303–313, 1993.

Karentz, D., and T. J. Smayda, Temporal patterns and variations in phytoplankton community organization and abundance in Narragansett Bay during 1959–1980, *J. Plankton Res., 20*, 145–168, 1998.

Keller, A. A., and G. Klein-MacPhee, Impact of elevated temperature on the growth, survival, and trophic dynamics of winter flounder larvae: a mesocosm study, *Can. J. Fish. Aquat. Sci., 57*, 2382–2392, 2000.

Keller, A. A., C. A. Oviatt, H. A. Walker, and J. D. Hawk, Predicted impacts of elevated temperature on the magnitude of the winter-spring phytoplankton bloom in temperate coastal waters: a mesocosm study, *Limnol. Oceanogr., 44*, 344–356, 1999.

Kenney, R. D., M. A. M. Hyman, R. E. Owen, G. P. Scott, and H. E. Winn, Estimation of prey densities required by western North Atlantic right whales, *Mar. Mamm. Sci., 2*, 1–13, 1986.

Klein-MacPhee, G., Synopsis of biological data for the winter flounder, *Pseudopleuronectes americanus* (Walbaum), *NOAA Tech. Rep.,* NMFS Circ., 414, 43 pp. 1978.

Knowlton, A. R., S. D. Kraus, and R. D. Kenney, Reproduction in North Atlantic right whales (*Eubalaena glacialis*), *Can. J. Zool., 72*, 1297–1305, 1994.

Kroencke, I., J. W. Dippner, H. Heyen, and B. Zeiss, Long-term changes in macrofaunal communities off Norderney (East Frisia, Germany) in relation to climate variability, *Mar. Ecol. Prog. Ser., 167*, 25–36, 1998.

Krovnin, A. S., A comparative study of climatic changes in the North Pacific and North Atlantic and their relation to the abundance of fish stocks, in *Climate Change and Northern Fish Populations*, edited by R. J. Beamish, pp. 181–198, *Can. Spec. Publ. Fish. Aquat. Sci., 121*, 1995.

Lindahl, O., A. Belgrano, L. Davidsson, and B. Hernroth. Primary production, climatic oscillations, and physico-chemical processes: The Gullmar Fjord time-series data set (1985-1996), *ICES J. Mar. Sci., 55*, 723–729, 1998.

Loeng, H., Features of the physical oceanographic conditions of the Barents Sea, *Polar Res., 10*, 5–18, 1991.

Loeng, H., J. Blindheim, B. Ådlandsvik, and G. Ottersen, Climatic variability in the Norwegian and Barents Seas, *ICES Mar. Sci. Symp., 195*, 52–61, 1992.

Mackenzie, B. R., and W. C. Leggett, Quantifying the contribution of small-scale turbulence to the encounter rates between larval fish and their zooplankton prey: effects of wind and tide, *Mar. Ecol. Prog. Ser., 73*, 149–160, 1991.

Maddock, L., D. S. Harbour, and G. T. Boalch, Seasonal and year-to-year changes in the phytoplankton from the Plymouth area, 1963–1986, *J. Mar. Biol. Ass. U.K., 69*, 229–244, 1989.

Magurran, A. E., *Ecological Diversity and its Measurement*, 135 pp., Princeton University Press, 1988.

Mann, K. H., and K. F. Drinkwater, Environmental influences on fish and shellfish production in the Northwest Atlantic, *Environ. Rev., 2*, 16–32, 1994.

Marak, R. R., Food habits of larval cod, haddock and coalfish in the Gulf of Maine and Georges Bank area, *J. Cons. Int. Explor. Mer, 25,* 147–157, 1960.

Marsh, R., B. Petrie, C. R. Weidman, R. R. Dickson, J. W. Loder, C. G. Hannah, K. Frank, and K. Drinkwater, The Middle Atlantic Bight tilefish kill of 1882, *Fish. Oceanogr., 8*, 39–49, 1999.

Mejuto, J., and J. M. de la Serna, *Updated Standardized Catch Rates by age for the Swordfish (Xiphias gladius) from the Spanish Longline Fleet in the Atlantic using Commercial Trips from the Period 1983-1995*, ICCAT Collec. Vol. Sci. Pap., SCRS/96/141, 1997.

MERCINA, Oceanographic responses to climate in the Northwest Atlantic, *Oceanogr., 14*, 76–82, 2001.

Myers, R. A., and N. G. Cadigan, Density-dependent juvenile mortality in marine demersal fish, *Can. J. Fish. Aquat. Sci., 50*, 1576–1590, 1993.

Nordberg, K., M. Gustafsson, and A. L. Krantz, Decreasing oxygen concentrations in the Gullmar Fjord, Sweden, as confirmed by benthic foraminifera, and the possible association with NAO, *J. Mar. Syst., 23*, 303–316, 2000.

O'Connell, S., Weather's wrong for right whale, *BBC Wildlife Magazine, 19*, 40, 2001.

Ottersen, G., Environmental impact on variability in recruitment, larval growth and distribution of Arcto-Norwegian cod, Ph. D. Thesis, 136 p, Geophysical Institute, University of Bergen, 1996.

Ottersen, G., and S. Sundby, Effects of temperature, wind and spawning stock biomass on recruitment of Arcto-Norwegian cod, *Fish. Oceanogr., 4*, 278–292, 1995.

Ottersen, G., and H. Loeng, Covariability in early growth and year-class strength of Barents Sea cod, haddock and herring: The environmental link, *ICES J. Mar. Sci., 57*, 339–348, 2000.

Ottersen, G., H. Loeng, and A. Raknes, Influence of temperature variability on recruitment of cod in the Barents Sea, *ICES Mar. Sci. Symp., 198*, 471–481, 1994.

Ottersen, G., K. Michalsen, and O. Nakken, Ambient temperature and distribution of north-east Arctic cod, *ICES J. Mar. Sci., 55*, 67–85, 1998.

Ottersen, G., B. Planque, A. Belgrano, E. Post, P. C. Reid, and N. C. Stenseth, Ecological effects of the North Atlantic Oscillation, *Oecologia, 128*, 1–14, 2001.

Ottersen, G., and N. C. Stenseth, Atlantic climate governs oceanographic and ecological variability in the Barents Sea, *Limnol. Oceanogr., 46*, 1774–1780, 2001.

Pearcy, W. G., Ocean ecology of North Pacific salmonids, Washington Sea Grant Program, 179 pp., University of Washington Press, Seattle, Washington, 1992.

Planque, B., and C. J. Fox, Interannual variability in temperature and the recruitment of Irish Sea cod, *Mar. Ecol. Prog. Ser., 172*, 101–105, 1998.

Planque, B., and T. Frédou, Temperature and the recruitment of Atlantic cod (*Gadus morhua), Can. J. Fish. Aquat. Sci., 56*, 1–9, 1999.

Planque, B., and P. C. Reid, Predicting *Calanus finmarchicus* abundance from a climatic signal, *J. Mar. Biol. Ass. U.K., 78*, 1015–1018, 1998.

Planque, B., and A. H. Taylor, Long-term changes in zooplankton and the climate of the North Atlantic, *ICES J. Mar. Sci., 55*, 644–654, 1998.

Reddin, D., and W. M. Shearer, Sea-surface temperature and distribution of Atlantic salmon in the Northwest Atlantic Ocean, *Amer. Fish. Soc. Symp., 1*, 262–275, 1987.

Reeves, R. R., and E. Mitchell, The Long Island, New York, right whale fishery: 1650-1924, *Rep. Int. Whal. Commn. (Spec. Issue), 10*, 201–220, 1986.

Reid, P. C., Continuous plankton records: Changes in the composition and abundance of the phytoplankton of the north-eastern Atlantic Ocean and North Sea, 1958–1974, *Mar. Biol., 40*, 337–339, 1977.

Reid, P. C., Continuous plankton records: Large-scale changes in the abundance of phytoplankton in the North Sea from 1958 to 1973, *Rapp. P.-v. Réun. Cons. int. Explor. Mer, 172*, 384–389, 1978.

Reid, P. C., M. Edwards, H. G. Hunt, and A. J. Warner, Phytoplankton change in the North Atlantic, *Nature, 391*, 546, 1998.

Reid, P. C., B. Planque, and M. Edwards, Is observed variability in the long-term results of the Continuous Plankton Recorder survey a response to climate change?, *Fish. Oceanogr., 7*, 282–288, 1998.

Rodionov, S. N., Atmospheric teleconnections and coherent fluctuations in recruitment to North Atlantic cod (*Gadus morhua*) stocks, in *Climate Change and Northern Fish Populations*, edited by R. J. Beamish, pp. 45–55, *Can. Spec. Publ. Fish. Aquat. Sci.*, 121, 1995.

Rogers, J. C., Atmospheric circulation changes associated with the warming over the North Atlantic in the 1920s, *J. Clim. Appl. Meteor,, 24*, 1303–1310, 1985.

Rose, G. A., B. de Young, D. W. Kulka, S. V. Goddard, and G. L. Fletcher, Distribution shifts and overfishing the northern cod (*Gadus morhua*): a view from the ocean, *Can. J. Fish. Aquat. Sci., 57*, 644–663, 2000.

Rothschild, B. J., and T. R. Osborn, Small-scale turbulence and plankton contact rates, *J. Plank. Res., 10*, 465–474, 1988.

Sakshaug, E., Biomass and productivity distributions and their variability in the Barents Sea, *ICES J. Mar. Sci., 54*, 341–350, 1997.

Sameoto, D., Decadal changes in phytoplankton color index and selected calanoid copepods in continuous plankton recorder data from the Scotian Shelf, *Can. J. Fish. Aquat. Sci., 58*, 749–761, 2001.

Santiago, J., *The North Atlantic Oscillation and Recruitment of Temperate Tunas*, ICCAT SCRS/97/40, 20 pp., 1997.

Schlosser, P., M. Bonisch, M. Rhein, and R. Bayer, Reduction of deep water formation in the Greenland Sea during the 1980s: evidence from tracer data, *Science, 251*, 1054–1056, 1991.

Sharp, G. D., Climate and fisheries: cause and effect or managing the long and short of it all, *S. African J. Mar. Sci., 5*, 811–838, 1987.

Sheperd, J. G., J. G. Pope, and R. D. Cousens, Variations in fish stocks and hypotheses concerning their links with climate, *Rapp. P.-v. Reun. Cons. int. Explor. Mer, 185*, 255–267, 1984.

Sherman, B. H., A prototype methodology for the assessment of multiple ecological disturbance in the Baltic Sea ecosystem, *Human and Ecological Risk Assessment, 7*, 1519–1540, 2001.

Sherman, B .H., and P. R. Epstein, Past anomalies as a diagnostic tool for evaluating multiple marine ecological disturbance events, *Human and Ecological Risk Assessment, 7*, 1493–1518, 2001.

Sherman, K., R. Maurer, R. Byron, and J. Green, Relationship between larval fish communities and zooplankton prey species in an offshore spawning ground, *Rapp. P.-v. Reun. Cons. int. Explor. Mer, 178*, 289–294, 1981.

Sherman, K., W. G. Smith, J. R. Green, E. B. Cohen, M. S. Berman, K. A. Marti, and J. R. Goulet, Zooplankton production and fisheries of the northeastern shelf, in *Georges Bank*, edited by R. H. Backus, 268–282, MIT Press, Cambridge, 1987.

Sherman K., A. R. Solow, J. W. Jossi, and J. Kane, Biodiversity and abundance of the zooplankton of the Northeast Shelf ecosystem, *ICES J. Mar. Sci., 55*, 730–738, 1998.

Shindell D. T., R. L. Miller, G. A. Schmidt, and L. Pandolfo, Simulation of recent northern winter climate trends by greenhouse gas forcing, *Nature, 399*, 452–455, 1999.

Smayda, T. J., Patterns of variability characterizing marine phytoplankton, with examples from Narragansett Bay, *ICES J. Mar. Sci., 55*, 562–573, 1998.

Stephens J. A., M. B. Jordan, A. H. Taylor, and R. Proctor, The effects of fluctuations in North Sea flows on zooplankton abundance, *J. Plank. Res., 20*, 943–956, 1998.

Sundby, S., H. Bjørke, A. V. Soldal, and S. Olsen, Mortality rates during the early life stages and year class strength of the Arcto-Norwegian cod (*Gadus morhua* L.), *Rapp. P.-v. Reun. Cons. Int. Explor. Mer, 191*, 351–358, 1989.

Sundby, S., B. Ellertsen, and P. Fossum, Encounter rates between first-feeding cod larvae and their prey during moderate to strong turbulent mixing, *ICES J. Mar. Sci. Symp., 198*, 393–405, 1994.

Sverdrup, H. U., On conditions of the vernal blooming of phytoplankton, *J. Cons. Int. Explor. Mer, 18*, 287–295, 1953.

Sætersdal, G., and H. Loeng, Ecological adaption of reproduction in Northeast Arctic cod, *Fish. Res., 5*, 253–270, 1987.

Taylor, A. H., North-south shifts of the Gulf Stream and their climatic connection with the abundance of zooplankton in the UK and its surrounding seas, *ICES J. Mar. Sci., 52*, 711–721, 1995.

Taylor, A. H. and J. A. Stephens, Latitudinal displacements of the Gulf Stream (1966 to 1977) and their relation to changes in temperature and zooplankton abundance in the NE Atlantic, *Oceanol. Acta, 3*, 145–149, 1980.

Taylor, A. H., and J. A. Stephens, The North Atlantic Oscillation and the latitude of the Gulf Stream, *Tellus, 50A*, 134–142, 1998.

Taylor, A. H., J. M. Colebrook, J. A. Stephens, and N. G. Baker, Latitudinal displacements of the Gulf Stream and the abundance of plankton in the north-east Atlantic, *J. Mar. Biolo. Ass. U.K., 72*, 919–921, 1992.

Templeman, W., Year-class success in some North Atlantic stocks of cod and haddock, *ICNAF Spec. Publ., 8*, 223–241, 1972.

Thompson, P. M., and J. C. Ollason, Lagged effects of ocean climate change on fulmar population dynamics, *Nature, 413*, 417–420, 2001.

Tiews, K., On the disappearance of bluefin tuna in the North Sea and its ecological implications for herring and mackerel, *Rapp. P.-v. Réun. Cons. Int. Explor. Mer, 172*, 301–309, 1978.

Tunberg, B. G., and W. G. Nelson, Do climatic oscillations influence cyclical patterns of soft bottom macrobenthic communities on the Swedish west coast?, *Mar. Ecol. Prog. Ser., 170*, 85–94, 1998.

van Loon, H., and J. C. Rogers, The seesaw in winter temperatures between Greenland and northern Europe. Part 1: General description, *Mon. Wea. Rev., 106*, 296–310, 1978.

Visbeck, M., E. Chassignet, R. Curry, T. Delworth, B. Dickson, and G. Krahmann, The ocean's response to North Atlantic Oscillation variability, this volume.

Warner A. J., and G. C. Hays, Sampling the Continuous Plankton Recorder survey, *Prog. Oceanog., 34*, 237–256, 1994.

Wen, D., and R. A. Laursen, Structure-function relationships in an antifreeze polypeptide, *J. Biol. Chem., 26*, 14102-14108, 1992.

Winn, H. E., C. A. Price, and P. W. Sorensen, The distribution biology of the right whale (*Eubalaena glacialis*) in the western North Atlantic, *Rep. Int. Whal. Commn. (Spec. Issue), 10*, 129–138, 1986.

Wishner, K. F., J. R. Schoenherr, R. Beardsley, and C. Chen, Abundance, distribution and population structure of the copepod *Calanus finmarchicus* in a springtime right whale feeding area in the southwestern Gulf of Maine, *Cont. Shelf. Res., 15*, 475–507, 1995.

Worthington, L. V., Anomalous conditions in the Slope Water area in 1959, *J. Fish. Res. Board Can., 21*, 327–333, 1964.

Andrea Belgrano, Department of Biology, University of New Mexico, 167 Castetter Hall, Albuquerque, NM, 87131-1091, U.S.A.
belgrano@unm.edu

Ángel Borja, Head of the Marine Environment Section, Department of Oceanography, AZTI, Herrera Kaia, Portualdea s/n, 20110 Pasaia, Spain.
aborja@pas.azti.es

A. Conversi, Via XX Settembre 15, 19032 Lerici (SP), Italy.
conversi@goased.msrc.sunysb.edu, conversi@area.ba.cap.it

Kenneth F. Drinkwater, Department of Fisheries and Oceans, Bedford Institute of Oceanography, Box 1006, Dartmouth, Nova Scotia, Canada, B2Y 4A2.
drinkwaterk@mar.dfo-mpo.gc.ca

Martin Edwards, Sir Alister Hardy Foundation for Ocean Science (SAHFOS), The Laboratory, Citadel Hill, Plymouth, U.K., PL1 2PB.
maed@mail.pml.ac.uk

C. H. Greene, Ocean Resources and Ecosystems Program, Dept. of Earth and Atmospheric Sciences, 2130 Snee Hall, Cornell University, Ithaca, NY, 14853-2701, U.S.A.
chg2@cornell.edu

Geir Ottersen, Institute of Marine Research, P.O. Box 1870 Nordnes, 5817 Bergen, Norway
Current address:
Department of Biology, Division of Zoology, University of Oslo, P.O. Box 1050 Blindern, N-0316 Oslo, Norway
geir.ottersen@bio.uio.no

Andrew J. Pershing, Ocean Resources and Ecosystems Program, Dept. of Earth and Atmospheric Sciences, 2130 Snee Hall, Cornell University, Ithaca, NY, 14853-2701, U.S.A.
ajp9@cornell.edu

Henry A. Walker, US Environmental Protection Agency, Office of Research and Development, National Health and Environmental Effects Research Laboratory, Atlantic Ecology Division, 27 Tarzwell Drive, Narragansett, RI, 02882, U.S.A.
walker.henry@epamail.epa.gov

SCOTT W. STARRATT

The Response of Terrestrial Ecosystems to Climate Variability Associated with the North Atlantic Oscillation

Atle Mysterud[1], Nils Chr. Stenseth[1], Nigel G. Yoccoz[1,2], Geir Ottersen[3], and Rolf Langvatn[4]

Climatic factors influence a variety of ecological processes determining patterns of species density and distribution in a wide range of terrestrial ecosystems. We review the effects of the NAO on processes and patterns of terrestrial ecosystems, including both plants and animals. In plants, the NAO index correlates with date of first flowering, tree ring growth and with quality of agricultural crops (wheat and wine grapes). Also, breeding dates are earlier after high NAO index winters for amphibians and birds in Europe. Population dynamical consequences of the NAO have also been reported for birds, and the differential impact of the NAO on two similar species may prevent competitive exclusion. Different effects of the NAO on large herbivore populations have been reported for different regions, depending on limiting factors and the correlation with local weather parameters. The NAO synchronizes population dynamics of lynx and some other carnivore populations in the eastern U.S. Most effects are on an ecological time scale; the evolutionary consequences of long term trends in the NAO are poorly documented. Important for predator and prey dynamics is (1) the disruption of phenology (the match-mismatch hypothesis), (2) that there may be delayed effects (cohort-effects), and (3) that effects of the NAO may interact with other factors such as density. We discuss the challenges related to nonlinearity, of using different climate indices, and how we can progress using these pattern-oriented NAO studies at coarse scales to conduct better process-oriented small-scale experiments.

1. INTRODUCTION

It has long been recognized that climatic factors may have a profound influence on a variety of ecological processes determining both species density and distribution in a wide range of terrestrial ecosystems. During the last few years there has been a dramatic increase in the number of studies documenting the impact of large-scale climatic variability, such as the El Niño-Southern Oscillation (ENSO) [*Philander*, 1990] and the North Atlantic Oscillation (NAO) [*Hurrell*, 1995]. For ENSO two reviews have recently been published [*Holmgren et al.,* 2001; *Jaksic*, 2001]. Here we review the reported effects of the NAO on processes and patterns of terrestrial ecosystems.

The impacts of climate on individuals and populations are through variations in local weather parameters, such as temperature, wind, rain and snow, and interactions among weather parameters and biotic factors. Such local weather variations are, however, often governed by climatological phenomena extending over large geographic areas. Interaction between the ocean and atmosphere may form dynamical systems, exhibiting complex patterns of variation, which may profoundly influence ecological processes in a number of ways [*Ottersen*

[1]Department of Biology, Division of Zoology, University of Oslo, Oslo, Norway

[2]Division of Arctic Ecology, Norwegian Institute for Nature Research, Polar Environmental Center, Tromsø, Norway

[3]Institute of Marine Research, Bergen, Norway. Current address: Department of Biology, Division of Zoology, University of Oslo, Oslo, Norway

[4]University Courses on Svalbard, Longyearbyen, Spitsbergen, Norway

The North Atlantic Oscillation:
Climatic Significance and Environmental Impact
Geophysical Monograph 134

10.1029/134GM11

et al., 2001]. A composed function involving a variety of climatic parameters over time and space may prove useful to improve our understanding of the ecological impacts of climate. The NAO may be considered to represent such an integrated measure of climate, not the least since it incorporates several climatological features in the North Atlantic region.

Furthermore, the increasing focus on global warming and its ecological consequences [e.g., *Hughes*, 2000; *McCarty*, 2001] provides additional reasons to relate the NAO to ecosystem functioning. If the recent increasing trend in the NAO index is somehow linked to global warming, as for instance suggested by *Hurrell et al.* [2001], the NAO might indeed be a better measure than more local climatic variables at predicting future effects of global climatic change, on the scale of continents or subcontinents. The effect of global change may increase temperatures locally in some regions, while leading to cooler conditions in other regions - differential effects that indeed may be linked up by the NAO. Some of the possible ecological effects of the NAO [*Post et al.*, 1999a; *Forchhammer*, 2001; *Ottersen et al.*, 2001] as well as ecological effects of climatic factors in general [*Sæther*, 1997; *Crawford*, 2000; *McCarty*, 2001; *Hughes*, 2000; *Weladji et al.*, 2002] have already been reviewed. None of these papers, however, are in-depth reviews on terrestrial ecosystems; rather they either provide general overviews of the field or are fairly selective in the material they review. Further, essentially all previous studies and reviews have only considered linear relationships between the NAO and ecological processes. Little focus has also been put on the many possible pitfalls when analysing time-series of data. A more in-depth review of the effects of the NAO on terrestrial ecosystems is thus warranted - hence, this review.

As described in other parts of this Monograph [e.g. *Hurrell et al.*, this volume], the NAO is a large-scale fluctuation in atmospheric pressure difference between the subtropical North Atlantic region (centered on the Azores) and the sub-polar North Atlantic region (centered on Iceland) mainly affecting winter climate. The increased pressure difference reflected in high (or positive) NAO index months corresponds to more and stronger winter storms crossing the Atlantic Ocean along a more northerly track leading to high temperatures in Western Europe and low temperatures in eastern coastal Canada [*Mann and Lazier*, 1991]. A low-index (or negative) NAO-phase leads to the opposite conditions. Ecologists must be aware of this geographically differential impact of the NAO. That is, the correlation of temperature, wind, and precipitation with the NAO varies between regions [*Dickson et al.*, 2000]. Therefore, the value of the NAO as a proxy varies regionally and has to be evaluated when studying specific areas. This is so on the full North Atlantic scale, regionally, and locally according to altitude [e.g., *Mysterud et al.*, 2000].

We first provide an overview of reported patterns relating to the effects of the NAO on terrestrial ecosystems. This part is organized as a typical natural historian would see it, with a marked focus on taxonomic groups rather than the mechanisms or specific population ecological phenomenon under study. This does not only reflect our genuine interest in patterns of different taxa, but also that studies relating the NAO to specific mechanisms at present is often, but not always, somewhat speculative. Although we aim at covering the reported patterns rather comprehensively, we do not aim at providing a full review of how (i.e., mechanisms) climate may affect terrestrial ecosystems. At most we try to suggest which mechanisms might plausibly be involved. At the very end we conclude with a discussion on emerging insights of the patterns observed and look to future challenges.

2. IMPACTS OF THE NAO ON TERRESTRIAL ECOSYSTEMS

An overview of the relationships between the NAO and different biological variables are listed in Table 1. In the following, we provide a closer description of these studies.

2.1. Plants

Predicted responses of plants to climatic warming include an earlier and longer annual growing season [*Menzel and Fabian*, 1999], changes in biomass production [*Myneni et al.*, 1997], increased distribution range [*Sturm et al.*, 2001], increased [*Sætersdal et al.*, 1998] or decreased [*Crawford*, 2000] species richness, increased rates of population growth, and altered timing of plant dynamics [*Post and Stenseth*, 1999]. From experimental work, it is known that increased temperature may advance key phenological events such as leaf bud burst and flowering [*Arft et al.*, 1999]. Changes in phenology (seasonal activity driven by environmental factors) from year to year may be a sensitive and easily observable indicator of changes in the biosphere [*Menzel and Fabian*, 1999]. As a result most work on plants as it relates to the NAO has been done on phenology.

2.1.1. Phenology. The most extensive data set on phenological development that has been analysed with regard to the effect of the NAO [see *Post and Stenseth*, 1999], is based on reports of annual dates of first flowering from 43 species of herbaceous and woody plants in 37 sites in Norway for the time period 1928 until 1977 [*Lauscher and Lauscher*, 1990]. Plants bloomed earlier following positive NAO index winters. Dates of flowering by *Anemone hepat-*

Table 1. A review of studies relating an effect of the NAO to different biological variables in terrestrial ecosystems. The review is restricted to papers explicitly using the NAO index.

Taxonomic group Species	Location	Time period or length	Biological variable	Relationship with NAO	Suggested mechanism	Reference
Plants						
Anemone hepatica, A. nemorosa, Convallaria majalis, Linnaea borealis, Epilobium angustifolium, Tussilago farfara, Caltha palustris, Oxalis acetosella, Primula officinalis, Trientalis europaea	Helle, Norway (and other locations)	1928-1977	Dates of flowering; flowering season lenght	- (lin), except two last species; + (lin) in 4 cases	Temperature	*Post and Stenseth*, 1999; *Post et al.*, 2001a; *Post et al.*, 2001b
Calluna vulgaris and *Vaccinium myrtillus*	Several locations, Norway	1928-1977	Dates of flowering	- (lin) on 50% of locations	Temperature	*Post and Stenseth*, 1999
Betula pubescens	Several locations, Norway	1928-1977	Date of leaf emergence	- (lin) on 50% of locations	Temperature	*Post and Stenseth*, 1999
European beech (*Fagus sylvatica*)			Diameter of tree rings	- (lin) in parts of time series	availability of water in soil in spring	*Piovesan and Schirone*, 2000
Several species (not given)	eastern North America and western Europe	1700-1979	Diameter of tree rings	sign. relationship	temperature/ precipitation depending on region	*Cook et al.*, 1998
Fir (*Abies balsamea*)	Isle Royale, USA		Diameter of tree rings	- (lin)		*Post et al.*, 1999d
Scots pine (*Pinus silvestris*); white spruce (*Picea glauca*)	Fennoscandia and Labrador		Diameter of tree rings			*D'Arrigo et al.*, 1993
Norway spruce (*Picea abies*)	Flatanger, central Norway	1898-1997 1920-1940	Diameter of tree rings	0 - (lin)	Temperature	*Solberg et al.*, 2002
Wheat	U.K.	1972-1996	3 quality indices	+ (lin) with 2 indices; no relationship with 3rd	precipitation in August	*Kettlewell et al.*, 1999
Wine	Spain	1964-1994	quality	0		*Rodó and Comín*, 2000
Wine	Portugal	33 yrs	quality	- (lin)	Precipitation	*Esteves and Orgaz*, 2001
Amphibians						
Bufo calamita, Rana esculenta and R. temporaria	U.K.	1970-1994	Spawning date	- (lin)	Temperature	*Forchhammer et al.*, 1998a
Triturus vulgaris, T. helveticus and T. cristatus	U.K.	1970-1994	First date of sighting	- (lin)	Temperature	*Forchhammer et al.*, 1998a
Birds						
Corn bunting (*Miliaria calandra*), Common chiffchaff (*Phylloscopus collybita*), and Magpie (*Pica pica*)		1970-1994	First egg-laying date	- (lin)		*Forchhammer et al.*, 1998a

Table 1. (continued)

Taxonomic group Species	Location	Time period or length	Biological variable	Relationship with NAO	Suggested mechanism	Reference
Sky lark (*Alauda arvensis*)	Norway	50 years	Arrival date	- (lin)		*Forchhammer et al.*, 1998a
Great tit (*Parus major*), blue tit (*Parus caeruleus*) and pied flycatcher (*Ficedula hypoleuca*)	northern Germany	1970-1995	Hatching date	- (lin)	not disc.	*Forchhammer and Post*, 2000
Golden plover (*Pluvialis apricaria*) and Sandpiper (*Actitis hypoleucos*)	U.K.	1970-1995	Abundance	+ (lin), - (lin)		*Forchhammer et al.*, 1998a
Dipper (*Cinclus cinclus*)	Agder, Norway	1978-1997	Abundance	+ (nonlin)	temp/icing	*Sæther et al.*, 2000
Pied flycatcher (*Ficedula hypoleuca*)	Dlouhá Loučka, Czech Republik	1985-1997	Breeding density	0	not disc.	*Sætre et al.*, 1999
Pied flycatcher (*Ficedula hypoleuca*)	Gotland, Sweden	1980-1995	Laying date, clutch size, fledgling success, number of recruits, tarsus length, wing length	- (lin), + (lin), 0, 0, 0, 0	not disc.	*Przybylo et al.*, 2000
Collared flycatcher (*Ficedula albicollis*)	Dlouhá Loučka, Czech Republik	1985-1997	Breeding density	+ (lin)	not disc.	*Sætre et al.*, 1999
Mammals						
Red deer (*Cervus elaphus*)	west coast of Norway	1965-1992	Body weight and skeletal size of adults	- (lin); not sign in males	Snow depth (direct while *in utero*)	*Post et al.*, 1997; *Post and Stenseth*, 1999
Red deer (*Cervus elaphus*)	west coast of Norway	1965-1998	Body weight of adults	nonlin	Snow depth (direct and indir. through plants)	*Mysterud et al.*, 2001b
Red deer (*Cervus elaphus*)	Sør-Trøndelag, Norway	1957-1996	Body weight of calves	+ in males; - in females	None	*Post et al.*, 1999c
Red deer (*Cervus elaphus*)	Sør-Trøndelag, Norway	1957-1996	Longevity	+ (lin)	None	*Post et al.*, 1999c
Red deer (*Cervus elaphus*)	Sør-Trøndelag, Norway	1977-1997	Body weight of calves	+ (lin)	Snow depth (direct)	*Loison et al.*, 1999b; *Loison et al.*, submitted
Red deer (*Cervus elaphus*)	Norway	1962-1992	Number of harvested animals	+ (lin)		*Forchhammer et al.*, 1998b; *Post and Stenseth*, 1999
Red deer (*Cervus elaphus*)	west coast of Norway	1968-1989	Prop. of females breeding as 2 yr. olds	+ (lin)	indir. through plants	*Post and Stenseth*, 1999
Red deer (*Cervus elaphus*)	west coast of Norway	1977-1993	Prop. of male calves harvested	+ lin	snow depth (direct)	*Post et al.*, 1999b
Red deer (*Cervus elaphus*)	west coast of Norway	1977-1998	Prop. of male calves harvested	+ lin	snow depth (direct)	*Mysterud et al.*, 2000
Domestic sheep (*Ovis aries*)	west coast of Norway	1989-1998	Body weight of lambs	nonlin	snow depth (indir. through plants)	*Mysterud et al.*, 2001b

Table 1. (continued)

Taxonomic group Species	Location	Time period or length	Biological variable	Relationship with NAO	Suggested mechanism	Reference
Red deer (*Cervus elaphus*)	Rum, Scotland	13 and 14 yrs	Male; female abundance	- (lin); + (lin)		*Post and Stenseth*, 1999
Red deer (*Cervus elaphus*)	Rum, Scotland	13 yrs	Calf winterand summer mortality	0		*Post and Stenseth*, 1999
Red deer (*Cervus elaphus*)	Rum, Scotland	16 yrs	Yearling mortality	- (lin)		*Post and Stenseth*, 1999
Red deer (*Cervus elaphus*)	Rum, Scotland	11 yrs	Adult winter mortality	+ (lin)		*Post and Stenseth*, 1999
Red deer (*Cervus elaphus*)	Rum, Scotland	20 yrs	Adult mass	0		*Post and Stenseth*, 1999
Red deer (*Cervus elaphus*)	Rum, Scotland	9-13 yrs	Calf birth mass	+ (lin)		*Post and Stenseth*, 1999
Red deer (*Cervus elaphus*)	Rum, Scotland	15-21 yrs	Cohort fecundity (3 yr olds)	- (lin), not sign in yeld hinds		*Post and Stenseth*, 1999
Soay sheep (*Ovis aries*)	Village Bay	6 yrs	Lamb birth mass	- (lin)		*Post and Stenseth*, 1999
Soay sheep (*Ovis aries*)	Hirta and Village Bay	13 and 16 yrs	Abundance	- (lin)		*Post and Stenseth*, 1999
Soay sheep (*Ovis aries*)	Village Bay		Abundance	+ (lin)		*Catchpole et al.*, 2000
Soay sheep (*Ovis aries*)	Village Bay	6 yrs	Lamb winter mortality	0		*Post and Stenseth*, 1999
Soay sheep (*Ovis aries*)	Hirta, Scotland		Mortality (lambs and adults)	+ (lin); more in lambs and old ind.; more at high density		*Milner et al.*, 1999; *Catchpole et al.*, 2000; *Coulson et al.*, 2001
Soay sheep (*Ovis aries*)	Hirta, Scotland	1986-1999	Abundance	- (lin); more at high density		*Coulson et al.*, 2001
Soay sheep (*Ovis aries*)	Hirta, Scotland	1985-	Weight at birth, time of birth, prop. twinning, maturation	- (lin)		*Forchhammer et al.*, 2001
Feral goats (*Hircus capra*)	Rum, Scotland	16 yrs	Abundance	- (lin)		*Post and Stenseth*, 1999
Moose (*Alces alces*)	Isle Royale, USA	1968-1985	Abundance	+ (lin)	Snow depth	*Post and Stenseth*, 1998; *Post and Stenseth*, 1999
Moose (*Alces alces*)	Norway (south and north)	14-20 yrs	Yearling mass	+ (lin)		*Post and Stenseth*, 1999
Moose (*Alces alces*)	Sweden	13 yrs	calf, yearling and adult mass	- (lin); not sign. for adult males		*Post and Stenseth*, 1999
Muskox (*Ovibos moschatus*)	East Greenland	28 yrs	Abundance	+ (lin)		*Post and Stenseth*, 1999
Reindeer (*Rangifer tarandus*)	Finland	13 yrs	Calf mass	0		*Post and Stenseth*, 1999
Reindeer (*Rangifer tarandus*)	Norway	11 yrs	Adult female fecundity	- (lin)		*Post and Stenseth*, 1999
Caribou (*Rangifer tarandus*)	West Greenland, Sisimiut and Paamiut	22 and 72 yrs	Abundance	+ (lin), 0		*Post and Stenseth*, 1999
Caribou (*Rangifer tarandus*)	Québec	8 yrs	Abundance	0		*Post and Stenseth*, 1999
Chamois (*Rupicapra rupicapra*)	Alps, France	1985-1997	Adult winter survival	- (lin)	indirect through vegetation	*Loison et al.*, 1999a

Table 1. (continued)

Taxonomic group Species	Location	Time period or length	Biological variable	Relationship with NAO	Suggested mechanism	Reference
Isard (*Rupicapra pyrenaica*)	Pyrénées, France	1985-1997	Adult winter survival	- (lin)	indirect through vegetation	*Loison et al.*, 1999a
White-tailed deer (*Odocoileus virgininaus*)	Minnesota	1975-1985	Abundance	+ (lin)	Snow	*Post and Stenseth*, 1998; *Post and Stenseth*, 1999
Clethrionomys glareolus and *Apodemus flavicollis*	Bialowieza National Park, Poland	1965-2000	Abundance	- (lin)	Seed production	*Stenseth et al.*, 2002
Muskrat (*Ondatra zibetthicus*)	Canada (3 climatic zones)	1925-1949	Population synchrony	+ (lin)	Spring conditions	*Haydon et al.*, 2001
Mink (*Mustela vison*)	Canada (3 climatic zones)	1925-1949	Population synchrony	+ (lin)	Spring conditions	*Haydon et al.*, 2001
Grey wolf (*Canis lupus*)	Isle Royale, USA	1959-1998	Pack size, moose remains in summer feces	+ (lin), + (lin),	snow depth	*Post et al.*, 1999d
Canadian lynx (*Lynx canadensis*)	Canada (3 climatic zones)	1821-1999	Abundance	-, +, 0 (lin) dep. on region	snow depth	*Stenseth et al.*, 1999

ica, *A. nemorosa*, *Tussilago farfara*, and *Caltha palustris* were negatively related to the NAO index at most sites. The median percentage of variation in annual dates of flowering that was explained by the NAO index was 28% for *Anemone hepatica*, 18% for *A. nemorosa*, 19% for *Tussilago farfara*, and 9% for *Caltha palustris*. Woody plants were less responsive to fluctuations in the NAO: flowering by *Calluna vulgaris* and *Vaccinium myrtillus*, and leaf emergence by *Betula pubescens*, covaried with the NAO on less than half of the sites. The mean number of days by which timing of flowering differed between the highest (1973: NAO = 2.18) and the lowest (1969: NAO = -4.76) NAO index years was 19.7 days for *Anemone nemorosa*, 26.4 days for *A. hepatica*, 25.5 days for *Tussilago farfara*, 15.5 days for *Vaccinium myrtillus* and 13.0 days for *Caltha palustris* [*Post and Stenseth*, 1999].

Timing of flowering varied more strongly with the NAO in southern than in northern Norway (for *Anemone nemorosa*, *A. hepatica*, and Vaccinium *myrtillus*), and more at lower elevations (for *Tussilago farfara*). Correlations between flowering dates of *Caltha palustris* and the NAO were unrelated to latitude, longitude and elevation [*Post and Stenseth*, 1999].

At Helle, southern Norway, timing of flowering by 10 of 12 species was negatively related to the NAO index. The flowering season for 4 of 11 species at Helle was prolonged following warm, wet winters by 18.8 days for *Anemone hepatica*, 17.1 days for *Tussilago farfara*, and 13.4 days for *Caltha palustris*. Within years, the spatial variability of timing of flowering across sites increased with the NAO index for *Anemone nemorosa*, *Calluna vulgaris*, *Vaccinium myrtillus*, *Tussilago farfara*, *Caltha palustris*, and, marginally, *Anemone hepatica*. Early-blooming plants were more strongly influenced by the NAO than late-blooming plants. Also, the length of flowering by early-blooming plants exhibited a greater degree of correlation with the NAO index than did the length of flowering by late-blooming plants [*Post and Stenseth*, 1999].

A comparison of correlation coefficients suggests that responses of plants to the NAO were slight compared to plant responses to local weather in other studies. Still, the advance in onset of flowering may increase the number of flowers produced, the number and size of seeds, and the survival of seedlings. Prolonged flowering season at the scale observed in response to the NAO enhanced flower production, seed set, and seedling recruitment in other studies [*Post and Stenseth*, 1999].

On the same dataset [*Lauscher and Lauscher*, 1990], annual dates of first flowering by *Anemone nemorosa*, *A. hepatica* and *Tussilago farfara* in 26 populations in Norway, a general autoregressive model of the timing of life-history events in relation to variation in the NAO was developed [*Post et al.*, 2001a]. Consistent with earlier studies, plants in most populations and all three species bloomed earlier following warmer winters. Earlier blooming may reflect increasing influences of resources and density-dependent population limitation with an increasing NAO index [*Post et al.*, 2001a]. The degree of reproductive asynchrony of these 3 species also increased with the magnitude of the large-scale environmental disturbance (including effects of the NAO [*Post et al.*, 2001b]].

2.1.2. Growth of tree-rings. The relationship between the NAO and tree rings in the European beech (*Fagus sylvatica*) has been reported [*Piovesan and Schirone*, 2000]. Prewhitened tree-ring chronologies respond mainly to summer precipitation and they do not correlate in a significant manner with the (winter) NAO. In this high-frequency pattern the NAO signs are only found on a small number of rings characterized by being very narrow or wide. By contrast, tree-ring width chronologies in which all the frequency components are conserved were significantly related to the NAO. The inverse correlation between actual measurements of ring width and NAO is probably a consequence of the availability of water in the soil at the beginning of the growing season. In the Mediterranean area the recharging of soil moisture depends on the amount of winter precipitation, which is inversely correlated with the NAO in this region. Strong signals of winter precipitation and the NAO are found in the low-frequency components of tree-ring growth [*Piovesan and Schirone*, 2000]. Scots pine (*Pinus sylvestris*) tree-rings in Fennoscandia were also significantly related to variations in the winter NAO [*D'Arrigo et al.*, 1993].

In Flatanger, central Norway, Norway spruce (*Picea abies*) tree rings were not related to the NAO during the period 1873-1997, except for the period 1920-1940 when there was a negative relationship between growth and the NAO [*Solberg et al.*, 2002]. Diameter of tree rings in fir (*Abies balsamea*) at Isle Royale, USA, was negatively related to the NAO of current and previous year [*Post et al.*, 1999d]. Furthermore, links between tree rings (a variety of chronologies) and the NAO index have been used to back-track the NAO [*Cook et al.*, 1998; see also *Jones et al.*, 2001]. It is likely that the climatic variable most strongly related to tree ring width will vary from location to location depending on growth at a particular site [*Solberg et al.*, 2002].

2.1.3. Agricultural crops. The prospect of using the NAO to predict effects on economically important crops is of great interest. The quality of the wheat in the U.K., as indexed by alpha-amylase activity with the Hagberg falling number (HFN) and specific weight, was positively correlated with the NAO [*Kettlewell et al.*, 1999], but there was no relationship with protein concentration (a third index of quality). These three measures of quality are used by the U.K. flour milling and baking industry to determine whether wheat grain is suitable for baking and should be purchased. The difference between the indices can be traced to the timing of the vegetative/ripening period, and therefore can indicate plausible mechanisms. The lack of relationship between the NAO index and grain protein concentration was expected, since the nitrogen supply to the crop from the soil and fertilizer is the main factor influencing protein concentration. Winter wheat is sown between September and December depending on the previous crop. The weather during the vegetative period before grain growth has relatively little influence on HFN and specific weight. The weather, especially rainfall, during the ripening period in August is the main determinant of HFN and specific weight. Both HFN and specific weight decrease with rain before harvest as a result of sprouting (germination), and also from weathering of the grain surface by alternate wetting and drying in the case of specific weight. The authors argue that an effect of the NAO on cereal quality may be widespread in Europe [*Kettlewell et al.*, 1999].

Rodó and Comín [2000] claimed that the NAO had no effect on wine quality in Spain, while the El Niño-Southern Oscillation (ENSO) did affect wine quality in the same region. A study from Portugal, however, indicated that there is a negative relationship between the NAO in April and wine quality [*Esteves and Orgaz*, 2001].

2.1.4. Desertification in Africa. Vegetation productivity and desertification in sub-Saharan Africa may be influenced by global climate variability attributable to the NAO [*Oba et al.*, 2001]. Combined and individual effects of the NAO and ENSO indices revealed that 75% of the interannual variation in the area of Sahara Desert was accounted for by the combined effects, with most variance attributable to the NAO. Effects were shown in the latitudinal variation on the 200 mm isocline, which was influenced mostly by the NAO. The combined indices explained much of the interannual variability in vegetation productivity in the Sahelian zone and southern Africa, implying that both the NAO and ENSO (quite naturally) may be useful for monitoring effects of global climate change in sub-Saharan Africa [*Oba et al.*, 2001].

2.2. Invertebrates

No study has yet been published on any specific analysis of the effects of the NAO on invertebrate populations, though such effects are likely. For instance, the effect of temperature on butterfly populations is well documented [*Roy and Sparks*, 2000; *Sparks and Yates*, 1997; *Sparks et al.*, 2000; *Parmesan et al.*, 1999; *Thomas et al.*, 2001a].

2.3. Amphibians

Populations of amphibians have been declining worldwide [*Morell*, 1999; *Alford et al.*, 2001; *Pounds et al.*, 1999; *Pounds*, 2001; *Kiesecker et al.*, 2001], an observation which has led to considerable interest regarding whether this is due to climate [e.g., *Alexander and Eischeid*, 2001]. All of the amphibians native to Britain undergo spring migrations to breeding ponds to reproduce, as is typical of most temperate species [*Beebee*, 1995]. The three native anurans, the toad (*Bufo calamita*) and the two species of frogs (*Rana esculata* and *R. temporaria*), demonstrate clines of spawning times, from the earliest in west or southwest of England to the latest in the cooler north and east. These clines are thought to be adaptive and climate-related. Over the 17 year period from 1978-1994, first spawning and sighting date of the three native anurans and urodeles (*Triturus vulgaris*, *T. helveticus* and *T. cristatus*) have been increasing in earliness (i.e., a significant time trend) [*Beebee*, 1995]. Average times for spawning were 2 to 3 weeks earlier during the last five compared to the first five years of study, while the first newts were arriving in the ponds an average of 5-7 weeks earlier. The NAO explained the year-to-year variation in both first day of spawning and sighting [Plate 1; *Forchhammer et al.*, 1998a]. The mechanism is probably directly related to temperature [*Beebee*, 1995]. Gametogenesis is temperature-dependent; in temperate amphibians it is completed either during the early spring after emergence from hibernation in the case of late breeders such as *B. calamita* and *R. esculenta*, or in winter for the remaining early-breeding species, and migration to breeding ponds requires threshold minimum temperatures and humidity levels [*Beebee*, 1995]. Spawning dates were negatively correlated with average minimum temperatures in March and April and maximum temperatures in March. Average maximum temperatures increased by an average of 0.11°C per year at one site and 0.24°C at another site [*Beebee*, 1995]. These observations suggest that amphibian reproductive cycles in at least one part of the North Atlantic region is closely linked to the dynamics of the NAO. No study has so far linked population dynamics of amphibians to the NAO.

2.4. Birds

It is well known that climate may affect populations of birds [e.g., *Moss et al.*, 2001]. For example, spring weather may be important for survival of chicks [*Moss et al.*, 2001] and the energetics of breeding in general [*Stevenson and Bryant*, 2000]. Birds have also been reported to extend their ranges northwards in recent years [*Thomas and Lennon*, 1999]. In the following, we focus on the reported effects of the NAO on as dif-

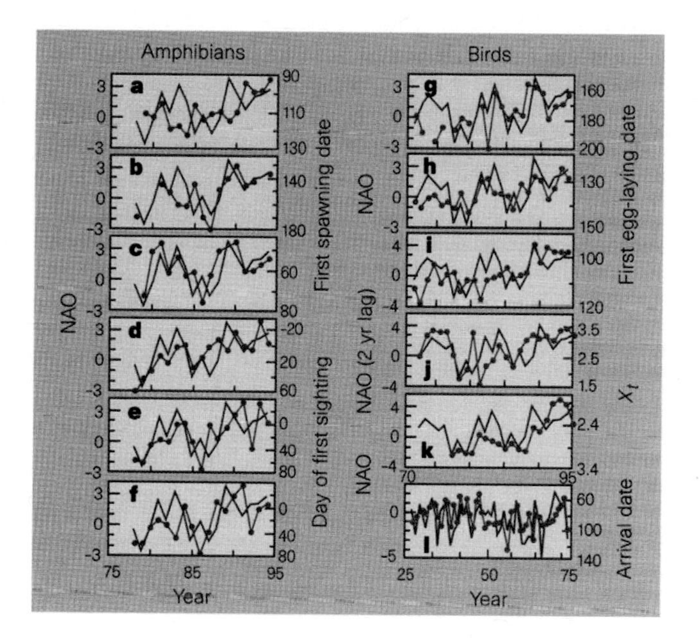

Plate 1. Variations in amphibian and bird breeding [after *Forchhammer et al.*, 1998a]. a-f, Temporal variation in first spawning date in *Bufo calamita* (a), *Rana esculenta* (b) and *R. temporaria* (c), and temporal variation in first day of sighting in *Triturus vulgaris* (d), *T. helveticus* (e) and *T. cristatus* (f). g-i, Temporal variation in first egg-laying date in *Miliaria calandra* (g), *Phylloscopus collybita* (h) and *Pica pica* (i). j-l, Temporal variation in log-transformed breeding numbers (Xt) in UK populations of *Pluvialis apricaria* (j) and *Actitis hypoleucos* (k), and in arrival date of a Norwegian population of *Alauda arvensis* (l). Superimposed on each figure is the NAO index (grey line). In coastal, northern Europe, high, positive values of the NAO indicate warm, moist winters, whereas unusually cold, dry winters are indicated by low, negative NAO index values. Pearson correlation cofficients between the NAO index and breeding characteristics are, for the non-detrended and detrended (in brackets) time series, as follows (italic type indicates significance, $P < 0.05$): a, -0.13 (0.23); b, *-0.52* (*-0.44*); c, *-0.57* (*-0.49*); d, *-0.59* (-0.33); e, *-0.50* (-0.25); f, *-0.41* (-0.39); g, *-0.50* (*-0.37*); h, *-0.56* (*-0.50*); i, *-0.34* (-0.17); j, *0.41* (*0.37*); k, *-0.68* (*-0.51*); l, *-0.59* (*-0.60*). Detrending was performed by linear regression. All correlations were tested for autocorrelation.

ferent topics as phenology, population dynamics and its effect on altering competitive interactions between bird species.

2.4.1. Breeding phenology. Egg-laying dates of corn bunting (*Miliaria calandra*), common chiffchaff (*Phylloscopus collybita*) and magpie (*Pica pica*) in the U.K. were all negatively correlated with the NAO, i.e. they were laying eggs earlier in years with a high NAO-index [Plate 1; *Forchhammer et al.,* 1998a]. This is likely due to an increase in spring temperatures [*Crick et al.,* 1997], as the relationship between breeding and spring temperatures has not changed [*McCleery and Perrins,* 1998]. Sky lark (*Alauda arvensis*) in Norway is also arriving earlier in years with a high value of the NAO index [Plate 1, *Forchhammer et al.,* 1998a]. Hatching dates of great tits (*Parus major*), blue tits (*Parus caeruleus*) and pied flycatchers in northern Germany between 1970 and 1995 are earlier after high NAO index winters (warm, wet winters) [*Forchhammer and Post,* 2000].

The breeding abundance of plover (*Pluvialis apricaria*) and sandpiper (*Actitis hypoleucos*) in the U.K. was, respectively, positively and negatively correlated with the NAO with a 2-year lag [Plate 1; *Forchhammer et al.,* 1998a].

2.4.2. Phenotypic plasticity vs. selection in flycatchers. Breeding performance (laying date, clutch size, fledging success and number of recruits produced) and morphological traits (tarsus and wing length) of collared flycatchers (*Ficedula albicollis*) in a population breeding on Gotland, Sweden, were studied over a period of 16 years (1980-95) [*Przybylo et al.,* 2000]. Effects on ecosystems may be investigated at short time scales (ecological effects; what is termed phenotypic plasticity) and at longer time scales (evolutionary effects; what is termed selection). Authors therefore performed both cross-sectional (transversal, each individual sampled once; affected by selection) and within-individual variation (longitudinal, following individuals over time; not affected by selection) correlations between the NAO-index and breeding performance and morphological traits. None of four measures of breeding performance (laying date, clutch size, fledging success and number of recruits produced) changed consistently over the study period, while tarsus length of males (and marginally females) decreased over time. Of the six traits investigated using cross-sectional data, only laying date was related to variation in the NAO-index. All characters investigated, showed significant repeatability within individuals among years, revealing the importance of factors specific to individuals in determining their value. However, within individuals, the NAO-index significantly affected laying date and clutch size such that females laid earlier and produced larger clutches after warmer, moister winters (high NAO index). The high degree of concordance between transversal and longitudinal analyses of the effect of NAO on laying date and clutch size suggests that the population-level response to the NAO-index may be explained as a result of phenotypic plasticity [*Przybylo et al.,* 2000].

2.4.3. Population ecology of the dipper. A particularly thorough study of the population ecology of the European dipper (*Cinclus cinclus*), a 50-60 g passerine species widely distributed in riparian habitats close to running water all over the Palearctic region, was done in southern Norway [*Sæther et al.,* 2000]. The dipper feeds on insects under water, and is therefore sensitive to any long-term freezing of the water surface. At northern latitudes the amount of ice strongly affects which areas have available winter-feeding habitats for the dipper. A large proportion of individuals were colour-ringed for individual recognition over a 20-year period (1978-1997). The population fluctuated markedly from a minimum of 27 pairs in 1982 to a maximum of 117 pairs in 1993. The recruitment rate decreased with increasing population size, mainly due to a reduction of the immigration rate with increasing population size. Fewer individuals were recruited after cold winters. Mean winter temperature was closely correlated with the annual variation in the number of days with ice cover in the study area. High winter temperature was correlated with high NAO index values. Cold winters had similar effects on the survival of both dispersing and non-dispersing individuals (Figure 1). A 2.5°C increase in winter temperature in this region is expected to increase carrying capacity by 58%. Most of this is due to effect on local dynamics, and not on immigration rates. Further, there was a weakly nonlinear relation between the carrying capacity and change in mean temperature and the NAO [*Sæther et al.,* 2000].

2.4.4. Sympatric populations of flycatchers. A rare case study of two sympatric flycatchers, the pied flycatcher (*Ficedula hypoleuca*) and the collared flycatcher (*F. albicollis*), derive from a subalpine mixed deciduous forest near Dlouhá Louka in the Czech Republic [*Sætre et al.,* 1999]. In this area, the winter temperatures over the past 130 years have covaried positively and relatively strongly with the NAO index [*Hurrell and van Loon,* 1997]. Annual estimates of breeding densities of both species of flycatchers from 1985 to 1997, as well as fledging success, intrinsic nestling mortality (i.e., starvation and/or disease), nest predation and adult disappearance was available. Nearly all interannual variation in breeding density of the pied flycatcher was explained by interspecific competition, intraspecific competition and intrinsic nestling mortality. Interspecific competition was significantly more important than intraspecific competition, while there was no effect of the NAO. In contrast, the collared flycatcher was less affected by interspecific competition, but more affected by intraspecific competition, intrinsic nestling mortality and the NAO. The collared flycatcher is both more abundant and a stronger interspecific competitor than the pied flycatcher. According to the competitive exclusion principle [*Hardin*, 1960], the pied flycatcher should not be able to coexist with the collared flycatcher in a stable equilibrium in this habitat. In the pied flycatcher, nestling mortality was explained by competition, and more so by interspecific competition, while intraspecific competition was most important for the collared flycatcher. There was an increase in the intensity of density dependence by nearly an order of magnitude, and a reduction in the intrinsic rate of increase by 10% when the influence of the NAO was removed [*Sætre et al.,* 1999]. This study suggests that the climatic fluctuations (in this case linked to the NAO), may prevent competitive exclusion.

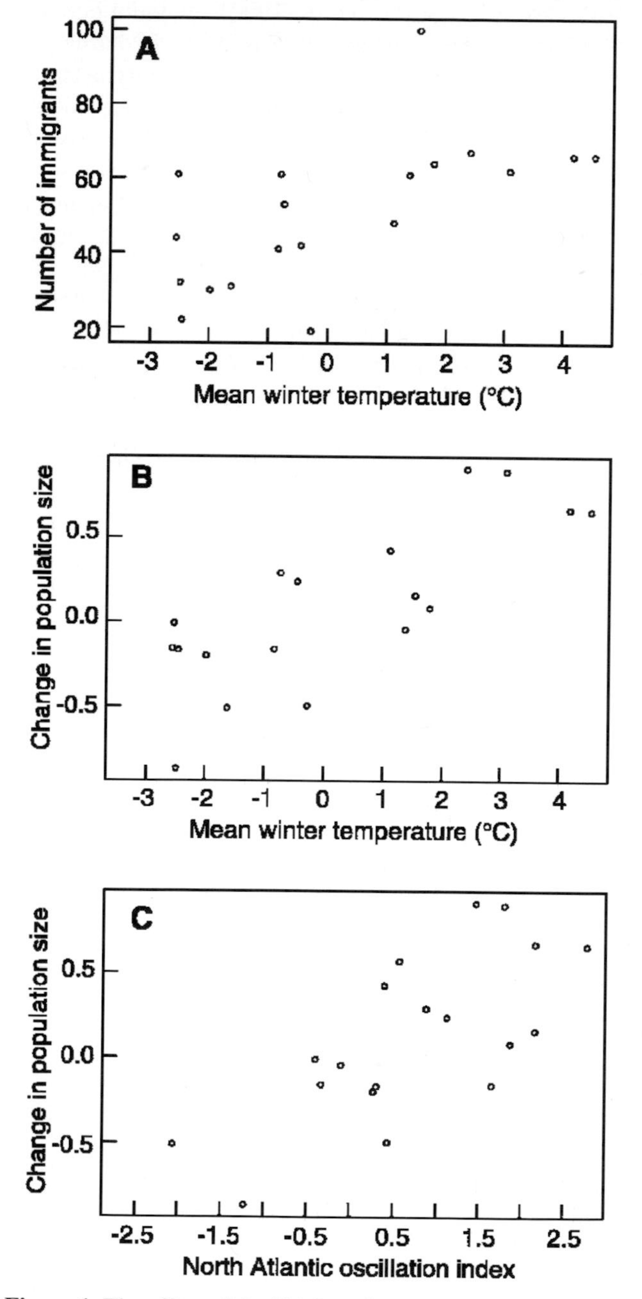

Figure 1. The effect of the NAO on population size of the dipper [after *Sæther et al.*, 2000]. (A) The number of immigrants plotted against mean winter temperature ($r = 0.62$, $n = 18$, $p = 0.003$), and the relative change in population size in relation to (B) mean winter temperatures and (C) the NAO index.

2.5. Mammals

Most studies on the effect of the NAO on mammals have been conducted on ungulates. Large animals, like most ungulates, have the capacity to switch habitats several times a day in response to variation in microclimate [*Parker and Robbins*, 1985; *Schmitz*, 1991; *Schwab and Pitt*, 1991 *Mysterud and Østbye*, 1995; *Mysterud*, 1996] and may also migrate over long distances due to seasonal changes in climate [*Albon and Langvatn*, 1992; *Mysterud*, 1999]. Nevertheless, climate is known to exert both direct effects through behavior and physiology - the indirect effects mainly through the quality of plants [*Albon and Langvatn*, 1992; *Lesage et al.*, 2000], such as variability of plant flowering dates [*Post and Stenseth*, 1999] and the content of phenolics [*Jonasson et al.,* 1986], but also through activity of parasitic insects [*Gunn and Skogland*, 1996; *Weladji et al.,* 2002].

The most thorough studies on the effect of the NAO on demography of large herbivores come from the case study of red deer on the west coast of Norway and on Rum, Scotland, as well as the Soay sheep (*Ovis aries*) in the St. Kilda archipelago, Scotland. Comparable studies relating the NAO to population ecology in North America is much needed.

2.5.1. Red deer on the west coast of Norway. The effect of the NAO on local climate is strong on the west coast of Norway [*Hurrell*, 1995] - the main distribution range of red deer in Norway. There exists extensive data on several demographic traits of Norwegian red deer, and the effects of the NAO have been reported on body weight of adults [*Post et al.,* 1997; *Mysterud et al.,* 2001b] and calves [*Post et al.,* 1999c; *Loison et al.,* 1999b], fecundity of females [*Post and Stenseth*, 1999], sex ratio of harvested calves [*Post et al.,* 1999b; *Mysterud et al.,* 2000] and also on number of red deer harvested [*Forchhammer et al.,* 1998b]. The red deer case can reveal also important methodological questions and highlights some important cautionary notes, since contradicting conclusions regarding the effect of the NAO on body weight have been presented based on the same data set [*Post et al.,* 1997; *Mysterud et al.,* 2001b], and the interpretation of the effect of the NAO on sex ratios of calves are also different [*Post et al.,* 1999b; *Mysterud et al.,* 2000].

The collection of data on body weights and fecundity of red deer on the west coast of Norway began in 1965 and is still ongoing [e.g., *Langvatn and Albon*, 1986; *Langvatn et al.,* 1996; *Mysterud et al.,* 2001c]. During this time span, the harvest of red deer in Norway has increased from 2484 in 1965 to 22534 in 2000 [*Statistics Norway*, 2001]. This increased harvest most importantly reflects an increase in density within the main distribution range, but the population has also expanded to the south, north and east from its main westerly initial point [*Mysterud et al.,* 2000; *Mysterud et al.,* 2001c]. There are thus significant local variations in red deer density and dynamics, both reflecting local weather conditions, pronounced variation in mountainous topography and the management practice (i.e., hunting licenses issued) [*Mysterud et al.,* 2000; *Mysterud et al.,* 2001a; *Mysterud et al.,* 2001c].

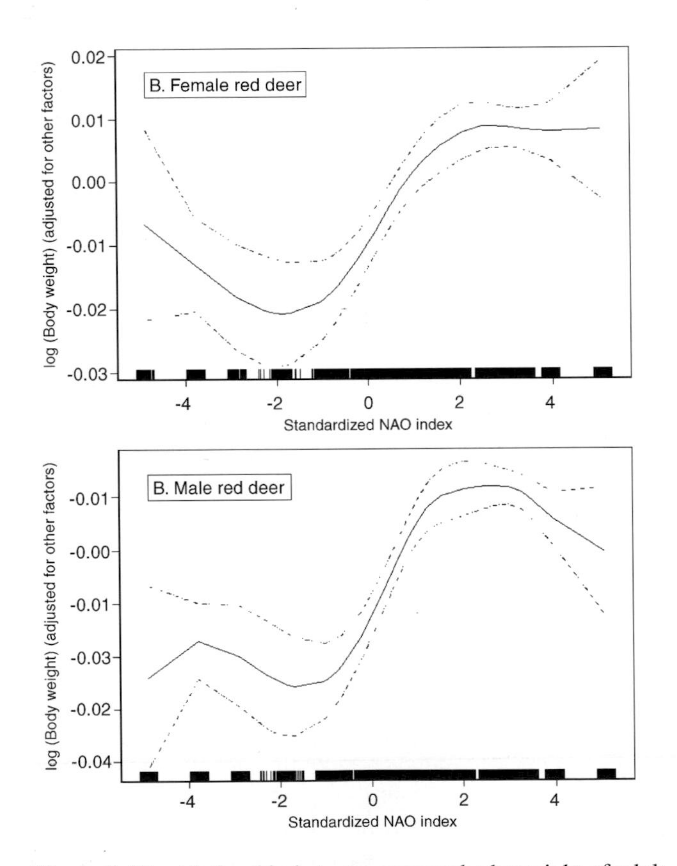

Figure 2. The relationship between autumn body weight of adult (1 yr) female (a) and male (b) red deer and the North Atlantic Oscillation. The dashed lines are 95% pointwise confidence intervals. The tick marks show the locations of the observations on that variable [see *Venables and Ripley*, 1994].

An initial study (using a coarse scale) found a negative effect of the NAO (when *in utero*) on body weight for adult females (not significant for males; *Post et al.,* 1997; *Post and Stenseth*, 1999]. A second study at a more local scale found a main positive effect, but the effect was reversed at very low NAO index values [Figure 2, *Mysterud et al.,* 2001b]. However, the *Post et al.* [1997] study did not correct for local density, which increased much during the study [*Mysterud et al.,* 2001c], and the seemingly negative effect of the NAO is probably due to the confounding increase in local density. This is consistent with other studies suggesting that the choice of spatial and temporal scale may be crucially important (as different levels of aggregation may result in opposite relationships; e.g., as for density-dependence [*Ray and Hastings*, 1996; *Donalson and Nisbet*, 1999; *Coulson et al.,* 1997; *Coulson et al.,* 1999]).

Such an additional relationship between the NAO and population ecology of red deer lends support to a positive effect of the NAO. Cohort fecundity of 2-year olds was positively correlated with the NAO index [*Post and Stenseth*, 1999], and there is usually a positive relationship between body weight and fecundity [e.g., *Reimers*, 1983; *Langvatn et al.,* 1996; *Hewison*, 1996]. Therefore, with this new insight, the positive phase of the NAO is associated with increased body weight [*Mysterud et al.,* 2001b], which in turn increases fecundity of females [*Post and Stenseth*, 1999] and probably leads to an increased harvest [*Forchhammer et al.,* 1998b; 2 year lag; *Post and Stenseth*, 1999].

Some of the reported patterns lack - as yet - a mechanistic basis. For example, one study reported a negative effect of the NAO on female calves body weight, while they found a positive effect on male calves body weight [*Post et al.,* 1999c]. The data set was very limited in size. In the same study, it was reported that longevity increased with 5 years for red deer born after high NAO index winters. As this is a harvested population in which most animals end up being shot, and this hunting pressure is not stable over time [*Langvatn and Loison*, 1999], we urge for caution in the interpretation of such correlations (and whether there truly are differential impacts of the NAO on males and females awaits further study). Indeed, a later study in the same subpopulation did not find any significant interaction between calf sex and the NAO [*Loison et al.,* submitted] - i.e., the sexes reacted similarly to the NAO.

2.5.2. Sex ratios of red deer. Variability in sex ratios (often measured as the proportion of male offspring) has been extensively studied within an evolutionary context to understand why and when sex-ratios should differ from 50:50 [for mammals, see *Williams*, 1979; *Clutton-Brock and*

Iason, 1986; *Clutton-Brock*, 1991; *Kojola*, 1998; *Hewison and Gaillard*, 1999]. Adaptive manipulation of the birth-sex ratio may be expected when costs and/or benefits differ for production of either sex [*Trivers and Willard*, 1973; *Maynard Smith*, 1980; *Hewison and Gaillard*, 1999]. Given the evidence for a relationship between maternal condition and offspring sex ratios in mammals at the individual level [*Clutton-Brock and Iason*, 1986; *Skogland*, 1986; *Cameron et al.,* 1999], it is plausible that all extrinsic factors affecting body condition, such as population density and climate, also may affect offspring sex ratio at a population level [*Kruuk et al.,* 1999; *Post et al.,* 1999b; *Mysterud et al.,* 2000]. Variation in population density and climate is over short time scales. It is therefore likely that selection may not act to change the relationship between body condition and sex ratio that was adaptive for the individual at former conditions. It is plausible that the individual response may be evident also at the population scale.

On the west coast of Norway, the NAO index is positively correlated to the proportion of male calves harvested in autumn in one red deer population [*Post et al.,* 1999b; *Mysterud et al.,* 2000], but not in 3 other populations [*Mysterud et al.,* 2000]. The NAO is a less important predictor of sex ratio than population density. The change in proportion of males harvested increased with as little as 5% from low to high NAO index values [*Mysterud et al.,* 2000]. Increased local winter precipitation (between November and January; i.e., similar to high NAO index winters [*Catchpole et al.,* 2000]) was associated with female-biased offspring sex ratios also in a population of red deer on Rum, Scotland [*Kruuk et al.,* 1999]. High winter rainfall increases juvenile mortality in the Rum population during winter, and then indicates that they are associated with nutritional stress [*Kruuk et al.,* 1999]. In both populations, sex ratio became more female-biased with severe conditions (more snow in wintering areas in Norway; more rain on Rum; also with increasing density in both populations), but high NAO index values gave male-biased sex ratios in Norway and female-biased sex ratios on Rum, Scotland.

The physiological mechanism by which sex ratio is determined in red deer remains unknown [*Hewison and Gaillard*, 1999]. Abortion is one possible mechanism. It is difficult to explain the effect of the NAO in populations in western Norway otherwise, since the NAO index values used are from winter (December through March, i.e. during gestation [*Mysterud et al.,* 2000]). Another possibility is that winter climate may have delayed effects and affect condition (at conception) through summer foraging conditions (see next section). However, the NAO the year prior to conception did not correlate with sex ratios, leaving this an unlikely mechanism [*Mysterud et al.,* 2000].

2.5.3. Comparing red deer and domestic sheep in Norway. In Norway, domestic sheep (*Ovis aries*) are free-ranging only from May/June to September/October [*Mysterud*, 2000]. During winter, sheep are fed indoors and, as a result, experience a stable interannual energy supply during that season. Thus, any relation between growth of lambs (during summer) and winter climate must operate through indirect effects of winter climate on forage quantity and quality during summer [*Mysterud et al.,* 2001b]. In Norway, sheep and red deer are sympatric in summer and have very similar diet composition and habitat use [*Mysterud*, 2000]. Through a comparison of the two species, direct (costs of thermoregulation and movement in snow) and indirect effects (delayed snow melt can affect foraging conditions during summer) of winter climate may be separated.

Using the red deer data described above, and data from 139,485 sheep from the same region, the relationship between the NAO and individual performance (as measured by body weight) was explored. Indeed, there was a similar response to variation in the NAO in both sheep and red deer [*Mysterud et al.,* 2001b]. Since both red deer (being outdoor during winter) and sheep (being indoor during winter) showed the same response to winter climate, it is likely that there is an indirect effect of the winter-NAO on summer foraging conditions. Forage quality of grasses and herbs decline with age since emergence [*Van Soest*, 1994]. Hence, it is commonly assumed that winter climate may also indirectly affect foraging conditions during summer, as deep snow may lead to an extended period of snow melt, and hence a prolonged period of access to newly emergent high quality forage [*Albon and Langvatn*, 1992; *Loison et al.,* 1999b; *Post and Stenseth*, 1999]. There is an increase in spatial variability in timing of flowering with an increasing NAO index [*Post and Stenseth*, 1999]. The prolonged period with access to newly emergent, high quality forage is favorable to both red deer and sheep. Even slight changes in forage quality may substantially affect body growth of ungulates [*White*, 1983]. This result therefore provides support for an indirect effect of winter climate on summer foraging conditions.

Red deer in Norway winter at low elevations [*Albon and Langvatn,* 1992]. Since temperatures in coastal, low land regions often are around 0°C during winter, and since temperature declines with altitude, altitude is a key factor determining whether precipitation comes as rain or snow, and thereby determines the ecological impact of the NAO [*Mysterud et al.,* 2000]. There was a negative correlation between the NAO index and snow depth at low altitude (below 400 m), whereas there was a positive correlation between the NAO index and snow depth at high altitude

[*Mysterud et al.,* 2000]. Forage in the field layer is thus probably more easily available to red deer in these coastal and lowland areas during winters with a high NAO index [*Loison et al.,* 1999b; *Mysterud et al.,* 2000; *Mysterud et al.,* 2001b], and the higher temperatures also represent less cold stress to the animals [*Simpson et al.,* 1978].

A trend of increasingly warm and wet winters (as is typical for years with high NAO index values) may therefore favor large herbivorous ungulates on the west coast of Norway through two separate mechanisms: (1) less snow in the low-elevation wintering areas will decrease energetic costs of thermoregulation and movement, and increasing access to forage in the field layer during winter [*Loison et al.,* 1999b; *Mysterud et al.,* 2000]; and (2) more snow in the high-elevation summer areas will lead to a prolonged period of access to high quality forage during summer [*Mysterud et al.,* 2001b]. When testing for effects of the NAO on sex ratios (see above), we also tested for an effect of the NAO the year before ovulation (i.e., lagged one year), but no significant effect was found [*Mysterud et al.,* 2000]. These results taken together suggest that both direct and indirect effects of the NAO may be important for wild red deer on the west coast of Norway.

2.5.4. Red deer on Rum, Scotland. The red deer on Rum, Scotland, is the best studied large mammal population in the world [e.g., *Clutton-Brock et al.,* 1982b; *Clutton-Brock et al.,* 1984; *Clutton-Brock et al.,* 1982a; *Clutton-Brock et al.,* 1983; *Coulson et al.,* 1997; *Kruuk et al.,* 1999]. Both male and female calf birth mass are positively related to the NAO index (year when *in utero*). Cohort fecundity (3-yr-olds and milk hinds) are negatively correlated with the NAO index (year when *in utero*), while there is no effect on cohort fecundity of non-lactating hinds. Adult mass of males, but not females, is positively related to the NAO index [*Post and Stenseth,* 1999]. Adult male abundance is negatively related to the NAO index (previous year), while female abundance was positively related to the NAO index (2 years previous). Neither calf winter nor summer mortality is related to the NAO index. Yearling survival was positively correlated with the NAO index (year when *in utero*), while adult male and female winter survival are negatively correlated with the NAO index (1 year previous). As most studies based on direct estimation of demographic rates have shown that prime-age adult survival is hardly affected by density or climate (as opposed to survival of young and old age classes) [*Gaillard et al.,* 1998; *Gaillard et al.,* 2000] the differential effects of the NAO on survival of yearling and adult red deer are highly unlikely, and awaits more detailed studies, preferably with a process-oriented approach [*Coulson et al.,* 2000].

2.5.5. Soay sheep in the St. Kilda archipelago, Scotland. The studies of Soay sheep on the St. Kilda archipelago, Scotland, provide another interesting assessment of the effect of the NAO, as both the demographic parameters and population dynamics of this species are well known and famous for its population cycles [e.g., *Jewell et al.,* 1974; *Clutton-Brock et al.,* 1991; *Clutton-Brock et al.,* 1992; *Clutton-Brock et al.,* 1996; *Grenfell et al.,* 1992; *Coulson et al.,* 1999; *Grenfell et al.,* 1998]. On St. Kilda, survival of Soay sheep is negatively correlated with March rainfall [*Catchpole et al.,* 2000]. The NAO index is positively correlated with March rainfall [*Catchpole et al.,* 2000]. The plausible sequence of causality is that the NAO influences March rainfall, which in turn quite likely increases mortality of sheep [*Catchpole et al.,* 2000]. This is most probably due to the combined negative effects on the energy balance of high winds and wetting of the pelage [*Parker,* 1988; *McIllroy,* 1989]. At least, a positive NAO index pattern negatively influenced survival of sheep [*Milner et al.,* 1999; *Catchpole et al.,* 2000], and more so in lambs [*Milner et al.,* 1999; not significant in the analysis of *Post and Stenseth,* 1999]. Lamb birth weight of Soay sheep is strongly negatively correlated with the NAO index in the population in Village Bay [*Post and Stenseth,* 1999]. Both populations in Village Bay and at Hirta, Scotland declined after high NAO index winters (previous) [*Post and Stenseth,* 1999], while population size was reported to increase with the NAO index in another study [*Catchpole et al.,* 2000], but this correlation was regarded to be spurious [*Catchpole et al.,* 2000].

The NAO induced marked cohort effects in this population [*Forchhammer et al.,* 2001]. Cohorts born after warm, wet and windy winters (high NAO index) were lighter at birth, born earlier, less likely to have a twin and matured later than cohorts born following cold and dry winters (low NAO index). High NAO index winters preceding birth depressed juvenile survival but increased adult survival and fecundity. The negative influence of high NAO index winters on juvenile survival are likely to be related to mothers compromised physical condition while cohorts were *in utero*, whereas the positive influence on adult survival and fecundity may relate to the improved post-natal forage conditions following high NAO index winters [*Forchhammer et al.,* 2001].

The interactions between density, climatic fluctuations and demography are important in order to understand the population fluctuations of Soay sheep [Figure 3, *Coulson et al.,* 2001]. Because sex and age structure of the population fluctuated independently of population size, and because animals of different age and sex respond in different ways to density and weather, identical weather conditions can result

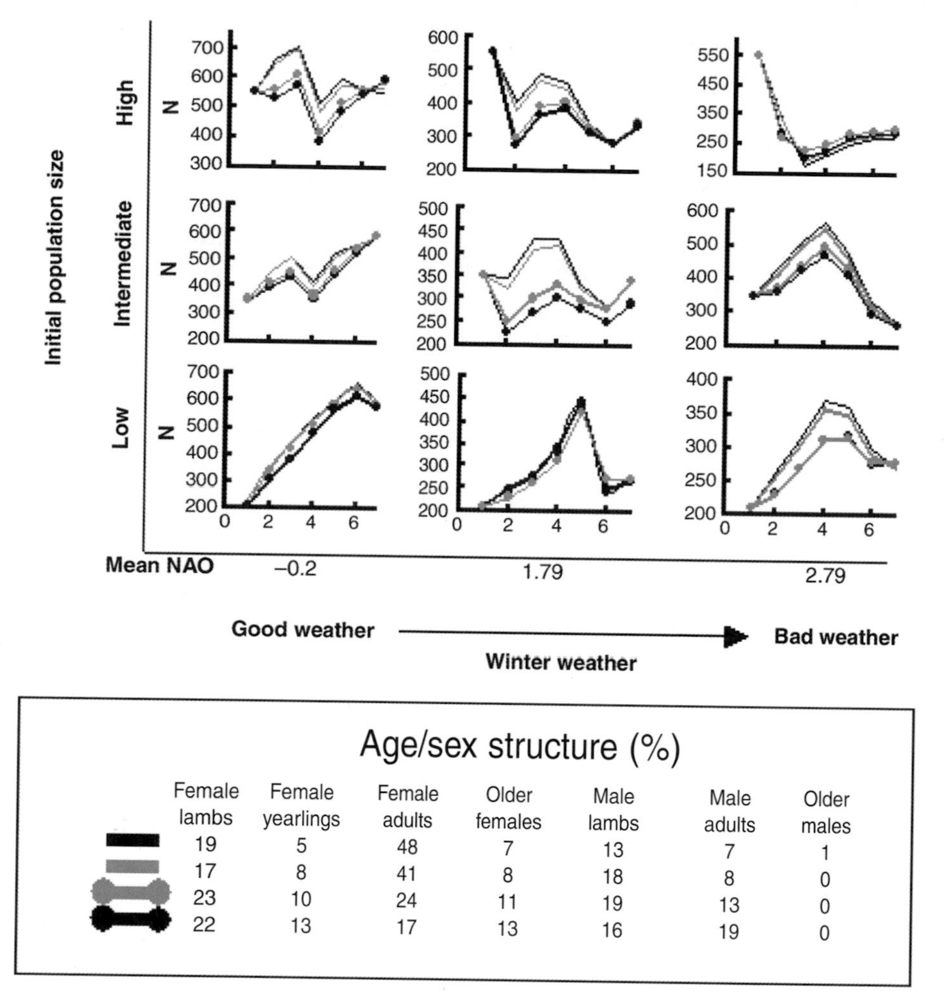

Figure 3. The importance of age and sex structure of a population of Soay sheep on the effects of the NAO [after *Coulson et al.*, 2001].

in different dynamics in populations of equal size. In each age/sex class, survival rates are lower in wet, windy winters (high NAO index values). Density and weather interact so that bad weather depressed survival rates relatively more at high density than at low density. The strength of the interaction is greatest in young and old animals [*Coulson et al.*, 2001]. As one feature of the NAO is decadal scale shifts between periods of cold, dry winters followed by wetter, stormier winters [*Hurrell*, 1995], these results suggests that the dynamics of the Soay sheep are likely to alternate between periods when the dynamics are over-compensatory (i.e., cyclic) and periods when they are not. This suggests that changes in climate associated with the NAO are likely to alter the dynamical properties of the system [*Coulson et al.*, 2001].

2.5.6. Other herbivore and granivore populations. The relationship between the NAO and various demographic variables have also been reported for several other eastern Atlantic populations [*Post and Stenseth*, 1999]. Yearling mass of moose (*Alces alces*) both in the south and north of Norway correlated positively with the NAO index (year when *in utero*). Calf mass, yearling mass, and adult female (but not male) mass covaried positively with the NAO index (3 years previous). There was no effect of the NAO (*in utero*) on reindeer (*Rangifer tarandus*) calf mass in Finland. Adult female fecundity was negatively related to the NAO index (1 year previous) in Norway. Abundance of feral goats (*Capra hircus*) on Rum, Scotland declined with increasing values of the NAO index (1 year previous) [*Post and Stenseth*, 1999].

The NAO has a different climatic impact in northern and central Europe [*Hurrell*, 1995]. A study of the effect of the NAO as well as local weather (winter, spring and autumn precipitation and temperature, duration of snow cover during winter, and day-degrees during spring) on adult survival of chamois (*Rupicapra rupicapra*) and isard (*R. pyrenaica*) during a 13-yr period, respectively, in the Alps and in the Pyrénées in France is therefore especially interesting [*Loison et al.,* 1999a]. In the Pyrénées, the NAO partly accounted for cohort effects in survival, while this was not the case in the Alps. Individuals born after positive NAO index winters (indicative of high temperature and low precipitation) exhibited a low adult winter survival in the Pyrénées. Furthermore, survival increased when precipitation was high and temperatures low during the preceding spring. In the Alps, survival decreased following a positive NAO index and low spring day-degree. A very interesting result of this study is the contrasting response to the effects of local weather (spring temperatures) in the two sites. There was no direct impact of winter weather on survival in either site. Hence, the NAO may operate indirectly through effects on vegetation also in these areas (see [*Mysterud et al.,* 2001b] for Norway).

Very few studies have been performed with regard to the effect of the NAO on the western side of the Atlantic. It was reported that muskoxen (*Ovibos moschatus*) in Greenland and caribou (*Rangifer tarandus*) in the Sisimiut herd of West Greenland declined after low NAO index winters [*Post and Stenseth,* 1999]. There is no effect of the NAO (1 year previous) on caribou in the Paamiut herd of West Greenland or in Quebec. Moose on Isle Royale, U.S., declined 2 years after low NAO index winters, and white-tailed deer in Minnesota declined 3 years after low NAO index winters [*Post and Stenseth,* 1998; *Post and Stenseth,* 1999]. All these reports use time-series analyses of abundance data, and also found evidence of negative density-dependence in all populations [*Post and Stenseth,* 1999]. There was no data on demographic traits to substantiate the results.

A study of the impact of the NAO on rodents comes from the Bialowieza forest in Poland [*Stenseth et al.,* 2002]. Using data on abundances of two species of forest rodents (*Clethrionomys glareolus* and *Apodemus flavicollis*), the annual density-dependent and density-independent structures were decomposed into their seasonal components. The NAO was incorporated significantly into both annual models (negative effect), but not if seed production is also incorporated. This suggests that the between-year variation primarily is accounted for by differences in seed-production, particularly oak seeds. The analysis further demonstrates that these between-year effects primarily operate during the winter [*Stenseth et al.,* 2002].

Muskrat (*Mustela vison*) populations in Canada were more synchronous in the eastern part of the country where the influence of the NAO is the strongest [*Haydon et al.,* 2001].

2.5.7. Mammalian carnivores. There is at present a very limited literature on the effect of the NAO on carnivorous species. This is partly because long-term studies on populations of carnivores are rare, and virtually nothing is available on temporal changes in demography. This is probably mainly due to low population densities and the practical difficulties of obtaining such data. It might of course be that carnivores are less affected directly by climatic factors, as they are at least not directly mechanistically linked through effects of climate on plants. Recent studies indicate that this may not be the case. One can easily identify two mechanistic links between climate and mammalian carnivores; (1) climate (in particular snow conditions) can affect hunting behavior, and (2) since the populations that carnivores feed on are affected, this can indirectly affect the carnivore populations. Carnivore foraging behavior is largely stereotypical within taxonomic families [*Murray et al.,* 1995]. Felids typically stalk or ambush prey from cover, while canids have a more variable hunting strategy, including hunting in packs (for wolves [*Macdonald,* 1983]). Fortunately, there have been studies conducted on each of these two important taxonomic groups, in addition to a study of a mustelid [*Haydon et al.,* 2001].

Grey Wolf. Isle Royale, U.S., is a protected national park and the site of the longest continuous study of an undisturbed, three-trophic-level system involving grey wolves (*Canis lupus*), moose and their primary winter forage, balsam fir (*Abies balsamea*) [*McLaren and Peterson,* 1994]. Moose play a pivotal role in ecosystem function on Isle Royale because they are the primary prey of wolves [*Peterson et al.,* 1984], and because their heavy browsing on balsam fir and other woody species, they determine several demographic traits of the fir. Wolf pack dynamics and hunting behavior on Isle Royale are recorded during aerial surveys initiated in 1959, shortly after the wolves colonized the island. Variation in the annual mean pack size between 1959 and 1998 showed a close correlation with the state of the NAO, which correlated negatively with snow depth [*Post et al.,* 1999d]. Pack size increased in response to winter snow depth (and therefore decreased with an increasing NAO index). As pack size increased, the number of moose killed per pack per day rose, particularly young and old moose. Increases in snow depth (decreasing NAO index values) also decreased the condition and survival of adult moose. Examination of summer wolf faeces revealed increases in both occurrence and biomass of adult moose following negative (snowy) NAO index winters. Annual mortality of wolves declined following winter in which wolves hunted in large packs [*Post et al.,* 1999d].

The food acquired by an individual wolf declines with increasing pack size [*Schmidt and Mech,* 1997], also on the Isle Royale [*Thurber and Peterson,* 1993]. One year after snowy winters, increases in wolf pack size and killing rates

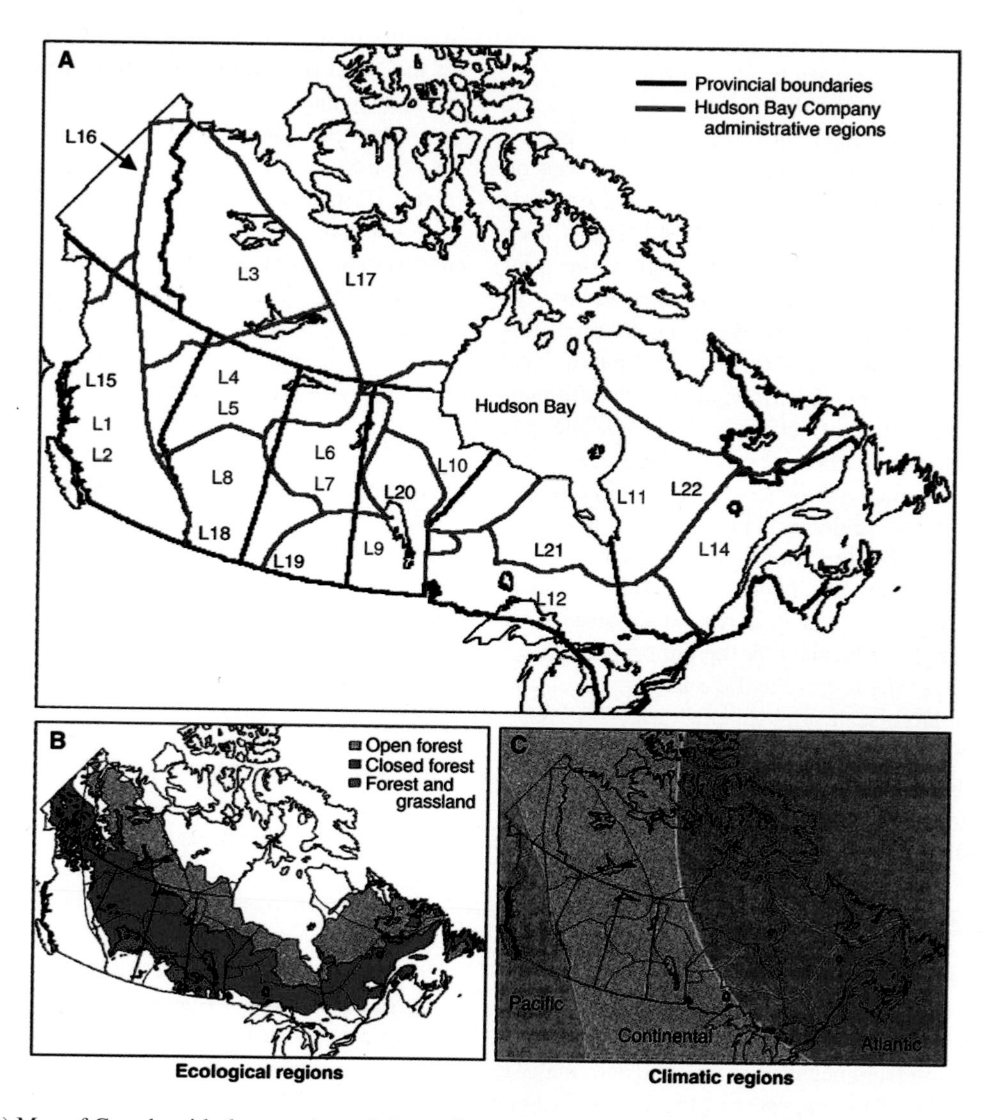

Plate 2. (A) Map of Canada with demarcations of the studied time series (red indicates the Hudson Bay Company time series and black indicates the recent series). (B) Ecological regions of Canada. (C) Climatic regions of Canada defined by the NAO. When surface pressures are lower than normal near Iceland and higher than normal over the subtropical Atlantic (the positive phase of the NAO), enhanced northerly flow over eastern Canada cools surface temperatures and enhanced southerly flow from the Gulf of Mexico into much of central Canada produces warm surface anomalies. Over the Pacific-maritime region, there is no significant NAO signature [after *Stenseth et al.,* 1999].

led directly to a decline in moose abundance, independently of the density-dependent limitation of moose. In turn, growth of fir, as estimated by width of tree rings, increased after snowy winters. This study suggests that there can be an effect of the NAO on the behavior of a top predator (in this case the wolves) that may cascade down to affect the secondary (the moose) and primary producers (fir trees).

The Canadian lynx. Across the boreal forest of Canada, lynx and snowshoe hare populations undergo regular density cycles [*Krebs et al.,* 2001]. Twenty one time-series on the population dynamics of lynx from 1821 and onwards are available [Plate 2; *Stenseth et al.,* 1999]. It has been examined whether the lynx populations display the most similar phase- and density-dependent structure within ecologically based habitat regions (northern, open boreal forest and southern, closed boreal forest) or climatological regions (Atlantic-maritime, Continental and Pacific-maritime) based on the spatial influences of the NAO. The lynx population cycles were found to be more alike (i.e., have the most closely related dynamic structure) within climatologically-based habitat regions, than within ecologically-based habitat regions [Plate 2; *Stenseth et al.,* 1999].

This, perhaps surprising, finding may be explained as follows. Over much of central and western Canada, surface climate is most strongly influenced by the atmospheric circulation upstream over the North Pacific and in particular by a natural mode of large-scale atmospheric variability known as the Pacific-North American (PNA) teleconnection pattern [*Wallace and Gutzler*, 1981; *Barnston and Livezey*, 1987]. However, the influence of the PNA on Canadian surface temperature is relatively spatially homogenous [*Trenberth and Hurrell*, 1994]. In contrast, the influence of the NAO on surface winter temperatures varies considerably from coast to coast and shows spatial variation corresponding well to the best grouping found for the lynx time series. Even though the effect of the NAO on lynx abundance is not strong, the lynx series fall along an east-west gradient progressing from negative to positive and finally to no effect of the NAO. This suggests that the climate in some way affects hunting behavior (and success) of the lynx, suggestively through snow depth and/or structure, but this remains to be measured [*Stenseth et al.,* 1999]. Further analysis of the snow depth and cover patterns have shown spatial patterns similar to the ones exhibited by the lynx dynamics [*Brown and Braaten*, 1998].

In light of the lynx study [*Stenseth et al.,* 1999], it is interesting that also mink (*Mustela vison*) populations in Canada were more synchronous in the eastern part of the country where the influence of the NAO is the strongest [*Haydon et al.,* 2001]. This may indicate that the NAO can act to synchronize populations.

3. EMERGING ECOLOGICAL INSIGHTS

Compared to the effect of El Niño events on terrestrial ecosystems [e.g., *Holmgren et al.,* 2001], the effect of the NAO seems less dramatic, probably since the influence is mostly during winter. As would be expected, the influence of the NAO on ecological patterns reflects this main influence on winter climate (and hence outside the vegetation growth-season). For example, early-blooming plants were more strongly influenced by the NAO than late-blooming plants [*Post and Stenseth*, 1999]. Although the effect of the NAO is mainly in the north Atlantic region, impacts in the north could extend far outside the direct impact area [*Danell et al.,* 1999], for example when it is affecting migratory birds.

3.1. Biases in the Literature

This review clearly demonstrates that there are marked geographic and taxonomic biases in the literature on effects of the NAO on terrestrial ecosystems. These biases cannot be explained based on present knowledge of the NAO as a climatic factor; we know that the NAO is known to correlate with local climate also in other areas. From current knowledge on general climatic effects, we would expect impacts of the NAO also in other systems. More likely they reflect that the study of ecological impacts of the NAO is recent, and that there are not always good data sets available. For example, the majority of studies have been conducted in western Europe and on large herbivores. This partly reflects that the effect of the NAO on terrestrial ecosystems was first shown on red deer in western Europe [*Post et al.,* 1997; it had, however, previously been demonstrated for tree rings; *D'Arrigo et al.,* 1993], and partly that long-term, high quality, data sets are available for this taxonomic group. Nothing has been published on the possible relationship between invertebrates and the NAO, and only one study of small mammals [*Stenseth et al.,* 2002]. There is also a bias in the features being studied. Clearly, much has been done on phenological developments of plants, amphibians and birds - much less on how it affects demography or distribution range of these [*Parmesan et al.,* 1999]. This reflects a lack of good, long-term data sets on these issues. For large herbivores we have the opposite picture. There we have very good data on demography, and little data on breeding phenology (i.e., calving time) has yet been analyzed (but see [*Forchhammer et al.,* 2001] for an exception in Soay sheep). Further, most studies have only been concerned with single species dynamics [but see *Sætre et al.,* 1999 for an exception in birds]. Within the field of ecology, in general, we still have limited insight into relative roles of, for instance, climate and interspecific competition [*Fox and Morin*, 2001].

3.2. Does Phenology Matter? The Match-Mismatch Hypothesis

There are already several case studies on plants, amphibians and birds showing that they may start to breed earlier with increasing positive NAO index winters in Europe (see below). But, does it really matter that spring begins a week or two earlier? Apparently it does in some cases, as species at different trophic levels (e.g., prey and predator) may have different responses to the same climatic change. Global warming may therefore disrupt tight multitrophic interactions involved in the timing of reproduction and growth. In the ocean, the match-mismatch hypothesis, referring to the fact that the growth and survival of cod larvae depends on synchronous production with their main food items, has long been recognized to represent a fundamental concept [*Cushing*, 1974; *Cushing*, 1990]. Recently, similar cases of match-mismatch have been reported for terrestrial ecosystems [*Thomas et al.*, 2001b; *Visser and Holleman*, 2001].

The synchrony of winter moth (Operophtera *brumata*) egg hatching and oak (*Quercus robur*) bud have been low in recent warm springs [*Visser and Holleman*, 2001]. This was due to an increase in spring temperatures without a decrease in the incidence of freezing spells in winter. Another example was from two populations of blue tits (*Parus caeruleus*) that breed at different dates relative to the local peak in prey abundance, due to differences in gene flow from adjacent habitats [*Thomas et al.*, 2001b]. In southern continental France the local landscape is dominated by deciduous Downy oak (*Quercus pubescens*) forest where spring leaf flush occurs in early May. However, the blue tit adapted to breeding in May, also overflows from deciduous oak forest into less common patches of evergreen oak, where pairs breed about 3 weeks too early relative to the local peak in caterpillar abundance. As food supply (caterpillar abundance) and demand become progressively mismatched, the increased cost of rearing young pushes the metabolic effort of adults beyond their apparent sustainable limit, drastically reducing the persistence of adults in the breeding population. The economics of parental foraging and limits to sustainable metabolic effort are key selective forces underlying synchronized seasonal breeding and long-term shifts in breeding date in response to climatic change. By advancing spring leaf flush and ensuing food availability, climatic warming results in a mismatch between the timing of peak food supply and nestling demand, shifting the optimal time for reproduction in birds [*Thomas et al.*, 2001b].

3.3. Ecological vs. Evolutionary Responses

It is therefore vital to understand whether the response of species to variation in the NAO is due to phenotypic plasticity (ecological effects at short time scales) or selection (evolutionary effects at long time scales). The effects of the NAO on the pied flycatcher were on phenotypic plasticity [*Przybylo et al.*, 2000, see above]. This is in agreement with evidence from Pleistocene glaciations which indicates that most species responded ecologically by shifting their ranges poleward and upward in elevation, rather than evolutionary through local adaptation [*Parmesan*, 2000; see also *Flagstad et al.*, 2001]. That species mainly respond with phenotypic plasticity may become a problem if changes are extensive. For long-distance migrants climate change may advance the phenology of their breeding areas, but the timing of some species' spring migration relies on endogenous rhythms that are not affected by climate change [*Both and Visser*, 2001]. Thus, the spring migration of these species will not advance even though they need to arrive earlier on their breeding grounds to utilize the peak in quality or availability of forage and breed at the appropriate time. The migratory pied flycatcher has advanced its laying date over the past 20 years [*Both and Visser*, 2001]. This temporal shift has been insufficient, however, as indicated by increased selection for earlier breeding over the same period. The shift is hampered by its spring arrival date, which has not advanced. Some of the numerous long-distance migrants will certainly suffer from climate change, because either their migration strategy is unaffected by climate change, or the climate in breeding and wintering areas are changing at different speeds, preventing adequate adaptation [*Both and Visser*, 2001]. However, this may vary among taxonomic groups. Both ecological and evolutionary processes were responsible for increased ranges in invertebrates in England [*Thomas et al.*, 2001a]. Range expansion was at least partly due to increased habitat breadth in two butterfly species (phenotypic plasticity), while range expansion was due to increased dispersal tendencies in two bush cricket species. The mechanisms in the latter case was higher fractions of longer-winged (dispersive) individuals (i.e., selection [*Thomas et al.*, 2001a; *Pimm*, 2001].

3.4. Long Term Effects and Migration

A central question is therefore whether species are able to find suitable new habitats with the climate they are adapted to. Snow depth is considered a main response to climate change in northern areas [*Danell et al.*, 1999], and hence a main ecological mediator of climate change. In Northern Europe the NAO index is generally positively correlated with both temperature and precipitation [*Hurrell*, 1995]. An increasing trend in the NAO index may thus lead to less snow in lowland areas and more snow at high altitude [*Mysterud et al.*, 2000], imposing differential forcing on high and low altitude ecological systems. Some animals may move between the different altitudinal belts and find the most suitable one for a given time/environment [*Danell et al.*, 1999; *Mysterud et al.*, 2001b], while other species are not able to do so [*Inouye et al.*, 2000]. Mountains have therefore often acted as refugia and past climatic changes have resulted in less loss of species in such regions than on the plains [*Danell et al.*, 1999]. Differential effects of altitude largely remain to be explored with regard to the effect of the NAO. Only one study reported that the effect of the NAO on first flowering of a plant (*Tussilago farfara*) was more variable at lower elevations [*Post and Stenseth*, 1999]. There is, however, an important study from high altitude in the Rocky Mountains that can shed light on a differential impact depending on altitude [*Inouye et al.*, 2000].

In the Colorado Rocky Mountains, calendar date of the beginning of the growing season at high altitude is variable but has not changed significantly over the past 25 years [*Inouye et al.*, 2000]. This result differs from the growing evidence from low altitudes that climate change is resulting in a longer growing season, earlier migrations, and earlier reproduction in a variety of taxa. At high altitude, the beginning of the growing season is controlled by melting of the previous winter's snowpack. Despite a trend towards warmer spring temperatures, the average date of snowmelt has not changed, perhaps because of the trend for increased winter precipitation which comes as snow at high altitude. This disjunction between phenology at low and high altitudes may create difficulties for several species, such as many birds, that migrate over altitudinal gradients. American robins (*Turdus migratorius*) arrive on average 14 days earlier than they did in 1981, and as a result experience an increase in the interval between arrival date and the first date of bare ground by about 18 days. Earlier formation of permanent snowpack and for a longer period of snow cover at high altitude has also implications for hibernating species. The yellow-bellied marmot (*Marmota flaviventris*) emerges about 38 days earlier than 23 years ago, apparently in response to warmer spring air temperatures [*Inouye et al.*, 2000]. The likely consequences of too early emergence may therefore be pronounced [cf. *Yom-Tov*, 2001].

3.5. Climate and Functional Plant Groups

The effect of the NAO on plant phenology may be quite different among species [*De Valpine and Harte*, 2001] or functional plant groups [*Arft et al.*, 1999]. At present, there appears to be insufficient data reported on the important, and possible large-scale effects of the NAO on the scale of vegetation communities. Perhaps most importantly, we do not know which role the NAO and climate in general may have on the location of tree line. Average climate as well as climatic variability influence the position of the tree line together with interactions with herbivores [e.g., *Davis and Shaw*, 2001; *Hofgaard*, 2001]. The easiest available long-term data on the demography of trees are from growth of tree-rings [*Holmgren et al.*, 2001]. However, recruitment will also prove crucial to understand forest expansion. Ability to resist or adapt to change is vital. Plants do not move (as much) as animals, and rapid changes in climate as we see it today may therefore possibly drive some plants to extinction. Some plants have historically moved fast enough so as to track climate change [*Pitelka et al.*, 1997]. These migrations have come as response to change that in some cases has been as fast as the rapid warming predicted for the next few decades. Species with low reproductive rates, such as trees and other perennials, that take longer to reach reproductive maturity, could have longer lag times and might be stranded in unsuitable habitat by rapid climate change [*Pitelka et al.*, 1997]. In the event of rapid climate change, weed species would be the ones that would have little trouble shifting their ranges. Also, successful new colonization may be the result of repeated attempts to disperse, and are only successful when colonizers persist long enough to germinate. The large impact of the NAO on desertification in the sub-saharan Africa [*Oba et al.*, 2001] underlines the importance of the possible effects of the NAO on large-scale vegetation patterns.

3.6. Delayed Effects Through Cohort Effects

Delayed effects of climate are also important in animal populations. Individuals born in a specific year may be larger or smaller than the average depending on the climatic conditions in the year of birth. Any substantial change in early environmental conditions affecting birth weight and/or early growth may have a considerable impact of later survival and reproductive performance - that is known as cohort effects [*Clutton-Brock et al.*, 1982b; *Albon et al.*, 1987; *Albon et al.*, 1992; *Gaillard et al.*, 1997; *Post et al.*, 1997; *Rose et al.*, 1998; *Lindström*, 1999; *Forchhammer et al.*, 2001]. For example, Soay sheep lambs born under poor conditions will often mature at a later age [e.g., *Forchhammer et al.*, 2001], and effects of the NAO in one year may thus have consequences on age of first reproduction several years later.

3.7. Confounding Effects and Interactions with Other Factors

Much of what is reported regarding patterns of change related to the NAO, implicitly or explicitly assumes that other factors affecting the populations remain unchanged. It is vitally important to consider also other ecological factors that can change over time, in particular population density. There are already reported cases both on birds and mammals of (negative) density-effects which may override (positive) effects of climatic change. For example, while the NAO index showed an upward trend during the study period for red deer in Norway (1965-1998) and positively correlated with body weights [*Mysterud et al.,* 2001b], the parallel increase in local density had a stronger impact on body weights and led to decreasing weights over time [*Mysterud et al.,* 2001c]. Similarly, although warm winters (high NAO indices) positively affected the abundance of golden plovers in the UK, direct density dependence strongly depressed their numbers over time despite the increasing trend in the NAO index [*Forchhammer et al.,* 1998a]. In other cases, the effect of density dependence (intra- or interspecifically) may be prevented by the NAO. For example, negative effects of the NAO on breeding density of the collared flycatcher allowed a less strong interspecific competitor (the pied flycatcher, which is not affected by the NAO) to live in sympatry [*Sætre et al.,* 1999]. For plants, changing land use and atmospheric nitrogen deposition have had greater impacts on plants (at least in the north) than climate change to date [T. Callaghan, personal communication].

It is also evident that many climatic effects depend on other ecological factors (i.e., interactions). For large mammalian populations it is commonly observed that effects of severe winter weather may have a stronger impact on survival at high density [*Sauer and Boyce,* 1983; *Portier et al.,* 1998]. Such interactions have also been reported between density dependence and the NAO [*Post and Stenseth,* 1999; *Coulson et al.,* 2001]. Similar relations apply also for plants. Rare and preferred plant species are often over-utilized by herbivores, which means that plants that have managed to invade new areas will be killed or at least suppressed, thereby reducing the predicted rate of their movement in response to global change. The active role of animals must be taken into account to a much greater extent in modelling work on the climate-driven changes of vegetation zones [*Danell et al.,* 1999; *Niemelä et al.,* 2001; *Holmgren et al.,* 2001], as well as the interaction between climate and other ecological factors. For plants, increasing levels of CO_2, UV-B and increasing fire in forests all interact with climate [T. Callaghan, personal communication].

Any impact of the NAO will therefore to a large degree depend on other factors affecting the population biology of the particular species in question. It is therefore too simplictic to predict population abundances and demographic rates only based on the phase of the NAO, particularly if the responses are nonlinear [e.g., *Mysterud et al.,* 2001b].

4. FUTURE CHALLENGES

This tour of relationships between the NAO and different terrestrial ecosystem parameters poses some important questions and challenging methodological problems.

Studies on the potential impact of climate – and NAO variations – belong to three general classes: i) statistical modelling of time-series of population or ecosystem state parameters (e.g., population size; pattern-oriented approach *sensu* [*Coulson et al.,* 2000]); ii) statistical modelling of parameters describing population processes (e.g., survival; process-oriented approach *sensu* [*Coulson et al.,* 2000]); and iii) experimental manipulations of climatic conditions. Experiments based on adequate replication and randomization provide the best evidence for establishing causal relationships [e.g., *Cox,* 1992], but for obvious reasons are difficult to implement at a scale consistent with climatic variability, let alone the NAO. Correlative studies, while being more realistic and leading to better replication, are on the other hand plagued with confounding issues, something particularly clear when studying the effects of the NAO: many pathways (e.g., increase in temperature, decrease in snow depth) are possible, different age and sex classes may be differentially affected by climate [*Coulson et al.,* 2001], and statistical considerations alone cannot unambiguously distinguish between different assumed models [e.g., *Yoccoz et al.,* 2001]. Process-oriented studies are an essential supplement to pattern studies as effects of climatic variables together with other parameters such as density are easier to assess directly given we adequately account for population structure with respect to age, sex and cohort [*Coulson et al.,* 2001; *Gaillard et al.,* 2001]. In plant science, there is a stronger tradition for doing experimental work relating to climate, in particular in greenhouses, but also long-term field manipulations have been performed [e.g., *Stenström et al.,* 1997; *De Valpine and Harte,* 2001]. It is commonly assumed that it is close to impossible to do experiments with climate when it comes to animals, as many use large areas. The recent work on Svalbard reindeer (*Rangifer tarandus platyrhynchus*) provides therefore a rare example of how it is possible to manipulate climate (in this case snow depth) at a scale that is relevant to the animals [*Van der Wal et al.,* 2000; see also *Walsh et al.,* 1997]. By adding or removing snow to plots, melting in spring was delayed or enhanced. This in turn affected both standing biomass and quality of the plants, and the subsequent selection of patches by reindeer. We urge for more small-scale experimental work of this kind, as some proposed mechanisms remains poorly understood. For example, the correlation between the NAO index and the quality of the wheat crop in the U.K. was suggestively related to August rainfall, which correlated with the NAO [*Kettlewell et al.,* 1999]. A problem related to this is therefore that, to our knowledge, no thorough climatological study has been done to try and relate the winter NAO to summer conditions. Further, the North Atlantic Oscillation is also a determinant of interannual variation in dust transport from Africa (which yearly is one billion tonnes) and possibly other aerosols [*Moulin et al.,* 1997], whereas the ecological impact on terrestrial ecosystems remains unexplored. Related to this is how series of specific events may become enhanced. Repeated periods of rain, melting and freezing can make foraging difficult for grazers and browsers [*Yoccoz and Ims*, 1999], as well as negatively affect plant survival [*Crawford*, 2000]. Surprisingly, milder winters may lead to more frost damage as not all part of the plants are equally prone to freezing. Mild periods of winter weather that encouraged premature spring growth caused severe dieback of non-hardy shoots [*Crawford*, 2000].

4.1. Methodological Challenges

The effects of large-scale climatic fluctuations on ecology have typically been assumed to be linear. With the eruption of new statistical methods such as generalized additive models (GAM) [*Hastie and Tibshirani*, 1990], nonlinearities may easily be assessed without any restrictive assumptions regarding the functional relationship between the predictor variables (e.g., the NAO) and the response variable. There are (at least) two different ways in which the effect of the NAO on ecological systems may be expected to be nonlinear. Firstly, the NAO may not be linearly related to local climatic variables, and animals and plants are indeed expected to respond to the climate they experience at a local scale. Secondly, the animals or plants response to changes in local climate may not always be linear. Linearity may occur only at some particular scale. At present, there is only one case study reporting the former [*Mysterud et al.,* 2001b], while the second awaits further studies. It should be emphasized that the issue of non-linear responses may at least in part be an issue of scale, as transforming the predictor and/or the response variable may linearize the relationships. Non-monotonous relationships, such as found in *Mysterud et al.* [2001b], cannot be linearized, however.

There is evidence that climate change may lead to a higher frequency of more extreme events [e.g. *Meehl et al.,* 2000; *Easterling et al.,* 2000]. It is generally regarded that the spatio-temporal distribution of extreme events are often ecologically more significant than seasonal mean values [*Seastedt and Knapp*, 1993; *Yoccoz and Ims*, 1999; *Danell et al.,* 1999; *Parmesan*, 2000]. Increased environmental variance has in general a negative effect on population growth, and non-linear relationships between climate and population processes imply that a change in climatic variability will affect the mean value of a process, even if the mean value of the climatic parameter remains the same [*May*, 1986; *Caswell*, 2001]. One example of such a nonlinear response to a linear increase in temperature is the effect of frost formation [see the review of *Inouye*, 2000]. For example, the physiological consequence of formation of ice crystals in plant tissue is often plant death, or at least damage of sensitive parts that can include flower buds, ovaries and leaves [*Inouye*, 2000]. During years of lower snow accumulation, an early-blooming herbaceous perennial (*Delphinium nelsonii*) experienced colder temperatures between the period of snowmelt and flowering [*Inouye and McGuire*, 1991]. Flowering was delayed, floral production was lower, and flowering curves were more negatively skewed in years of low snow accumulation [*Inouye and McGuire*, 1991]. Corroborative evidence derives from experimental work in the Rocky Mountains, were a major effect of warming was to protect the plant *Helianthella quinquenervis* from frost damage [*De Valpine and Harte*, 2001] - illustrating the importance of extreme weather events. Another example of a nonlinear response to climate is the eggs of the autumnal moth *Epirrita autumnata* that die at ca. -36°C [*Niemelä*, 1979; *Tenow and Nilssen*, 1990; *Danell et al.,* 1999]. The result of low winter temperatures will be seen along the mountain slopes during the next summer if there is an outbreak year. In the valley bottoms, if the winter temperature passes the critical freezing tolerance of the eggs, there will be almost no defoliated birches while heavily defoliated trees will occur at higher, warmer (in winter) altitudes. Warming and a changing frequency of extreme cold events may lead to expansion of the outbreak areas whereas refuge areas with no outbreaks will decrease in size [*Danell et al.,* 1999; *Niemelä et al.,* 2001]. However, warming may also lead to an increase in parasitoid abundances, which are likely to play an important role in the regulation of moths populations [*Ruohomäki et al.,* 2000]. This again emphasizes the importance of considering different pathways when analyzing the consequences of climatic change.

4.2. Capturing the Ecology of Climate Fluctuations: Which NAO Index to use?

Several indices for the NAO have been defined [notably those by *Walker and Bliss*, 1932; *Rogers*, 1984; *Hurrell*,

1995; *Jones et al.*, 1997]. Ecologists have mainly used Hurrell's winter (December through March) index based on the difference of normalized sea level pressures (SLP) between Lisbon, Portugal and Stykkisholmur, Iceland. However, other locations are also used [*Jones et al.*, this volume]. *Mysterud et al.* [2001b] studied the effect of the NAO on sheep and red deer body weight and the results using December-March NAO indices with the southern station from either the Azores, Lisbon or Gibraltar were very similar. Indeed the correlations between these NAO winter-indices are so high [*Jones et al.*, this volume] that it should generally not be of any concern to ecologists studying the effects of the NAO. The seasonal aspect is of greater importance. Since the NAO is most pronounced during winter, when it accounts for more than one third of the variability in sea-level pressure [*Cayan*, 1992], an index for this period should generally be most suited for studying ecological effects [but see *Post et al.*, 1999c; *Esteves and Orgaz*, 2001], also when these effects are measured later on in the year. However, *Portis et al.* [2001] have recently introduced a "mobile" NAO index (NAOm) where the locations of the northern and southern nodes are allowed to shift geographically with season. The NAOm is defined as the difference between normalized SLP anomalies at the locations of maximum negative correlation between the subpolar and subtropical nodes. This new NAO index could be useful for studies involving other seasons than winter, including many ecological investigations.

The various NAO indices thus defined are useful in their ability to represent the main patterns of weather variability over a large area (e.g., sea-level atmospheric pressures over the North Atlantic), as well as with respect to their simplicity (atmospheric pressures at two locations, or a leading empirical orthogonal function of atmospheric pressure). An index that is best at representing weather patterns may not necessarily be the best or even appropriate at capturing the climatic effects on terrestrial ecosystems. Indeed, the examples presented above show clearly that the links between the NAO and ecosystems occur through many different pathways, and therefore a better local index could perhaps be constructed from local climate variables. Another limitation is that the relationship between the NAO and local climatic variables may change over time [i.e., nonstationarity; *Jones et al.*, this volume], and hence limit the time period for which effects on ecosystems can be predicted from known relationships with the NAO (for an example of a different impact of the NAO depending on time period, see [*Solberg et al.*, 2002]). However, a NAO index has some clear advantages, such as stressing the correlations between different climatic components (e.g., precipitation and temperature on the west coast of Norway). An interesting question is to understand to what extent the NAO represents climatic effects on terrestrial ecosystems, and if there is any systematic pattern in the unexplained part that could be related to other large-scale circulation patterns (e.g., the second empirical orthogonal function). Recent analyses of weather modelling (especially of sea level pressure) are based on nonlinear Principal Component Analysis (PCA) and show a pattern of variation somewhat different from the NAO (being based on linear PCA) [*Monahan et al.*, 2000; *Monahan et al.*, 2001]. In the future, we might therefore obtain improved climatic functions that may be better suited than the NAO in predicting temporal and spatial variation in climate.

The interactive role of biologists, climatologists and statisticians will certainly be useful in order to further advance the insight on the effect of the NAO on terrestrial ecosystems. It is evident to us that ideas for climatologically interesting work have already surfaced from the work on terrestrial ecosystems (e.g., the nonlinear relations between the NAO and local climate) [*Mysterud et al.*, 2001b]. There are further issues and questions raised by ecologists that need a climatological investigation. We wish to emphasize that insights from detailed biological studies provide not only ideas about small-scale variations in climate that may profoundly influence ecosystems, but also how a large-scale climatic phenomenon such as the NAO influences local climate.

Acknowledgements. We thank Terry Callaghan and Anne Loison for comments that helped us improve the manuscript, and the Research Council of Norway (NFR) for economic support.

REFERENCES

Albon, S. D., T. H. Clutton-Brock, and F. E. Guinness, Early development and population dynamics in red deer. II. Density-independent effects and cohort variation, *Journal of Animal Ecology, 56*, 69-81, 1987.

Albon, S. D., T. H. Clutton-Brock, and R. Langvatn,. Cohort variation in reproduction and survival: implications for population demography, in *The Biology of Deer*, edited by R. D. Brown, pp. 15-21, Springer Verlag, New York, 1992.

Albon, S. D., and R. Langvatn, Plant phenology and the benefits of migration in a temperate ungulate, *Oikos, 65*, 502-513, 1992.

Alexander, M. A., and J. K. Eischeid, Climate variability in regions of amphibian declines, *Conservation Biology, 15*, 930-942, 2001.

Alford, R. A., P. M. Dixon, and J. H. K. Pechmann, Ecology: Global amphibian population declines, *Nature, 412*, 499-500, 2001.

Arft, A. M., et al., Responses of tundra plants to experimental warming: meta-analysis of the International Tundra Experiment, *Ecological Monographs, 69*, 491-511, 1999.

Barnston, A. G., and R. E. Livezey, Classification, seasonality and persistence of low-fequency atmospheric circulation patterns, *Mon. Wea. Rev., 115*, 1083-1126, 1987.

Beebee, T. J. C., Amphibian breeding and climate, *Nature, 374*, 219-220, 1995.

Both, C., and M. E. Visser, Adjustment to climate change is constrained by arrival date in a long-distance migrant bird, *Nature, 411*, 296-298, 2001.

Brown, R. D., and R. O. Braaten, Spatial and temporal variability of Canadian monthly snow depths, 1946-1995, *Atmospheric-Ocean, 36*, 37-54, 1998.

Cameron, E. Z., W. L. Linklater, K. J. Stafford, and C. J. Veltman, Birth sex ratios relate to mare condition at conception in Kaimanawa horses, *Behavioral Ecology, 10*, 472-475, 1999.

Caswell, H., *Matrix Population Models*, 722 pp., Sinauer Associated Inc., Sunderland, MA, 2001.

Catchpole, E. A., B. J. T. Morgan, T. N. Coulson, S. N. Freeman, and S. D. Albon, Factors influencing Soay sheep survival, *Applied Statistics, 49*, 453-472, 2000.

Cayan, D. R., Latent and sensible heat flux anomalies over the northern oceans: the connection to monthly atmospheric circulation, *J. Climate, 5*, 354-369, 1992.

Clutton-Brock, T. H., *The Evolution of Parental Care*, 352 pp., Princeton University Press, Princeton, N.J., 1991.

Clutton-Brock, T. H., S. D. Albon, and F. E. Guinness, Competition between female relatives in a matrilocal mammal, *Nature, 300*, 178-180, 1982a.

Clutton-Brock, T. H., S. D. Albon, and F. E. Guinness, Maternal dominance, breeding success and birth sex ratios in red deer, *Nature, 308*, 358-360, 1984.

Clutton-Brock, T. H., F. E. Guinness, and S. D. Albon, *Red Deer. Behavior and Ecology of Two Sexes*, 378 pp., Edinburgh University Press, Edinburgh, 1982b.

Clutton-Brock, T. H., F. E. Guinness, and S. D. Albon, The costs of reproduction to red deer hinds, *Journal of Animal Ecology, 52*, 367-383, 1983.

Clutton-Brock, T. H., and G. R. Iason, Sex ratio variation in mammals, *Quarterly Review of Biology, 61*, 339-374, 1986.

Clutton-Brock, T. H., O. F. Price, S. D. Albon, and P. A. Jewell, Persistent instability and population regulation in Soay sheep, *Journal of Animal Ecology, 60*, 593-608, 1991.

Clutton-Brock, T. H., O. F. Price, S. D. Albon, and P. A. Jewell, Early development and population fluctuations in Soay sheep, *Journal of Animal Ecology, 61*, 381-396, 1992.

Clutton-Brock, T. H., I. R. Stevenson, P. Marrow, A. D. MacColl, A. I. Houston, and J. McNamara, Population fluctuations, reproductive costs and life-history tactics in female Soay sheep, *Journal of Animal Ecology, 65*, 675-689, 1996.

Cook, E. R., R. D. D'Arrigo, and K. R. Briffa, A reconstruction of the North Atlantic Oscillation using tree-ring chronologies from North America and Europe, *Holocene, 8*, 9-17, 1998.

Coulson, T., S. Albon, J. Pilkington, and T. Clutton-Brock, Small-scale spatial dynamics in a fluctuating ungulate population, *Journal of Animal Ecology, 68*, 658-671, 1999.

Coulson, T., E. A. Catchpole, S. D. Albon, B. J. T. Morgan, J. M. Pemberton, T. H. Clutton-Brock, M. J. Crawley, and B. T. Grenfell, Age, sex, density, winter weather, and population crashes in Soay sheep, *Science, 292*, 1528-1531, 2001.

Coulson, T., E. J. Milner-Gulland, and T. Clutton-Brock, The relative roles of density and climatic variation on population dynamics and fecundity rates in three contrasting ungulate species, *Proceedings of the Royal Society of London, Series B, 267*, 1771-1779, 2000.

Coulson, T. N., S. D. Albon, F. E. Guinness, J. M. Pemberton, and T. H. Clutton-Brock, Population substructure, local density, and calf winter survival in red deer (*Cervus elaphus*), *Ecology, 78*, 852-863, 1997.

Cox, D. R., Causality: some statistical aspects, *Journal of the Royal Statistical Society A, 155*, 291-301, 1992.

Crawford, R. M. M., Ecological hazards of oceanic environments, *New Phytologist, 147*, 257-281, 2000.

Crick, H. Q. P.,C. Dudley, D. E. Glue, and D. L. Thomson, UK birds are laying eggs earlier, *Nature, 388*, 526, 1997.

Cushing, D. H., *The Natural Regulation of Fish Populations. Sea Fisheries Research*, edited by H. Jones, pp. 399-412, F. R. Elek Science, London, 1974.

Cushing, D. H., Plankton production and year-class strength in fish populations: an update of the match/mismatch hypothesis, *Advances in Marine Biology, 26*, 249-293, 1990.

D'Arrigo, R. D., E. R. Cook, G. C. Jacoby, and K. R. Briffa, NAO and sea surface temperature signatures in tree-ring records from the North Atlantic sector, *Quarternary Science Reviews, 12*, 431-440, 1993.

Danell, K., A. Hofgaard, T. V. Callaghan, and J. P. Ball, Scenarios for animal responses to global change in Europe's cold regions: an introduction, *Ecological Bulletins, 47*, 8-15, 1999.

Davis, M. B., and R. G. Shaw, Range shifts and adaptive responses to Quaternary climate change, *Science, 292*, 673-679, 2001.

De Valpine, P., and J. Harte, Plant responses to experimental warming in a montane meadow, *Ecology, 82*, 637-648, 2001.

Dickson, R. R., T. J. Osborn, J. W. Hurrell, J. Meincke, J. Blindheim, B. Adlandsvik, T. Vinje, G. Alekseev, and W. Maslowski, The Arctic Ocean response to the North Atlantic Oscillation, *J. Climate, 13*, 2671-2696, 2000.

Donalson, D. D., and R. M. Nisbet, Population dynamics and spatial scale: effects of system size on population persistence, *Ecology, 80*, 2492-2507, 1999.

Easterling, D. R., G. A. Meehl, C. Parmesan, S. A. Changnon, T. R. Karl, and L. O. Mearns, Climate extremes: observations, modeling, and impacts, *Science 289*, 2068-2074, 2000.

Esteves, M. A., and M. D. M. Orgaz, The influence of climatic variability on the quality of wine, *International Journal of Biometeorology, 45*, 13-21, 2001.

Flagstad, Ø., P. O. Syvertsen, N. C. Stenseth, and K. S. Jakobsen, Environmental change and rates of evolution: the phylogeographic pattern within the hartebeest complex as related to climatic variation, *Proceedings of the Royal Society of London, Series B, 268*, 667-677, 2001.

Forchhammer, M. C., Terrestrial ecological responses to climate change in the Northern Hemisphere, in *Climate Change Research – Danish Contributions*, edited by A. M. K. Jørgensen, J. Fenger, and K. Halsnæs, pp. 219-236, GEC GAD, Copenhagen, 2001.

Forchhammer, M. C., T. H. Clutton-Brock, J. Lindström, and S. D. Albon, Climate and population density induce long-term cohort variation in a northern ungulate, *Journal of Animal Ecology, 70*, 721-729, 2001.

Forchhammer, M. C., and E. Post, Climatic signatures in ecology, *Trends in Ecology and Evolution*, 15, 286, 2000.

Forchhammer, M. C., E. Post, and N. C. Stenseth, Breeding phenology and climate, *Nature, 391*, 29-30, 1998a.

Forchhammer, M. C., N. C. Stenseth, E. Post, and R. Langvatn, Population dynamics of Norwegian red deer: density-dependence and climatic variation, *Proceedings of the Royal Society of London, Series B, 265*, 341-350, 1998b.

Fox, J. W., and P. J. Morin, Effects of intra- and interspecific interactions on species responses to environmental change, *Journal of Animal Ecology 70*, 80-90, 2001.

Gaillard, J.-M., J.-M. Boutin, D. Delorme, G. Van Laere, P. Duncan, and J.-D. Lebreton, Early survival in roe deer: causes and consequences of cohort variation in two contrasted populations, *Oecologia, 112*, 502-513, 1997.

Gaillard, J.-M., M. Festa-Bianchet, and N. G. Yoccoz, Population dynamics of large herbivores: variable recruitment with constant adult survival, *Trends in Ecology and Evolution, 13*, 58-63, 1998.

Gaillard, J.-M., M. Festa-Bianchet, and N. G. Yoccoz, Not all sheep are equal, *Science, 292*, 1499-1500, 2001.

Gaillard, J.-M., M. Festa-Bianchet, N. G. Yoccoz, A. Loison, and C. Toigo, Temporal variation in fitness components and population dynamics of large herbivores, *Annual Review of Ecology and Systematics, 31*, 367-393, 2000.

Grenfell, B. T., O. F. Price, S. D. Albon, and T. H. Clutton-Brock, Overcompensation and population cycles in an ungulate, *Nature, 355*, 823-826, 1992.

Grenfell, B. T., K. WilsonB. F. Finkenstädt, T. N. Coulson, S. Murray, S. D. AlbonJ. M. Pemberton, T. H. Clutton-Brock, and M. J. Crawley, Noise and determinism in synchronized sheep dynamics, *Nature, 394*, 674-677, 1998.

Gunn, A., and T. Skogland, Responses of caribou and reindeer to global warming, *Ecological Studies, 124,* 189-200, 1996.

Hardin, G., The competitive exclusion principle, *Science* 131, 1292-1297, 1960.

Hastie, T., and R. Tibshirani, *Generalized Additive Models*, 335 pp., Chapman and Hall, London, 1990.

Haydon, D. T., N. C. Stenseth, M. S. Boyce, and P. E. Greenwood, Phase coupling and synchrony in the spatiotemporal dynamics of muskrat and mink populations across Canada, *Proceedings of the National Academy of Sciences, USA, 98*, 13149-13154, 2001.

Hewison, A. J. M., Variation in the fecundity of roe deer in Britain: effects of age and body weight, *Acta Theriologica, 41*, 187-198, 1996.

Hewison, A. J. M., and J.-M.Gaillard, Successful sons or advantaged daughters? The Trivers-Willard model and sex-biased maternal investment in ungulates, *Trends in Ecology and Evolution, 14*, 229-234, 1999.

Hofgaard, A., Inter-relationships between treeline position, species diversity, land use and climate change in the central Scandes Mountains of Norway, *Global Ecology and Biogeography Letters, 6*, 419-429, 2001.

Holmgren, M., M. Scheffer, E. Ezcurra, J. R. Gutiérrez, and G. M. J. Mohren, El Niño effects on the dynamics of terrestrial ecosystems, *Trends in Ecology and Evolution, 16*, 89-94, 2001.

Hughes, L., Biological consequences of global warming: is the signal already apparent?, *Trends in Ecology and Evolution, 15*, 56-61, 2000.

Hurrell, J. W., Decadal trends in the North Atlantic Oscillation: regional temperatures and precipitation, *Science, 269*, 676-679, 1995.

Hurrell, J. W., and H. van Loon, Decadal variations in climate associated with the North Atlantic Oscillation. Climatic Change, 36, 310-326, 1997.

Hurrell, J. W., Y. Kushnir, and M. Visbeck, The North Atlantic Oscillation, *Science, 291*, 603-604, 2001.

Hurrell, J. W., Y. Kushnir, G. Ottersen, and M. Visbeck, An overview of the North Atlantic Oscillation, this volume.

Inouye, D. W., The ecological and evolutionary significance of frost in the context of climate change, *Ecology Letters 3*, 457-463, 2000.

Inouye, D. W., B. Barr, K. B. Armitage, and B. D. Inouye, Climate change is affecting altitudinal migrants and hibernating species, *Proceedings of the National Academy of Sciences, USA, 97*, 1630-1633, 2000.

Inouye, D. W., and A. D. McGuire, Effects of snowpack on timing and abundance of flowering in *Delphinium nelsonii* (*Ranunculaceae*): implications for climate change, *American Journal of Botany, 78*, 997-1001, 1991.

Jaksic, F. M., Ecological effects of El Niño in terrestrial ecosystems of western South America, *Ecography*, *24*, 241-250, 2001.

Jewell, P. A., C. Milner, and J. M. Boyd, *Island Survivors. The Ecology of the Soay Sheep of St Kilda*, 386 pp., The Athlone Press of the University of London, London, 1974.

Jonasson, S., J. P. Bryant, F. S. I. Chapin, and M. Andersson, Plant phenols and nutrients in relation to variations in climate and rodent grazing, *American Naturalist, 128*, 394-408, 1986.

Jones, P. D., T. Jónsson, and D. Wheeler, Extension using early instrumental pressure observations from Gibraltar and SW Iceland to the North Atlantic Oscillation, *Int. J. Climatol.*, *17*, 1433-1450, 1997.

Jones, P. D., T. J. Osborn, and K. R. Briffa, The evolution of climate over the last millennium, *Science*, *292*, 662-667, 2001.

Jones, P. D., T. J. Osborn, and K. R. Briffa, Pressure-based measures of the North Atlantic Oscillation (NAO): A comparison and an assessment of changes in the strength of the NAO and in its influence on surface climate parameters, this volume.

Kettlewell, P. S., R. B. Sothern, and W. L. Koukkari, U.K. wheat quality and economic value are dependent on the North Atlantic Oscillation, *Journal of Cereal Science*, *29*, 205-209, 1999.

Kiesecker, J. M., A. R. Blaustein, and L. K. Belden, Complex causes of amphibian population declines. *Nature*, *410*, 681-684, 2001.

Kojola, I., Sex ratio and maternal investment in ungulates, *Oikos, 83*, 567-573, 1998.

Krebs, C. J., R. Boonstra, S. Boutin, and A. R. E. Sinclair, What drives the 10-year cycle of the snowshoe hares? *BioScience*, *51*, 25-35, 2001.

Kruuk, L. E. B., T. H. Clutton-Brock, S. D. Albon, J. M. Pemberton, and F. E. Guinness, Population density affects sex ratio variation in red deer, *Nature*, *399*, 459-461, 1999.

Langvatn, R. and S. D. Albon, Geographic clines in body weight of Norwegian red deer: a novel explanation of Bergmann's rule? *Holarctic Ecology, 9*, 285-293, 1986.

Langvatn, R., S. D. Albon, T. Burkey, and T. H. Clutton-Brock, Climate, plant phenology and variation in age at first reproduction in a temperate herbivore, *Journal of Animal Ecology, 65*, 653-670, 1996.

Langvatn, R., and A. Loison, Consequences of harvesting on age structure, sex ratio and population dynamics of red deer *Cervus elaphus* in central Norway, *Wildlife Biology, 5*, 213-223, 1999.

Lauscher, A., and F. Lauscher, *Phänologie Norwegens, teil IV. Eigenverlag*, F. Lauscher, 119 pp., Vienna, Austria, 1990.

Lesage, L., M. Crête, J. Huot, and J.-P. Ouellet, Quality of plant species utilized by northern white-tailed deer in summer along a climatic gradient, *Ecoscience*, *7*, 439-451, 2000.

Lindström, J., Early development and fitness in birds and mammals, *Trends in Ecology and Evolution, 14*, 343-348, 1999.

Loison, A., J.-M. Jullien, and P. Menaut, Relationship between chamois and isard survival and variation in global and local climate regimes: contrasting examples from the Alps and Pyrenees, *Ecological Bulletins*, *47*, 126-136, 1999a.

Loison, A., R. Langvatn, and E. J. Solberg, Body mass and winter mortality in red deer calves: Disentangling sex and climate effects, *Ecography, 22*, 20-30, 1999b.

Macdonald, D. W., The ecology of carnivore social behavior, *Nature, 301*, 379-384, 1983.

Mann, K. H., and J. R. N. Lazier, *Dynamics of Marine Ecosystems*, 466 pp., Blackwell, Cambridge, 1991.

May, R. M., When two and two do not make four: nonlinear phenomena in ecology, *Proceedings of the Royal Society of London, Series B, 228*, 241-266, 1986.

Maynard Smith, J., A new theory of sexual investment, *Behavioral Ecology and Sociobiology*, *7*, 247-251, 1980.

McCarty, J. P., Ecological consequences of recent climate change, *Conservation Biology, 15*, 320-331, 2001.

McCleery, R. H., and C. M. Perrins, Temperature and egg-laying trends, *Nature*, *391*, 30-31, 1998.

McIllroy, S. H., Rain and windchill as factors in the occurrence of pneumonia in sheep, *The Veterinary Record*, *125*, 79-84, 1989.

McLaren, b. E., and R. O. Peterson, Wolves, moose and tree rings on Isle Royale, *Science*, *266*, 1555-1558, 1994.

Meehl, G. A., et al., An introduction to trends in extreme weather and climate events: Observations, socioeconomic impacts, terrestrial ecological impacts, and model projections, *Bull. Amer. Meteor. Soc.*, *81*, 413-416, 2000.

Menzel, A., and P. Fabian, Growing season extended in Europe, *Nature*, *397*, 659, 1999.

Milner, J. M., D. A. Elston, and S. D. Albon, Estimating the contributions of population density and climatic fluctuations to interannual variation in survival of Soay sheep, *Journal of Animal Ecology*, *68*, 1235-1247, 1999.

Monahan, A. H., J. C. Fyfe, and G. M. Flato, A regime view of northern hemisphere atmospheric variability and change under global warming, *Geophys. Res. Lett.*, *27*, 1139-1142, 2000.

Monahan, A. H., L. Pandolfo, and J. C. Fyfe, The preferred structure of variability of the Northern Hemisphere atmospheric circulation, *Geophys. Res. Lett.*, *28*, 1019-1022, 2001.

Morell, V., Are pathogens felling frogs? *Science*, *284*, 728-731, 1999.

Moss, R., J. Oswald, and D. Baines, Climate change and breeding success: decline of the capercaillie in Scotland, *Journal of Animal Ecology, 70*, 47-61, 2001.

Moulin, C., C. E. Lambert, F. Dulac, and U. Dayan, Control of atmospheric export of dust from North Africa by the North Atlantic Oscillation, *Nature*, *387*, 691-694, 1997.

Murray, D. L., S. Boutin, M. O'Donoghue, and V. O. Nams, Hunting behavior of a sympatric felid and canid in relation to vegetative cover, *Animal Behavior*, *50*, 1203-1210, 1995.

Myneni, R. B., C. D. Keeling, C. J. Tucker, G. Asrar, and R. R. Nemani, Increased plant growth in the northern high latitudes from 1981-1991, *Nature, 386*, 698-702, 1997.

Mysterud, A., Bed-site selection by adult roe deer *Capreolus capreolus* in southern Norway during summer, *Wildlife Biology, 2*, 101-106, 1996.

Mysterud, A., Seasonal migration pattern and home range of roe deer (*Capreolus capreolus*) in an altitudinal gradient in southern Norway, *Journal of Zoology*, *247*, 479-486, 1999.

Mysterud, A., Diet overlap among ruminants in Fennoscandia, *Oecologia, 124*, 130-137, 2000.

Mysterud, A., R. Langvatn, N. G. Yoccoz, and N. C. Stenseth, Plant phenology, migration and geographic variation in body weight of a large herbivore: the effect of a variable topography, *Journal of Animal Ecology*, *70*, 915-923, 2001a.

Mysterud, A., N. C. Stenseth, N. G. Yoccoz, R. Langvatn, and G. Steinheim, Nonlinear effects of large-scale climatic variability on wild and domestic herbivores, *Nature, 410*, 1096-1099, 2001b.

Mysterud, A., N. G. Yoccoz, N. C. Stenseth, and R. Langvatn, Relationships between sex ratio, climate and density in red deer: the importance of spatial scale, *Journal of Animal Ecology*, *69*, 959-974, 2000.

Mysterud, A., N. G. Yoccoz, N. C. Stenseth, and R. Langvatn, The effects of age, sex and density on body weight of Norwegian red deer: evidence of density-dependent senescence, *Proceedings of the Royal Society of London, Series B, 268*, 911-919, 2001c.

Mysterud, A., and E. Østbye, Bed-site selection by European roe deer (*Capreolus capreolus*) in southern Norway during winter, *Canadian Journal of Zoology*, *73*, 924-932, 1995.

Niemelä, P., Topographical delimitation of Oporinia-damages: experimental evidence of the effect of winter temperature, *Reports Kevo Subarctic Research Station*, *15*, 33-36, 1979.

Niemelä, P., F. S. I. Chapin, K. Danell, and J. P. Bryant, Herbivory-mediated responses of selected boreal forests to climatic change, *Clim. Change*, *48*, 427-440, 2001.

Oba, G., E. Post, and N. C. Stenseth, Sub-saharan desertification and productivity are linked to hemisphere climate variability, *Global Change Biology*, *7*, 241-246, 2001.

Ottersen, G., B. Planque, A. Belgrano, E. Post, P. C. Reid, and N. C. Stenseth, Ecological effects of the North Atlantic Oscillation, *Oecologia*, *128*, 1-14, 2001.

Parker, K. L. Effects of heat, cold, and rain on coastal black-tailed deer, *Canadian Journal of Zoology*, *66*, 2475-2483, 1988.

Parker, K. L., and C. T. Robbins, Thermoregulation in ungulates, in *Bioenergetics of wild herbivores*, edited by R. J. Hudson, and R. G. White, pp. 161-182, CRC Press, Inc., 1985.

Parmesan, C., Impacts of extreme weather and climate on terrestrial biota, *Bull. Am. Meteor. Soc.*, *81*, 443-450, 2000.

Parmesan, C., et al., Poleward shifts in geographical ranges of butterfly associated with regional warming, *Nature*, *399*, 579-583, 1999.

Peterson, R. O., R. E. Page, and K. M. Dodge, Wolves, moose, and the allometry of population cycles, *Science*, *224*, 1350-1352, 1984.

Philander, S. G., *El Niño, La Niña, and the Southern Oscillation*, 293 pp., Academic Press, New York, 1990.

Pimm, S. L., Entrepreneurial insects, *Nature*, *411*, 531-532, 2001.

Piovesan, G., and B. Schirone, Winter North Atlantic oscillation effects on the tree rings of the Italian beech (*Fagus sylvatica L.*), *International Journal of Biometeorology*, *44*, 121-127, 2000.

Pitelka, L. F. et al., Plant migration and climate change, *American Scientist*, *85*, 464-473, 1997.

Portier, C., M. Festa-Bianchet, J.-M. Gaillard, J. T. Jorgenson, and N. G. Yoccoz, Effects of density and weather on survival of bighorn sheep lambs (*Ovis canadensis*), *Journal of Zoology*, *245*, 271-278, 1998.

Portis, D. H., J. E. Walsh, M. El Hamly, and P. J. Lamb, Seasonality of the North Atlantic Oscillation, *J. Climate*, *14*, 2069-2078, 2001.

Post, E., M. C. Forchhammer, and N. C. Stenseth, Population ecology and the North Atlantic Oscillation (NAO), *Ecological Bulletins*, *47*, 117-125, 1999a.

Post, E., M. C. Forchhammer, N. C. Stenseth, and T. V. Callaghan, The timing of life-history events in a changing climate, *Proceedings of the Royal Society of London, Series B*, *268*, 15-23, 2001a.

Post, E., M. C. Forchhammer, N. C. Stenseth, and R. Langvatn, Extrinsic modification of vertebrate sex ratios by climatic change, *American Naturalist*, *154*, 194-204, 1999b.

Post, E., R. Langvatn, M. C. Forchhammer, and N. C. Stenseth, Environmental variation shapes sexual dimorphism in red deer, *Proceedings of the National Academy of Sciences, USA*, *96*, 4467-4471, 1999c.

Post, E., S. A. Levin, Y. Iwasa, and N. C. Stenseth, Reproductive asynchrony increases with environmental disturbance, *Evolution*, *55*, 830-834, 2001b.

Post, E., R. O. Peterson, N. C. Stenseth, and b. E. McLaren, Ecosystem consequences of wolf behavioral response to climate, *Nature*, *401*, 905-907, 1999d.

Post, E., and N. C. Stenseth, Large-scale climatic fluctuations and population dynamics of moose and white-tailed deer, *Journal of Animal Ecology*, *67*, 537-543, 1998.

Post, E., and N. C. Stenseth, Climatic variability, plant phenology, and northern ungulates, *Ecology*, *80*, 1322-1339, 1999.

Post, E., N. C. Stenseth, R. Langvatn, and J.-M. Fromentin, Global climate change and phenotypic variation among red deer cohorts, *Proceedings of the Royal Society of London, Series B*, *264*, 1317-1324, 1997.

Pounds, J. A., Climate and amphibian declines, *Nature*, *410*, 639-640, 2001.

Pounds, J. A., M. P. L. Fogden, and J. H. Campbell, Biological response to climate change on a tropical mountain, *Nature*, *398*, 611-615, 1999.

Przybylo, R., B. C. Sheldon, and J. Merilä, Climatic effects on breeding and morphology: evidence for phenotypic plasticity, *Journal of Animal Ecology*, *69*, 395-403, 2000.

Ray, C., and A. Hastings, Density dependence: are we searching at the wrong spatial scale? *Journal of Animal Ecology*, *65*, 556-566, 1996.

Reimers, E., Reproduction in wild reindeer in Norway, *Canadian Journal of Zoology*, *61*, 211-217, 1983.

Rodó, X., and F. A. Comín, Links between large-scale anomalies, rainfall and wine quality in the Iberian Peninsula during the last three decades, *Global Change Biology*, *6*, 267-273, 2000.

Rogers, J. C., The association between the North Atlantic Oscillation and the Southern Oscillation in the Northern Hemisphere, *Mon. Wea. Rev.*, *112*, 1999-2015, 1984.

Rose, K. E., T. H. Clutton-Brock, and F. E. Guinness, Cohort variation in male survival and lifetime breeding success in red deer, *Journal of Animal Ecology*, *67*, 979-986, 1998.

Roy, D. B., and T. H. Sparks, Phenology of British butterflies and climate change, *Global Change Biology*, *6*, 407-416, 2000.

Ruohomäki, K., M. Tanhuanpää, M. P. Ayres, P. Kaitaniemi, T. Tammaru, and E. Haukioja, Causes of cyclicity of *Epirrita autumnata* (Lepidoptera, Geometridae): grandiose theory and tedious practice, *Population Ecology*, *42*, 211-223, 2000.

Sauer, J. R., and M. S. Boyce, Density dependence and survival of elk in northwestern Wyoming, *Journal of Wildlife Management*, *47*, 31-37, 1983.

Schmidt, P. A., and L. D. Mech, Wolf pack size and food acquisition, *American Naturalist*, *150*, 513-517, 1997.

Schmitz, O. J., Thermal constraints and optimization of winter feeding and habitat choice in white-tailed deer, *Holarctic Ecology*, *14*, 104-111, 1991.

Schwab, F. E., and M. D. Pitt, Moose selection of canopy cover types related to operative temperature, forage, and snow depth, *Canadian Journal of Zoology*, *61*, 3071-3077, 1991.

Seastedt, T. R., and A. K. Knapp, Consequences of nonequilibrium resource availability across multiple time scales: the transient maxima hypothesis, *American Naturalist*, *141*, 621-633, 1993.

Simpson, A. M., A. J. F. Webster, J. S. Smith, and C. A. Simpson, Energy and nitrogen metabolism of red deer (*Cervus elaphus*) in cold environments; a comparison with cattle and sheep, *Journal of Comparative Biochemistry and Physiology*, *60*, 251-256, 1978.

Skogland, T., Sex ratio variation in relation to maternal condition and parental investment in wild reindeer Rangifer t. tarandus, *Oikos*, *46*, 417-419, 1986.

Solberg, B. Ø., A. Hofgaard, and H. Hytteborn, Shifts in radial growth responses of coastal *Picea abies* induced by climatic change during the 20th century, central Norway, *Ecoscience*, *9*, 79-88, 2002.

Sparks, T. H., E. P. Jeffree, and C. E. Jeffree, An examination of the relationship between flowering times and temperature at the national scale using long-term phenological records from the UK, *International Journal of Biometeorology*, *44*, 82-87, 2000.

Sparks, T. H., and T. J. Yates, The effect of spring temperature on the appearance dates of British butterflies 1883-1993, *Ecography*, *20*, 368-374, 1997.

Statistics, Norway, *Official hunting statistics of Norway*, Statistics Norway, Oslo and Kongsvinger, 2001.

Stenseth, N. C., et al., Common dynamic structure of Canada lynx populations within three climatic regions, *Science*, *285*, 1071-1073, 1999.

Stenseth, N. C., H. Viljugrein, W. Jedrzejewski, A. Mysterud, and Z. Pucek, Population dynamics of *Clethrionomys glareolus* and *Apodemus flavicollis:* seasonal components of density-dependence and density-independence. *Acta Theriologica, 47,* Supplement 1, in press, 2002.

Stenström, M., F. Gugerli, and G. H. R. Henry, Response of *Saxifraga oppositifolia* L. to simulated climate change at three contrasting latitudes, *Global Change Biology*, *3*, 44-54, 1997.

Stevenson, I. R., and D. M. Bryant, Climate change and constraints on breeding, *Nature*, *406*, 366-367, 2000.

Sturm, M., C. Racine, C., and K. Tape, Increasing shrub abundance in the Arctic, *Nature*, *411*, 546-547, 2001.

Sætersdal, M., H. J. B. Birks, and S. M. Peglar, Predicting changes in Fennoscandian vascular-plant species richness as a result of future climatic change, *Journal of Biogeography 25*, 111-122, 1998.

Sæther, B.-E., Environmental stochasticity and population dynamics of large herbivores: a search for mechanisms, *Trends in Ecology and Evolution*, *12*, 143-149, 1997.

Sæther, B.-E., J. Tufto, S. Engen, K. Jerstad, O. W. Røstad, and J. E. Skåtan, Population dynamical consequences of climate change for a small temperate songbird, *Science*, *287*, 854-856, 2000.

Sætre, G-P., E. Post, and M. Král, Can environmental fluctuation prevent competitive exclusion in sympatric flycatchers? *Proceedings of the Royal Society of London, Series B, 266*, 1247-1251, 1999.

Tenow, O., and A. Nilssen, Egg cold hardiness and topoclimatic limitations to outbreaks of *Epirrita autumnata* in northern Fennoscandia, *Journal of Applied Ecology*, *27*, 723-734, 1990.

Thomas, C. D., E. J. Bodsworth, R. J. Wilson, A. D. Simmons, Z. G. Davies, M. Musche, and L. Conradt, Ecological and evolutionary processes at expanding range margins, *Nature*, *411*, 577-581, 2001a.

Thomas, C. D., and J. J. Lennon, Birds extend their ranges northwards, *Nature*, *399*, 213, 1999.

Thomas, D. W., J. Blondel, P. Perret, M. M. Lambrechts, and J. R. Speakman, Energetic and fitness costs of mismatching resource supply and demand in seasonally breeding birds, *Science, 291*, 2598-2600, 2001b.

Thurber, J. M., and R. O. Peterson, Effects of population density and pack size on the foraging ecology of gray wolves, *Journal of Mammalogy*, *74*, 879-889, 1993.

Trenberth, K. E., and J. W. Hurrell, Decadal atmosphere-ocean variations in the Pacific, *Clim. Dyn.*, 9, 303-319, 1994.

Trivers, R. L., and D. E. Willard, Natural selection of parental ability to vary the sex ratio of offspring, *Science*, *179*, 90-92, 1973.

Van der Wal, R., N. Madan, S. van Lieshout, C. Dormann, R. Langvatn, and S. D. Albon, Trading forage quality for quantity? Plant phenology and patch choice by Svalbard reindeer, *Oecologia*, *123*, 108-115, 2000.

Van Soest, P. J., *Nutritional Ecology of the Ruminant*, 476 pp., Cornell University Press, New York, 1994.

Venables, W. N. and B. D. Ripley, *Modern Applied Statistics with S-plus*, 462 pp., Springer Verlag, New York, 1994.

Visser, M. E., and L. J. M. Holleman, Warmer springs disprupt the synchrony of oak and winter moth phenology, *Proceedings of the Royal Society of London, Series B*, *268*, 289-294, 2001.

Walker, G. T., and E. W. Bliss, World weather V, *Mem. Royal Meteorological Society*, *4*, 53-84, 1932.

Wallace, J. M., and D. S. Gutzler, Teleconnections in the geopotential height field during the Northern Hemisphere winter, *Mon. Wea. Rev.*, *109*, 784-812, 1981.

Walsh, N. E., T. R. McCabe, J. M. Welker, and A. N. Parsons, Experimental manipulations of snow-depth: effects on nutrient content of caribou forage, *Global Change Biology*, *3*, 158-164, 1997.

Weladji, R. B., D. R. Klein, Ø Holand, and A. Mysterud, Comparative response of *Rangifer tarandus* and other northern ungulates to climatic variability, *Rangifer, 22*, 33-50, 2002.

White, R. G., Foraging patterns and their multiplier effects on productivity of northern ungulates, *Oikos*, *40*, 377-384, 1983.

Williams, G. C., The question of adaptive sex ratio in outcrossed vertebrates, *Proceedings of the Royal Society of London, Series B*, *205*, 567-580, 1979.

Yoccoz, N. G., and R. A. Ims, Demography of small mammal in cold regions: the importance of environmental variability, *Ecological Bulletins*, *47*, 137-144, 1999.

Yoccoz, N. G., J. D. Nichols, and T. Boulinier, Monitoring of biological diversity in space and time, *Trends in Ecology and Evolution*, *16*, 446-453, 2001.

Yom-Tov, Y., Global warming and body mass decline in Israeli passerine birds, *Proceedings of the Royal Society of London, Series B*,I, 947-952, 2001.

Rolf Langvatn, University Courses on Svalbard, BP 156, N-9170 Longyearbyen, Spitsbergen, Norway
rolf.langvatn@ninatrd.ninaniku.no

A. Mysterud, Department of Biology, Division of Zoology, University of Oslo, P.O. Box 1050 Blinden, N-0316 Oslo, Norway
atle.mysterud@bio.uio.no

Geir Ottersen, Institute of Marine Research, P.O. Box 1870 Nordnes, 5817 Bergen, Norway

Current address:

Department of Biology, Division of Zoology, University of Oslo, P.O. Box 1050 Blindern, N-0316 Oslo, Norway
geir.ottersen@bio.uio.no

N. C. Stenseth, Department of Biology, Division of Zoology, University of Oslo, P.O. Box 1050 Blinden, N-0316 Oslo, Norway
n.c.stenseth@bio.uio.no

Nigel G. Yoccoz, Department of Arctic Ecology, Norwegian Institute for Nature Research, Polar Environmental Center, N-9296 Tromsø, Norway
nigel.yoccoz@ninatos.ninaniku.n

The Response of Freshwater Ecosystems to Climate Variability Associated with the North Atlantic Oscillation

Dietmar Straile[1], David M. Livingstone[2], Gesa A. Weyhenmeyer[3], D. Glen George[4]

The North Atlantic Oscillation (NAO) affects the physics, hydrology, chemistry and biology of freshwater ecosystems over a large part of the Northern Hemisphere. Physical impacts of the NAO include effects on lake temperature profiles, lake ice phenology, river runoff and lake water levels. These physical and hydrological responses influence the chemistry and biology of fresh waters by affecting the leaching of nutrients from the soil and by altering the distribution of nutrients and oxygen in lakes. Finally, the population dynamics of freshwater organisms on several trophic levels—including autotrophs, herbivores and vertebrate predators—are directly and indirectly linked to the NAO via food-web interactions. As a result, the effects of mild winters associated with the positive index phase of the NAO can influence the food-web characteristics of lakes in summer. A considerable body of evidence documents the importance of these indirect and food-web mediated effects of the NAO, which might even result in ecosystem regime shifts. Owing to the large-scale impact of the NAO, lakes exhibit spatial coherence over large areas with respect to both physical and biological properties. This coherence is modified by geographical factors such as altitude and latitude, and by lake-specific characteristics such as depth and trophic status.

1. INTRODUCTION

Freshwater ecosystems have been repeatedly shown to respond significantly to large-scale climatic fluctuations [e.g., *Strub et al.,* 1985; *Firth and Fisher,* 1992; *George and Taylor,* 1995; *Anderson et al.,* 1996; *Schindler*, 1997]. Because human demand for high-quality fresh water is continually increasing, the practical relevance of such responses for the large-scale management of inland waters is now gradually beginning to be realized. Specifically, the impact of the North Atlantic Oscillation on lacustrine systems, especially in Europe, is becoming a topic of some importance. Briefly summarizing the information given more extensively in other parts of this volume, the North Atlantic Oscillation (NAO) represents a large-scale fluctuation in the air pressure difference between the Azores High and the Iceland Low [e.g., *Hurrell et al.*, this volume]. It dominates much of the atmospheric behavior in the North Atlantic region, and is known to influence air temperature and precipitation over large areas of the Northern Hemisphere in winter [e.g., *Hurrell,* 1995; *Hurrell and van Loon*, 1997]. It is commonly represented in terms of an index based on the difference of the sea-level air pressure measured at a meteorological station close to the center of the Azores High and that measured at a station in Iceland [e.g., *Jones et al.*, this volume]. Positive winter values of an NAO index correspond to a strong meridional pressure gradient that results in strong westerly winds transporting warm, moist maritime air across Europe, giving rise to warm, wet winters there. In contrast, low NAO index values correspond to weak westerlies and cold, dry winters in Europe.

[1]Limnological Institute, University of Konstanz, Konstanz, Germany
[2]Water Resources Department, Swiss Federal Institute of Environmental Science and Technology (EAWAG), Dübendorf, Switzerland
[3]Department of Environmental Assessment, Swedish University of Agricultural Sciences, Uppsala, Sweden
[4]Centre for Ecology and Hydrology, Windermere, U.K.

The North Atlantic Oscillation:
Climatic Significance and Environmental Impact
Geophysical Monograph 134

10.1029/134GM12

Here we will discuss the response of freshwater ecosystems to NAO-mediated changes in temperature regimes and water balances. Impacts of the NAO on lakes and rivers have been documented within a region stretching from Lake Mendota in central North America to Lake Baikal in eastern Siberia, and from Lake Kallavesi in Finland to the Caspian Sea (Figure 1). They can affect the physical, chemical and biological characteristics of these systems. The biological characteristics affected include plankton populations, community structure, and the food web.

Our review will not be one of individual case studies. Instead, we adopt an ecosystem approach to the impacts of the NAO, starting with its immediate effects on the physical characteristics of lakes, and continuing with the consequences of these physical responses for lake chemistry and biology. Finally, we consider more indirect biologically mediated chemical and food-web responses to the NAO.

2. PHYSICAL RESPONSES

2.1. Lake Temperatures

The heat balance of a lake is governed predominantly by five heat exchange processes involving four meteorological driving variables; viz. air temperature, cloud cover, water vapor pressure and wind speed [*Edinger et al.,* 1968]. Of these driving variables, air temperature is generally regarded to have the most significant influence on lake temperature variability [e.g., *Henderson-Sellers,* 1988; *Hondzo and Stefan,* 1992; 1993]. Air temperature is involved explicitly in three of the five heat exchange processes (convective sensible heat exchange, evaporative heat exchange and the atmospheric emission of long-wave radiation incident on the lake), and the surface temperature of a lake tends asymptotically towards an equilibrium temperature that is often close to the ambient air temperature despite deviations induced by radiative heat exchange and wind mixing [*Dingman,* 1972; *Arai,* 1981]. Thus, despite the fact that air temperature is only partially responsible for the heat balance of a lake, empirical studies show that lake surface temperatures are closely related to local and regional air temperatures on both short and long time scales [*McCombie,* 1959; *Shuter et al.,* 1983; *Livingstone and Lotter,* 1998; *Livingstone et al.,* 1999; *Livingstone and Dokulil,* 2001].

Air temperatures over large areas of the Northern Hemisphere are known to be influenced significantly by the NAO [e.g., *Hurrell,* 1995]. Thus, even neglecting a possible NAO influence on cloud cover, water vapor pressure and wind speed, the NAO influence on air temperature alone might well suffice to generate an NAO signature in lake surface temperature via the relevant air-water heat exchange processes.

The response of a lake to meteorological forcing is most immediate at the lake surface. Because of the relative ease of obtaining surface data, historical time-series of surface temperature tend to go back further in time than time-series of temperatures at depth, facilitating long-term comparison with meteorological data and with indices of the NAO. Based on 40 years of data, *George et al.* [2000] showed a highly significant correlation between the winter NAO and the winter surface temperatures of Esthwaite Water and Windermere in the English Lake District, with as much as 40–50% shared variance. Likewise, winter (DJF) temperature in Black Brows Beck, a small stream in the English Lake District, was highly correlated with the winter NAO, with over 60% shared variance [*Elliott et al.,* 2000]. Because of its proximity to the North Atlantic, a strong

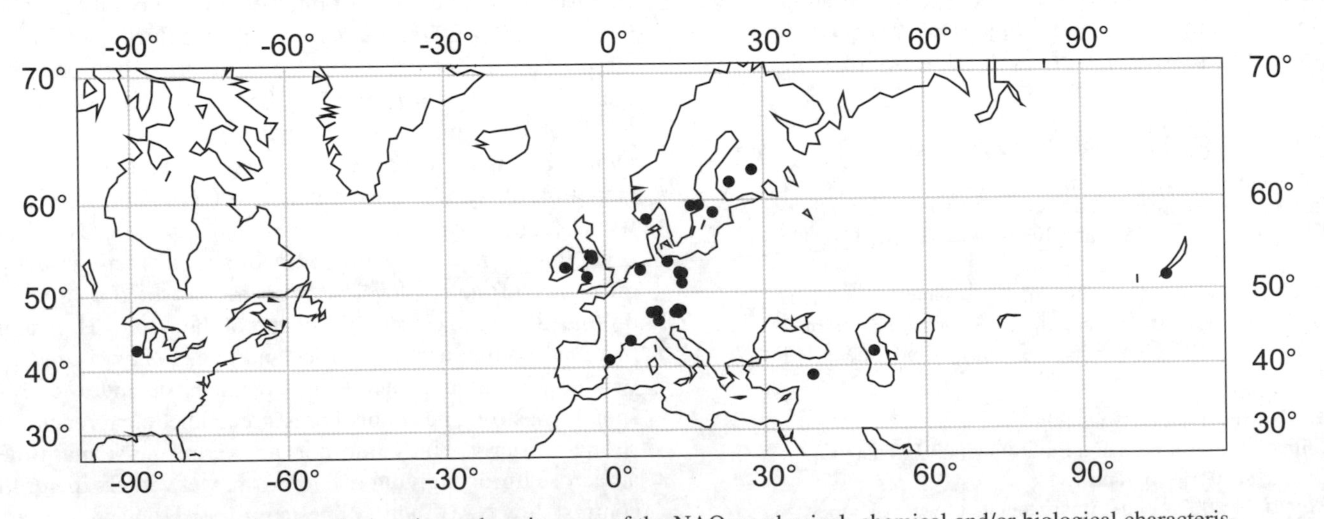

Figure 1. Map showing the locations where impacts of the NAO on physical, chemical and/or biological characteristics of freshwater ecosystems have been demonstrated.

NAO influence might be expected in the western maritime region of Europe. However, studies in the European peri-alpine region, in which a strong NAO influence might not be expected *a priori*, have also revealed a significant NAO influence on surface or near-surface lake water temperatures. Based on 16 yr of data, *Straile* [2000] detected a significant NAO influence on the water temperature at 8 m depth in Lake Constance, with over 30% shared variance with the water temperature in March. Average winter (JFM) and spring (AMJ) temperatures in Lake Veluwe, in the Netherlands, from 1959 - 1999 were found to be highly correlated with the NAO index ($r = 0.75$, $p < 0.01$ and $r = 0.49$, $p < 0.01$ respectively [*Scheffer et al.*, 2001a]). In an analysis of 80 years of monthly mean surface temperatures from eight lakes in Austria, *Livingstone and Dokulil* [2001] found surface temperatures to be highly spatially coherent among lakes in all seasons, and to reflect much of the temporal structure inherent in the regional air temperature. Comparison of an average Austrian lake surface temperature with *Hurrell's* [1995] winter NAO index revealed a significant correlation from January to May, with over 25% shared variance in March. In each of the months January to May, the NAO signal in the average Austrian lake surface temperature was found to be even stronger than the NAO signal in regional air temperature, suggesting that although air temperature is probably the most important meteorological driving variable transmitting an NAO signal to the lakes, it is unlikely to be the only one. An additional result concerns the blocking of the NAO influence by ice cover: in general, the NAO signal detected in lake surface temperature was found to be strongest in lowland lakes with infrequent or short periods of ice cover, becoming successively weaker with altitude, as the frequency and duration of ice cover events increased.

Fluctuations in the surface temperature of a lake are transmitted downwards by vertical mixing processes. However, although the response of lake surface temperature to meteorological forcing is fairly immediate, this is not true of the temperature of the deeper water. A lake tends to filter out high-frequency temperature fluctuations with increasing depth, implying that, although short-period fluctuations in air temperature may be reflected only in water temperatures in the uppermost few meters of a lake, longer-period fluctuations may well be reflected in some way in the water temperature in the depths of the lake [*Livingstone*, 1993]. These fluctuations, however, are only transmitted to the deep water of the lake when thermal stratification is weak, i.e., during the winter half-year. In late winter and early spring, when the NAO influence on lake surface temperature is at its greatest, holomictic lakes (lakes that can undergo mixing down to the lowest depths without hindrance from chemical stabilization of the water column) usually become homothermic, with vigorous vertical mixing occurring during the so-called spring turnover. Thus at this time of year, not only surface temperatures, but also deep-water temperatures, are directly determined to a large extent by the prevailing meteorological conditions. Any NAO signal present in the meteorological variables that determine the lake heat balance at this time of year will therefore be transmitted not only to the upper layers of the lake, but to the whole water body. Because the thermal stratification that subsequently ensues effectively isolates the hypolimnion (the region below the thermocline) from the epilimnion (the region above the thermocline), and thus from any further climatic forcing mediated by the epilimnion, deep-water temperatures during the summer half-year are determined by, and highly correlated with, the deep-water temperature attained at the end of spring turnover [*Robertson and Ragotzkie*, 1990; *Hondzo and Stefan*, 1993; *Livingstone*, 1993]. Thus, any climatic signal captured in the hypolimnetic temperature of a lake during spring turnover is likely to persist for some months afterwards.

With respect to the NAO, this was analyzed by *Gerten and Adrian* [2001] in three different lakes in northern Germany. They were able to show that a signal from the winter NAO was present in the temperatures of all three lakes at all depths, but that the persistence of this signal differed significantly from lake to lake. In the epilimnion of all three lakes, the NAO signal was confined to late winter and early spring, implying that the advent of thermal stratification in late spring effectively eliminates any influence of the previous winter's NAO above the thermocline in summer (the effect of the summer NAO is negligible). In the hypolimnion, however, the response varied from lake to lake. In shallow, polymictic Müggelsee, the NAO signal was confined to spring. In shallow, dimictic Heiligensee it persisted, albeit with a weakening tendency, throughout much of the summer. Finally, in much deeper, dimictic Stechlinsee, the NAO signal persisted throughout the whole of summer and autumn, only being eliminated by vertical mixing in December. Thus, although an NAO signal is likely to be present to some extent in the temperature of all lakes within the NAO's area of influence, individual lake characteristics can cause major modifications of this signal.

Using a superposed epoch analysis, Figure 2 illustrates the effect the NAO can have on the longer-term development of the water temperature in an even deeper lake, Lake Zurich (136 m deep), located in Switzerland. In this analysis, the 50 years from 1948 to 1997 were first divided into two 25-yr groups according to whether their winter NAO index was less than or greater than the median value (0.42). For each group, the monthly mean lake water temperatures

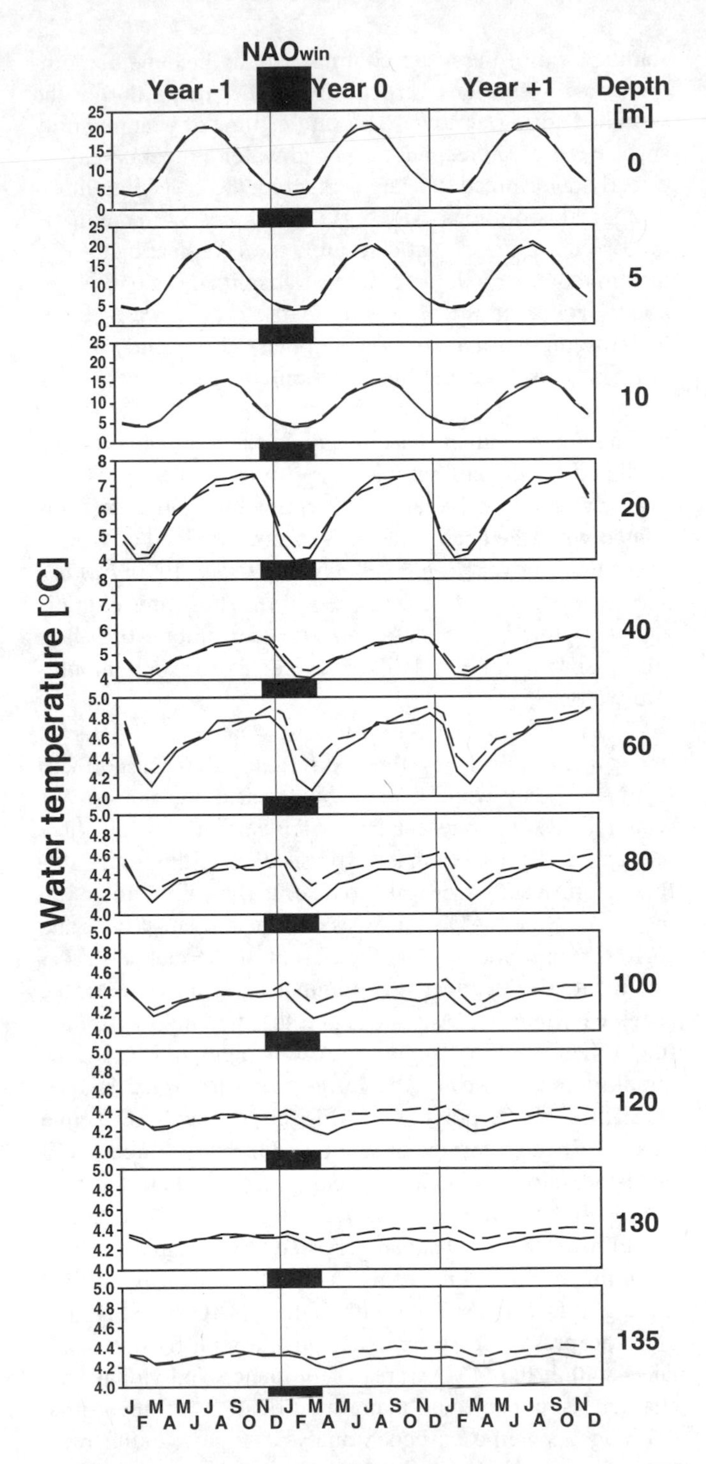

Figure 2. Superposed epoch analysis of water temperatures at 11 depths in Lake Zurich, Switzerland, with respect to the occurrence of high and low winter indices of the NAO. The period covered by the winter NAO index NAOwin (December–March) is shaded. The figure shows water temperatures averaged over all years in which NAOwin was greater (dashed lines) or less (solid lines) than the median value. See text for detailed explanation.

at a particular depth were then averaged over all 25 yr for the 3-yr "epoch" encompassing the year of the NAO index taken ("Year 0") and the years immediately before ("Year – 1") and after ("Year 1"). This analysis shows up the effect of high and low NAO indices on the development of the temperature regime of the lake. Lake Zurich does not lie particularly close to the center of action of the NAO. Nevertheless, a significant difference is apparent in the way the thermal structure of the lake develops following winters with a high or low winter NAO index. High (low) winter NAO indices tend to result in high (low) temperatures in the hypolimnion in late winter and early spring, which persist not only throughout the following summer stratification period (Year 0), but carry over through the subsequent winter into the next summer stratification period (Year 1). In the year previous to the occurrence of either a high or low NAO index (Year -1), no significant difference in lake temperature is observed at any depth, indicating that the causative event does indeed occur at the beginning of Year 0. The persistence of the NAO effect from one year to the next is possible because Lake Zurich, although holomictic, does not always circulate fully each year. During warm winters (high NAO index) stratification in Lake Zurich may persist uninterruptedly from one summer to the next [*Livingstone*, 1993; 1997a]. A similar pattern has been observed in Lake Constance, where it has also been shown that the thermal stability of the water column in winter and spring is positively related to the winter NAO index [*Straile et al.*, in prep.].

In general, epilimnetic temperatures of all lakes within the region of influence of the NAO are likely to show a large degree of spatial coherence in winter and spring that is induced by spatial coherence in meteorological forcing related to the winter NAO. In the hypolimnion, however, the persistence of the winter NAO signal can vary substantially from lake to lake, depending on individual lake characteristics such as lake morphometry and degree of susceptibility to meteorological forcing in summer.

2.2. *Lake Ice Cover*

Historical ice phenology records from lakes distributed throughout the Northern Hemisphere are known to reflect the general rise in global air temperature that has occurred over approximately the last 150 years [*Magnuson et al.*, 2000]. Such records are therefore not only of local interest, but also are valuable indicators of large-scale climate forcing such as that associated with the NAO. As in the case of lake surface temperatures, the timing of freeze-up (ice-on) and break-up (ice-off) of a lake, although influenced by several meteorological variables, appears to be related most strongly to air temperature [*Palecki and Barry*, 1986;

Ruosteenoja, 1986; *Robertson et al.*, 1992; *Vavrus et al.*, 1996; *Livingstone*, 1997b]. Because the NAO affects air temperatures over large parts of the Northern Hemisphere [*Hurrell*, 1995; *Hurrell*, 1996; *Hurrell and van Loon*, 1997], it also leaves a signal in the break-up dates of lakes distributed widely over the Northern Hemisphere [*Weyhenmeyer et al.*, 1999; *George*, 2000a; *Livingstone*, 2000]. However, these signals are not always constant in time [*Livingstone*, 2000].

The strongest NAO influence on Eurasian winter air temperatures appears to occur north of about 60°N [*Hurrell*, 1996; *Hurrell and van Loon*, 1997]. Finland, which lies almost entirely north of 60°N, contains many lakes with long historical time-series of ice phenology observations [*Palecki and Barry*, 1986]. An analysis of break-up data from two of these lakes (Kallavesi and Näsijärvi) from the 1830s to the present reveals a fluctuating, but always strong, NAO signal in the entire record [*Livingstone,* 2000, Figure 3a]. The signal strength appears to vary with a periodicity of about 30 yr, with the proportion of shared variance generally ranging between 10% and 30%, but achieving a maximum of 43% - surprisingly high in view of the complexity involved in the thawing of lake ice and the simplicity of the parameterization of the climatic effects of the NAO by the winter NAO index. Apart from the fluctuations in signal strength, no longer-term trends or abrupt alterations in the signal strength are apparent (Figure 3a).

Air temperatures in Siberia also exhibit a substantial NAO influence [*Hurrell*, 1996; *Hurrell and van Loon*, 1997]. The largest lake on earth with respect to volume, Lake Baikal, is located in Siberia and might therefore also be expected to be subject to the influence of the NAO. The analysis of a 128-yr ice record did indeed reveal the timing of break-up of Lake Baikal to be related not only to air temperatures over an extensive area of northern Asia encompassing most of Siberia and parts of northern China, but also to the winter NAO [*Livingstone*, 1999]. However, there are distinct differences between Lake Baikal and the Finnish lakes with respect to the NAO signal in the ice phenology data. In the Lake Baikal data, a significant NAO signature can be detected only in the latter part of the series, beginning in the 1918-1967 50-yr data window and becoming generally stronger as time progresses [*Livingstone*, 1999; 2000]. An NAO signature detected in the timing of break-up of Lej da San Murezzan (Lake St. Moritz), a high-altitude lake in Switzerland, was found to behave similarly to that detected in the Lake Baikal data, with a significant signature detected only in the latter part of the series [*Livingstone*, 2000; Figure 3b]. The most interesting behavior, however, is that exhibited by Lake Mendota, Wisconsin. In this lake, break-up during the second half of the 20th century has been influenced more by the El-Niño/Southern Oscillation (ENSO) phenomenon than by the NAO [*Robertson*, 1989; *Anderson et al.*, 1996; *Livingstone*, 2000]. This is to be expected in view of the fact that air temperatures in central North America from 1935 to 1994 appear to have been influenced much more by the Southern Oscillation (SO) than by the NAO [*Hurrell*, 1996]. However, the ice phenol-

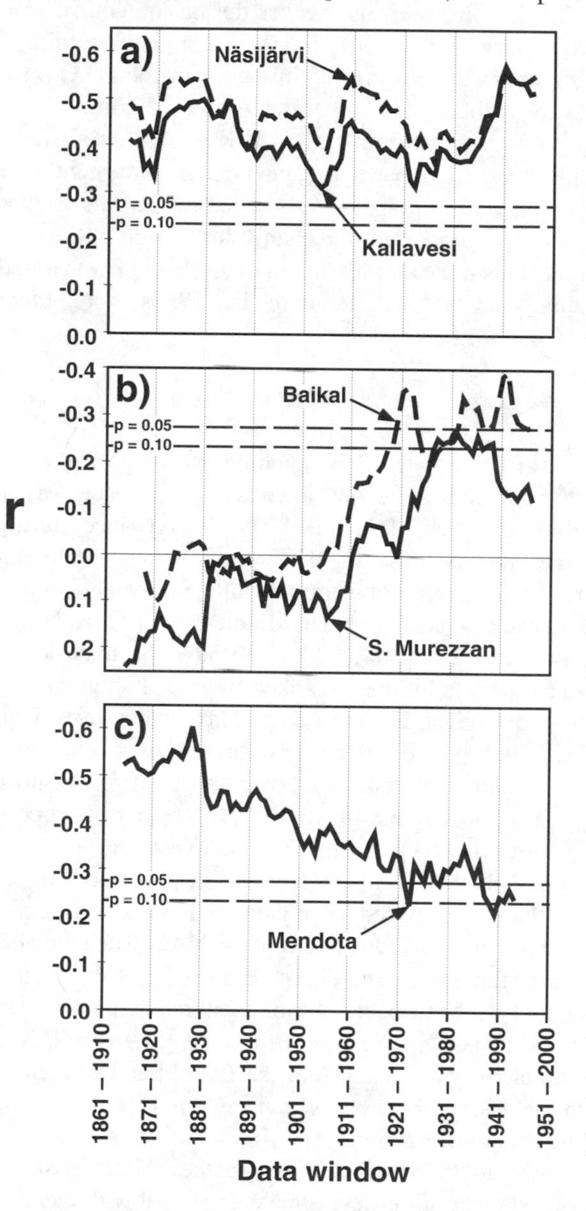

Figure 3. Running correlation coefficients (r) between the Julian day of break-up and the 3-monthly (DJF) winter index of the preceding NAO, based on 50-year data windows, for (a) Kallavesi and Näsijärvi (Finland); (b) Baikal (Russia) and San Murezzan (Switzerland); and (c) Mendota (USA). The p=0.05 and p=0.10 significance levels (n = 50, 2-tailed t-test) are as shown (based on *Livingstone* [2000]).

ogy data from Lake Mendota suggest that this situation may have been different in the recent past. During the last part of the 19th century and the first part of the 20th century, there is no evidence that breakup on Lake Mendota was influenced by El Niño [*Robertson*, 1989] or the SO [*Livingstone*, 2000], but there is evidence of a strong correlation with the winter NAO [*Livingstone*, 2000; Figure 3c]. This correlation has become steadily weaker during the course of the 20th century (Figure 3c). Taken in conjunction with the simultaneous increase in the influence of the NAO on the breakup of the Swiss and Russian lakes (Figure 3b), this may point to a shift in Northern Hemisphere quasi-stationary planetary wave patterns having occurred during the first half of the 20th century [see also *Jones et al.,* this volume]. This may be connected to a change in the sign of the correlation between the air pressure in the Iceland Low and in the Aleutian Low that occurred in the late 1930s [*van Loon and Madden*, 1983].

2.3 Lake Level Changes and River Discharge

Like air temperature, precipitation over large areas of the Northern Hemisphere is known to be influenced significantly by the NAO [*Hurrell*, 1995]. For instance, during a positive index phase of the NAO, winter precipitation increases in Northern Europe while it decreases in the Mediterranean area and the Middle East [*Hurrell*, 1995; *Cullen and deMenocal*, 2000]. Because the hydrology of inland waters is intimately linked to precipitation, an NAO influence on river discharge and on lake water levels would be expected, which has been confirmed by several studies on rivers and lakes distributed over various parts of Eurasia.

The runoff of rivers in western and northern Europe has increased with increasing NAO index. *Kiely* [1999], for example, showed that the increase in the runoff of Irish rivers that has occurred since the mid-1970s corresponds to the recent positive index phase of the NAO. The annual discharge of the Hafren and Gwy rivers in central Wales shared 41% and 31%, respectively, of their interannual variance with the winter NAO index [*Bradley and Ormerod*, 2001]. The total river discharge into the Baltic Sea has increased with increasing NAO index, causing a reduction in salinity [*Hänninen et al.*, 2000].

In the Mediterranean region and the Middle East, river runoff and lake water levels have decreased with increasing NAO index. *Lloret et al.* [2001] showed that interannual fluctuations in the runoff of the Rhône (France) and the Ebro (Spain) were synchronous and negatively correlated with an NAO index ($r = -0.4$ and $r = -0.26$, respectively, both $p < 0.05$), as was the Euphrates streamflow between 1938 and 1972 ($r = -0.42$, $p < 0.02$) [*Cullen and deMenocal*, 2000].

In an analysis of the climatic factors responsible for fluctuations in the level of the Caspian Sea, *Rodionov* [1994] showed that the NAO plays a dominant role, with a decline in the surface level of the Caspian Sea being linked to a positive NAO index and vice versa. The Caspian Sea surface level is determined mainly by the rate of water supply to the sea from the Volga basin to the north (modified, of course, by anthropogenic factors). About a year before a decline in the Caspian Sea surface level, a noticeable decrease in air pressure at higher latitudes over the North Atlantic tends to occur, coupled with an increase at lower latitudes (i.e., the positive index phase of the NAO). As a result, mid-latitude westerlies over the North Atlantic intensify, and storm tracks shift northwards. The transport of heat and moisture from the North Atlantic to the extreme north of European Russia increases, resulting in decreased ice cover on the Barents Sea. Simultaneously, the establishment of a high-pressure cell over the Volga basin results in lower precipitation rates, lower runoff, and consequently a decline in the Caspian Sea surface level.

A rise in the Caspian Sea surface level tends to be preceded by the reverse climatic situation, in which air pressure increases at high latitudes and decreases at low latitudes (i.e., the negative index phase of the NAO), leading to weakened westerlies and more meridional atmospheric circulation. The establishment of a blocking ridge over the North Atlantic and an upper atmospheric trough over eastern Europe forces moisture-laden southerly cyclones to travel northeastward along the edge of the trough, resulting in heavy precipitation in the Volga basin and a consequent rise in the Caspian Sea surface level.

3. CHEMICAL RESPONSES

Physical responses to the NAO might result in secondary chemical responses, due for instance to changes in river inflow or in mixing or upwelling processes. Several studies point to the importance of changes in the leaching of nutrients into lakes and rivers in response to changes in the NAO. Alterations in winter temperature and precipitation result in changes in nutrient release from soil in the catchment area. For instance, *Weyhenmeyer* [in prep.] shows that total phosphorus, phosphate phosphorus, total nitrogen and nitrate nitrogen in March in the Galten Basin, a shallow basin of Lake Mälaren in Sweden, are all positively related to the nutrient transport from inflowing waters, winter air temperature and the winter NAO index. She suggests that in the colder winters that occur in low NAO index years, nutrient release from the frozen ground in the catchment area is reduced. Additionally, alkalinity and conductivity in surface waters throughout the year show a strong dependence on winter meteorological conditions [*Weyhenmeyer*, in prep.]. The effect of warmer winters on these two variables is persistent since catchment processes predominantly determine both during winter rather than internal lake processes, so seasonal variability is slight.

In contrast to the Swedish results, *Monteith et al.* [2000] observed that March nitrate nitrogen concentrations in nine lakes and six streams in the UK are strongly negatively correlated with the winter NAO index. In this case, low winter temperatures and freezing conditions causing frost damage to plants and roots were suggested to result in an increase in nitrate leaching. The seeming contradiction between the two studies might be related to the difference in winter severity and average March temperatures between Sweden and the UK. In the catchment area of the Galten Basin, March is still winter and the ground is normally frozen. In cold winters, nitrogen leaching in March is therefore negligible. In warm winters, however, nitrogen leaching begins in March, resulting in a positive relationship between the NAO and March nitrate nitrogen concentrations. In contrast, nitrogen leaching at the UK sites is at its maximum in winter and early spring, when biological demand and uptake in the catchment area are lowest. In this region, which experiences milder winters on average, rare freezing events increase nitrate leaching due to their damaging effects on plants, roots and microorganisms.

The input of nutrients into lakes might also vary with NAO-related changes in precipitation. For example, in small Blelham Tarn in the English Lake District, the winter concentration of dissolved reactive phosphorus is positively related to winter precipitation ($r = 0.61$, $p < 0.01$). As precipitation in the UK is linked to the NAO, winter concentrations of dissolved reactive phosphorus in this lake are also positively correlated with the winter NAO index ($r = 0.53$, $p < 0.01$) [*George*, 2002]. Similar effects have been recorded in nearby Esthwaite Water [*George et al.*, in prep.]. Increased precipitation in high NAO index years appears also to result in an increase in the wet deposition of chloride and other marine-derived ions at several sites in the UK [*Evans et al.*, 2001]. At lake sites in north Wales, the Lake District and the Galloway district of southwest Scotland, concentrations of chloride and other marine-derived ions follow an approximately decadal cycle that is suggested to be in phase with the winter NAO. Marine-derived cations cause the displacement of absorbed acid cations such as hydrogen and labile aluminum in the soil, which can lead to acidification of runoff [*Evans et al.*, 2001]. Acidification and recovery from acidification in these systems might therefore also be influenced by the NAO.

The patterns reported above stem mostly from observations on small, shallow lakes. Such systems often have short water residence times, and are therefore more likely to react rapidly to changes in inflow than are large water bodies. For instance, in the Lake District, a relationship between nutrients and the NAO was observed in comparatively small lakes such as Blelham Tarn and Esthwaite Water (see above), but not in the two large basins of Lake Windermere [*George et al.*, in prep.]. In Lake Mälaren, the water chemistry is strongly linked to recent meteorology and the NAO only in the small, shallow Galten Basin, but not in other larger and deeper parts of the lake [*Weyhenmeyer*, in prep.]. Hence, the sensitivity of, for instance, nutrient concentrations to NAO variability seems to depend on the size of the lake.

The NAO has been shown to affect mixing processes in lakes (see above) and the ocean [e.g., *Williams and McLaren*, 2000; *Oschlies*, 2001]. This is important because mixing affects the distribution of nutrients and other particulate and dissolved substances in the water column. For example, spring turnover in lakes is an important process for the transport of oxygen into deeper water layers, and inhibition of spring turnover can result in deep-water anoxia. In high NAO index years, warm winters can indeed inhibit spring turnover, resulting in extreme deep-water oxygen depletion in some deep perialpine lakes [*Livingstone*, 1997a]. In Lake Constance, near-bottom oxygen concentrations during spring turnover and during the following summer have been shown to be negatively related to the winter NAO index [*Straile et al.*, in prep.]. Climatic variations associated with the NAO can therefore have a major impact on the occurrence of oxygen deficiencies in lakes. The remediation of oxygen deficiency in the deep water of lakes is therefore only partially under the control of lake managers, and might be a problem in many deep lakes in the future despite reduced nutrient loadings [*Livingstone and Imboden*, 1996].

A reduction in the vigor of spring turnover additionally implies that nutrients (e.g., silicate and phosphorus) that have accumulated in the hypolimnion during the previous stratification period will not be homogeneously distributed vertically over the whole water column. Because of such climate-dependent differences in nutrient upwelling from year to year, nutrient concentrations in the epilimnion of Lake Constance tend to be lower in high NAO years than in low NAO years, whereas in the hypolimnion the opposite is the case [*Straile et al.*, in prep.].

4. BIOLOGICAL RESPONSES

4.1. Phytoplankton

Many experimental studies have shown the growth and species composition of algae to depend strongly on light availability, temperature, and nutrient availability [e.g., *Sommer*, 1989; *Reynolds*, 1989; *Huisman et al.*, 1999]. Because the NAO can exert a strong influence on these factors (see above), it is also likely to affect the phytoplankton.

The presence of ice cover on lakes will reduce the light available for phytoplankton photosynthesis, resulting in the earlier (later) occurrence of peaks in algal growth with earlier (later) ice-off. The timing of the phytoplankton peak in Lake Erken (Sweden) and in Müggelsee (Germany) was found to be related to the timing of ice-off [*Weyhenmeyer et al.*, 1999] and to the duration of ice-cover [*Adrian et al.*, 1999], respectively. As the timing of ice-off is linked to the NAO in both lakes [*Weyhenmeyer et al.*, 1999; R. Adrian, personal communication], so is the timing of the spring bloom, as well as the late winter and early spring phytoplankton biomass in Müggelsee [*Weyhenmeyer et al.*, 1999; *Gerten and Adrian*, 2000; *Straile and Adrian*, 2000]. Additionally, reduced ice cover in Plußsee (Germany) following mild winters has been suggested to be responsible for lower transparency in early spring, indicating a higher phytoplankton biomass [*Güss et al.*, 2000].

In addition to overall phytoplankton biomass, phytoplankton species composition can also be affected by ice dynamics, and hence by the NAO. High NAO index years have been found to favor the growth of diatoms relative to other phytoplankton species in both Lake Erken and Müggelsee. This is because extensive diatom blooms develop only under ice-free conditions, as diatom species need turbulent conditions to prevent them sinking out of the euphotic zone. In contrast, the dinoflagellate *Peridinium* is able to grow well under the ice of Lake Erken when snow cover is low and the ice is clear [*Weyhenmeyer et al.*, 1999]. In Müggelsee, earlier ice-off results in an earlier diatom peak, whereas the timing of the biomass peaks of other phytoplankton taxa is not affected by ice duration [*Adrian et al.*, 1999]. Hence in both Lake Erken and Müggelsee, the importance of diatoms relative to other phytoplankton is greatest in high NAO index (early ice-off) years. Obviously, changes in ice phenology due to the NAO cannot have impacts on phytoplankton in lakes that do not normally freeze during winter. In Lake Constance, for example, no influence of the NAO on the phytoplankton dynamics in early spring could be detected. Using a one-dimensional numerical hydrodynamical model, *Gaedke et al.* [1998] showed that phytoplankton growth in this deep lake was only possible when no deep mixing (below 20 m depth) of algal cells occurred, either because of the onset of stratification or the absence of wind.

Studies so far have documented the importance of the NAO to phytoplankton growth and species composition only via its effect on ice-cover dynamics. Regarding the potential importance of water temperature, it should be noted that phytoplankton population growth has been observed at quite low water temperatures, e.g. under the ice in Lake Erken [*Weyhenmeyer et al.*, 1999] or during late winter in the absence of mixing in Lake Constance [*Gaedke et al.*, 1998]. This suggests that the NAO may be influencing lake phytoplankton in spring primarily through its effect on light availability (via ice cover), rather than its effect on water temperature. Although it is clear that water temperature plays an important role in controlling phytoplankton growth rates [*Raven and Geider*, 1988], an influence of the NAO on lake phytoplankton populations via such a temperature control mechanism would seem to be of only secondary importance. However, it has been suggested that an increase in water temperatures was responsible for the earlier occurrence of cyanobacteria in the 1990s in Sweden's largest lakes (Mälaren, Vänern and Vättern) in late spring and early summer [*Weyhenmeyer*, 2001]. Up to now, no studies, either in limnetic or marine environments, have documented the consequences of NAO-related changes in nutrient availability on phytoplankton dynamics. For food-web mediated effects of the NAO on phytoplankton populations, see section 5.2 below.

4.2. Zooplankton

The spring succession in temperate lakes is characterized by high growth rates of several zooplankton species in response to increased water temperature and algal food supply [*Sommer et al.*, 1986]. For instance, the biomass of *Daphnia* in Lake Constance increased on average almost 1000-fold from less than 10 mg C m^{-2} to 1000 mg C m^{-2} within approximately 6 weeks [*Straile*, 2000]. Interannual variability in the *Daphnia* growth rate during this period—and, as a result, in the *Daphnia* biomass in May—is strongly related to water temperature, and consequently to the NAO. This correlation is most probably due to a physiological response of *Daphnia* egg development and growth rates to increased water temperatures, and is not mediated by an increase in the food supply. During late winter, the growth of *Daphnia* is restricted by a low food supply and low temperatures. However, after the onset of stratification this "co-limitation" switches to a temperature control of *Daphnia* growth, because water temperature increases more slowly than phytoplankton biomass [*Gaedke et al.*, 1998; *Straile*, 2000]. Hence, during the period of exponential *Daphnia* population increase, it is water temperature, and not food supply, that controls *Daphnia* growth. Subsequent studies in different lakes have shown that in addition to daphnids, the abundances of other cladoceran species, and also of rotifer species, are related to the NAO. In Müggelsee, spring abundances of *Daphnia sp., Bosmina sp.* and of the rotifer *Keratella* have been found to be higher in high NAO index years than in low index years [*Gerten and Adrian*, 2000; *Straile and Adrian*, 2000]. Furthermore, the timing of the peak abundance of several rotifer species (*Keratella cochlearis, Keratella quadrata, Brachionus angularis, Brachionus calyciflorus, Polyarthra dolichoptera*) in this lake is significantly related to the duration of ice cover [*Adrian et al.*, 1999], which suggests a link to the NAO.

In Bautzen Reservoir, Germany, high water temperatures during late winter/early spring are associated with high average *Daphnia* biomass in April/May [*Benndorf et al.*, 2001]. The *Daphnia* biomass was not directly linked to the NAO by *Benndorf et al.* [2001]; however, the geographical location of this reservoir (51°10'N, 14°26'E) suggests a possible relationship with the NAO. To examine this, we digitized their Fig 3b, estimated the mean *Daphnia* biomass for the years 1983-1996, and correlated it with the winter NAO index. We found the mean *Daphnia* biomass in Bautzen Reservoir during April/May to be related significantly to the NAO ($r = 0.52$, $p < 0.05$). The impact of the NAO on *Daphnia* spring dynamics in two lakes and one reservoir that are more than 700 km apart, and that differ strongly in many respects limnologically, suggests that many lakes within central Europe might be similarly affected by the NAO.

Increased water temperatures might also have negative impacts on zooplankton species, since enhanced metabolic requirements in warmer water might not be met by corresponding increases in phytoplankton production. This mechanism has been suggested for the negative relationship between the NAO and *Daphnia* abundance in winter in Esthwaite water [*George*, 2000b]. Additionally, competition between *Daphnia* and a calanoid copepod, *Eudiaptomus gracilis*, whose abundance was positively related to the NAO during this period, might have contributed to the decline of *Daphnia* in mild winters [*George and Hewitt*, 1999].

Up to now the only freshwater copepod species whose abundance is known to be related to the NAO is the calanoid copepod *Eudiaptomus* in Esthwaite water [*George and Hewitt*, 1999]. In addition, *Adrian* [1997] showed that the abundance of copepod species in Heiligensee (Berlin, Germany) is associated with winter temperatures. Although copepod abundances were not linked directly to the NAO in this work, the proximity of Heiligensee to Müggelsee suggests that an NAO influence is likely. All other reported links between zooplankton and the NAO concern cladoceran and rotifer species. This suggests that the life history of zooplankton species may be an important factor governing their response to increased water temperatures and to the NAO. In contrast to copepods, cladocerans and rotifers reproduce parthenogenetically, which enables their populations to grow exponentially during spring, thus allowing them to double their population sizes within a short period. Hence, even a small change in temperature - and consequently in growth and development rates - during the period of exponential increase can result in a difference in population size large enough to be detected easily during long-term sampling programs. During a period in which cladocerans and rotifers reproduce rapidly, more slowly developing copepods might complete just one life cycle. An increase in temperature will certainly speed up the ontogenetic development of copepods also; however, it would take a long time for temperature differences to show up as differences in overall abundance. This suggests that changes in temperature are more likely to have a large impact on copepod populations when they result in changes in food availability and/or predation pressure that alter the copepods' survival chances. Supporting this hypothesis, changes in copepod abundance were indeed found to be associated not only with changes in water temperature but also with changes in food availability in both Esthwaite Water and Heiligensee [*Adrian*, 1997; *George and Hewitt*, 1999].

4.3. Zoobenthos

The response of freshwater benthos to the NAO has been less thoroughly investigated than that of the plankton. The only two studies reporting a relationship were conducted in the UK [*George*, 2000a; *Bradley and Ormerod*, 2001]. Interannual variability in the emergence of alder-flies, *Sialis lutaria*, from the littoral zone of Lake Windermere was studied by *Elliott* [1996] from the 1960s to the 1990s and shown to be related to the NAO by *George* [2000a]. Year-to-year variations of the date of the first emergence of adults shared 20% variance with the winter NAO index and 92% variance with average spring water temperatures, suggesting a rather direct control of the timing of emergence by water temperature.

During 1984-1998, *Bradley and Ormerod* [2001] studied macroinvertebrate communities in 8 streams in central Wales that differed in chemistry and catchment land use. The persistence of these communities between pairs of successive years was analyzed with Spearman's rank correlation coefficient and Jaccard's coefficient of similarity. Invertebrate communities in all different stream types switched significantly from being highly persistent during negative phases of the NAO to unstable during positive phases. The overall abundance of macroinvertebrates was not related to the NAO, at least partially because of opposing relationships of individual species with the NAO; e.g. the relationship of *Nemurella picteti, Elmis aenea, Hydropsyche siltalai,* and *Paraleptophlebia submarginata* to the phase of the NAO was found to be negative, while that of *Chloroperla tripunctata* was found to be positive. However, the actual mechanisms linking the abundance of individual species and community persistence in the Welsh streams to the NAO are unclear.

4.4. Vertebrate Predators

Evidence for an impact of the NAO on freshwater fish species is rare. This contrasts with marine systems, where the recruitment of various fish species has been shown to be linked to the NAO [e.g., *Fromentin et al.*, 1998; *Stenseth et al.*, 1999; *Ottersen and Loeng*, 2000; *Drinkwater et al.*, this volume]. In most cases this relationship has been attributed to the direct effects of temperature, although the actual mechanisms that have been suggested to underlie the climate-recruitment links often remain speculative [*Ottersen et al.*, 2001]. As the temperature of many lakes and rivers is closely related to the phase of the NAO, we might also expect a link between the NAO and freshwater fishes. This link has been suggested to exist for sea trout (*Salmo trutta*) in Black Brows Beck, a small stream in the English Lake District [*Elliott and Hurley*, 1998] The authors developed an individual-based model to predict sea trout fry emergence dates. The model predictions were validated by 8 years of field data. In a subsequent study, *Elliott et al.* [2000] correlated the predicted emergence dates with the NAO index and found 41% shared variance. *Elliott et al.* [2000] suggested that water temperature was the mediating factor, as water temperature in this stream was correlated with the phase of the NAO and is the main driving factor in the model of *Elliott and Hurley* [1998]. Water temperature has also been suggested to be the most important factor determining the recruitment of whitefish (*Coregonus lavaretus*) in Lake Constance [*Eckmann et al.*, 1988] and perch (*Perca fluviatilis*) in Bautzen Reservoir [*Mehner et al.*, 1998]. Nevertheless, the link between the NAO and recruitment was not investigated in these studies, and more work needs to be done on this topic in the future.

In addition to fish, birds may also be important predators in freshwater environments. For example, dippers (*Cinclus cinclus*) are important predators of macrobenthos in small streams. Change in the population size of dippers in southern Norway has been found to be positively related to winter temperature and to the winter NAO index [*Sæther et al.*, 2000]. Mild winters changed the local dynamics of this dipper population by increasing the carrying capacity of the habitat, but they also increased the immigration rate [see also *Mysterud et al.*, this volume]. A second study, in which breeding phenology along the riparian habitat of streams flowing into Lake Zurich was investigated from 1992 to 2000, supports the importance of winter temperature for dippers [*Hegelbach*, 2001]. Dippers started to breed around February and March, and earlier breeding was observed in mild winters; i.e., the date of first breeding was significantly related to water and air temperature in February. A comparison with earlier studies revealed that on average, the start of the breeding season in the 1990s was earlier than in previous decades. *Hegelbach* [2001] suggests that a faster development of invertebrate prey at warmer water temperatures links the breeding phenology of dippers with winter temperatures.

5. BIOLOGICALLY MEDIATED RESPONSES

5.1 Water Chemistry

The impact of the NAO on phytoplankton populations (see above) has indirect consequences for water chemistry. An increase in phytoplankton biomass, and probably also primary production, in response to mild winters is likely to result in changes in chemical variables such as pH and the concentrations of O_2 and nutrients. In Plußsee [*Güss et al.*, 2000] and in the Galten Basin of Lake Mälaren [*Weyhenmeyer*, in prep.], higher O_2 concentrations as a consequence of increased O_2 production, as well as higher pH and reduced concentrations of SRSi and ammonia resulting from an increased uptake of CO_2 and nutrients, were observed to follow mild winters. In both lakes, some of these indirect phytoplankton-mediated effects of the NAO last until April and May [*Güss et al.*, 2000; *Weyhenmeyer*, in prep.].

In addition, NAO-related changes in summer water chemistry can be traced back to changes in summer phytoplankton abundance. For example, increased epilimnetic O_2 concentrations and increased pH correspond to decreased transparency (indicating a more intense phytoplankton bloom) in Plußsee in July and August [*Güss et al.*, 2000]. In autumn, O_2 concentrations are negatively related to the conditions in the previous winter, indicating an increase in the degradation of particulate organic matter originating from more intense phytoplankton blooms [*Güss et al.*, 2000]. These results suggest that meteorological conditions in winter, and hence also the NAO, play a central role in explaining the chemistry of freshwater systems throughout the year.

5.2 Food Web Interactions

Populations interact with each other in food webs. The sensitivity and response of one particular population to increased water temperatures, for example, might therefore alter its interactions with competitors, mutualists, predators, prey, and pathogens. These possible indirect effects of climate forcing related to the NAO might be especially important if critical species within a food web are affected. Small changes in a factor like water temperature could alter food-web dynamics and generate system-wide ecological changes [*Sanford*, 1999; *Scheffer et al.*, 2001b].

Clearly, *Daphnia*, which is both an important component of the diet of many fish species and an important consumer of phytoplankton and protozoa, is a critical species in lake food webs [*Jürgens*, 1994]. Grazing by *Daphnia* during its mass development in late spring/early summer suppresses phytoplankton populations, resulting in a so-called clear-water phase [*Sommer et al.*, 1986]. In Lake Constance, higher spring water temperatures tend to result in higher *Daphnia* growth rates and biomass compared to years with lower water temperatures, resulting in turn in earlier occurrence of the clear-water phase [*Straile*, 2000]. Consequently, the timing of the clear-water phase in Lake Constance has been shown to be significantly related to the phase of the NAO, with a shared variance of 30% (Figure 4). Given the indirect nature of the relationship, which involves several intermediate steps (Figure 4) that are all significantly related to each other [*Straile*, 2000], this high percentage of shared variance is indeed remarkable.

In a subsequent study, *Straile and Adrian* [2000] showed that the observed patterns were not specific to Lake Constance, but could also be found in Müggelsee [but see *Gerten and Adrian*, 2000], a small polytrophic lake in Berlin that differs strongly from Lake Constance limnologically. Within both lakes, interannual variability in vernal warming and *Daphnia* population growth was found to be related to the interannual variability of the NAO. Furthermore, the timing of the clear-water phase in the two lakes was highly correlated ($r = 0.63$, $p < 0.05$), indicating synchrony in the occurrence of successional events despite the great distance (> 700 km) between the lakes [*Straile and Adrian*, 2000]. In addition, a link between the NAO (or winter temperatures) and the timing of the clear-water phase was also found in Plußsee, Germany [*Müller-Navarra et al.*, 1997] and Lake Geneva, France/Switzerland [*Anneville et al.*, 2002].

The spatial coherence in plankton succession across large regions due to the synchronizing impact of the NAO was

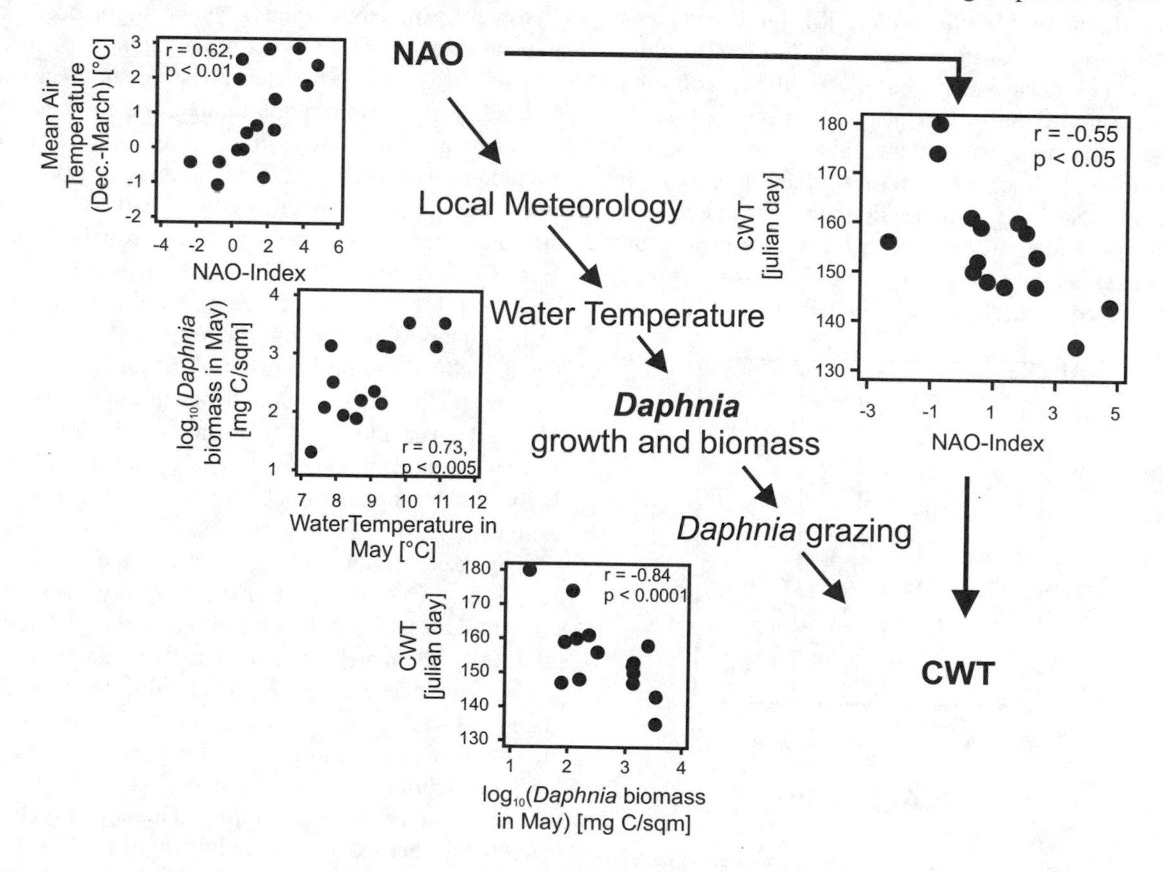

Figure 4. Diagram illustrating the propagation of the effects of the NAO on winter meteorological conditions through water temperature, *Daphnia* population growth and biomass on clear-water timing (CWT) in Lake Constance (modified from *Straile* [2000]).

further investigated by *Straile* [2002]. He used the timing of the clear-water phase as a phenological marker and the link between it and the NAO as a proxy for the occurrence of the chain of effects depicted in Figure 4. He investigated the influence of the NAO on vernal warming and *Daphnia* population growth in 28 lakes across central Europe, encompassing an area of $>10^5$ km^2 and an altitude range of almost 1000 m. On average, the timing of the clear-water phase in these lakes was found to have advanced by 0.5 days per year (corresponding to approximately 2 weeks over the last three decades), and to have been strongly influenced by the NAO (Figure 5). The NAO has therefore been responsible for a synchronization of the plankton succession among central European lakes during the last few decades, giving rise to a large-scale coherence in successional patterns. Such coherence has hitherto only been found for abiotic variables in lake ecosystems [e.g., *Magnuson et al.*, 1990; *Kratz et al.*, 1998; *Baines et al.*, 2000; *Livingstone and Dokulil*, 2001].

Further evidence for the large-scale coherence of successional events in central European lakes is given by the link between the phase of the NAO and the timing of the clear-water phase found by *Scheffer et al.* [2001a] in Dutch lakes. In contrast to the lakes chosen by *Straile* [2002], which show a distinct clear-water phase, the highly eutrophic Dutch lakes only rarely experience a clear-water phase. In those rare cases, early clear-water phases are associated with a positive phase of the NAO ($r^2 = 0.21$, $p < 0.01$). In addition, multiple logistic regression (the dependent variable is binary) shows that the probability of a clear-water phase in the Dutch lakes decreased with increasing average chlorophyll *a* concentration ($r = -0.067$, $p < 0.001$), but was

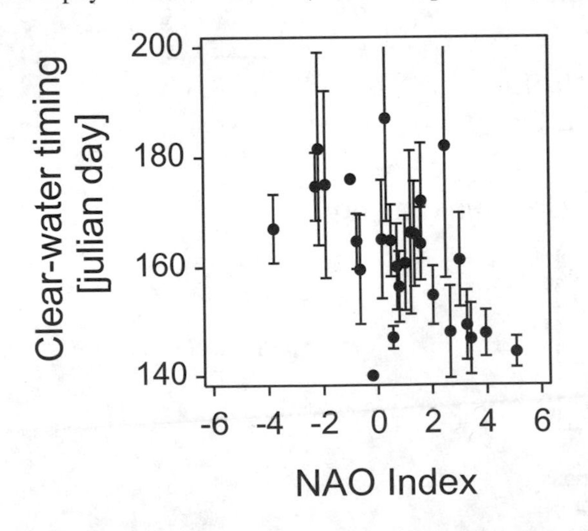

Figure 5. NAO impact on the timing of the clear-water phase in central European lakes. The average date of occurrence of the clear-water phase in a specific year is significantly related to the winter NAO index ($p < 0.05$) (see *Straile* [2002] for details).

positively related to the NAO index ($r = 0.335$, $p < 0.005$). *Scheffer et al.* [2001a] show that the observed changes in the timing and occurrence of a clear-water phase can be reproduced remarkably well using a simple periodically forced algae-zooplankton model with temperature-dependent growth and mortality terms. Although some important mechanisms are not considered in this model (see below), it does show that even small temperature changes might crucially affect the dynamics of a lake system. In shallow lakes, a clear-water phase is not only a distinct phase of plankton succession but also represents an important window of opportunity for the growth and recolonization of submerged macrophytes. Hence it can also function as a possible switch between alternative stable states, from a system dominated by phytoplankton to one dominated by macrophytes [*Scheffer et al.*, 1993]. This switch is regarded as a key process in the restoration of eutrophic lakes and has profound consequences for the entire food-web structure. Subtle changes in temperature associated with the NAO might therefore result in ecosystem regime shifts [*Scheffer et al.*, 2001a].

However, high NAO index years might not always be associated with the suppression of algal growth and the establishment of a clear-water phase in summer, but rather with a *Daphnia* summer minimum and a consequent reduction in the grazing of algae. In Plußsee, summer phytoplankton blooms were found to be more intense after mild winters, and it was suggested that this could be due to a decrease in the abundance and grazing pressure of zooplankton in summer [*Güss et al.*, 2000]. In the biomanipulated Bautzen Reservoir, high daphnid biomass during spring associated with a high NAO index (see above) was found to be the best single predictor for the occurrence of a *Daphnia* midsummer decline and low *Daphnia* abundance in summer [*Benndorf et al.*, 2001]. The authors of this study even conclude that climate warming is a potential threat to successful biomanipulation. Low zooplankton abundance in summers following mild winters could in principle result from a beneficial effect of increased winter and early spring temperature on the recruitment (see above), growth and prey consumption of fish [*Hill and Magnuson*, 1990]. According to *Benndorf et al.* [2001], the match or mismatch *sensu Cushing* [1990] of *Daphnia* population development with peak predation of juvenile fishes is especially crucial for *Daphnia* population dynamics during summer [for examples on the importance of the match/mismatch mechanism in terrestrial ecosystems see *Mysterud et al.*, this volume]. This again highlights the need for further studies on the impact of the NAO on freshwater fishes.

Why do mild winters increase the probability of a clear-water phase in Dutch lakes on the one hand, but result in lower *Daphnia* summer abundance in Bautzen Reservoir and more intense phytoplankton blooms in Plußsee on the other? A possible explanation might be related to differences in the mean depth of the lakes. All Dutch lakes are very shallow, and a clear-water phase can be stabilized by macrophytes providing a refuge for *Daphnia* and suppressing phytoplankton by allelochemicals [*Scheffer et al.*, 1993; *Benndorf et al.*, 2002]. In contrast, in Bautzen Reservoir and in Plußsee, the stabilization of a clear-water phase by macrophytes is not possible as the lakes are too deep to allow macrophyte growth over enough of the lake bottom. Consequently, the effect of climate variability on plankton dynamics might be different in shallow and deep lakes, partially because of the different opportunities they offer for macrophyte growth.

Clearly, the discussion of the indirect effects of the NAO must remain to some extent speculative. As they are "indirect", these effects are often difficult to detect and to understand. Nevertheless, an increasing body of evidence suggests that indirect, food-web mediated effects of the NAO (and indeed of winter warming in general) are very important. These effects can propagate far into the following growing season, possibly resulting in ecosystem-wide regime shifts [*Scheffer et al.*, 2001a].

6. SUMMARY AND CONCLUSIONS

Research on the influence of the NAO on freshwater ecosystems is still in its early stages. However, the tremendous impact of the NAO on the physics, chemistry and biology of many Northern Hemisphere lakes is already apparent. Of course, much of the variability in any one freshwater ecosystem is not related to the NAO. However, the impact of the NAO is strong enough to result in a striking coherence in many of the physical and biological characteristics of freshwater ecosystems over a very large area. Large-scale coherent responses to the NAO have been observed, for instance, in the case of ice phenology [*Livingstone*, 2000], spring surface temperatures [*Livingstone and Dokulil*, 2001] and the successional dynamics of lake plankton [*Straile*, 2002; *Weyhenmeyer et al.*, 2002]. This high degree of coherence, which is observed even among lakes which differ strongly limnologically - for example with respect to trophic status - gives aquatic ecologists a tool which they can use to disentangle the effects of climatic changes on lakes from the effects of direct human impact (e.g. eutrophication).

To a large extent, the strong, coherent impact of the NAO on European lakes is a result of the time of year during which the NAO most strongly affects the European climate; viz. winter and early spring. This time of year, during which both spring turnover and the onset of stratification occur, is critical for lakes. Because of this, the NAO can have a major impact on the distribution and seasonal development of temperature and nutrients, and is able to influence both the time of onset and the rate of plankton succession.

Although the signature of the winter NAO is detectable in regional air temperatures in central Europe only up until March, it can persist in lake surface water temperatures until May [*Livingstone and Dokulil*, 2001]. Because deep-water temperatures are generally determined by the temperature attained during partial or total homothermy in early spring and change comparatively little after the onset of stratification [*Livingstone*, 1993], the signature of the winter NAO present in the lake temperature in spring can persist much longer in the hypolimnion than in the epilimnion [Figure 2; *Straile and Adrian*, 2000; *Gerten and Adrian*, 2001]. This has implications for respiration rates in the hypolimnion, and hence for the rate of oxygen depletion there [*Livingstone*, 1997a].

Food-web interactions result in a prolongation of the effects of the NAO on lake biology. The impact of the NAO is detectable, for instance, in the timing of the clear-water phase in early summer [*Straile and Adrian*, 2000; *Straile*, 2002]. Likewise, evidence suggests that the NAO affects phytoplankton populations and lake chemistry in summer and autumn, although the mechanisms linking winter meteorology to lake chemistry and biology in summer are complex and not yet clear [*Güss et al.*, 2000; *Weyhenmeyer*, in prep.]. Finally, in shallow lakes the switch from domination by phytoplankton to domination by macrophytes in summer may be influenced by the climatic conditions prevailing in winter and early spring, and hence also by the NAO [*Scheffer et al.*, 2001a].

The impact of the NAO on a freshwater ecosystem depends critically on its size and geographical location. As the NAO affects air temperature and precipitation differently in different regions [*Hurrell*, 1995], the response of lakes and rivers will obviously also be different. For instance, increased precipitation in northern and western Europe in high NAO index years results in increased river discharge in these regions [e.g., *Kiely*, 1999; *Hänninen et al.*, 2000], while decreased precipitation in high NAO index years lowers river discharge in southern Europe and the Middle East [*Cullen and deMenocal*, 2000; *Lloret et al.*, 2001]. However, even within one particular climatic region, different lake responses might occur at different altitudes or latitudes. The impact of the NAO on the leaching of nutrients from soil and subsequent March nutrient concentrations in lakes differs between lakes in Sweden and the UK because of differences in average winter severity, which affects whether the soil is frozen or not, although air temperatures in both Sweden and the UK are positively correlated with the winter NAO index. The response of lake surface water temperature to the NAO depends on the altitude of the lake, because high-altitude lakes tend to be ice-covered longer, protecting the lake from the impact of the winter NAO [*Livingstone and Dokulil*, 2001]. Lake temperatures in high-altitude or high-latitude lakes that thaw in early summer instead of spring are unlikely to show a strong NAO signal, because the climatic influence of the NAO at this time of year is comparatively weak. However, the NAO will affect the timing of ice-out of such lakes [*Livingstone*, 2000].

In addition to altitude and latitude, lake size and depth also matter. Small lakes with low water turnover times are more susceptible to fluctuations in inflow associated with the NAO than are large lakes. In contrast, only deep lakes are sensitive to NAO-related effects on spring turnover, as shallow lakes mix completely after every winter regardless of the state of the NAO. The depth of a lake further determines the response of its hypolimnetic water temperatures to the NAO and the persistence of the NAO signal produced. As the response of lacustrine plant and animal populations to the NAO is mediated by lake physics and chemistry, the population responses are likely to differ among freshwater systems depending on altitude, latitude, size and depth.

Recent internationally-based analyses of the impact of the NAO on freshwater ecosystems across large areas of the Northern Hemisphere are providing an improved understanding of the large-scale mechanisms responsible for spatially coherent interannual fluctuations in physical, chemical and biological lake variables. As the current tendency for the NAO to remain in its positive phase may be stabilized by increased concentrations of greenhouse gases [*Paeth et al.*, 1999; see also *Gillett et al.*, this volume], the current series of unusually warm winters in Europe may be set to continue. A better understanding of the large-scale role played by the NAO in determining the functioning of freshwater ecosystems is urgently needed to predict the consequences of such a development.

Acknowledgements. We thank John J. Magnuson and two anonymous referees for comments and suggestions. Funding was provided by the European Union (Contract No. ENV4-CT97-0453) and the Swiss Federal Office of Education and Science (BBW; Contract No. 97.0344) within the framework of the European Union Environment and Climate project REFLECT ("Response of European Freshwater Lakes to Environmental and Climatic Change").

REFERENCES

Adrian, R., Calanoid-cyclopoid interactions: evidence from an 11-year field study in a eutrophic lake, *Freshw. Biol., 38*, 315-325, 1997.

Adrian, R., N. Walz, T. Hintze, S. Hoeg, and R Rusche, Effects of ice duration on the plankton succession during spring in a shallow polymictic lake, *Freshw. Biol., 41*, 621-623, 1999.

Anderson, W. L., D. M. Robertson, and J. J. Magnuson, Evidence of recent warming and El-Niño-related variations in ice breakup of Wisconsin lakes, *Limnol. Oceanogr., 41*, 815-821, 1996.

Anneville, O., S. Soussi, F. Ibañez, V. Ginot, J.-C. Druart, and N. Angeli, Temporal mapping of phytoplankton assemblages in Lake Geneva: annual and interannual changes in their patterns of succession, *Limnol. Oceanogr., 47,* 1355-1366, 2002.

Arai, T., Climatic and geomorphological influences on lake temperature, *Verh. Internat. Verein. Limnol., 21*, 130-134, 1981.

Baines, S. B., K. E. Webster, T. K. Kratz, S. R. Carpenter, and J. J. Magnuson, Synchronous behavior of temperature, calcium, and chlorophyll in lakes of northern Wisconsin, *Ecology, 81*, 815-825, 2000.

Benndorf, J., W. Böing, J. Koop, and I. Neubauer, Top-down control of phytoplankton: the role of time-scale, lake depth and trophic state, *Freshw. Biol., in press*, 2002.

Benndorf, J., J. Kranich, T. Mehner, and A. Wagner, Temperature impact on the midsummer decline of *Daphnia galeata*: an analysis of long-term data from the biomanipulated Bautzen Reservoir (Germany), *Freshw. Biol., 46*, 199-211, 2001.

Bradley, D. C., and S. J. Ormerod, Community persistence among stream invertebrates tracks the North Atlantic Oscillation, *J. Anim. Ecol., 70*, 987-996, 2001.

Cullen, H. M., and P. B. deMenocal, North Atlantic Influence on Tigris-Euphrates streamflow, *Int. J. Climatol., 20*, 853-863, 2000.

Cushing, D. H., Plankton production and year-class strength in fish populations: an update of the match/mismatch hypotheses, *Adv. Mar. Biol., 26*, 249-293, 1990.

Dingman, S. L., Equilibrium temperatures of water surfaces as related to air temperature and solar radiation, *Water Resour. Res., 8*, 42-49, 1972.

Drinkwater, K. F., A. Belgrano, A. Borja, A. Conversi, M. Edwards, C. H. Greene, G. Ottersen, A. J. Pershing, and H. Walker, The response of marine ecosystems to climate variability associated with the North Atlantic Oscillation, this volume.

Eckmann, R., U. Gaedke, and H. J. Wetzlar, Effects of climatic and density-dependent factors on year-class strength of *Coregonus lavaretus* in Lake Constance, *Can. J. Fish. Aquat. Sci., 45*, 1088-1093, 1988.

Edinger, J. E., D. W. Duttweiler, and J. C. Geyer, The response of water temperatures to meteorological conditions, *Water Resour. Res., 4*, 1137-1143, 1968.

Elliott, J. M., Temperature-related fluctuations in the timing of emergence and pupation of Windermere alder-flies over 30 years, *Ecol. Entomol., 21*, 241-247, 1996.

Elliott, J. M., and M. A. Hurley, An individual-based model for predicting the emergence period of sea trout fry in a Lake District stream, *J. Fish. Biol., 53*, 414-433, 1998.

Elliott, J. M., M. A. Hurley, and S. C. Maberly, The emergence period of sea trout fry in a Lake District stream correlates with the North Atlantic Oscillation, *J. Fish. Biol., 56*, 208-210, 2000.

Evans, C. D., D. T. Monteith, and R. Harriman, Long-term variability in the deposition of marine ions at west coast sites in the

UK acid waters monitoring network: impacts on surface water chemistry and significance for trend determination. *The Science of the total Environment, 265*, 115-129, 2001.

Firth, P., and S. G. Fisher (eds.), *Global climate change and freshwater ecosystems*, 321 pp., Springer, Berlin, 1992.

Fromentin, J. M., N. C. Stenseth, J. Gjosaeter, T. Johannessen, and B. Planque, Long-term fluctuations in cod and pollack along the Norwegian Skagerrak coast, *Mar. Ecol. Prog. Ser., 162*, 265-278, 1998.

Gaedke, U., D. Ollinger, E. Bäuerle, and D. Straile, The impact of the interannual variability in hydrodynamic conditions on the plankton development in Lake Constance in spring and summer, *Arch. Hydrobiol. Spec. Issues Advanc. Limnol., 53*, 565-585, 1998.

George, D. G., Using 'climate indicators' to monitor patterns of change in freshwater lakes and reservoirs, in *Water in the Celtic world: managing resources for the 21st century*, edited by J. A. A. Jones, K. Gilman, A Jigorel, and J. Griffin, pp. 93-102, British Hydrological Society, Occasional Papers 11, 2000a.

George, D. G., The impact of regional-scale changes in the weather on the long-term dynamics of *Eudiaptomus* and *Daphnia* in Esthwaite Water, Cumbria, *Freshw. Biol., 45*, 111-121, 2000b.

George, D. G., Regional-scale influences on the long-term dynamics of lake plankton, in *Phytoplankton Productivity: Carbon Assimilation in Marine and Freshwater Ecosystems*, edited by P. J. le B Williams, D. N. Thomas, and C. S. Reynolds, Blackwell Science, 265-290, 2002.

George, D. G., and D. P. Hewitt, The influence of year-to-year variations in winter weather on the dynamics of *Daphnia* and *Eudiaptomus,* in Estwaite Water, Cumbria, *Funct. Ecol., 13(Suppl.1)*, 45-54, 1999.

George, D. G., J. F. Talling, and E. Rigg, Factors influencing the temporal coherence of five lakes in the English Lake Distict, *Freshw. Biol., 43*, 449-461, 2000.

George, D. G., and A. H. Taylor, UK lake plankton and the Gulf stream, *Nature, 378*, 139-1995.

Gerten, D., and R. Adrian, Climate-driven changes in spring plankton dynamics and the sensitivity of shallow polymictic lakes to the North Atlantic Oscillation, *Limnol. Oceanogr., 45*, 1058-1066, 2000.

Gerten, D., and R. Adrian, Differences in the persistency of the North Atlantic Oscillation signal among lakes, *Limnol. Oceanogr., 46*, 448-455, 2001.

Gillett, N. P., H. F. Graf, and T. J. Osborn, Climate change and the North Atlantic Oscillation, this volume.

Güss, S., D. Albrecht, H.-J. Krambeck, D. C. Müller-Navarra, and H. Mumm, Impact of weather on a lake ecosystem, assessed by cyclo-stationary MCCA of long-term observations, *Ecology, 81*, 1720-1735, 2000.

Hänninen, J., I Vuorinen, and P. Hjelt, Climatic factors in the Atlantic control the oceanographic and ecological changes in the Baltic Sea, *Limnol. Oceanogr., 45*, 703-710, 2000.

Hegelbach, J., Wassertemperatur und Blütenphänologie als Anzeiger des früheren Brutbeginns der Wasseramsel (*Cinclus cinclus*) im schweizerischen Mittelland, *J. Ornithol.* , *142*, 284-294, 2001.

Henderson-Sellers, B., Sensitivity of thermal stratification models to changing boundary conditions, *Appl. Math. Mod., 12*, 31-43, 1988.

Hill, D. K., and J. J. Magnuson, Potential effects of global climate warming on the growth and prey consumption of Great Lakes fishes, *Trans. Am. Fish. Soc., 119*, 265-275, 1990.

Hondzo, M., and H. G. Stefan, Propagation of uncertainty due to variable meteorological forcing in lake temperature models, *Water Resour. Res., 28*, 2629-2638, 1992.

Hondzo, M., and H. G. Stefan, Regional water temperature characteristics of lakes subjected to climate change, *Clim. Change, 24*, 187-211, 1993.

Huisman, J., R. R. Jonker, C. Zonneveld, and F. J. Weissing, Competition for light between phytoplankton species: experimental tests of mechanistic theory, *Ecology, 80*, 211-222, 1999.

Hurrell, J. W., Decadal trends in the North Atlantic Oscillation: regional temperatures and precipitation, *Science, 269*, 676-679, 1995.

Hurrell, J. W., Influence of variations in extratropical wintertime teleconnections on Northern Hemisphere temperatures, *Geophys. Res. Lett., 23*, 665-668, 1996.

Hurrell, J. W., and H. van Loon, Decadal variations in climate associated with the North Atlantic Oscillation, *Clim. Change, 36*, 301-326, 1997.

Hurrell, J. W., Y. Kushnir, G. Ottersen, and M. Visbeck, An Overview of the North Atlantic Oscillation, this volume.

Jones, P. D., T. J. Osborn, and K. R. Briffa, Pressure-based measures of the North Atlantic Oscillation (NAO): A comparison and an assessment of changes in the strength of the NAO and in its influence on surface climate parameters, this volume.

Jürgens, K., Impact of *Daphnia* on planktonic microbial food webs - A review, *Mar. Microb. Food Webs, 8*, 295-324, 1994.

Kiely, G., Climate change in Ireland from precipitation and streamflow observations, *Adv. Water Res., 23*, 141-151, 1999.

Kratz, T. K., P. A. Soranno, S. B. Baines, B. J. Benson, J. J. Magnuson, T. M. Frost, and R. C. Lathrop, Interannual synchronous dynamics in north temperatue lakes in Wisconsin, USA, in *Management of Lake and Reservoirs during Global Climate Change*, edited by D. G. George, J. G. Jones, P. Puncochar, C. S. Reynolds, and D. W. Sutcliffe, pp. 273-287, Kluwer Academic Publishers, 1998.

Livingstone, D. M., Temporal structure in the deep-water temperature of four Swiss lakes: a short-term climatic change indicator? *Verh. Internat. Verein. Limnol., 25*, 75-81, 1993.

Livingstone, D. M., An example of the simultaneous occurrence of climate-driven "sawtooth" deep-water warming/cooling episodes in several Swiss lakes, *Verh. Internat. Verein. Limnol., 26*, 822-826, 1997a.

Livingstone, D. M., Break-up dates of Alpine lakes as proxy data for local and regional mean surface air temperatures, *Clim. Change, 37*, 407-439, 1997b.

Livingstone, D. M., Ice break-up on southern Lake Baikal and its relationship to local and regional air temperatures in Siberia and to the North Atlantic Oscillation, *Limnol. Oceanogr., 44*, 1486-1497, 1999.

Livingstone, D. M., Large-scale climatic forcing detected in historical observations of lake ice break-up, *Verh. Internat. Verein. Limnol.*, 27, 2775-2783, 2000.

Livingstone, D. M., and M. T. Dokulil, Eighty years of spatially coherent Austrian lake surface temperatures and their relationship to regional air temperature and the North Atlantic Oscillation, *Limnol. Oceanogr., 46*, 1220-1227, 2001.

Livingstone, D. M., and D. M. Imboden, The prediction of hypolimnetic oxygen profiles: a plea for a deductive approach, *Can. J. Fish. Aquat. Sci., 53*, 924-932, 1996.

Livingstone, D. M., and A. F. Lotter, The relationship between air and water temperatures in lakes of the the Swiss plateau: a case study with paleolimnological implications, *J. Paleolimnol., 19,* 181-198, 1998.

Livingstone, D. M., A. F. Lotter, and I. R. Walker, The decrease in summer surface water temperature with altitude in Swiss alpine lakes: a comparison with air temperature lapse rates, *Arct. Antarct. Alpine Res., 31*, 341-352, 1999.

Lloret, J., J. Lleonart, I Solé, and J. M. Fromentin, Fluctuations of landings and environmental conditions in the north-western Mediterranean Sea, *Fish. Oceanogr., 10*, 33-50, 2001.

Magnuson, J. J., B. J. Benson, and T. K. Kratz, Temporal coherence in the limnology of a suite of lakes in Wisconsin, U.S.A., *Freshw. Biol., 23*, 145-159, 1990.

Magnuson, J. J., et al., Historical trends in lake and river ice cover in the Northern Hemisphere, *Science*, *289*, 1743-1746, 2000.

McCombie, A. M., Some relations between air temperature and the surface water temperatures of lakes. *Limnol. Oceanogr.*, *4*, 252-258, 1959.

Mehner, T., H. Dörner, and H. Schultz, Factors determining the year-class strength of age-0 Eurasian perch (*Perca fluviatilis L.*) in a long-term biomanipulated reservoir. *Archive of Fishery and Marine Research*, *46*, 241-251, 1998.

Monteith, D. T., C. D. Evans, and B. Reynolds, Are temporal variations in the nitrate content of UK upland freshwaters linked to the North Atlantic Oscillation? *Hydrological Processes, 14,* 1745-1749, 2000.

Müller-Navarra, D. C., S. Güss, and H. von Storch, Interannual variability of seasonal succession events in a temperate lake and its relation to temperature variability, *Global Change Biology*, *3*, 429-438, 1997.

Mysterud, A., N. C. Stenseth, N. G. Yoccoz, G. Ottersen, and R. Langvatn, The response of terrestrial ecosystems to climate variability associated with the North Atlantic oscillation, this volume.

Oschlies, A., NAO-induced long-term changes in nutrient supply to the surface waters of the North Atlantic, *Geophys. Res. Lett.*, *28*, 1751-1754, 2001.

Ottersen, G., and H. Loeng, Covariability in early growth and year-class strength of Barents Sea cod, haddock and herring: the environmental link, *ICES Journal of Marine Science*, 57, 339-348, 2000.

Ottersen, G., B. Planque, A. Belgrano, E. Post, and N. C. Stenseth, Ecological effects of the North Atlantic Oscillation, *Oecologia*, *128*, 1-18, 2001.

Paeth, H., A. Hense, R. Glowienka-Hense, R. Voss, and U Cubasch, The North Atlantic Oscillation as an indicator for greenhouse-gas induced regional climate change, *Clim. Dynamics*, *15*, 953-960, 1999.

Palecki, M. A., and R. G. Barry, Freeze-up and break-up of lakes as an index of temperature changes during the transition seasons: a case study for Finland, *J. Clim. Appl. Meteor.*, *25*, 893-902, 1986.

Raven, J. A., and R. J. Geider, Temperature and algal growth. *New Phytol.*, 110, 1988.

Reynolds, C. S., Physical determinants of phytoplankton succession, in *Plankton ecology: succession in plankton communities*, edited by U. Sommer, pp. 9-56, Springer, Berlin, 1989.

Robertson, D. M., *The use of lake water temperature and ice cover as climatic indicators,* PhD diss., University of Wisconsin, Madison, Wisconsin. 265 pp. 1989.

Robertson, D. M., and R. A. Ragotzkie, Changes in the thermal structure of moderate to large sized lakes in response to changes in air temperature, *Aquat. Sci.*, *52*, 360-380, 1990.

Robertson, D. M., R. A. Ragotzkie, and J. J. Magnuson, Lake ice records used to detect historical and future climatic changes. *Clim. Change, 21*, 407-427, 1992.

Rodionov, S. N., *Global and regional climate interaction: the Caspian Sea experience*, Kluwer, Dordrecht, 1994.

Ruosteenoja, K., The date of break-up of lake ice as a climatic index, *Geophysica, 22*, 89-99, 1986.

Sanford, E., Regulation of keystone predation by small changes in ocean temperature, *Science, 283*, 2095-2097, 1999.

Sæther, B.-E., J. Tufto, S. Engen, K. Jerstad, O. W. Røstad, and J. E Skåtan, Population dynamical consequences of climate change for a small temperate songbird, *Science*, *287,* 854-856, 2000.

Scheffer, M., S. Carpenter, C. Foley, and B. Walker, Catastrophic shifts in ecosystems, *Nature*, *413*, 591-596, 2001b.

Scheffer, M., S. H. Hosper, M. L. Meijer, B. Moss, and E. Jeppesen, Alternative equilibria in shallow lakes, *Trends Ecol. Evol., 8*, 275-279, 1993.

Scheffer, M., D. Straile, E. H. van Nes, and H. Hosper, Climatic warming causes regime shifts in lake food webs, *Limnol. Oceanogr., 46*, 1780-1783, 2001a.

Schindler, D. W., Widespread effects of climatic warming on freshwater ecosystems in North America, *Hydrol. Process., 11*, 1043-1067, 1997.

Shuter, B. J., D. A. Schlesinger, and A. P. Zimmerman, Empirical predictors of annual surface water temperature cycles in North American Lakes, *Can. J. Fish. Aquat. Sci., 40*, 1838-1845, 1983.

Sommer, U., The role of competition for resources in phytoplankton succession, in *Plankton ecology: succession in plankton communities*, edited by U. Sommer, pp. 57-106, Springer, Berlin, 1989.

Sommer, U., Z. M. Gliwicz, W. Lampert, and A. Duncan, The PEG-model of seasonal succession of planktonic events in fresh waters, *Arch. Hydrobiol., 106*, 433-471, 1986.

Stenseth, N. C., O. N. Bjørnstad, W. Falck, J. M. Fromentin, J. Gjosaeter, and J. S. Gray, Dynamics of coastal cod populations: intra- and intercohort density dependence and stochastic processes, *Proc. R. Soc. Lond. B*, *266*, 1645-1654, 1999.

Straile, D., Meteorological forcing of plankton dynamics in a large and deep continental European lake, *Oecologia, 122*, 44-50, 2000.

Straile, D., North Atlantic Oscillation synchronizes food-web interactions in central European lakes, *Proc. R. Soc. Lond. B*, *269,* 391-395, 2002.

Straile, D., and R. Adrian, The North Atlantic Oscillation and plankton dynamics in two European lakes - two variations on a general theme, *Global Change Biology*, *6*, 663-670, 2000.

Strub, P. T., T. Powell, and C. R. Goldman, Climatic forcing: effects of El Niño on a small, temperate lake, *Science, 227*, 55-57, 1985.

van Loon, H., and R. A. Madden, Interannual variations of mean monthly sea-level pressure in January, *J. Clim. Appl. Meteor.*, *22*, 687-692, 1983.

Vavrus, S. J., R. H. Wynne, and J. A. Foley, Measuring the sensitivity of southern Wisconsin lake ice to climate variations and lake depth using a numerical model, *Limnol. Oceanogr.*, *41*, 822-831, 1996.

Weyhenmeyer, G. A., Warmer winters–are planktonic algal populations in Swedens largest lakes affected? *Ambio, 30*, 565-571, 2001.

Weyhenmeyer, G. A., R. Adrian, U. Gaedke, D. M. Livingstone, and S. C. Maberly, Response of phytoplankton in European lakes to a change in the North Atlantic Oscillation, *Verh. Internat. Verein. Limnol.,* in press, 2002.

Weyhenmeyer, G. A., T. Blenckner, and K. Pettersson, Changes of the plankton spring outburst related to the North Atlantic Oscillation, *Limnol. Oceanogr., 44*, 1788-1792, 1999.

Williams, R. G., and A. J. McLaren, Estimating the convective supply of nitrate and implied variability in export production over the North Atlantic, *Global Biogeochem. Cycles, 14*, 1299-1313, 2000.

Dietmar Straile, Limnological Institute, University of Konstanz, 78457 Konstanz, Germany.
Dietmar.Straile@uni-konstanz.de

David M. Livingstone, Water Resources Department, Swiss Federal Institute of Environmental Science and Technology (EAWAG), Überlandstrasse 133, CH-8600 Dübendorf, Switzerland.
living@eawag.ch

Gesa A. Weyhenmeyer, Department of Environmental Assessment, Swedish University of Agricultural Sciences, P.O. Box 7050, 75007 Uppsala, Sweden.
Gesa.Weyhenmeyer@ma.slu.se

D. Glen George, Centre for Ecology and Hydrology, The Ferry House, Far Sawrey, Ambleside, Cumbria LA22 0LP, U.K.
DGG@ceh.ac.uk